Structure-Based Mechanics of Tissues and Organs

Ghassan S. Kassab • Michael S. Sacks
Editors

Structure-Based Mechanics of Tissues and Organs

Springer

Editors
Ghassan S. Kassab
California Medical Innovations Institute
San Diego, CA, USA

Michael S. Sacks
The University of Texas at Austin
Austin, Texas, USA

ISBN 978-1-4899-7629-1 ISBN 978-1-4899-7630-7 (eBook)
DOI 10.1007/978-1-4899-7630-7

Library of Congress Control Number: 2015959792

Springer New York Heidelberg Dordrecht London

Printed on acid-free paper

Springer Science+Business Media LLC New York is part of Springer Science+Business Media (www.springer.com)

Autobiographical Postscript

My research career follows closely the development of Biomedical Engineering (BME) in Israel, a goal to which I devoted a significant portion of my time and efforts. This turn of events was not planned ahead. Growing up during the first years of the State of Israel, like many youngsters at that time I was a member of a youth movement whose goal was to prepare us for life in the Kibbutz, with the mission to settle and build the country. These personal plans had to be changed due to a sport accident. I decided to go for higher education related to agriculture, so that my education could be of use to my future Kibbutz. Since my inclination was towards exact sciences, I chose agricultural engineering at the Technion in Haifa, the only engineering school in Israel at that time. In the third year of study, I added civil engineering to my curriculum. It was during these Technion years that I realized I had a great deal of curiosity, often asking myself "why" or "what is the underlying mechanistic reason"? It was then that I first entertained the thought of a possible research career.

Upon graduation in 1962, I joined a Kibbutz in the dry southern part of Israel—the Negev. It soon became clear that the small Kibbutz at that time could not benefit from my engineering skills, so I took a job as a water system engineer in the nearby town of Beer Sheba. My responsibility was to design and oversee construction of water supply networks to the new farms in the dry southern Negev area which occupies nearly half of the area of Israel and extends from Beer Sheva down south to the Red Sea, and to the Dead Sea in the east. Although the job provided me with a great deal of practical engineering experience, it required little of advanced engineering skills.

In 1964 I accepted an offer to work in Benin City, Nigeria, on a large-scale project aimed to supply treated fresh water to each village and town in the mid-eastern region of that country. In addition to satisfying my professional curiosity, the work was an opportunity to get to know Nigeria, its people and culture. I was attracted by the importance of the job which sought to reduce the rate of infectious diseases, especially the rate of infant mortality, estimated at that time to be above 80 %. This was caused primarily by consumption of untreated water from nearby rivers. The project was completed a couple of years after I left Nigeria. I was proud to learn that it achieved its goal: infant mortality dropped down to 10–15 %, still too high, but nonetheless a significant improvement.

After 1 year in Nigeria I was recruited by my Kibbutz to take the position of R&D engineer in the newly established drip irrigation factory—Netafim. Drip irrigation was conceived by an Israeli water expert and proved to be well suited for dry areas due to the substantial saving in water and the significant increase in yields compared to other methods such as sprinkle irrigation. It faced, however, a number of technological problems which I had to attend to. Today, drip irrigation is well established. Netafim is a highly successful multinational consortium, well known in countries using irrigation. The work was interesting and involved some research work. It became however clear that serious research can be done only in a well-equipped facility of an academic institution. Hence, after 2 years in the job I returned to the Technion in 1968 and took a job in the Hydraulics Laboratory as an R&D engineer. The renewed interaction with an academic environment was highly fulfilling, and just a few weeks into the job, I decided to pursue an academic career.

M.Sc. training at the Technion required a full research thesis. While searching for a suitable topic, I was informed of a new research field, Biomedical Engineering (BME). New, interdisciplinary, and with endless perspective, it seemed like an attractive field. In the late 1960s, one of the hottest topics in BME research was the pulsatile blood flow in arteries. Research in that topic flourished after the breakthrough work of John Womersley, who showed that for an oscillatory flow of a Newtonian fluid in an elastic tube, the flow is characterized by a parameter, which today bears his name (the Womersley number). In looking for a specific thesis topic, I browsed the literature and discussed various possibilities with Uri Dinnar and Hillel Rubin. The question came up on how the flow would be affected if one considered the blood vessels' viscoelasticity and the non-Newtonian nature of blood. It seemed an interesting extension of the Womersley theory and one that could be experimentally tested in the laboratory. While developing the theory, it turned out that Womersley analysis can be generalized to Maxwellian fluids (Holzinger and Rubin 1970) and to general linear viscoelastic fluids in elastic-viscous tubes (Lanir and Rubin 1971, 1972), by replacing the Newtonian Womersley number with a complex viscoelastic one, and the tube elastic parameters with its complex, elastico-viscous counterpart. In the experimental investigation, I used viscoelastic CMC polymer solutions of various concentrations. The theory and experiments were in good agreement and showed an unexpected result: unlike in Newtonian fluids in which the wave velocity increases monotonically with frequency, in viscoelastic fluids, it attains a maximum level at a low frequency and decreases thereafter.

I submitted my M.Sc. thesis in the summer of 1970. Two weeks later I was surprisingly notified that since my work was of a level and scope of a Ph.D. research, the Technion decided to grant me a Ph.D. degree. Although satisfying and exciting, this rather unusual and unexpected development presented a problem: I was just accepted for Ph.D. training at the University of California at San Diego (UCSD) under the supervision of Prof. Y.C. Fung. Determined to work with Dr. Fung, I decided to take my chance. My wife Suzi and I landed in San Diego shortly after my Ph.D. defense, with the intention to embark on postdoctoral training. Dr. Fung, after reading my Ph.D. papers, was satisfied with my credentials. However, due to severe recent cuts in his NASA research funding, he could not support me. In the face of this news, I decided to wait for 1 month before looking for another job. Luckily, a small leftover of a US Air Force research budget was found. I embarked on a work on fiber composite materials.

Fiber composites were extensively studied under tension and shear loading. The question at hand was how they would respond to compression in the fiber direction, especially when the matrix Poisson's ratio exceeded that of the fibers, a situation which could lead to de-bonding and fiber buckling. My analysis showed that although fibers may indeed buckle in a mode which increases with compression, in the range of infinitesimal deformations, this has insignificant effect on the composite response (Lanir and Fung 1972). For a while, I believed that this research and its results would be of no special significance, only to discover much later that the work is of interest in nanotubes research (Lourie et al. 1998).

My next project was in tissue biomechanics. This research occupied me for the rest of my postdoctoral training and shaped much of my future research. In this I could not have wished for a better mentor than Dr. Fung—"the father of modern biomechanics" (Kassab 2004). I was asked by Dr. Fung to develop the first biaxial large deformation tissue tester and apply it to study the properties of flat tissues, specifically the skin. Three criteria directed the setup design. For reliable characterization, the stretched specimen should be under uniform deformation; the strain measurement should be a no-contact one so as not to affect the soft tissue response; in view of tissue viscoelasticity, the rate of loading should be adjustable and controlled. The developed setup fulfilled all three criteria (Lanir and Fung 1974a). The test protocols included biaxial stress relaxation tests, constant rate of stretch tests at different rates, and tests of temperature effects (Lanir and Fung 1974b). The data clearly demonstrated the skin's anisotropy, nonlinearity, time dependence, and preconditioning adaptation. The effects of temperature were found to depend on the imposed rate of temperature change and on the tissue stretch: at low stretch levels the skin has a negative thermoelastic response (it contracts with increasing temperature, as rubber-like materials), while at high stretch levels it is positive (expanding with temperature, as in crystallized materials such as metals). Later, I learned that this is due to the transition from the amorphic, elastin-dominated low stretch range to the high stretch one dominated mainly by the structured collagen.

In 1972, I joined the Technion as the first full-time BME faculty (also first in Israel). The BME unit was an interdisciplinary program which offered solely graduate courses but no research facilities. Courses were conducted by secondary affiliated faculty members who lectured on their field of expertise. Yet the curriculum lacked a clear plan or structure. I was asked to head this unit with the mission to develop the research activity and introduce coherence and clear tracking to the graduate program. This was a challenging task for a newly appointed faculty member. After 1 year in the job I realized that although I was able to move the department in the right direction and recruited new faculty members, the job was too time consuming and thus unsuitable for a new, young faculty and could risk my academic progress towards tenure. After 1 year as department head, I resigned to resume my research work.

In Fung's laboratory at UCSD, I enjoyed experimental research in a well-equipped and properly staffed setting. These conditions were unavailable in the Technion BME laboratory at that time. I opted to invest my efforts in theoretical research. My first project was aimed at characterizing the skin mechanical properties based on the data collected at UCSD. Tissue characterization was up to that time exclusively phenomenological, culminating in Fung's exponential law (Fung 1967, 1972) and in the material laws of polymers such as the Mooney–Rivlin one. It turned out, however, that none of these hyper-elastic phenomenological laws adequately fit the skin's biaxial response. In looking for a better representation, I wondered if something can be learned from looking at the internal processes which give rise to the global tissue response. In searching for the relevant information, I was fortunate to come across a few histological studies which were published around that time on the response of tissue microstructure to uniaxial and biaxial stretch (Gibson et al. 1965; Viidik 1969, 1972; Millington and Brown 1970; Chu et al. 1972; Brown 1973). Three distinct processes could be identified from the published images: First, in the uniaxially stretched tendon, there is a gradual straightening (recruitment) of collagen fibers with stretch (Viidik 1972, 1978) and this is accompanied by an increase in the tissue rigidity. Second, in addition to collagen, flat tissues such as the skin and the mesentery consist also of elastin fibers which become straight at lower stretch levels than the collagen (Gibson et al. 1965; Chu et al. 1972; Evans et al. 1980). Third, in addition to straightening, in flat tissues there is a process of fiber rotation towards the direction of highest stretch (Brown 1973; Evans et al. 1980). The challenge was to find a way by which these processes could be incorporated into a general hyper-elastic constitutive formulation. After testing a number of possibilities, the most promising one proved to be the stochastic approach—to assign distribution functions to the fibers' straightening strains and orientations, for both the collagen and elastin fibers, and sum up all the fibers' contributions. The formulation is based on four assumptions: (a) affine deformation field, i.e., each embedded fiber responds kinematically as if it were an element in the tissue continuum; (b) the tissue's total strain energy equals the algebraic sum of its fibers' strain energies; (c) the strain energy of a slender fiber is solely due to its axial stretch and vanishes under compression due to buckling; and (d) the fluid-like ground substance matrix renders the tissue incompressible and contributes hydrostatic pressure to the global tissue response. The natural outcomes of these ideas were that the observed nonlinear and anisotropic properties of tissues result from the fibers' gradual recruitment (straightening) and their nonuniform orientation distributions.

This is how the microstructural approach to tissue constitutive modeling was born. If I have to identify a common trait to my research work since then, including studies unrelated to tissue mechanics, it is the strong inclination to look at the micro and to link the micro to the macro function. My experience has shown this to be a rewarding approach.

Today, the structural approach is widely used in tissue mechanics and applied in a variety of tissues (e.g., Belkoff and Haut 1991; Billiar and Sacks 2000; Chandran and Barocas 2006; Cortes et al. 2010; Crabb et al. 2006; Dahl et al. 2008; Driessen et al. 2005; Engelmayr and Sacks 2006; Federico et al. 2005; Gasser et al. 2006; Grytz and Meschke 2009; Hansen et al. 2002; Hollander et al. 2011a; Holzapfel et al. 2004; Horowitz et al. 1988a; Jhun et al. 2009; Lake et al. 2011; Lokshin and Lanir 2009a; Martufi and Gasser 2011; Nevo and Lanir 1989; Raz and Lanir 2009; Sacks et al. 2004; Sverdlik and Lanir 2002; Zulliger et al. 2004). Comparison with data of 1D (Raz and Lanir 2009; Sverdlik and Lanir 2002), 2D (Billiar and Sacks 2000; Lokshin and Lanir 2009a; Sacks et al. 2004) and 3D (Horowitz et al. 1988a; Hollander et al. 2011b) responses showed excellent agreement. The case of the arterial wall (Hollander et al. 2011a) is a recent example of the power of microstructural modeling. The arterial media is a 3D layered structure consisting of concentric lamellae, interlamellar thin elastin struts, and smooth muscle cells. The lamellae are composed of helical-oriented elastin and collagen fibers (Clark and Glagov 1985; Rhodin 1980; Wasano and Yamamoto 1983). Based on these structural features, together with David Durban and our student Yaniv Hollander and in collaboration with Ghassan Kassab we developed a microstructural constitutive model for the coronary media and validated it against 3D data of inflation/extension/twist tests carried out in Ghassan laboratory. By sensitivity analysis it was found that a reduced form of the model having only four parameters provided excellent fit to the entire 3D database (the model descriptive power). The model predictive power (fit to data not used in estimating the model constitutive parameters) was validated as well. In particular, characterization based on just inflation/extension data provides reliable estimates of the model parameters (thus saving the need to perform the more difficult twist test) and very good fit to the entire 3D data. Previous models of the media were either phenomenological or phenomenological–structural hybrids. Comparison of the structural with these models against the 3D data demonstrated the superior reliability of the structural approach in both its descriptive and predictive performances (Hollander et al. 2011b).

The theory of constitutive characterization imposes restrictions on material laws. One of these requires that the constitutive law must be convex under any deformation scheme. In essence, this restriction guarantees the existence and stability of the material response (Holzapfel et al. 2004; Truesdell and Noll 1965). I was able to prove that the structural constitutive formulation automatically satisfies the convexity conditions under any deformation (Lanir 1996). This distinct advantage stems from the convexity of the fibers' stress–strain relationships. Hence, in structural characterization there is no need to check the plausibility restrictions.

A daunting problem in tissue mechanics is the difficulty of finding a hyper-elastic law that is valid under any deformation scheme, i.e., while the response under a single protocol can be adequately represented by a variety of constitutive models, when attempting to fit data of multiple protocols with a single set of material parameters, no model proved to be adequate (Tong and Fung 1976). In studying this problem, I realized from the onset that since soft tissues are inelastic, it is unrealistic to expect that any hyper-elastic model can fit their inelastic response. An inelastic constitutive formulation was needed. In looking for a suitable inelastic formulation, it turned out that the stochastic structural approach is well suited for such generalization. Since the inelastic properties of tissues derive from the inelastic properties of their fibers, generalization to inelastic cases (such as viscoelasticity and pre-conditioning) can be readily achieved by replacing the fibers' elastic stress–strain laws with their viscoelastic and viscoplastic counterparts. The resulting nonlinear inelastic structural constitutive models provided excellent fit to both 1D (tendon and skin (Raz and Lanir 2009; Sverdlik and Lanir 2002; Lokshin and Lanir 2009b)) and 2D (skin (Lokshin and Lanir 2009a)) data under multiple and different protocols. For example, with my student Einat Raz we found that in the structural formulation for the viscoelastic case, the predicted responses of the tendon under creep and stress relaxation tests are mutually compatible and interlinked by a relationship that depends on the protocol and constitutive law (Raz and Lanir 2009). Moreover, both the 1D and 2D investigations found that for reliable tissue representation, preconditioning must be included as an integral part of the constitutive formulation, in addition to viscoelasticity.

During my sabbatical leave at Michigan in 1979 I met a young undergraduate student who showed a keen interest in Biomechanics. We had a couple of discussions on the experimental and theoretical aspects of tissue characterization. Our discussions may have impressed him enough to embark on a career in Tissue Biomechanics. He is Michael Sacks, a prominent researcher in the field and former editor of the ASME Journal of Biomechanical Engineering. Mike "blames" me for recruiting him to Biomechanics, a charge which I do not deny.

In the early 1980s I initiated the recruitment of Alice Maroudas to our BME department at the Technion. Alice's well-known experimental work on the articular cartilage and on intervertebral disc showed that in these tissues, due to the high concentration of their interstitial negatively charged proteoglycans, a major support against compressive loading is due to the osmotic-induced hydrostatic pressure of the ground substance (Maroudas and Bannon 1981). My own interest in this topic developed following a meeting with Richard Skalak of Columbia University who encountered a theoretical dilemma relating to the number of equations needed to solve a boundary value problem in biphasic materials. Two modeling approaches were known at that time: poroelasticity (McCutchen 1982) and biphasic theory (Armstrong et al. 1984; Kwan et al. 1984). Both considered elastic and viscous forces due to fluid filtration relative to the solid phase, but ignored the osmotic effects. Hence, both theories predicted that under equilibrium with external compression, it is the collagen fibers which bear the compressive load, while the fluid pressure vanishes, predictions which are in obvious contradiction to Alice and coworkers' experimental results. There was a need for a new constitutive approach. The challenge in developing a constitutive swelling theory for these tissues was to incorporate osmotic effects into a mechanical model. The solution which emerged following a discussion with Alex Silberberg of the Weizmann Institute was to integrate mixture theory (Truesdell 1962) with nonequilibrium thermodynamics (Silberberg 1982). The developed bi-component theory (Lanir 1987) extended the previous biphasic theories by introducing concentration (osmotic) forces into the chemical potentials of the solid and fluid components, subject to the Gibbs–Duhem condition. Three important new results emerged: First, a self-consistent bi-component constitutive theory was developed for swelling tissues. Second, the new theory was in agreement with experimental observations as to both the role of hydrostatic pressure in bearing compressive loadings and the stiffening effect of osmotic forces on the tissue response. Third, the driving force for fluid filtration was identified as the "swelling stress" (the difference between the solid stress and osmotic pressure).

Although proteoglycan concentrations in soft tissues such as skin, blood vessels, and muscle are lower than in cartilage and disc, osmotic swelling still has significant functional and biological importance in these tissues as well. Since osmotic swelling is counterbalanced by tension in the tissue fibers, it follows that the unloaded tissue is not stress free, but is internally loaded by residual stress. In an earlier study by Skalak and coworkers (1996), it was proposed that residual stress stems from incompatible growth of tissue elements which is balanced by incompatible elastic strain, to produce a combined compatible deformation. It is that elastic strain which produces residual stress. It turns out, however, that growth is not the only mechanism of residual stress. Based on our combined experimental and theoretical studies involving controlled manipulation of the tissue osmotic pressure, my coworkers and I were able to show that in the left ventricle (Lanir et al. 1996b) and in the aorta (Guo et al. 2007), osmotic loading may have a dramatic effect on the residual stress. A parallel microstructural stress analysis in the left ventricle (Lanir et al. 1996b) supported this notion, indicating that a significant portion of the tissue's residual stress stems from its osmotic swelling. In two recent publications, this link between osmotic swelling and residual stress was explored in detail and shown to be valid in the cartilage and the disc as well (Lanir 2009, 2012).

The cardiovascular system (CVS) was of special interest to me since my Ph.D. training. It presents unique and interesting challenges: it is subjected to periodic loading by the beating heart; its function is primarily mechanical (to pump the blood, transport it to the periphery, and collect it); it is constructed of hollow organs; its tissues have 3D microstructure; and most CVS tissues contain muscle cells which render them active properties. In addition, CVS research is of important clinical value due to the major impact of vascular diseases (in particular, coronary arterial diseases) on health in the western world. The coronary circulation functions within the myocardium and is substantially affected by the cyclic contraction of the heart, resulting in a unique dynamic flow features. Progress in this field of research is impeded by the difficulties in measurements within the moving heart. Thus, modeling simulation is an attractive research alternative and is widely applied. Simulation of the coronary flow requires knowledge of the input loading conditions imposed by the myocardium on the coronary vessels.

Since the coronary vessels are externally loaded by stress in the surrounding myocardium, a first step in the research is to establish the dynamic stress field within the contracting heart. Stress analysis in the heart is a special challenge due to its irregular shape, the complex passive and active properties of the myocardium, the effects of the dynamic volume changes due to blood sloshing in and out of the embedded coronary vessels, and its loading by the papillary-mitral valve system during systole. The myocardium constitutive properties derive from its two major constituents, the collagen network and the myocytes. Following the success of the structural approach in other tissues, it was only natural to apply micromechanical analysis in the myocardium as well. My work was carried out in three stages. First, based on the histological data on the collagen network of the heart which became available at that time (Borg and Caulfield 1979, 1981; Borg et al. 1981; Caulfield and Borg 1979; Robinson et al. 1983; Streeter et al. 1969), a 3D structural constitutive equation for the passive myocardium was developed and validated against mechanical data (Horowitz et al. 1988a), and then implemented in constructing a 2D finite element for myocardial strips (Horowitz et al. 1988b) and a 3D one for the full thickness LV wall (Horowitz et al. 1988c). In the second stage, together with my student Erez Nevo we developed a 3D passive and active constitutive law for the myocardium and applied it in stress analysis of the cylindrically shaped equatorial region of the beating LV along the entire cardiac cycle, first without (Nevo and Lanir 1989) and then with (Nevo and Lanir 1994) the effect of residual stress. In the third stage, a full 3D structural passive and active myocardial finite element was developed and implemented in dynamic stress analysis in a truncated ellipsoid-shaped LV (Imanuel 1996).

A major determinant of the coronary flow is the network structure. The network consists of millions of vessel segments. The network flow and the myocardium loading on the vessels within it are inhomogeneous. Hence reliable flow analysis can only be carried out based on realistic model of the network structure. Up to the early 1990s, pertinent information was scarce and insufficient. As a result, most flow modeling studies were based on lumped concepts in which the whole network or large portions of it were represented as compartments. Unfortunately, such an approach does not consider the effect of the network structure on the flow, nor does it account for the local nature of the myocardium–vessel interaction (MVI) and for its variation across the heart wall. It was obvious that detailed structural information was in pressing need and that no significant progress can be achieved without it. Fortunately, this need was met in the early 1990s.

In 1990, Y.C. Fung initiated and chaired the First World Congress in Biomechanics held on the UCSD in La-Jolla, CA. The last presentation in the last day of the congress was by Fung's Ph.D. student Ghassan Kassab. The title related to the detailed morphometry of the coronary network. In his work, Ghassan was able to provide detailed statistics of the network morphometry. Our meeting marked the beginning of a long collaboration and friendship. Ghassan's data relate to the coronary arterial, venous, and capillary networks (Kassab and Fung 1994; Kassab et al. 1993, 1994, 1997). These data have served several investigators in reconstructing portions of the coronary arterial network based on some simplifications. Our collaboration, together with my student Benny Kaimovitz, resulted in reconstruction of the entire arterial system and embedding it within a prolate spheroid heart model (Kaimovitz et al. 2005). This was achieved by solving the network geometry as a large-scale multistep optimization problem. We then continued to the venous system, which was reconstructed as an optimization problem subject to a boundary restriction that arterial and venous capillaries must be joined (Kaimovitz et al. 2010). The resulting network consists of close to ten million vessels. Its rendering (Wischgoll et al. 2007) highly resembles images of the native network.

Following the progress in the network reconstruction, it was possible to embark on realistic distributive coronary flow analysis. Three "mysteries" have perplexed the research community for many years, being subjects of confusion and debates. The first related to the question of the true nature of the dynamic myocardium–vessel interaction (MVI) in the beating heart (Westerhof et al. 2006). Together with my student Dotan Algranati and in collaboration with Ghassan Kassab, we developed an analytic/numeric platform which incorporated the coronary network structure and included a module for analyzing the in situ vessel nonlinear pressure–diameter relationship (PDR) and a module for network flow analysis based on our previously validated single vessel nonlinear flow model (Jacobs et al. 2008). With this platform, five MVI mechanisms were tested against published dynamic flow and diameter data of endocardial and epicardial microvessels. The results revealed (Algranati et al. 2010) that the only interaction mechanism which fits all the data consists of the combined effects of interstitial pressure (derived by the left ventricle pressure) and intramyocyte pressure (which develops as the myocytes contract). The second mystery related to the underlying reasons for the sub-endocardium higher vulnerability to hypo-perfusion ischemia, in spite of the fact that stenosis which induces ischemia occurs exclusively in epicardial vessels (Hoffman 1987). A detailed sensitivity analysis revealed (Algranati et al. 2011) that the basic reason for subendocardium vulnerability is the nonlinear nature of the vessels' PDR, coupled with the differences in the pressure work points between vessels in these two layers. From the clinical aspects, the analysis revealed that this vulnerability of subendocardial vessels can be moderated by lowering the heart rate and the left ventricle pressure. A third mystery was a clinical question relating to the reliability of indices used during catheterization in assessing functional stenosis severity, or its functional reciprocal—the predicted post treatment flow improvement (Spaan et al. 2006). Our analysis (Algranati 2010) focused on three commonly used indices, fractional flow reserve (FFR), percent stenosis area (%AS), and hyperemic stenosis resistance (HSR). In particular, the investigation related to the extent by which the predictive performance of these indices was affected by interpersonal variability in coronary hemodynamic and by mechanical factors. The results showed that while predictions of the true flow improvement of %AS and HSR are significantly affected by aortic blood pressure, hematocrit, and vessels' stiffness, FFR predictions are robust to changes in heart rate, hematocrit, and variability in aortic, venous, and left ventricle pressures. FFR predictions are, however, sensitive to changes in the vessels' stiffness, which may be affected by age and by pathological vessel remodeling due to smoking, diabetics, and hypertension.

During the years since my first administrative experience as a young department head, I was assigned a number of administrative duties which were associated with developing of the BME academic activities in the Technion and outside of it. I declined, however, requests to accept major administrative responsibilities knowing that it would seriously distract from and impede my research. There was one important exception: starting in 1995, I initiated and led the efforts to establish the BME undergraduate program, the first in Israel. In 1999 our department opened its gate to the first class of undergraduate BME students. Within a few years other universities and colleges followed suit.

My recent research focuses on two main topics. The first relates to soft tissues growth and remodeling (G&R). In contrast to inert materials, biological tissues have the unique ability to grow and remodel. Soft tissues adapt to altered mechanical environment by changing their size, structure, and mechanical properties. G&R occurs since in order to survive and proliferate, tissue cells strive for a homeostatic mechanical environment. They do so by adapting their extracellular matrix (ECM) via turnover (production and/or degradation) of ECM constituents. Modeling G&R can be of great value by unifying a collection of seemingly unrelated facts into a general scientific framework, thus providing insight into the processes involved. In practice, models serve for quantitative prediction and design (e.g., tissue engineering) where mechanical conditioning stimulates matrix production and plays a key role in the evolution of the constructs towards targeted microstructure and mechanical properties. Previous G&R models relate mostly to cardiovascular tissues. Earlier models focused mainly on the manifestation of G&R assuming that tissue dimension and structure adapt to the global stress or strain. For example, in blood vessels, the diameter and wall thickness remodel to maintain the luminal shear stress and wall hoop stress at their homeostatic range. However, since tissues G&R derives from loading-dependent local turnover events in the fiber level, a microstructural mechanistic G&R theory can mimic real adaptation events with high realism. To materialize this idea I developed the theory which incorporates the specific mechanical properties and turnover kinetics of each constituent, thereby establishing a general framework which can serve for integration of additional constituents and processes involved in G&R (Lanir 2015). The theory predictions show qualitative agreement with a number of well-known features of tissues including the fiber's nonuniform recruitment density distribution, the associated tissue convex nonlinear stress–stretch relationship, and the development of tissue pre-stretch and pre-stress states. In my ongoing research I attempt to extend the theory to multi-dimension tissues and to tissues with multiple fiber types, each characterized by its own turnover kinetics and mechanical response.

Another topic of my current research is the control of the coronary circulation. It is a natural extension of our previous coronary circulation research. In collaboration with Ghassan Kassab, we emphasize the combined modeling/experimental investigation of the detailed manner by which the coronaries regulate the flow to meet metabolic demand under a wide range of physical activity (the coronary reserve). Flow regulation is achieved through local control of the vessel's diameter (and associate resistance) by contraction/relaxation of smooth muscle cells in the vessel wall. Of particular interest are the yet unresolved questions on the effects of activation on the vessel/myocardium dynamic interaction, and of the mutual interactions between the three independent control mechanisms known to regulate the flow (pressure-induced myogenic control, shear control, and metabolic one). There are important clinical questions relating to the consequences of acute cardiac changes (such as heart rate and perfusion pressure) on the spatial distribution of ischemic regions, and why and which conditions cause hypo-perfusion-associated failure of subendocardial tone regulation before subepicardial one, and the failure of profound ischemia to maximally dilate the vessels. In the first phase of the work, together with Jon Young and based on micromechanical modeling of the vessel interaction with its surrounding myocardium, we were able to show that effective flow regulation requires both an interstitial gap between vessel and tissue and slackness in their tethering, so that the vessel can easily contract requiring little energy. In parallel to the theoretical analysis, experimental work by Kassab group verified the existence of a gap and tethering slackness (Young et al. 2012).

In closing, I wish to reflect on possible future directions of research in tissue and organ biomechanics, with emphasis on the microstructural approach:

Experimental tissue-structure characterization: Tissue morphometry is the cornerstone of the structural approach. Previous structural characterization was for the most part indirect: general forms of orientation and waviness distributions were assigned based on qualitative histological observations, and parameters of these distributions were subsequently estimated to best fit given response data. Attempts to directly determine the distribution functions based on morphological observations have been impeded by computational difficulties in reliable extraction of the 3D complex structural features of tissues and by the inherent experimental inaccuracies due to the need to process the specimen, or by the use of differential digestions which likely distorts the tissue's microstructure. Recent developments in nonlinear optics (such as multiphoton microscopy and coherent anti-Stokes Raman scattering microscopy), when coupled with computer-interfaced sequential optical sectioning, hold great promise for 3D structural characterization. The advantage of these methodologies is that tissue samples are maintained in their native state (i.e., unstained and unprocessed). In addition, the application of two types of nonlinear laser/tissue interactions—two photon excited fluorescence (TPEF) and second-harmonic generation (SHG), allows retrieval of distinct structural data for the elastin and collagen type I fibers (Chen et al. 2011, 2013; Rezakhaniha et al. 2012; Zoumi et al. 2004), respectively.

Efficient microstructure representation: Implementation of the structural approach in finite element stress analysis presents a heavy computational load due to the need to integrate the contributions of fibers over all their 3D orientation distribution function. Although modern fast computing facilities render this task tractable (e.g., in the case of the left ventricle (Imanuel 1996)), attempts are being made to develop a computationally more efficient representation. Although promising first steps have been achieved, a methodology which is general enough for any tissue structure and fibers' material law is still in need. The quest for such representation is a challenging theoretical/numerical undertaking. One current approach relies on the application of the generalized structural tensor (Gasser et al. 2006; Freed et al. 2005) which represents the tissue's 3D structure. In application, this tensor is multiplied by a weighted average of the fibers strain. Comparison with the exact solution revealed, however, that this method is valid only when all the fibers are in tension and when the fiber distribution span is small (Cortes et al. 2010;

Federico and Herzog 2008). Moreover, I have recently shown that the structure tensor is a reliable descriptor of tissue structure only under very limited cases (Lanir and Namani 2015). A second class of methodologies attempts to represent the tissue properties by a discrete set of fiber bundles. There are three variants: One was developed for amorphous polymers and represents the tissue properties by an eight-chain 3D rectangle unit cell whose shape determines the tissue orthotropy (Bischoff et al. 2002). While this unit cell is a structure, it is unrelated to the real tissue structure. The two other variants represent the tissue's continuous 3D orientation distribution by a discrete set of fibers. In the first, the fibers in the set have fixed orientations, and weight is assigned to each of them to optimize the fit to the tissue response (Elata and Rubin 1994). This approach was shown, however, to introduce undesirable anisotropy to an isotropic material (Bazant 1986; Ehret et al. 2010). The other variant applies the spherical t-design (Delsarte et al. 1977; Hardin and Sloane 1996). The latter is a set of N points on a sphere (or their equivalent fiber orientations) such that the average value of any polynomial f of degree t or less on this set equals the integral of f over the sphere. Hence, integration over the 3D orientation distribution is replaced by a sum of discrete values of that function. The design degree t depends on the orientation distribution function and on the fiber material law (Federico and Gasser 2010). Hence, t must be established for each case (Martufi and Gasser 2011) and may change as parameter estimates evolve during iterative search for optimal parameters during the process of tissue material characterization. In summary, it is seen that this challenge of efficient representation is yet to be met.

References

Algranati D. Physical determinants of normal and obstructed coronary flow. PhD thesis. Technion—I.I.T., Haifa; 2010.

Algranati D, Kassab GS, Lanir Y. Mechanisms of myocardium-coronary vessel interaction. Am J Physiol Heart Circ Physiol. 2010;298(3):H861–73.

Algranati D, Kassab GS, Lanir Y. Why is the subendocardium more vulnerable to ischemia? A new paradigm. Am J Physiol Heart Circ Physiol. 2011;300(3):H1090–100.

Armstrong CG, Lai WM, Mow VC. An analysis of the unconfined compression of articular cartilage. J Biomech Eng. 1984;106(2):165–73.

Bazant ZP. Efficient numerical integration on the surface of a sphere. Z Angew Math Mech. 1986;66(1):37–49.

Belkoff SM, Haut RC. A structural model used to evaluate the changing microstructure of maturing rat skin. J Biomech. 1991;24(8):711–20.

Billiar KL, Sacks MS. Biaxial mechanical properties of the native and glutaraldehyde-treated aortic valve cusp: part II—a structural constitutive model. J Biomech Eng. 2000;122:327–35.

Bischoff JE, Arruda EA, Grosh K. A microstructurally based orthotropic hyperelastic constitutive law. J Appl Mech. 2002;69:570–9.

Borg TK, Caulfield JB. Collagen in the heart. Tex Rep Biol Med. 1979;39:321–33.

Borg TK, Caulfield JB. The collagen matrix of the heart. Fed Proc. 1981;40(7):2037–41.

Borg TK, Ranson WF, Moslehy FA, Caulfield JB. Structural basis of ventricular stiffness. Lab Invest. 1981;44(1):49–54.

Brown IA. A scanning electron microscope study of the effects of uniaxial tension on human skin. Br J Dermatol. 1973;89(4):383–93.

Caulfield JB, Borg TK. The collagen network of the heart. Lab Invest. 1979;40(3):364–72.

Chandran PL, Barocas VH. Affine versus non-affine fibril kinematics in collagen networks: theoretical studies of network behavior. J Biomech Eng. 2006;128(2):259–70.

Chen H, Liu Y, Slipchenko M, Zhao X, Cheng J-X, Kassab G. The layered structure of coronary adventitia under mechanical load. Biophys J. 2011;101(11):2555–62. doi:10.1016/j.bpj.2011.10.043.

Chen H, Slipchenko MN, Liu Y, Zhao X, Cheng JX, Lanir Y, Kassab GS. Biaxial deformation of collagen and elastin fibers in coronary adventitia. J Appl Physiol. 2013;115(11):1683–93.

Chu BM, Frasher WG, Wayland H. Hysteretic behavior of soft living animal tissue. Ann Biomed Eng. 1972;1(2):182–203.

Clark JM, Glagov S. Transmural organization of the arterial media. The lamellar unit revisited. Arteriosclerosis. 1985;5:19–34.

Cortes DH, Lake SP, Kadlowec JA, Soslowsky LJ, Elliott DM. Characterizing the mechanical contribution of fiber angular distribution in connective tissue: comparison of two modeling approaches. Biomech Model Mechanobiol. 2010;9(5):651–8.

Crabb RA, Chau EP, Evans MC, Barocas VH, Hubel A. Biomechanical and microstructural characteristics of a collagen film-based corneal stroma equivalent. Tissue Eng. 2006;12(6):1565–75.

Dahl SL, Vaughn ME, Hu JJ, Driessen NJ, Baaijens FP, Humphrey JD, Niklason LE. A microstructurally motivated model of the mechanical behavior of tissue engineered blood vessels. Ann Biomed Eng. 2008;36(11):1782–92.

Delsarte P, Goethals JM, Seidel JJ. Spherical codes and designs. Geometriae Dedicata. 1977;6:363–88.

Driessen NJ, Bouten CV, Baaijens FP. A structural constitutive model for collagenous cardiovascular tissues incorporating the angular fiber distribution. J Biomech Eng. 2005;127(3):494–503.

Ehret AE, Itskov M, Schmid H. Numerical integration on the sphere and its effect on the material symmetry of constitutive equations—a comparative study. Int J Numer Meth Eng. 2010;81(2):189–206.

Elata D, Rubin M. Isotropy of strain energy functions which depend only on a finite number of directional strain measures. J Appl Mech. 1994;61(2):284–9.

Engelmayr Jr GC, Sacks MS. A structural model for the flexural mechanics of nonwoven tissue engineering scaffolds. J Biomech Eng. 2006;128(4):610–22.

Evans JH, Barbenel JC, Steel TR, Ashby AM. Structure and mechanics of tendon. Symp Soc Exp Biol. 1980;34:465–9.

Federico S, Gasser TC. Nonlinear elasticity of biological tissues with statistical fiber orientation. J R Soc Interface. 2010;7(47):955–66.

Federico S, Herzog W. Towards an analytical model of soft biological tissues. J Biomech. 2008;41(16):3309–13.

Federico S, Grillo A, La Rosa G, Giaquinta G, Herzog W. A transversely isotropic, transversely homogeneous microstructural-statistical model of articular cartilage. J Biomech. 2005;38(10):2008–18.

Freed AD, Einstein DR, Vesely I. Invariant formulation for dispersed transverse isotropy in aortic heart valves: an efficient means for modeling fiber splay. Biomech Model Mechanobiol. 2005;4(2–3):100–17.

Fung YC. Elasticity of soft tissues in simple elongation. Am J Physiol. 1967;213(6):1532–44.

Fung YC. Stress-strain-history relations of soft tissues in simple elongation. In: Perrone N, Anliker M, Fung YC, editors. Biomechanics—its foundations and objectives. Englewood Cliffs: Prentice-Hall, Inc.; 1972. p. 181–208.

Gasser TC, Ogden RW, Holzapfel GA. Hyperelastic modelling of arterial layers with distributed collagen fibre orientations. J R Soc Interface. 2006;3:15–35.

Gibson T, Kenedi RM, Craik JE. The mobile micro-architecture of dermal collagen: a bioengineering study. Br J Surg. 1965;52(10):764–70.

Grytz R, Meschke G. Constitutive modeling of crimped collagen fibrils in soft tissues. J Mech Behav Biomed Mater. 2009;2(5):522–33.

Guo X, Lanir Y, Kassab GS. Effect of osmolarity on the zero-stress state and mechanical properties of aorta. Am J Physiol Heart Circ Physiol. 2007;293(4):H2328–34.

Hansen KA, Weiss JA, Barton JK. Recruitment of tendon crimp with applied tensile strain. J Biomech Eng. 2002;124(1):72–7.

Hardin RH, Sloane NJA. McLaren's improved snub cube and other new spherical designs in three dimensions. Discrete Comput Geom. 1996;15:429–41.

Hoffman JI. Transmural myocardial perfusion. Prog Cardiovasc Dis. 1987;29(6):429–64.

Hollander Y, Durban D, Lu X, Kassab GS, Lanir Y. Experimentally validated microstructural 3D constitutive model of coronary arterial media. J Biomech Eng. 2011a;133(3):031007.

Hollander Y, Durban D, Lu X, Kassab GS, Lanir Y. Constitutive modeling of coronary arterial media—comparison of three model classes. J Biomech Eng. 2011b;133(6):061008.

Holzapfel G, Gasser T, Ogden R. A new constitutive framework for arterial wall mechanics and a comparative study of material models. In: Cowin S, Humphrey J, editors. Cardiovascular soft tissue mechanics. New York: Kluwer Academic; 2004. p. 1–48.

Holzinger Y, Rubin H. Oscillatory flow of Maxwellian fluid in thin-walled elastic circular tube. Israel J Technol. 1970;8:335–43.

Horowitz A, Lanir Y, Yin FC, Perl M, Sheinman I, Strumpf RK. Structural three-dimensional constitutive law for the passive myocardium. J Biomech Eng. 1988a;110(3):200–7.

Horowitz A, Sheinman I, Lanir Y, Perl M, Sideman S. Nonlinear incompressible finite element for simulating loading of cardiac tissue—part I: two dimensional formulation for thin myocardial strips. J Biomech Eng. 1988b;110(1):57–61.

Horowitz A, Sheinman I, Lanir Y. Nonlinear incompressible finite element for simulating loading of cardiac tissue—part II: three dimensional formulation for thick ventricular wall segments. J Biomech Eng. 1988c;110(1):62–8.

Imanuel O. Stress analysis in the left ventricle of the heart. PhD thesis. Technion—I.I.T., Haifa; 1996.

Jacobs J, Algranati D, Lanir Y. Lumped flow modeling in dynamically loaded coronary vessels. J Biomech Eng. 2008;130(5):054504.

Jhun CS, Evans MC, Barocas VH, Tranquillo RT. Planar biaxial mechanical behavior of bioartificial tissues possessing prescribed fiber alignment. J Biomech Eng. 2009;131(8):081006.

Kaimovitz B, Lanir Y, Kassab GS. Large-scale 3-D geometric reconstruction of the porcine coronary arterial vasculature based on detailed anatomical data. Ann Biomed Eng. 2005;33(11):1517–35.

Kaimovitz B, Lanir Y, Kassab GS. A full 3-D reconstruction of the entire porcine coronary vasculature. Am J Physiol Heart Circ Physiol. 2010;299(4):H1064–76.

Kassab GS. Y.C. "Bert" Fung: the father of modern biomechanics. Mech Chem Biosyst. 2004;1(1):5–22.

Kassab GS, Fung YC. Topology and dimensions of pig coronary capillary network. Am J Physiol. 1994;267(1 Pt 2):H319–25.

Kassab GS, Rider CA, Tang NJ, Fung YC. Morphometry of pig coronary arterial trees. Am J Physiol. 1993;265(1 Pt 2):H350–65.

Kassab GS, Lin DH, Fung YC. Morphometry of pig coronary venous system. Am J Physiol. 1994;267(6 Pt 2):H2100–13.

Kassab GS, Pallencaoe E, Schatz A, Fung YC. Longitudinal position matrix of the pig coronary vasculature and its hemodynamic implications. Am J Physiol. 1997;273(6 Pt 2):H2832–42.

Kwan MK, Lai WM, Mow VC. Fundamentals of fluid transport through cartilage in compression. Ann Biomed Eng. 1984;12(6):537–58.

Lake SP, Cortes DH, Kadlowec JA, Soslowsky LJ, Elliott DM. Evaluation of affine fiber kinematics in human supraspinatus tendon using quantitative projection plot analysis. Biomech Model Mechanobiol. 2011;11(1–2):197–205.

Lanir Y. Constitutive equations for the lung tissue. J Biomech Eng. 1983;105(4):374–80.

Lanir Y. Biorheology and fluid flux in swelling tissues. I. Bicomponent theory for small deformations, including concentration effects. Biorheology. 1987;24(2):173–87.

Lanir Y. Mechanisms of residual stress in soft tissues. J Biomech Eng. 2009;131(4):044506.

Lanir Y. Osmotic swelling and residual stress in cardiovascular tissues. J Biomech. 2012;45:780–9.

Lanir Y. Mechanistic micro-structural theory of soft tissues growth and remodeling: tissues with unidirectional fibers. Biomech Model Mechanobiol. 2015;14(2):245–66.

Lanir Y, Fung YCB. Fiber composite columns under compression. J Compos Mater. 1972;6:387–401.

Lanir Y, Fung YC. Two-dimensional mechanical properties of rabbit skin. I. Experimental system. J Biomech. 1974a;7(1):29–34.

Lanir Y, Fung YC. Two-dimensional mechanical properties of rabbit skin. II. Experimental results. J Biomech. 1974b;7(2):171–82.

Lanir Y, Namani R. Reliability of structure tensors in representing soft tissues structure. J Mech Behav Biomed Mater. 2015;46:222–8.

Lanir Y, Rubin H. Oscillatory flow of linear viscoelastic fluids in thin-walled elastico-viscous tubes. Rheol Acta. 1971;10:467–72.

Lanir Y, Rubin H. Pulsatile flow of linear viscoelastic fluids in elasticoviscous tubes. J Mec. 1972;11(2):295.

Lanir Y, Lichtenstein O, Imanuel O. Optimal design of biaxial tests for structural material characterization of flat tissues. J Biomech Eng. 1996a;118(1):41–7.

Lanir Y, Hayam G, Abovsky M, Zlotnick AY, Uretzky G, Nevo E, Ben-Haim SA. Effect of myocardial swelling on residual strain in the left ventricle of the rat. Am J Physiol. 1996b;270(5 Pt 2):H1736–43.

Lokshin O, Lanir Y. Micro and macro rheology of planar tissues. Biomaterials. 2009a;30(17):3118–27.

Lokshin O, Lanir Y. Viscoelasticity and preconditioning of rat skin under uniaxial stretch: microstructural constitutive characterization. J Biomech Eng. 2009b;131(3):031009.

Lourie O, Cox D, Wagner H. Buckling and collapse of embedded carbon nanotubes. Phys Rev Lett. 1998;81(8):1638–41.

Maroudas A, Bannon C. Measurement of swelling pressure in cartilage and comparison with the osmotic pressure of constituent proteoglycans. Biorheology. 1981;18(3–6):619–32.

Martufi G, Gasser TC. A constitutive model for vascular tissue that integrates fibril, fiber and continuum levels with application to the isotropic and passive properties of the infrarenal aorta. J Biomech. 2011;44(14):2544–50.

McCutchen CW. Cartilages is poroelastic, not viscoelastic (including an exact theorem about strain energy and viscous loss, and an order of magnitude relation for equilibration time). J Biomech. 1982;15(4):325–7.

Millington PF, Brown IA. Scanning electron microscope studies of some internal surfaces in human skin. Z Zellforsch Mikrosk Anat. 1970;106(2):209–19.

Nevo E, Lanir Y. Structural finite deformation model of the left ventricle during diastole and systole. J Biomech Eng. 1989;111(4):342–9.

Nevo E, Lanir Y. The effect of residual strain on the diastolic function of the left ventricle as predicted by a structural model. J Biomech. 1994;27(12):1433–46.

Raz E, Lanir Y. Recruitment viscoelasticity of the tendon. J Biomech Eng. 2009;131(11):111008.

Rezakhaniha R, Agianniotis A, Schrauwen JT, Griffa A, Sage D, Bouten CV, van de Vosse FN, Unser M, Stergiopulos N. Experimental investigation of collagen waviness and orientation in the arterial adventitia using confocal laser scanning microscopy. Biomech Model Mechanobiol. 2012;11(3–4):461–73.

Rhodin JAG. Architecture of the vessel wall. In: Bohr DF, Somlyo AD, Sparks HV, editors. Handbook of physiology—the cardiovascular system, vol. 2. Bethesda: American Physiological Society; 1980. p. 1–31.

Robinson TF, Cohen-Gould L, Factor SM. Skeletal framework of mammalian heart muscle. Arrangement of inter- and pericellular connective tissue structures. Lab Invest. 1983;49(4):482–98.

Sacks MS, Lam TV, Mayer Jr JE. A structural constitutive model for the native pulmonary valve. In: Conf Proc IEEE Eng Med Biol Soc, vol. 5; 2004. p. 3734–6.

Silberberg A. The mechanics and thermodynamics of separation flow through porous, molecularly disperse, solid media—the Poiseuille Lecture 1981. Biorheology. 1982;19(1/2):111–27.

Skalak R, Zargaryan S, Jain RK, Netti PA, Hoger A. Compatibility and the genesis of residual stress by volumetric growth. J Math Biol. 1996;34(8):889–914.

Spaan JA, Piek JJ, Hoffman JI, Siebes M. Physiological basis of clinically used coronary hemodynamic indices. Circulation. 2006;113(3):446–55.

Streeter Jr DD, Spotnitz HM, Patel DP, Ross Jr J, Sonnenblick EH. Fiber orientation in the canine left ventricle during diastole and systole. Circ Res. 1969;24(3):339–47.

Sverdlik A, Lanir Y. Time-dependent mechanical behavior of sheep digital tendons, including the effects of preconditioning. J Biomech Eng. 2002;124(1):78–84.

Tong P, Fung YC. The stress-strain relationship for the skin. J Biomech. 1976;9(10):649–57.

Truesdell C. Mechanical basis of diffusion. J Chem Phys. 1962;37:2336–44.

Truesdell C, Noll W. The non-linear field theories of mechanics. In: Flugge S, editor. Encyclopedia of physics. Berlin: Springer; 1965. p. 1–579.

Viidik A. On the rheology and morphology of soft collagenous tissue. J Anat. 1969;105(Pt 1):184.

Viidik A. Simultaneous mechanical and light microscopic studies of collagen fibers. Z Anat Entwicklungsgesch. 1972;136(2):204–12.

Viidik A. (1978) On the correlation between structure and mechanical function of soft connective tissues. Verh Anat Ges (72):75–89.

Wasano K, Yamamoto T. Tridimensional architecture of elastic tissue in the rat aorta and femoral artery—a scanning electron microscope study. J Electron Microsc (Tokyo). 1983;32(1):33–44.

Westerhof N, Boer C, Lamberts RR, Sipkema P. Cross-talk between cardiac muscle and coronary vasculature. Physiol Rev. 2006;86(4):1263–308.

Wischgoll T, Meyer J, Kaimovitz B, Lanir Y, Kassab GS. A novel method for visualization of entire coronary arterial tree. Ann Biomed Eng. 2007;35(5):694–710.

Young JM, Choy JS, Kassab GS, Lanir Y. Slackness between vessel and myocardium is necessary for coronary flow reserve. Am J Physiol Heart Circ Physiol. 2012;302(11):H2230–42.

Zoumi A, Lu X, Kassab GS, Tromberg BJ. Imaging coronary artery microstructure using second-harmonic and two-photon fluorescence microscopy. Biophys J. 2004;87(4):2778–86.

Zulliger MA, Fridez P, Hayashi K, Stergiopulos N. A strain energy function for arteries accounting for wall composition and structure. J Biomech. 2004;37(7):989–1000.

Dr. Lanir: A Personal Perspective

In 1979 as a very impressionable undergraduate in the Engineering Mechanics Program (now part of the Department of Mechanical Engineering) at Michigan State University, I was seeking some type of research experience and was very interested in the then new field of biomedical engineering that applied mechanics in some fashion. I was awarded a summer undergraduate research scholarship from the Program to work in the Department of Biomechanics located in East Fee Hall in the School of Osteopathic Medicine. There, I was exposed to all kinds of interesting and novel things: what living tissues are, how mechanics played a role in the understanding of their function, advanced computer-controlled equipment, and mathematical modeling. Heady stuff for a mere Junior. At that time the Department Chair, Dr. Robert Soutas-Little, would invite both domestic and international visitors to work for a time in the then new labs. One day Dr. Lanir showed up for what turned out to be a 2-year sabbatical. I recall his very friendly demeanor, and that he had a box of 3×5 cards for manuscript references that was very small when he arrived and that expanded to several large boxes before he left (there was no Google Scholar way back then). I often worked with Dr. Lanir, assisting in several projects and engaging in various discussions. Sometime later we were awarded an NSF summer undergraduate student research grant where, as part of a dedicated group of undergraduate students, we worked on the mechanics of skin grafts. Dr. Lanir was our faculty advisor and was very helpful in so many ways. He was never averse to meeting with us and was incredibly patient. The project was quite a success (at least to us), and we owed much to Dr. Lanir's detailed knowledge of skin biomechanics.

During his time at Michigan State University, I was introduced to Dr. Lanir's focus on tissue microstructure as a means to understand its function. I recall doing several things, including analyzing biomechanical data of spinal ligaments, measuring fiber crimp during carefully controlled uniaxial experiments on tendons. It should be noted that, at that time, there was an emphasis on developing phenomenological constitutive models of soft tissues based on related approaches to polymeric materials, which were the closest known material to soft tissues behavior. Most approaches were based on the pseudo-hyperelastic approach pioneered by Y.C. Fung. While very powerful in terms of relating the stress and strain behaviors, these approaches were limited to the ranges in which their parameters were obtained. Moreover, there were major challenges in understanding how to interpret the complex, often highly coupled behaviors. One thing that I recall vividly was a growing appreciation for the incredible diversity of functions that soft tissues exhibit, even though they were built upon a comparatively limited set of materials (collagens, elastin, proteoglycans, glycosaminoglycans, etc.). Such diversity was built upon variations in the underlying structure rather than intrinsic changes in underlying materials. To me, this was the essence of Dr. Lanir's approach. After 2 years, Dr. Lanir headed back to the Department of Biomedical Engineering at the Technion and I went on to complete my degrees and pursued an academic career.

Yet, in the time since I realized that it was during my 2-year experience with Dr. Lanir that he had taught me so many things, both biomechanical and otherwise, that have stayed with me. I have retained a strong scientific interest in biosolid micromechanics and continue to be strongly influenced by his substantial oeuvre. For example, when Kristen Billiar and I were first observing the complex biaxial mechanical responses of the aortic heart valve, my first approach for constitutive model development was to develop a complex complimentary strain energy density function, since we were utilizing stress-controlled experiments. However, I then performed some basic inverse simulations based on a variation of Dr. Lanir's structural model, and we discovered how subtle changes in fiber alignment completely captured the observed responses with only three model parameters. Observations like this made us feel that we were on the right track. Above all, Dr. Lanir showed me what it meant to do sophisticated scientific research, with all its ups and downs, at a level that I had not seen before as an undergraduate and rarely since.

It is thus with great pleasure, on behalf of my co-Editor Dr. Ghassan Kassab and all the contributing authors, to present this book as an expression of our affections and gratitude for all of Dr. Lanir's contributions to our field.

Austin, TX Michael S. Sacks

Sari Lanir, Michael Sacks, and Yoram Lanir in younger days (circa. 1996)

Happy Birthday to Yoram: Professional and Personal Reflections

The first World Congress of Biomechanics in 1990 was Chaired by Dr. Y.C. Fung and held in San Diego. It was then and there that I first met Dr. Yoram Lanir. I had just completed my Ph.D. training with Dr. Fung and I was thrilled to meet Yoram who had previously been a postdoctoral fellow with Dr. Fung. I had read many of Yoram's publications on biomechanics of blood flow and solid mechanics of soft tissue with great interest. The depth of his focus and the mystery of biomechanics was obvious and his conclusions rigorous. His works were unhurried, systematic, and definitive. He had the talent to capture the heart of the subject and simplify the analysis without loss of realism. Yoram embodied the character of a great engineer who captured the essence of a problem. He was clearly a master modeler who believed in structure-based analysis.

The first meeting with Yoram in 1990 was very memorable both academically and personally. He was extremely approachable and very humble. Despite my junior status, Yoram treated me as a colleague and was very interested in my Ph.D. thesis work on morphometry of coronary vasculature. Given his interest in structure-based analysis, he recognized the utility of the morphometric data we had labored over for several years. He advocated the use of measured geometry (anatomy, microstructure, etc.), and mechanical properties as a foundation for realistic analysis of organ function based on laws of physics with as few ad hoc assumptions as possible. The imprints of the structural-based analysis approach can be found throughout his contributions to the mechanics literature on skin, lung, tendon, heart, and vessels. He is clearly a pioneer in microstructure-based biomechanical analysis. Yoram's early work on skin mechanics and its relation to microstructure (i.e., elastin and collagen) have set the stage for similar analysis on many other organs.

The early encouragement and support I received from Yoram were long lasting. He exemplifies a great citizen of science. He is open armed and encouraging of younger scientists. Yoram has a mentoring character and draws great affection and respect from all who work with him. Our groups have reaped the rewards of numerous collaborations and friendship over the years.

In the Fall of 2010, I had the pleasure of visiting Yoram in Haifa to serve in Dr. Benjamin Kaimovitz's (one of Yoram's Ph.D. students) thesis defense. I very

Left to right: Hanna Kaimovitz (Benjamin Kaimovitz's wife), Yoram, Sari Lanir (Yoram's wife), Jawhara Kassab (Ghassan's sister), and Ghassan.

much enjoyed Yoram's hospitality and tour of Jerusalem. He managed to show me Jerusalem in a day. I was not surprised by his depth of knowledge in history, religion, and cultures nor his high energy as a nonstop tour guide. This exemplifies his character as an energetic doer with both depth and breadth. Yoram does everything with passion and completeness. As seen in the picture, the intensity wore us out and we finally rested (only for a few minutes) under an olive tree.

On behalf of my coeditor, Dr. Michael Sacks, and all the contributing authors, I wish Dr. Yoram Lanir a wonderful 70th birthday. This book is composed of our affections and gratitude for all of Yoram's contribution to our field.

Contents

Contributors

Dotan Algranati Faculty of Biomedical Engineering, Technion, Haifa 3200003, Israel

Rouzbeh Amini Department of Biomedical Engineering, Center for Cardiovascular Simulation, Institute for Computational Engineering and Science, University of Texas, Austin, TX, USA

Lynne E. Bilston Neuroscience Research Australia and University of New South Wales, Sydney, Australia

Huan Chen California Medical Innovations Institute, Roselle St, San Diego, CA 92121, USA

Daniel H. Cortes Department of Mechanical and Nuclear Engineering, Pennsylvania State University, PA

Department of Biomedical Engineering, University of Delaware, Newark, DE, USA

J. Crawford Downs Ocular Biomechanics Laboratory, Devers Eye Institute, Portland, Oregon

David Durban Faculty of Aerospace Engineering, Technion, Israel Institute of Technology, Haifa 32000, Israel

Dawn M. Elliott Department of Mechanical and Nuclear Engineering, Pennsylvania State University, PA

Department of Biomedical Engineering, University of Delaware, Newark, DE, USA

Bahar Fata Department of Biomedical Engineering, Center for Cardiovascular Simulation, Institute for Computational Engineering and Science, University of Texas, Austin, TX, USA

Cormac Flynn Department of Electrical and Computer Engineering, University of British Columbia, Vancouver, BC, Canada

Jeffrey J. Fredberg Program in Molecular and Integrative Physiological Sciences, School of Public Health, Harvard University, Boston, MA, USA

Thiranja P. Babarenda Gamage Auckland Bioengineering Institute, The University of Auckland, Auckland, New Zealand

T. Christian Gasser Department of Solid Mechanics, School of Engineering Sciences, The Royal Institute of Technology (KTH), Stockholm, Sweden

Amit Gefen Faculty of Engineering, Department of Biomedical Engineering, Tel Aviv University, Tel Aviv, Israel

Rafael Grytz Ocular Biomechanics Laboratory, Devers Eye Institute, Portland, Oregon

Julius M. Guccione Department of Surgery, University of California, San Francisco, CA 94143, Israel

Yaniv Hollander Faculty of Aerospace Engineering, Technion, Israel Institute of Technology, Haifa 32000, Israel

J.D. Humphrey Department of Biomedical Engineering, Yale University, New Haven, CT, USA

Peter J. Hunter Auckland Bioengineering Institute, The University of Auckland, Auckland, New Zealand

Yunlong Huo Peking University, China

Frank G. Jacobitz Mechanical Engineering Program, University of San Diego, San Diego, CA, USA

Jost B. Jonas Department of Ophthalmology, Medical Faculty Mannheim, Ruprecht-Karls-University Heidelberg, Mannheim, Germany

Adam M. Jones Mechanical Engineering Program, University of San Diego, San Diego, CA, USA

Jessica W.Y. Jor Auckland Bioengineering Institute, The University of Auckland, Auckland, New Zealand

Benjamin Kaimovitz IUPUI, Indianapolis, IN, USA

Ghassan S. Kassab California Medical Innovations Institute, San Diego, CA, USA

IUPUI, Indianapolis, IN, USA

Doron Klepach Department of Mechanical or Industrial Engineering, Altair, Tel Aviv University, Multiscale Design Systems, LLC, UCSF, UC Berkeley, University of California, Berkeley, USA

Yoram Lanir IUPUI, Indianapolis, IN, USA

Chung-Hao Lee Department of Biomedical Engineering, Center for Cardiovascular Simulation, Institute for Computational Engineering and Sciences (ICES), The University of Texas at Austin, Austin, TX, USA

Lik Chuan Lee Department of Mechanical Engineering, Michigan State University, East Lansing, MI 48824, USA

Xiao Lu California Medical Innovations Institute, San Diego, CA 92121, USA

Arnab Majumdar Department of Biomedical Engineering, Boston University, Boston, MA, USA

Günther Meschke Institute for Structural Mechanics, Ruhr-University Bochum, Bochum, Germany

Martyn P. Nash Department of Engineering Science, Auckland Bioengineering Institute, The University of Auckland, Auckland, New Zealand

Thao D. Nguyen Department of Mechanical Engineering, The Johns Hopkins University, Baltimore, MD, USA

Poul M.F. Nielsen Department of Engineering Science, Auckland Bioengineering Institute, The University of Auckland, Auckland, New Zealand

Cees Oomens Eindhoven University of Technology, Eindhoven, The Netherlands

Gerrit Peters Eindhoven University of Technology, Eindhoven, The Netherlands

Sonia R. Rosner Program in Molecular and Integrative Physiological Sciences, School of Public Health, Harvard University, Boston, MA, USA

M.B. Rubin Faculty of Mechanical Engineering, Technion—Israel Institute of Technology, Haifa, Israel

Michael S. Sacks, Ph.D. Department of Biomedical Engineering, Center for Cardiovascular Simulation, Institute for Computational Engineering and Sciences, The University of Texas at Austin, Austin, TX, USA

Geert W. Schmid-Schönbein Department of Bioengineering, University of California, San Diego, La Jolla, CA, USA

Naama Shoham Faculty of Engineering, Department of Biomedical Engineering, Tel Aviv University, Tel Aviv, Israel

Robert L. Steward Jr. Program in Molecular and Integrative Physiological Sciences, School of Public Health, Harvard University, Boston, MA, USA

Béla Suki Department of Biomedical Engineering, Boston University, Boston, MA, USA

Jonathan Wenk Department of Mechanical Engineering and Surgery, University of Kentucky, Lexington, KY 40506, USA

Niki L. Yamamura Mechanical Engineering Program, University of San Diego, San Diego, CA, USA

Will Zhang Department of Biomedical Engineering, Center for Cardiovascular Simulation, Institute for Computational Engineering and Science, University of Texas, Austin, TX, USA

Xuefeng Zhao California Medical Innovations Institute, San Diego, CA 92121, USA

Chapter 1
The Influence of Microstructure on Neural Tissue Mechanics

Lynne E. Bilston

Abstract Neural tissues have a complex microstructure, and this is reflected in their mechanical properties. Both brain and spinal cord tissues are heterogeneous, with white and grey matter regions having different constituents and structural arrangements. This gives rise to the complex, non-linearly viscoelastic mechanical behaviour of these tissues.

1.1 Introduction

The link between microstructure and mechanical behaviour of soft biological tissues has been the subject of much research over recent decades, led by the pioneering work of Yoram Lanir and his colleagues. Their work has laid out how, particularly in collagenous tissues, the non-linear stress–strain response, whereby soft tissues become stiffer as they are loaded, arises from progressive straightening and recruitment of fibres under stretch. Viscoelastic responses of the tissues, including stress-relaxation, preconditioning and creep can be described by constitutive models that are based on microstructural considerations. Some of these models have included mechanical anisotropy arising from structural anisotropy. Such models have been developed for skin, tendons, lung tissue, passive myocardium, brainstem, and arteries (e.g. Hollander et al. 2011a, b; Horowitz et al. 1988; Lanir 1976, 1978, 1979, 1980, 1983a, b, 1996; Lokshin and Lanir 2009; Ning et al. 2006; Wang et al. 2006).

1.2 Microstructure of Neural Tissues

The tissues of the nervous system are structurally heterogeneous. The tissues of the central nervous system (CNS), encompassing the brain and spinal cord are anatomically and mechanically distinct from those of the peripheral nervous system,

L.E. Bilston (✉)
Neuroscience Research Australia and University of New South Wales, Sydney, Australia
e-mail: l.bilston@neura.edu.au

G.S. Kassab, M.S. Sacks (eds.), *Structure-Based Mechanics of Tissues and Organs*,
DOI 10.1007/978-1-4899-7630-7_1

which we define here as the nerve roots, ganglia, and peripheral nerves. The CNS tissues are bathed in cerebrospinal fluid, and are protected by a series of collagenous membranes, the meninges, made up of the pia mater, arachnoid mater, and dura mater. The peripheral neural tissues are also protected by an external collagenous membrane, known as a nerve sheath. We consider here only the mechanics of the neural tissues themselves, excluding the meninges and neural sheaths.

1.3 CNS Tissues

1.3.1 Macrostructure of the Brain

The human brain is a highly complex structure, with structure adapted to function. The bulk of the brain is divided into two cerebral hemispheres, which sit above the brainstem and the cerebellum (Fig. 1.1). The cerebral hemispheres have a characteristic undulating surface (the gyri and sulci). The two cerebral hemispheres are joined by the corpus callosum, a highly aligned band of neurons. The brainstem connects the brain to the spinal cord, relaying efferent and afferent signals. It is

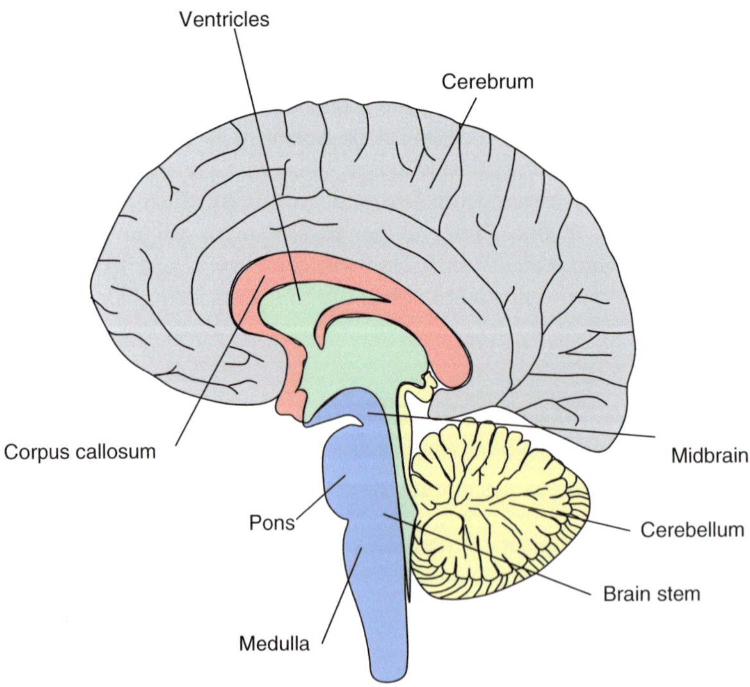

Fig. 1.1 The human brain in the mid-sagittal plane, showing the major structures. © L. Bilston, used by permission

made up of the midbrain, pons, and medulla. As well as being a major relay station, several key bodily functions reside in the brainstem, including regulation of cardiac and respiratory function. The cerebellum sits posteriorly at the base of the skull, and plays a major role in motor control.

1.3.2 Macrostructure of the Spinal Cord

The spinal cord is an elongated structure, from which two pairs of nerve roots exit at each spinal level. At the caudal end, the spinal cord tapers and a bundle of nerve roots, the cauda equina, continue to the lumbar and sacral levels.

1.3.3 Microstructure of the CNS Tissues

Brain tissue consists of neurons, glia, and other cells. The tissues of the brain are typically classified as either white or grey matter. White matter gets its colour from the white myelin sheaths (oligodendrocytes) around the axons that make it up, while grey matter neurons are largely unmyelinated. Most neural cell bodies reside in the grey matter. Both contain axons, but the grey matter axons tend to be shorter. Grey matter is more vascularised, and contains the dense dendritic networks that join neurons together and make up the functional connections of the brain.

White matter is often structurally anisotropic, due to aligned axons within specific fibre tracts, but it is not homogeneous, and many white matter regions contain multiple tracts, including crossing fibre tracts. Diffusion tensor imaging (DTI) is often used to map the detailed microstructure of the white matter in humans (e.g. Assaf and Pasternak 2008; Tsuchiya et al. 2003). DTI can measure how water molecules diffuse within the tissue in vivo, allowing directions of preferential diffusion to be identified, which are associated with white matter tracts with aligned axons.

Grey matter is also inhomogeneous, and can be either structurally isotropic or anisotropic. The grey matter of the cerebral cortex contains layers, some of which have aligned unmyelinated axons, and others which are more randomly oriented.

The brain is a mixture of white and grey matter, with the bulk of the deeper structures made up of white matter, with smaller grey matter subregions. The surface of the cerebral hemispheres (cerebral cortex) is made up of grey matter. The spinal cord consists of grey matter core, arranged in a ‘H-shaped’ cross section, surrounded by white matter (Fig. 1.2).

All of these structural and microstructural features are thought to influence the mechanical behaviour of neural tissue.

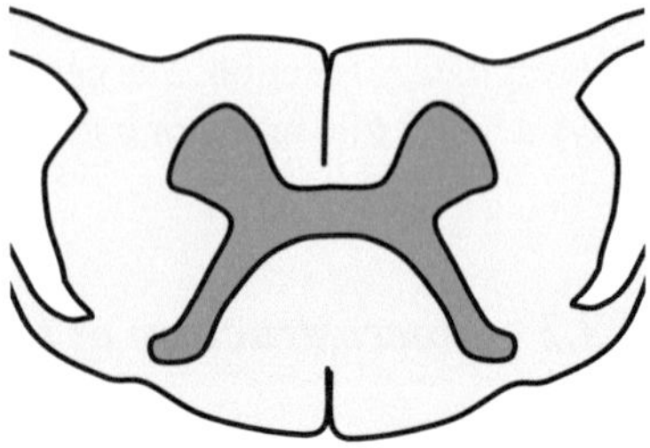

Fig. 1.2 Cross section of the spinal cord, showing grey matter 'core' surrounded by white matter

1.4 Peripheral Nervous System Tissues

Peripheral nerves leave the spinal cord through a nerve root. Pairs of nerve roots join to form a nerve at the exit from the spine, from whence they divide into smaller nerves in the periphery of the body. Each nerve is a hierarchical structure, made up of nerve fascicles, surrounded by a perineurial membrane, which are themselves made up of axons. Nerves may be either myelinated or unmyelinated. In peripheral nerves, Schwann cells are the principal support cell, and form the insulating myelin sheath in myelinated axons.

Nerve roots are also made up of a collection of axons, and the cell bodies of these axons are contained in an enlargement of the exiting nerve, called the dorsal root ganglion.

The key tissues of the peripheral nervous system are structurally anisotropic, due to the alignment of axons within nerve fascicles, and fascicles within the nerves.

1.5 Mechanical Behaviour of Neural Tissues

The mechanical behaviour of the tissues of the brain have been extensively studied at a macroscopic level over the last half-century, although their heterogeneous, non-linear viscoelastic behaviour and substantial changes to tissue behaviour post-mortem has made such studies difficult to perform reliably and data difficult to interpret. The spinal cord and peripheral nerves have been less well studied, although more recently their mechanical behaviour has begun to be elucidated.

1.5.1 CNS Tissues

Brain Tissue

Brain tissue is a non-linearly viscoelastic, multiphase tissue. Its complex mechanical behaviour arises from the inherently non-linearly viscoelastic components (neurons, glia) and also from interactions between extracellular fluid and these components. Since brain tissue behaves non-linearly at strains above 0.1–0.3 % (Bilston et al. 1997, 2001), for most practical purposes, non-linear viscoelastic approaches must be used. One exception to this are in vivo measurements of brain tissue properties performed using wave propagation methods, such as during magnetic resonance elastography (MRE), where the displacements are exceedingly small ($<10\ \mu m$) and linear viscoelasticity can be used (Green et al. 2008; Sack et al. 2008). Few studies have carefully tested the principle of strain-time separability for brain tissue, but one study suggests that it is not valid for brain tissue because the shape of the relaxation modulus varies with strain (Bilston et al. 2001), although this has often been ignored in constitutive modelling studies (see below). Once the linear viscoelastic limit is exceeded, brain exhibits shear thinning in oscillatory studies, with the shear modulus decreasing with increasing strain (Bilston et al. 1997; Nicolle et al. 2005). No studies have directly addressed the underlying structural rearrangements during such loading, and thus the microstructural mechanisms of shear thinning in brain tissue are not well understood. The observed effect is similar to that observed in polymer melts and some suspensions, however, where this has been attributed to rearrangement of the microstructure by the flow in a manner that provides less resistance to shear, including aggregation of particles and alignment of long chain polymers (Ferry 1980). In relaxation tests, brain tissue typically exhibits a power-law relaxation (e.g. Bilston et al. 2001) with no clear plateau modulus observed within the time scale of most experiments (typically hundreds of seconds). The relaxation behaviour is thought to be due to a combination of physical rearrangement of the cells and their processes, and also from interstitial fluid flow in this highly hydrated tissue. In tension, brain tissue has been shown to increase in stiffness with increasing strain, similar to many other soft biological tissues (Miller and Chinzei 2002; Miller et al. 2000). At a microstructural level, this has been assumed to be due to progressive recruitment of crimped fibres and their cross-links.

Brain white matter has typically been found to be slightly softer than grey matter (Green et al. 2008; Prange and Margulies 2002), but the microstructural reasons for this are not clear. It may relate to the presence of myelin, and fewer neural cell bodies and dendrites. The latter may act as 'cross-links' between cellular structures. Little is known about precisely how the interconnections affect tissue mechanics. Neuronal kinematics during elongation suggest that non-linear elasticity arises from similar mechanisms as in collagenous tissues—progressive straightening and recruitment of fibres (Bain et al. 2003). There is considerable heterogeneity in the mechanical behaviour of neurons and glial cells, cell bodies have been found to be stiffer than neural processes, and glial cells are softer than neurons (Lu et al. 2006). Selective removal of myelin and glial cells in embryonic spinal cords resulted in a reduction in tensile stiffness, confirming their mechanical contribution to overall CNS tissue behaviour (Shreiber et al. 2009). Studies have also shown that brain tissue mechanical behaviour is heterogeneous across the brain, with differences between the corpus callosum and corona radiata in white matter, and in different grey matter regions (Coats and Margulies 2006), between the mean behaviour of the cerebral hemispheres and the cerebellum (Zhang et al. 2011), and even within local structures such as the hippocampus (Elkin et al. 2007). These differences have been attributed to differences in the local microstructure of the tissue. The cerebellum, for example, has a much finer microstructure than the cerebrum, with delicate branching structures and lower proportion of glial cells than the cerebrum, which may explain its less stiff mechanical behaviour (Zhang et al. 2011).

The spinal cord has been much less well studied than the brain, but has also been shown to be a non-linear viscoelastic material (Bilston and Thibault 1996; Clarke et al. 2009; Fiford and Bilston 2005; Ichihara et al. 2001). Comprehensive rheological studies of shear properties have not been done due to the difficulties of sample preparation, and most reported data relies on tensile tests or compression (Bilston and Thibault 1996; Clarke et al. 2009; Fiford and Bilston 2005; Ichihara et al. 2001). Similarly to brain tissue, white matter has been shown to be slightly softer than grey matter (Ichihara et al. 2001). The spinal cord is widely assumed to be mechanically anisotropic, due to highly aligned white matter fibres (Tsuchiya et al. 2003), although formal studies confirming this are lacking. One unpublished study (Fiford 2006) has shown that the non-linear stress–strain curve observed in tensile tests of spinal cord tissue (Bilston and Thibault 1996; Clarke et al. 2009; Fiford and Bilston 2005) is likely to be due to progressive uncrimping and straightening of axons during stretch, similar to collagenous tissues. For a more detailed review of CNS tissue mechanical behaviour, see Bilston (2011) and Cheng et al. (2008).

Age can influence CNS tissue mechanical behaviour, although there are only a few studies documenting these changes, none of which directly link the changes to microstructural alterations during development and ageing. Porcine brain tissue of young animals, matching infant ages in humans, has been shown to be approximately twice as stiff as adult and older juvenile animal brain tissue (Prange and Margulies 2002). This is possibly due to reduced amounts of myelination in the youngest animal tissues. One recent MR elastography study has shown that the mean shear modulus of the human brain in vivo reduces with age in adults (Sack et al. 2009), and although the structural underpinnings of these changes have not been directly established, this may be due to cell loss as part of the ageing process.

Alterations in CNS mechanical properties as a result of disease processes are a growing area of research, at least in part because of the emerging technologies that allow measurement of in vivo CNS tissue properties, including MR elastography (Green et al. 2008, 2009; Sack et al. 2008; Kruse et al. 2008). To date, a small number of preliminary studies have examined mean brain shear modulus changes in normal pressure hydrocephalus (Streitberger et al. 2011), Alzheimer's disease (Murphy et al. 2011), demyelination (Schregel et al. 2012), and also shown that cancerous tissues can have different elastic and viscous properties to the surrounding healthy tissues (Jamin et al. 2011). In the latter work, the changes in viscous behaviour are thought to be due to differences in the vascular structure within the tissue, whereby cancerous tissue has a disorganised vascular network that acts to scatter (and thus attenuate) the mechanical waves used to probe tissue mechanical behaviour. Wave attenuation results in an apparent viscosity in the tissue in these methods.

1.5.2 Peripheral Nervous System Tissues

Peripheral nerves have largely been studied in tension. They exhibit the 'j-shaped' stress–strain curve typical of soft biological tissues, with an initial softer toe region, a linear region as strain increases, followed by yield and failure at larger strains (ranging from 20 to 70 %). The toe region of peripheral nerves is very soft, with little tension developed up to 15 % of tensile strain (Kwan et al. 1992). Many early studies report the average elastic modulus for the linear region of the stress–strain curve, which ranges from 0.5 to 28.5 MPa, depending on the nerve tested and the species. Peripheral nerves are also viscoelastic, with significant strain-rate dependent behaviour (Singh et al. 2006), and exhibiting substantial relaxation if elongated. In compression, similar overall behaviour is seen, with an initial toe region followed by a stiffening response, although the only detailed study is of nerve roots (Hubbard et al. 2008). Peripheral nerve mechanics have recently been reviewed in detail elsewhere (Nicholson and Winkelstein 2011).

1.6 Structure–Function Relationships

One of the most challenging and interesting aspects of researching neural tissue biomechanics is defining the link between structure, mechanics, and neural function. There is a plethora of work in the neurosciences and neurological domains that aims to link structure and function, using both traditional neuropathological and histological techniques and modern in vivo imaging techniques. Research to link neurological function to mechanical behaviour was initially driven by studies of traumatic brain injury, where neurological deficits were observed as a result of mechanical loading, despite the apparent macroscopic integrity of tissue. It is now well known that much of the damage to neural tissues arising from trauma occurs because of secondary damage that occurs hours later than the primary mechanical insult (Povlishock and Christman 1995). More recently, there has been interest in studying many other situations and medical disorders in which mechanical loading of neural tissues are thought to play a key role, including structural neurological disorders (e.g. hydrocephalus, syringomyelia), neurosurgical simulation, understanding how the nervous system develops in the embryo and foetus, and more. The challenge of developing mathematical models that can predict neural function as a function of mechanical loading is still an open one, although biomechanics researchers continue to work intensely on the problem. Starting at the cellular level and working up to the whole organism level, some of the particular challenges include those outlined in Fig. 1.3.

Meeting each of these challenges requires both reliable experimental data and robust validated mathematical models that can predict the tissue response across a broad range of loading conditions. The vast majority of research covering the functional response of neural tissues to mechanical loading has been conducted using

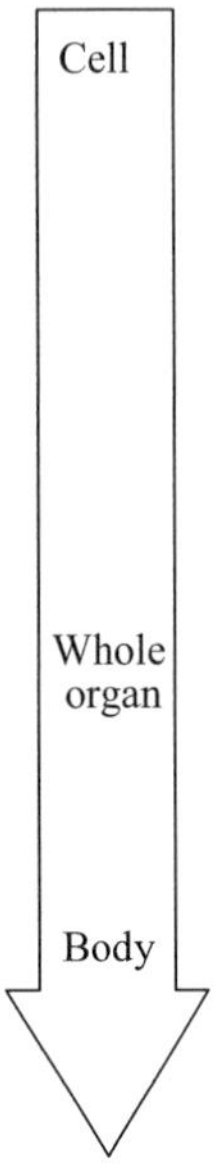

1. Defining the precise loading conditions (strain, strain rate, duration) that result in axonal functional and structural injury;
2. Determining how the population of nerve cells are loaded as a result of local tissue-level loading;
3. Determining how these relationships vary across the different regions of the nervous system (brain, spinal cord, peripheral nerves);
4. Determining how local regions of the nervous system are deformed within the overall loading of the organ;
5. Determining how the neural tissues are loaded when the body is loaded.

Fig. 1.3 Challenges in understanding neural tissue biomechanics and the relationship to microstructure

animal models, ranging from rodents to primates. Human studies of traumatic brain and spinal cord injury are largely limited to retrospective reconstructions of injury circumstances, often using finite element models to estimate tissue deformation occurring as a result of the injurious situation, and comparison with the known tissue injuries. This finite element modelling work has recently been reviewed elsewhere (Yang et al. 2011). Cadaveric studies can provide some insight into structural tissue failure, but not functional outcomes. Biomechanical modelling studies of structural neurological disorders are also typically based on retrospective analysis, for example, attempting to reproduce the dilation of the ventricles observed in a hydrocephalus patient as observed from MRI scans (Pena et al. 1999). Controlled studies where the precise loading conditions are known and the neural response is recorded, however, are rare in humans, for both practical and ethical reasons, even at physiological loading levels.

In the CNS, primate studies of brain injury (Gennarelli and Thibault 1982; Gennarelli et al. 1982) and spinal cord injury (e.g. Clarke and Bilston 2008; Clarke et al. 2008; Fiford et al. 2004) have suggested that both the amount of tissue deformation and the rate of deformation are important for predicting neural dysfunction. More recently, in studies aimed at linking microstructure to injury in brain tissue suggest that the role of strain rate is less clear (Cater et al. 2006).

In peripheral nerves, neurological deficits are associated with either compression (Nicholson and Winkelstein 2011; Clarke et al. 2007; Han et al. 2010) or stretch (e.g. Kwan et al. 1989, 1992). Neurological function can be disrupted by macroscopic mechanical loading either directly or indirectly via mechanically induced ischemia (Tanoue et al. 1996). Responses depend on the degree of deformation, the loading rate (Hubbard et al. 2008; Clarke et al. 2007; Rothman et al. 2010; Rothman and Winkelstein 2007), and the duration of loading, reflecting the viscoelastic nature of these tissues. The threshold for abnormal function can be altered by inflammation (Dilley et al. 2005). Precise details of how this relates to the local microstructure are unclear, although one study suggests that the perineurium fails before the nerve sheath (Kwan et al. 1992) and there are differences in compliance in nerves along their length (Phillips et al. 2004). One study of nerve roots under tension reported decreases in nerve conduction velocity with increasing strain and strain rate, and above a strain of 20 %, nerve conduction was blocked entirely (Singh et al. 2009).

1.7 Modelling of Neural Tissue Mechanical Behaviour

There have been many constitutive models proposed for CNS tissues, particularly for brain tissue. The majority of these are phenomenological, with only a few based on microstructural considerations. Early constitutive models attempted to use linear viscoelastic models (e.g. Fallenstein et al. 1969; Galford and McElhaney 1970), typically spring and dashpot-based models, but these are not valid for modelling the non-linear viscoelastic behaviour of CNS tissue for most practical applications, and have been superseded by non-linear models. The non-linear models tend to fall into two camps, those based on quasilinear viscoelasticity (e.g. Miller 1999) and those that are more general (e.g. Bilston et al. 2001; Brands et al. 2004; Darvish and Crandall 2001). The quasilinear viscoelastic models typically use a hyperelastic formulation (most commonly an Ogden model) to describe the non-linear elasticity, together with a linear viscoelastic relaxation modulus (most commonly a Prony series, but other formulations also exist) to match the viscous behaviour. As noted above, the strain-time separability of brain tissue has been questioned, which suggests that these models may be inappropriate for use to describe brain tissue, however, the deviations from strain-time separability are not large at moderate strains (Bilston et al. 1997, 2001; Darvish and Crandall 2001), and given the complexity of the mechanical response of brain tissue, the errors introduced by this assumption may be acceptable, depending on the application. Indeed, for neurosurgical simulation, the details of the constitutive model have been suggested to be less critical than the numerical formulation used to model incompressibility (Wittek et al. 2009). Models that attempt to capture the full range of non-linear behaviour of CNS tissue have included those that model the variation in relaxation behaviour with strain, rate-dependent viscosity (Bilston et al. 2001; Hrapko et al. 2006) and yield behaviour. Few of these models have directly considered the microstructure as the basis for modelling, although Yoram Lanir's work has shown that hyperelastic models can incorporate microstructural information, including fibre bundle waviness and recruitment with strain, through choice of an appropriate strain energy density function (Lanir 2009).

Mechanical anisotropy of brain tissue appears to vary across the brain (as does the degree of structural anisotropy as determined by DTI (Pierpaoli et al. 1996)), and constitutive models describing the anisotropy have been suggested. Velardi et al. (2006) used a transversely isotropic hyperelastic (not viscoelastic) model to describe the response of porcine corpus callosum and corona radiata tissue samples. Cloots et al. (2011) used multiscale finite element modelling to simulate the axonal, tissue, and macroscopic level behaviour, including anisotropy. None of these models fully capture the anisotropic non-linear viscoelastic behaviour of brain tissue. There is still much work to be done in developing comprehensive microstructurally based constitutive models for neural tissue.

Recent work combining DTI with MR elastography is a promising method of identifying subject-specific anisotropic mechanical properties in neural tissues (Qin et al. 2013). While this data is, so far, limited to the small strain linear viscoelastic regime, it could be combined with other forms of deformation (e.g. Clarke et al. 2011) to provide a comprehensive picture of the complex mechanical behaviour of neural tissues, and used for structurally based constitutive equations.

For some applications, such as the study of hydrocephalus (and the related spinal cord condition, syringomyelia) and those involving tissue oedema, alterations in the distribution of fluid within neural tissues play a key role. A small number of experimental studies (e.g. Cheng and Bilston 2007) and modelling studies have described brain tissue as a two-phase material, using poroelastic-type models. The simplest poroelastic (or biphasic) models have been shown to be unable to reproduce the strain-rate sensitivity observed in brain tissue (Chinzei and Miller 1996), however a poro-hyper-viscoelastic model has been more successful (Cheng and Bilston 2007) and has been used to model the time-dependent response of the brain leading to enlarged ventricles (Cheng and Bilston 2010). This model incorporates an intrinsically viscoelastic porous matrix through which an inviscid fluid permeates. The model has not been adapted to incorporate white matter anisotropy to date, although the mathematical framework to do so is relatively straightforward. Franceschini et al. (2006) modelled brain tissue using a poro-hyperelastic model, and García and Smith (2009) developed a biphasic hyperelastic model, focussed on interstitial transport of solutes (such as drugs) in the brain.

1.8 Future Directions

It is clear from this overview of the microstructure, mechanics, and functional responses of neural tissues that robust, validated constitutive models for neural tissue that link mechanical behaviour to tissue microstructure remain a considerable research challenge. Key challenges exist across the experimental, microstructural, functional, and modelling domains. Specifically, while there are a plethora of studies testing neural tissues in specific loading conditions, there are very few well-conducted experimental studies that characterise tissue response across a broad range of loading regimes, including tension, compression, shear orientations, under relaxation, oscillation and more complex loading histories, at large and small deformations and deformation rates, and across different anatomical regions. There is also a need to directly match the microstructure to observed responses, particularly during loading, to better understand how tissue structure influences observed mechanical behaviour, and how loading influences tissue structure. Given the importance of the functional activity of neural tissues, linking mechanical response to neural activity also requires further work, particularly delineating how loading of individual tissue components (neurons, glial, vasculature) interact to affect neural activity at different time scales ranging from high loading rates typical of trauma to the slow loading rates typical of structural neurological disorders, and peripheral nerve compression pathologies such as carpal tunnel syndrome.

Acknowledgment Lynne Bilston is supported by a National Health and Medical Research Council of Australia Senior Research Fellowship.

References

Assaf Y, Pasternak O. Diffusion tensor imaging (DTI)-based white matter mapping in brain research: a review. J Mol Neurosci. 2008;34(1):51–61.

Bain AC, Shreiber DI, Meaney DF. Modeling of microstructural kinematics during simple elongation of central nervous system tissue. J Biomech Eng. 2003;125:798.

Bilston LE, editor. Neural tissue biomechanics. Studies in mechanobiology, tissue engineering and biomaterials. Berlin: Springer; 2011.

Bilston LE, Thibault LE. The mechanical properties of the human cervical spinal cord in vitro. Ann Biomed Eng. 1996;24(1):67–74.

Bilston LE, Liu Z, Phan-Thien N. Linear viscoelastic properties of bovine brain tissue in shear. Biorheology. 1997;34(6):377–85.

Bilston LE, Liu Z, Phan-Thien N. Large strain behaviour of brain tissue in shear: some experimental data and differential constitutive model. Biorheology. 2001;38(4):335–45.

Brands DWA, Peters GWM, Bovendeerd PHM. Design and numerical implementation of a 3-D non-linear viscoelastic constitutive model for brain tissue during impact. J Biomech. 2004;37(1):127–34.

Cater HL, Sundstrom LE, Morrison B. Temporal development of hippocampal cell death is dependent on tissue strain but not strain rate. J Biomech. 2006;39(15):2810–8.

Cheng S, Bilston LE. Unconfined compression of white matter. J Biomech. 2007;40(1):117–24.

Cheng S, Bilston LE. Computational model of the cerebral ventricles in hydrocephalus. J Biomech Eng. 2010;132:054501.

Cheng S, Clarke EC, Bilston LE. Rheological properties of the tissues of the central nervous system: a review. Med Eng Phys. 2008;30(10):1318–37.

Chinzei K, Miller K. Compression of swine brain tissue: experiment in vitro. J Mech Eng Lab. 1996;50(4):106–15.

Clarke EC, Bilston LE. Contrasting biomechanics and neuropathology of spinal cord injury in neonatal and adult rats following vertebral dislocation. J Neurotrauma. 2008;25(7):817–32.

Clarke EC, McNulty PA, Macefield VG, Bilston LE. Mechanically evoked sensory and motor responses to dynamic compression of the ulnar nerve. Muscle Nerve. 2007;35(3):303–11.

Clarke EC, Choo AM, Liu J, Lam CK, Bilston LE, Tetzlaff W, et al. Anterior fracture-dislocation is more severe than lateral: a biomechanical and neuropathological comparison in rat thoracolumbar spine. J Neurotrauma. 2008;25(4):371–83.

Clarke EC, Cheng S, Bilston LE. The mechanical properties of neonatal rat spinal cord in vitro, and comparisons with adult. J Biomech. 2009;42(10):1397–402.
Clarke E, Cheng S, Green M, Sinkus R, Bilston L. Using static preload with magnetic resonance elastography to estimate large strain viscoelastic properties of bovine liver. J Biomech. 2011;44(13):2461–5.
Cloots RJH, van Dommelen JAW, Nyberg T, Kleiven S, Geers MGD. Micromechanics of diffuse axonal injury: influence of axonal orientation and anisotropy. Biomech Model Mechanobiol. 2011;10(3):413–22.
Coats B, Margulies SS. Material properties of porcine parietal cortex. J Biomech. 2006;39(13):2521–5.
Darvish KK, Crandall JR. Nonlinear viscoelastic effects in oscillatory shear deformation of brain tissue. Med Eng Phys. 2001;23(9):633–45.
Dilley A, Lynn B, Pang SJ. Pressure and stretch mechanosensitivity of peripheral nerve fibres following local inflammation of the nerve trunk. Pain. 2005;117(3):462–72.
Elkin BS, Azeloglu EU, Costa KD, Morrison 3rd B. Mechanical heterogeneity of the rat hippocampus measured by atomic force microscope indentation. J Neurotrauma. 2007;24(5):812–22.
Fallenstein GT, Hulce VD, Melvin JW. Dynamic mechanical properties of human brain tissue. J Biomech. 1969;2(3):217–26.
Ferry J. Viscoelastic properties of polymers. New York: Wiley; 1980.
Fiford RJ. Biomechanics of spinal cord injury in a novel rat model. PhD thesis, University of Sydney; 2006.
Fiford RJ, Bilston LE. The mechanical properties of rat spinal cord in vitro. J Biomech. 2005;38(7):1509–15.
Fiford RJ, Bilston LE, Waite P, Lu J. A vertebral dislocation model of spinal cord injury in rats. J Neurotrauma. 2004;21(4):451–8.
Franceschini G, Bigoni D, Regitnig P, Holzapfel GA. Brain tissue deforms similarly to filled elastomers and follows consolidation theory. J Mech Phys Solids. 2006;54(12):2592–620.
Galford JE, McElhaney JH. A viscoelastic study of scalp, brain, and dura. J Biomech. 1970;3: 211–21.
García J, Smith J. A biphasic hyperelastic model for the analysis of fluid and mass transport in brain tissue. Ann Biomed Eng. 2009;37(2):375–86.
Gennarelli TA, Thibault LE. Biomechanics of acute subdural hematoma. J Trauma. 1982;22(8):680–6.
Gennarelli TA, Thibault LE, Adams JH, Graham DI, Thompson CJ, Marcincin RP. Diffuse axonal injury and traumatic coma in the primate. Ann Neurol. 1982;12(6):564–74.
Green MA, Bilston LE, Sinkus R. In vivo brain viscoelastic properties measured by magnetic resonance elastography. NMR Biomed. 2008;21(7):755–64.
Green MA, Bilston LE, van Houten E, Sinkus R, editors. In-vivo brain viscoelastic anisotropic properties using DTI and MR-elastography. In: Proceedings of the 17th ISMRM; Honolulu, HI; 2009.
Han SE, Lin CSY, Boland RA, Bilston LE, Kiernan MC. Changes in human sensory axonal excitability induced by focal nerve compression. J Physiol. 2010;588(10):1737–45.
Hollander Y, Durban D, Lu X, Kassab GS, Lanir Y. Experimentally validated microstructural 3D constitutive model of coronary arterial media. J Biomech Eng. 2011a;133(3):031007.
Hollander Y, Durban D, Lu X, Kassab GS, Lanir Y. Constitutive modeling of coronary arterial media—comparison of three model classes. J Biomech Eng. 2011b;133(6):061008.
Horowitz A, Lanir Y, Yin FC, Perl M, Sheinman I, Strumpf RK. Structural three-dimensional constitutive law for the passive myocardium. J Biomech Eng. 1988;110(3):200–7.
Hrapko M, van Dommelen JAW, Peters GWM, Wismans JSHM. The mechanical behaviour of brain tissue: large strain response and constitutive modelling. Biorheology. 2006;43(5): 623–36.
Hubbard RD, Chen Z, Winkelstein BA. Transient cervical nerve root compression modulates pain: load thresholds for allodynia and sustained changes in spinal neuropeptide expression. J Biomech. 2008;41(3):677–85.

Ichihara K, Taguchi T, Shimada Y, Sakuramoto I, Kawano S, Kawai S. Gray matter of the bovine cervical spinal cord is mechanically more rigid and fragile than the white matter. J Neurotrauma. 2001;18(3):361–7.
Jamin Y, Boult JK, Bamber JC, Sinkus R, Robinson SP, editors. High resolution magnetic resonance elastography of orthotopic murine glioma in vivo. Montreal: ISMRM; 2011.
Kruse SA, Rose GH, Glaser KJ, Manduca A, Felmlee JP, Jack Jr CR, et al. Magnetic resonance elastography of the brain. Neuroimage. 2008;39(1):231–7.
Kwan MK, Wall EJ, Weiss JA, Rydevik BL, Garfin SR, Woo SL-Y. Biomechanical analysis of rabbit peripheral nerve: in situ stresses and strains. ASME Biomech Symp AMD. 1989;98: 109–12.
Kwan MK, Wall EJ, Massie J, Garfin SR. Strain, stress and stretch of peripheral nerve rabbit experiments in vitro and in vivo. Acta Orthop. 1992;63(3):267–72.
Lanir Y. Biaxial stress-relaxation in skin. Ann Biomed Eng. 1976;4(3):250–70.
Lanir Y. Structure-strength relations in mammalian tendon. Biophys J. 1978;24(2):541–54.
Lanir Y. A structural theory for the homogeneous biaxial stress-strain relationships in flat collagenous tissues. J Biomech. 1979;12(6):423–36.
Lanir Y. A microstructure model for the rheology of mammalian tendon. J Biomech Eng. 1980;102(4):332–9.
Lanir Y. Constitutive equations for the lung tissue. J Biomech Eng. 1983a;105(4):374–80.
Lanir Y. Constitutive equations for fibrous connective tissues. J Biomech. 1983b;16(1):1–12.
Lanir Y. Plausibility of structural constitutive equations for swelling tissues—implications of the C-N and S-E conditions. J Biomech Eng. 1996;118(1):10–6.
Lanir Y. Physical mechanisms of soft tissues rheological properties. In: Chien S, editor. Biomechanics: from molecules to man: tributes to Yuan-Cheng Fung on his 90th birthday. Singapore: World Scientific; 2009. p. 1–12.
Lokshin O, Lanir Y. Viscoelasticity and preconditioning of rat skin under uniaxial stretch: microstructural constitutive characterization. J Biomech Eng. 2009;131(3):031009.
Lu Y-B, Franze K, Seifert G, Steinhäuser C, Kirchhoff F, Wolburg H, et al. Viscoelastic properties of individual glial cells and neurons in the CNS. Proc Natl Acad Sci. 2006;103(47):17759–64.
Miller K. Constitutive model of brain tissue suitable for finite element analysis of surgical procedures. J Biomech. 1999;32(5):531–7.
Miller K, Chinzei K. Mechanical properties of brain tissue in tension. J Biomech. 2002;35(4): 483–90.
Miller K, Chinzei K, Orssengo G, Bednarz P. Mechanical properties of brain tissue in-vivo: experiment and computer simulation. J Biomech. 2000;33(11):1369–76.
Murphy MC, Huston J, Jack CR, Glaser KJ, Manduca A, Felmlee JP, et al. Decreased brain stiffness in Alzheimer's disease determined by magnetic resonance elastography. J Magn Reson Imaging. 2011;34(3):494–8.
Nicholson KJ, Winkelstein BA. Nerve and nerve root biomechanics. In: Bilston LE, editor. Neural tissue biomechanics. Studies in mechanobiology, tissue engineering and biomaterials. Springer: Berlin; 2011. p. 203–29.
Nicolle S, Lounis M, Willinger R, Palierne JF. Shear linear behavior of brain tissue over a large frequency range. Biorheology. 2005;42(3):209–23.
Ning X, Zhu Q, Lanir Y, Margulies SS. A transversely isotropic viscoelastic constitutive equation for brainstem undergoing finite deformation. J Biomech Eng. 2006;128(6):925–33.
Pena A, Bolton MD, Whitehouse H, Pickard JD. Effects of brain ventricular shape on periventricular biomechanics: a finite-element analysis. Neurosurgery. 1999;45(1):107.
Phillips JB, Smit X, De Zoysa N, Afoke A, Brown RA. Peripheral nerves in the rat exhibit localized heterogeneity of tensile properties during limb movement. J Physiol. 2004;557(3):879–87.
Pierpaoli C, Jezzard P, Basser PJ, Barnett A, Di Chiro G. Diffusion tensor MR imaging of the human brain. Radiology. 1996;201(3):637–48.
Povlishock JT, Christman CW. The pathobiology of traumatically induced axonal injury in animals and humans: a review of current thoughts. J Neurotrauma. 1995;12(4):555–64.

Prange MT, Margulies SS. Regional, directional, and age-dependent properties of the brain undergoing large deformation. J Biomech Eng. 2002;124(2):244–52.
Qin EC, Sinkus R, Geng G, Cheng S, Green M, Rae CD, Bilston LE. Combining MR elastography and diffusion tensor imaging for the assessment of anisotropic mechanical properties: a phantom study. J Magn Reson Imaging. 2013;37:217–26.
Rothman SM, Winkelstein BA. Chemical and mechanical nerve root insults induce differential behavioral sensitivity and glial activation that are enhanced in combination. Brain Res. 2007;1181:30–43.
Rothman SM, Nicholson KJ, Winkelstein BA. Time-dependent mechanics and measures of glial activation and behavioral sensitivity in a rodent model of radiculopathy. J Neurotrauma. 2010;27(5):803–14.
Sack I, Beierbach B, Hamhaber U, Klatt D, Braun J, Sack I, et al. Non-invasive measurement of brain viscoelasticity using magnetic resonance elastography. NMR Biomed. 2008;21(3): 265–71.
Sack I, Beierbach B, Wuerfel J, Klatt D, Hamhaber U, Papazoglou S, et al. The impact of aging and gender on brain viscoelasticity. Neuroimage. 2009;46(3):652–7.
Schregel K, Wuerfel née Tysiak E, Garteiser. P, Gemeinhardt I, Prozorovski T, Aktas O, et al. Demyelination reduces brain parenchymal stiffness quantified in vivo by magnetic resonance elastography. Proc Natl Acad Sci U S A. 2012;109(17):6650–5.
Shreiber D, Hao H, Elias R. Probing the influence of myelin and glia on the tensile properties of the spinal cord. Biomech Model Mechanobiol. 2009;8(4):311–21.
Singh A, Lu Y, Chen C, Cavanaugh JM. Mechanical properties of spinal nerve roots subjected to tension at different strain rates. J Biomech. 2006;39(9):1669–76.
Singh A, Kallakuri S, Chen C, Cavanaugh JM. Structural and functional changes in nerve roots due to tension at various strains and strain rates: an in-vivo study. J Neurotrauma. 2009;26(4): 627–40.
Streitberger K-J, Wiener E, Hoffmann J, Freimann FB, Klatt D, Braun J, et al. In vivo viscoelastic properties of the brain in normal pressure hydrocephalus. NMR Biomed. 2011;24(4):385–92.
Tanoue M, Yamaga M, Ide J, Takagi K. Acute stretching of peripheral nerves inhibits retrograde axonal transport. J Hand Surg Br. 1996;21(3):358–63.
Tsuchiya K, Katase S, Fujikawa A, Hachiya J, Kanazawa H, Yodo K. Diffusion-weighted MRI of the cervical spinal cord using a single-shot fast spin-echo technique: findings in normal subjects and in myelomalacia. Neuroradiology. 2003;45(2):90–4.
Velardi F, Fraternali F, Angelillo M. Anisotropic constitutive equations and experimental tensile behavior of brain tissue. Biomech Model Mechanobiol. 2006;5(1):53–61.
Wang C, Garcia M, Lu X, Lanir Y, Kassab GS. Three-dimensional mechanical properties of porcine coronary arteries: a validated two-layer model. Am J Physiol Heart Circ Physiol. 2006;291(3):H1200–9.
Wittek A, Hawkins T, Miller K. On the unimportance of constitutive models in computing brain deformation for image-guided surgery. Biomech Model Mechanobiol. 2009;8(1):77–84.
Yang KH, Mao H, Wagner C, Zhu F, Chou CC, King AI. Modeling of the brain for injury prevention. In: Bilston LE, editor. Neural tissue biomechanics. Studies in mechanobiology, tissue engineering and biomaterials. Berlin: Springer; 2011. p. 69–120.
Zhang J, Green MA, Sinkus R, Bilston LE. Viscoelastic properties of human cerebellum using magnetic resonance elastography. J Biomech. 2011;44(10):1909–13.

Chapter 2
Modeling of Collagenous Tissues Using Distributed Fiber Orientations

Daniel H. Cortes and Dawn M. Elliott

Abstract Collagen is the most abundant protein in mammals and is the major component of load-bearing tissues including tendons, ligaments, cartilage, and others. The mechanical behavior of collagenous tissues depends on the relative collagen content and its organization. Fiber orientation plays a crucial role in the mechanical behavior of these tissues. Several mechanical properties such as anisotropy and Poisson's ratio are mostly determined by fiber organization. Additionally, mechanical models that include fiber orientation distributions better predict the mechanical behavior of collagenous tissues. Dr. Lanir proposed a pioneering formulation to model the mechanics of collagenous tissues that includes fiber nonlinearity, buckling, and distributed orientations. This formulation had been used to model a variety of tissues and is considered the gold standard for the analysis of distributed fibers. The objective of this chapter is to describe the methods to analyze the mechanical behavior of tissues with fiber orientation distributions. This chapter includes methods to measure fiber orientation, a detailed description of Lanir's formulation, simplified versions of Lanir's approach, and applications to several collagenous tissues.

2.1 Introduction

Collagen is the most abundant protein in mammals and is the major component of load-bearing tissues including tendons, ligaments, cartilage, disc, skin, arteries, and valves (Fratzl 2008). The fibrillar collagens, particularly types I, II, and III are primarily present in these tissues that are responsible for mechanical load transmission (Hulmes 2008). These load-bearing tissues are also composed of other structural proteins such as elastin and proteoglycans. The mechanical behavior of

D.H. Cortes • D.M. Elliott, Ph.D. (✉)
Department of Mechanical and Nuclear Engineering, Pennsylvania State University, 329 Leonhard Building University Park, PA 16802

Department of Biomedical Engineering, University of Delaware, 161 Colburn Laboratory, 150 Academy Street, Newark, DE 19716, USA
e-mail: delliott@udel.edu

G.S. Kassab, M.S. Sacks (eds.), *Structure-Based Mechanics of Tissues and Organs*,
DOI 10.1007/978-1-4899-7630-7_2

collagenous tissues depends on the relative collagen content and its organization. Fiber orientation plays a crucial role in the mechanical behavior of these tissues. For example, tendons and ligaments are largely composed of collagen type I, and collagen fibers have a preferred orientation in the direction where load is transmitted. Conversely, tissues loaded in multiple directions, such as the annulus fibrosus of the intervertebral disc or the adventitia of arteries, have two or more preferred fiber orientations. The degree of anisotropy and both the modulus and Poisson's ratio are influenced by the fiber organization (Ateshian et al. 2009).

Mechanical models of collagenous tissues explicitly consider fibers as one of their main components. For simplicity, fibers are often considered to be aligned. However, several studies have shown that mechanical models that include the orientation distribution of fibers better describe the experimental mechanical behavior (Ateshian et al. 2009; Gasser et al. 2006; Sacks 2003; Federico and Herzog 2008). For instance, a model of articular cartilage that considers fiber orientation distribution is able to predict several experimental observations such as tension-compression nonlinearity, high Poisson's ratio in tension, and stiffening after proteoglycan depletion (Ateshian et al. 2009). Therefore, modeling the fiber orientation distribution is crucial for the accurate prediction of the mechanical behavior of collagenous tissues.

The objective of this chapter is to describe the methods to analyze the mechanical behavior of collagenous tissues with fiber orientation distributions. Techniques to measure fiber orientation distributions are presented in Sect. 2.2. Two approaches to model the mechanical behavior of fibers with distributed orientations, Angular Integration and Structural Tensors, are presented in Sect. 2.3. Lanir proposed one of these methods, Angular Integration, in 1983 and it is still considered the gold standard formulation for the mechanics of distributed fibers. The other method to analyze fiber distributions, Structural Tensors, is a simplification of Lanir's formulation that is numerically more efficient. A brief comparison between those approaches is also presented. To show the advantages of considering fiber orientation distributions, applications of these modeling approaches to different tissues are presented in Sect. 2.4. Finally, concluding remarks are presented in Sect. 2.5.

2.2 Experimental Measurements of Fiber Orientation Distribution

The mechanical behavior of collagenous tissues is greatly influenced by the collagen fiber organization. Fiber orientation is a structural parameter that defines the anisotropy of the tissue. For instance, the modulus of tendons is greater in the direction of the fibers than in the transverse direction. For some collagenous tissues, such as Achilles tendon, the orientation of the fibers is predominantly in one direction. These tissues can be accurately modeled using a fiber population with a single orientation. However, collagen fibers in tissues such arteries, mitral valve, and articular cartilage have a wide range of fiber orientations. To accurately model the mechanical behavior of these tissues is important to consider the distribution of fiber orientation. In this section, methods to measure fiber orientation distributions are discussed.

2.2.1 Small Angle Light Scattering

Small angle light scattering (SALS) is an optical technique where a laser light is passed through a thin specimen and a portion of the incident light is scattered due to different refraction indices of the fibers and the surrounding matrix (Sacks and Chuong 1992). The scattered pattern is measured by rotating a linear array of photo-detectors around the optical axis. The pattern is related to the 2D Fourier transformation of the transmitted light (Yang et al. 1987). The general features of a typical SALS scattered pattern are shown using a surface plot when the height represents the intensity of transmitted light (Fig. 2.1a). The contour plot (Fig. 2.1b) shows two preferred directions for this example. Since light scatters in a direction perpendicular to the long axis of a fiber, there is a 90° shift between the scattered light and the fiber orientation. A plot of the angular distribution of light intensities also shows two preferred orientations (Fig. 2.1c). From this plot the main fiber orientations, the spread of the fiber orientation distribution (e.g., variance), and the volume fraction of fibers can be obtained. The preferred orientation is obtained by locating the angle where light intensity reaches a maximum. The variance can be obtained by fitting a distribution function to each of the fiber populations. Finally, the volume fraction of each fiber population can be calculated as the ratio of the area under the distribution function to the total area of the scan. This technique has been applied to pericardium (Sacks 2003), aortic valves (Billiar and Sacks 2000), and other tissues (Sacks and Chuong 1992; Gilbert et al. 2008; Waldman et al. 1999).

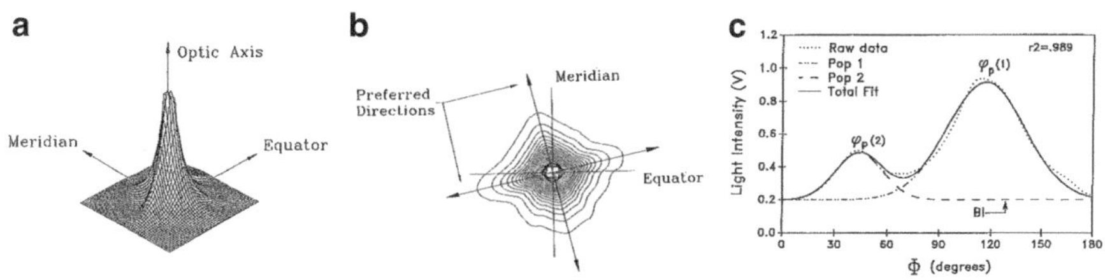

Fig. 2.1 Features of a typical small angle light scattering (SALS) experiment. (**a**) Surface plot shows the intensity of scattered laser light; (**b**) a contour plot reveals preferred fiber directions. (**c**) Fiber-orientation parameters can be calculated by fitting distribution functions to the angular distribution of light intensity. Adapted from Sacks and Chuong (1992)

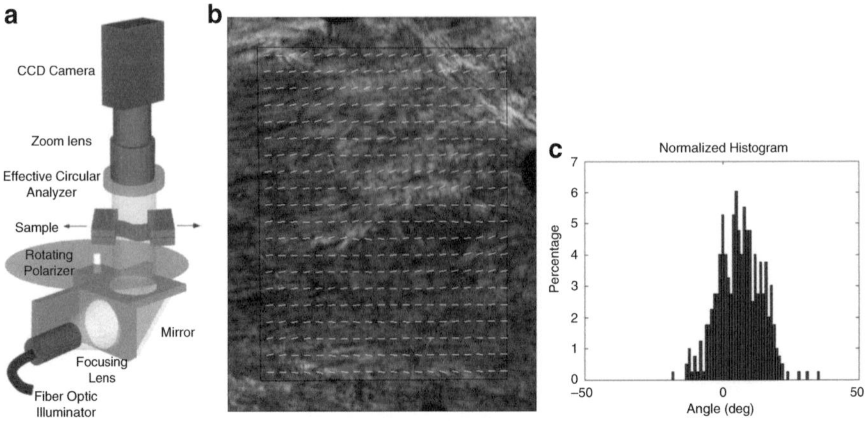

Fig. 2.2 Quantitative polarized light-mechanical testing system. (**a**) Imaging and testing setup (adapted from Tower et al. (2002)), (**b**) Orientation map on a supraspinatus tendon sample, (**c**) normalized histogram of fiber orientations calculated from the orientation map shown in (**b**)

2.2.2 *Quantitative Polarized Light*

This technique uses the birefringence properties of collagen fibers to determine their orientation. In this method, a rotating linear polarizing film, the sample, and a circular polarizing analyzer (a combination of quarter wave and linear polarizing filters) are placed in-between a light source and a CCD camera (Fig. 2.2). When an isotropic sample is tested, rotation of the linear polarizer results in a constant nonzero intensity beam. However in an anisotropic fibrous tissue, the polarized beam is split into two, one faster than the other, resulting in an elliptically polarized beam. If the polarization axis of the beam is rotated, a sinusoidal variation of the intensity is recorded. The oscillation's phase and amplitude are related to the sample's alignment and retardation, respectively. Since a phase value can be measured per pixel, a fiber orientation map can be calculated for the sample (Tower et al. 2002). This technique has several advantages: the components are inexpensive and the data can be acquired fast enough to measure in real-time fiber realignment during a mechanical test. Additionally, the entire sample can be imaged at once. This technique has been applied to supraspinatus tendon (Lake et al. 2009), facet capsule ligament (Quinn et al. 2010; Quinn and Winkelstein 2011), and collagen gels (Lake et al. 2011; Lake and Barocas 2011; Raghupathy et al. 2011).

2.2.3 *Fast Fourier Transformation*

This method calculates the distribution of fiber orientations by applying a fast Fourier transformation (FFT) to an optical image illustrating the fiber population

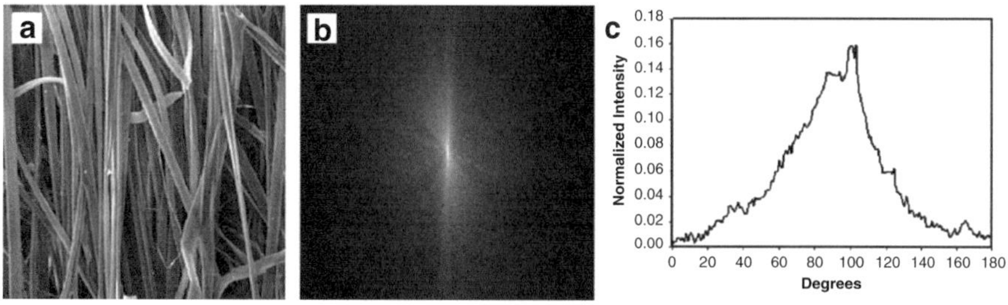

Fig. 2.3 SEM Image (**a**), power spectrum (rotated 90°) (**b**), and discrete orientation distribution for an electrospun scaffold (**c**). (Adapted from Ayres et al. (2006))

of the tissue. The rationale behind this technique is that in an image of perfectly aligned fibers the spatial variation (frequency) of the image intensity is higher in the perpendicular direction than along the fiber. Consequently, a 2D FFT of this image will easily identify the orientation of the maximum frequency and this will correspond to the direction perpendicular to the fibers. In an image of a tissue with distributed fibers, the power spectrum of a 2D FFT converts this image into a 2D map where, at each pixel, the intensity represents the amplitude of the intensity variation; the frequency is represented by the distance from the origin; and the direction represents the orientation in the original image (Fig. 2.3). A discrete orientation distribution can be calculated from the power spectrum by summing the pixel intensities along the radii of the power spectrum. A continuous orientation distribution function can be obtained by fitting the normalized discrete orientation distribution with a continuous probability function. Since this technique can be applied to images obtained from a variety of imaging modalities, it has been applied to variety of tissues including human annulus fibrosus (Guerin and Elliott 2006), chordae tendineae (Vidal Bde and Mello 2009), and electrospun scaffolds (Ayres et al. 2006, 2008).

2.2.4 Second-Harmonic-Generation

Second harmonic generation is a nonlinear effect where the light scattered by a material with non-centrosymmetric structural features has a component with twice the frequency (second harmonic) (Stoller et al. 2002; Chang and Deng 2010). This phenomenon was first recognized in crystals by Franken et al. (1961) shortly after the demonstration of laser. Since collagen molecules are organized naturally into structures with a lack of center of inversion symmetry, they are able to generate second harmonic light (Fine and Hansen 1971). By rotating the polarization axis of the incident beam, the amplitude and phase of the modulated fundamental and harmonic signals are recorded. The information about fiber orientation is contained in the phase of these signals. The preferred fiber orientation at a specific point is calculated

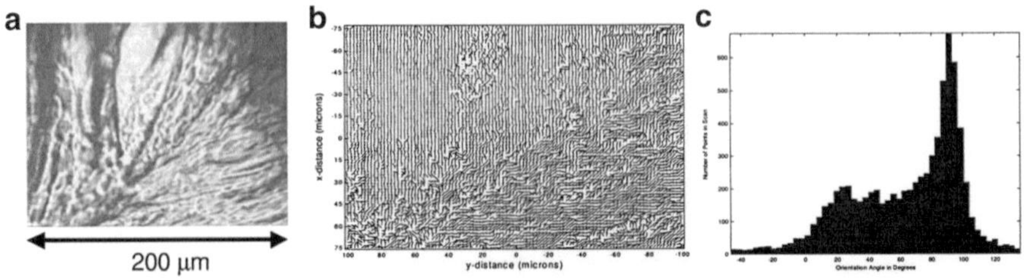

Fig. 2.4 (**a**) Light micrograph of a small region of intervertebral disc showing two different regions of fiber orientation. (**b**) Orientation image of approximately the same region. (**c**) Histogram showing the frequency of fiber orientation in the intervertebral disc (**b**). Adapted from Stoller et al. (2002)

using this technique. By scanning the sample a spatial map of fibers orientations can be obtained. A histogram of fiber orientations shows the overall fiber distribution of the scanned area (Fig. 2.4). This technique has been applied to a variety of tissues such as rat tail tendon, bovine fascia, porcine cornea, and human intervertebral disc (Stoller et al. 2002; Chang and Deng 2010; Fine and Hansen 1971).

2.2.5 *Magnetic Resonance Imaging*

Two MRI methods have been used to quantify fiber orientation: Diffusion Tensor Imaging (DTI) and ^{1}H NMR of multipolar spin states. Both techniques are based on the fact that motion of hydrogen atoms (^{1}H spin) of water molecules in the vicinity of collagen fibrils is anisotropic and has a preferred direction along the collagen molecule. In DTI, the fiber orientation can be determined by direction with maximum flow rate (Merboldt et al. 1985). In ^{1}H NMR of multipolar spin states, the distribution of the orientation of collagen fibrils can be estimated by measuring the anisotropy of a spin property called residual dipolar coupling (Berendsen 1962; Fechete et al. 2003). These techniques have been used to measure the distribution of fibril orientations in tendons (Fechete et al. 2003) and cartilage (de Visser et al. 2008; Deng et al. 2007; Pierce et al. 2010; Raya et al. 2011).

2.2.6 *Continuous Functions of Fiber Orientation Distribution*

Once the fiber orientation is experimentally measured, it can be used directly in a model (see next section) or more often, the fiber orientation distribution is described as continuous functions of the fiber angle. Probability functions are commonly used since they represent the spread (variance) of fiber orientation around one or several preferred orientation without changing the "amount" of fibers, i.e., the integration of the area under curve over the entire range is always a constant value. This is important to uncouple the anisotropy of the fiber distribution and the stiffness of the fiber population (this will be discussed in more detail in the next section).

The distribution functions can be divided in planar (2D), transversely isotropic (3D) and orthotropic (3D) distributions. Planar distributions are used when the majority of the fibers lay in a single plane, e.g., bovine pericardium (Sacks 2003) and supraspinatus tendon (Lake et al. 2009). Normal and von Mises distribution have been used to represent planar distributions with a single preferred orientation (Bischoff 2006; Holzapfel and Ogden 2010) ((2.1) and (2.2), respectively):

$$R(\theta) = \frac{1}{\sigma\sqrt{2\pi}} \mathrm{e}^{\left[-\frac{(\theta-\mu)^2}{2\sigma^2}\right]}, \tag{2.1}$$

$$R(\theta) = \frac{\mathrm{e}^{b\,\cos(\theta-\mu)}}{2\pi I_0(b)}, \tag{2.2}$$

where $R(\theta)$ is the fiber density function, μ is the preferred fiber orientation, σ is the standard deviation of the distribution, I_0 is first-kind zeroth-order Bessel function, and b is a parameter related to the circular variance as follows:

$$\sigma^2 = 1 - \frac{I_1(b)}{I_0(b)}, \tag{2.3}$$

where I_0 and I_1 are Bessel functions of first kind of zeroth and first order, respectively. Other distribution functions have been proposed to represent fiber distribution with two preferred fiber orientations (bimodal distribution) (Bischoff 2006):

$$R(\theta) = \frac{1}{\sigma\sqrt{2\pi}} \left\{ \mathrm{e}^{\left[-\frac{(\theta-\mu)^2}{2\sigma^2}\right]} + \mathrm{e}^{\left[-\frac{(\theta+\mu)^2}{2\sigma^2}\right]} \right\}. \tag{2.4}$$

This distribution is symmetric about $\theta = 0°$, the means of the two modes are $\pm\mu$ and each mode has a standard deviation σ. A transversely isotropic distribution is a 3D distribution which is symmetric around the axis of the mean fiber orientation. This type of distributions have been applied to describe the fiber distribution in the arterial wall (Gasser et al. 2006), aortic heart valves (Freed et al. 2005), and supraspinatus tendons (Thomopoulos et al. 2006). Gasser et al. (2006) proposed a transversely isotropic function that can be interpreted as the axis-symmetric version of the von Mises distribution:

$$R(\theta) = 4\sqrt{\frac{b}{2\pi}} \frac{\mathrm{e}^{[b(\cos(2\theta)+1)]}}{\mathrm{erfi}\left(\sqrt{2b}\right)}, \tag{2.5}$$

where b is a concentration parameter associated with the von Mises distribution, $\mathrm{erfi}(x) = -i \times \mathrm{erf}(x)$ denotes the imaginary error function, and θ is the "nutation" angle around the axis of symmetry. An ellipsoidal distribution has been used to describe orthotropic symmetries (Ateshian et al. 2009; Nagel and Kelly 2012):

$$R(\theta, \varphi) = \frac{1}{\xi}\left[\frac{(\cos\theta \sin\varphi)^2}{\xi_1^2} + \frac{(\sin\theta \sin\varphi)^2}{\xi_2^2} + \frac{(\cos\varphi)^2}{\xi_3^2}\right]^{-1/2}, \tag{2.6}$$

where θ and φ are Eulerian angles, ξ_1, ξ_2, and ξ_3 are the semi-axes of the ellipsoid, and

$$\xi = \frac{1}{4\pi}\int_0^{2\pi}\int_0^{\pi}\left[\frac{(\cos\theta \sin\varphi)^2}{\xi_1^2} + \frac{(\sin\theta \sin\varphi)^2}{\xi_2^2} + \frac{(\cos\varphi)^2}{\xi_3^2}\right]^{-1/2} \sin\theta d\theta d\varphi. \tag{2.7}$$

2.3 Modeling Fiber Mechanics Using a Continuous Distribution of Fiber Orientation

Collagenous tissues are often modeled as a fiber-reinforced composite material (Spencer 1984). In this approach, the tissue is analyzed as a mixture of matrix and fibers, where the matrix represents the non-fibrillar components. In these models, the matrix is usually considered isotropic and the anisotropy of the tissue is characterized by the fiber orientation. For instance, tendons and ligaments are usually modeled using fibers aligned in a single orientation; therefore, the stiffness along the fibers is much higher than the transverse direction (Lanir 1978). Another important characteristic of collagen fibers is that they confer the nonlinearity of the tissue via fiber crimping and buckling. Collagen fibers have different degree of waviness (crimping). Therefore, when load is applied to the tissue, not all fibers are stretched simultaneously. Instead, fibers are progressively recruited increasing the stiffness of the tissue with deformation (Lanir 1978, 1983; Comninou and Yannas 1976; Diamant et al. 1972). The other mechanism by which fibers contribute to the nonlinearity of the tissue is fiber buckling. Due to their high slenderness ratio, fibers buckle at negligible compressive forces (Holzapfel et al. 2004). Thus, collagenous tissues exhibit direction-dependent tension-compression nonlinearity (Soltz and Ateshian 2000). The significant influence of the orientation of collagen fibers on the mechanical behavior of the tissue demands a careful consideration of the orientation distribution for the accurate modeling of collagenous tissues. In the pioneering work of Lanir, a formulation, based on angular integrals, was proposed to include fiber nonlinearity, buckling, and distributed orientations. These integrals represent the addition of the contribution of infinitesimal fractions of fibers oriented in a given direction. This section describes Lanir's Angular Integration formulation and subsequent simplifications for applications to various tissues.

2.3.1 Angular Integration

In Lanir's formulation (Lanir 1983), Angular Integration, the fiber distribution was described by a spatial density distribution function, which quantifies the volumetric fraction of fibers oriented in a particular direction. The total strain energy and stresses were calculated as the integration of the energy and stresses of fibers in all directions. This general formulation was later simplified for the case of planar tissues under biaxial testing (Lanir et al. 1996). Similar approaches were presented by (Ateshian et al. 2009; Sacks 2003; Billiar and Sacks 2000; Girard et al. 2009; Nguyen et al. 2008). In general, the strain energy (Ψ) and the second Piola–Kirchhoff stress tensor ($\mathbf{S}_{\mathrm{f}}$) for a family of distributed fibers can be expressed as

$$\Psi_{\mathrm{f}} = \int_0^{2\pi} \int_0^{\pi} R(\theta, \varphi)\, \overline{\Psi}(\lambda) \sin\theta d\theta d\varphi, \tag{2.8}$$

$$\mathbf{S}_{\mathrm{f}} = 2\frac{\partial \Psi_{\mathrm{f}}}{\partial \mathbf{C}} = 2\int_0^{2\pi} \int_0^{\pi} R(\theta, \varphi)\, \frac{\partial \overline{\Psi}(\lambda)}{\partial \mathbf{C}} \sin\theta d\theta d\varphi, \tag{2.9}$$

where λ is the stretch of a fiber, $\overline{\Psi}$ is the strain energy of a fiber, $\mathbf{C}$ is the right Green-Cauchy strain tensor. In (2.8) and (2.9), it is assumed that the fiber will buckle under any compressive deformation; therefore, $\overline{\Psi}(\lambda) = 0$ and $\partial\overline{\Psi}(\lambda)/\partial\mathbf{C} = 0$ for $\lambda \leq 1.0$. The partial derivative in (2.8) can be rewritten using the chain rule as

$$\frac{\partial \overline{\Psi}}{\partial \mathbf{C}} = \frac{\partial \overline{\Psi}}{\partial \lambda}\frac{\partial \lambda}{\partial \mathbf{C}} = \frac{1}{2\lambda}\frac{\partial \overline{\Psi}}{\partial \lambda}\mathbf{M} \otimes \mathbf{M}, \tag{2.10}$$

where $\mathbf{M}$ is a unit vector representing the average direction of an infinitesimal fraction of fibers, and the stretch, in the direction $\mathbf{M}$, is related to the strain tensor by $\lambda^2 = \mathbf{C} : \mathbf{M} \otimes \mathbf{M}$. Therefore, (2.9) can be rewritten as

$$\mathbf{S}_{\mathrm{f}} = 2\frac{\partial \Psi_{\mathrm{f}}}{\partial \mathbf{C}} = 2\int_0^{2\pi} \int_0^{\pi} R(\theta, \varphi)\, \frac{1}{2\lambda}\frac{\partial \overline{\Psi}}{\partial \lambda}\mathbf{M} \otimes \mathbf{M} \sin\theta d\theta d\varphi. \tag{2.11}$$

Equation (2.11) is valid for a general 3D fiber distribution. However, it can be simplified for the case of planar fiber distributions as

$$\mathbf{S}_{\mathrm{f}} = 2\int_0^{\pi} R(\theta)\, \frac{1}{2\lambda}\frac{\partial \overline{\Psi}}{\partial \lambda}\mathbf{M} \otimes \mathbf{M}\, d\theta. \tag{2.12}$$

Although (2.12) has been solved for a particular choice of $R(\theta)$ and $\overline{\Psi}$ (Raghupathy and Barocas 2009), in general, it is very difficult to obtain a closed-form solution; therefore, numerical integration is the usual alternative.

2.3.1.1 Numerical Solution of the Angular Integration Formulation

In this section, some aspects of the numerical solution required for Lanir's Angular Integration formulation are discussed. Equations (2.8)–(2.12) show that the calculation of the stress and strain energy requires an angular integration. In very few simplified cases the integral can be solved analytically (Raghupathy and Barocas 2009; Cortes et al. 2010). However, in the majority of practical cases a numerical integration must be executed. In the case of planar (2D) distributions, such that of (2.12), a trivial 1D integration is necessary. However, 3D distributions, such as those in (2.5)–(2.7), require an integration over the unit sphere. This integral cannot be solved just by dividing the range of the variables θ and φ in equally spaced segments since there would be an accumulation of integration points around the poles of the unit sphere causing numerical inaccuracies (Fig. 2.5a). Instead, an icosahedron-based method has been used to solve this problem (Ateshian et al. 2009; Nagel and Kelly 2012). In this method, the triangular faces of an icosahedron are divided using equally spaced points and then mapped onto the unit sphere producing a more uniform distribution of the area of surface elements. It was shown that the accuracy of this method can be further improved by rotating the cloud integration points so that one point coincides with the orientation of the peak value of the orientation distribution function (Nagel and Kelly 2012). In this way, the numerical integration method is able to exactly capture the maximum radius of the distribution function independent of the number of elements used to divide the sphere.

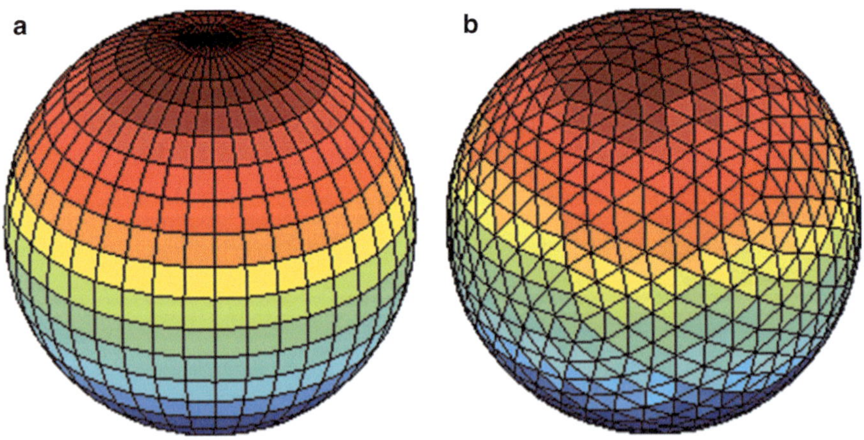

Fig. 2.5 Discretization of the unit sphere using (**a**) equally spaced segments over the range of the variables θ and φ and (**b**) icosahedron-based method

2.3.2 Generalized Structure Tensors: A simplified Approach for Fiber Distribution

Structure tensors are an alternative approach to formulate the constitutive relations for fiber-reinforced materials. A generalized structure tensor represents the anisotropy of the fiber population and, as shown below, the stress of the fibers is proportional to this tensor. For instance, a material with aligned fibers is represented by a structure tensor defined as the dyadic product $\mathbf{a}_0 \otimes \mathbf{a}_0$, where $\mathbf{a}_0$ is a unit vector in the direction of the fibers in the reference configuration. The strain energy of a composite material with aligned fibers can be expressed as a function of the invariants of $\mathbf{C}$ and $\mathbf{a}_0 \otimes \mathbf{a}_0$:

$$I_1 = \mathbf{C} : \mathbf{I}, \quad I_2 = 1/2\left\{\mathrm{tr}^2\mathbf{C} - \mathrm{tr}\mathbf{C}^2\right\}, \quad I_3 = \det\,\mathbf{C}, \quad I_4 = \mathbf{C} : \mathbf{a}_0 \otimes \mathbf{a}_0. \tag{2.13}$$

A general discussion on the use of these invariants and tensors can be found in (Spencer 1984). Recently, Freed et al. (2005) and Gasser et al. (2006) formulated structure tensors which consider the effect of fiber angular distribution. The angular distribution tensor ($\mathbf{H}$) is defined as

$$\mathbf{H} = \int_0^{2\pi}\int_0^{\pi} R(\theta,\varphi)\,\mathbf{M}\otimes\mathbf{M}\,\sin\theta d\theta d\varphi. \tag{2.14}$$

Again, (2.14) can be simplified for planar fiber distributions as

$$\mathbf{H} = \int_0^{\pi} R(\theta)\,\mathbf{M}\otimes\mathbf{M}\,d\theta. \tag{2.15}$$

The pseudo-invariant $\bar{I}_4 = \mathbf{C} : \mathbf{H}$ can be expressed in an integral form, using (2.14), as

$$\begin{aligned}\bar{I}_4 &= \mathbf{C} : \int_0^{2\pi}\int_0^{\pi} R(\theta,\varphi)\,\mathbf{M}\otimes\mathbf{M}\sin\theta d\theta d\varphi \\ &= \int_0^{2\pi}\int_0^{\pi} R(\theta,\varphi)\,\mathbf{C} : \mathbf{M}\otimes\mathbf{M}\,\sin\theta d\theta d\varphi \\ &= \int_0^{2\pi}\int_0^{\pi} R(\theta,\varphi)\,\lambda^2\sin\theta d\theta d\varphi = \bar{\lambda}^2.\end{aligned} \tag{2.16}$$

A similar expression can be easily obtained for planar distributions. From (2.16), it can be observed that $\bar{I}_4$ is the weighted average of λ^2. Notice that in the definition of $\bar{I}_4$, there is not a condition to exclude fibers under compression. Therefore, if one of the principal values of $\mathbf{C}$ is lower than 1.0, a fraction of compressed fibers will be included in the average $\bar{\lambda}^2$. Uniaxial tension, biaxial tension, and simple shear are examples of deformation states for which at least one of the principal values of

C is less than one. Consequently, error will be introduced when the structure tensor formulation is considered in these loading scenarios.

For the generalized structural tensor, the strain energy (Ψ_f) and the second Piola–Kirchhoff ($\mathbf{S}_\mathrm{f}$) stress tensor can be expressed as

$$\Psi_\mathrm{f} = \overline{\Psi}\left(\overline{\lambda}\right), \tag{2.17}$$

$$\mathbf{S}_\mathrm{f} = 2\frac{\partial\overline{\Psi}}{\partial\mathbf{C}} = \frac{1}{\overline{\lambda}}\frac{\partial\overline{\Psi}}{\partial\overline{\lambda}}\frac{\partial\overline{\lambda}}{\partial\mathbf{C}} = \frac{1}{\overline{\lambda}}\frac{\partial\overline{\Psi}}{\partial\overline{\lambda}}\mathbf{H}. \tag{2.18}$$

An important advantage of this approach is that the angular integrals are evaluated just once during the definition of the tensor **H**. After that, the stresses are obtained by algebraic operations. This is important in numerical methods such as finite elements since the number of calculation is greatly reduced. Notice that the strain energy function and the stresses are calculated using the average stretch rather than the actual stretch in the fibers; therefore, for nonlinear fibers, lower stresses are obtained. In this formulation the buckling condition establishes that the whole fiber distribution is neglected when the mean of stretch is lower than one (Gasser et al. 2006): $\Psi_\mathrm{f} = 0$, $\mathbf{S}_\mathrm{f} = \mathbf{0}$ for $\overline{\lambda} < 1.0$ (or equivalently $\overline{I}_4 < 1.0$). However, since this criterion is based on the average value of the stretch, compressed fibers may be included in the calculation of $\overline{\lambda}$ when $\overline{\lambda} > 1.0$; and conversely, fibers in tension may be disregarded when $\overline{\lambda} < 1.0$. This will cause a difference between the angular integration (Lanir's formulation) and structure tensors. The differences between these formulations are discussed in the next section.

2.3.3 Comparison Between the Angular Integration and Structure Tensor Approaches

Although the structure tensor approach greatly reduces the amount of calculations in numerical methods such as finite elements, several studies have shown that these methods lead to different results for the same fiber distribution and applied strains (Federico and Herzog 2008; Cortes et al. 2010; Pandolfi and Vasta 2012). In general, the difference between the formulations increases with deformation and the spread of the orientation distribution. In this section, a numerical comparison between the angular integration (Lanir's formulation) and structure tensors is presented for a von Mises and a transversely isotropic distributions (Cortes et al. 2010). The following popular strain energy has been chosen as a constitutive relation to describe the mechanical behavior of the collagen fibers (Holzapfel et al. 2000)

$$\overline{\Psi} = \frac{c_1}{2c_2}\left[\mathrm{e}^{c_2\left(\lambda^2-1\right)^2} - 1\right], \tag{2.19}$$

where c_1 is an elastic constant and c_2 a is non-dimensional parameter associated with the degree of nonlinearity. A single set of material properties is used throughout this section: $c_1 = 5$ MPa, $c_2 = 30$. These properties describe the behavior of the supraspinatus tendon (Kadlowec 2009).

The generalized structure tensor associated with the distribution function shown in (2.5) can be expressed as

$$\mathbf{H} = \kappa\mathbf{I} + (1 - 3\kappa)\,\mathbf{a}_0 \otimes \mathbf{a}_0, \tag{2.20}$$

where $\kappa = 2\pi \int_0^{\pi} R(\theta)\sin^3\theta \, d\theta$ ranges from 0 to 1/3 for perfectly aligned and isotropic distributed fibers, respectively. An expression similar to (2.20) can be obtained for a von Mises distribution

$$\mathbf{H}_2 = \kappa_{2\mathrm{D}}\mathbf{I}_2 + (1 - 2\kappa_{2\mathrm{D}})\,\mathbf{a}_{0_2} \otimes \mathbf{a}_{0_2}, \tag{2.21}$$

where $\mathbf{H}_2$, $\mathbf{I}_2$, and $\mathbf{a}_{0_2} \otimes \mathbf{a}_{0_2}$ are 2D versions of the tensors shown in (2.20), the distribution parameter is now defined as $\kappa_{2\mathrm{D}} = \int_0^{\pi} \rho(\theta)\sin^2\theta \, d\theta$ and ranges from 0 to 1/2. The parameters κ and $\kappa_{2\mathrm{D}}$ are associated with the degree of anisotropy of the fiber distribution (Fig. 2.6). For instance, a value of $\kappa = \kappa_{2\mathrm{D}} = 0$ corresponds to aligned fibers, whereas $\kappa_{2\mathrm{D}} = 1/2$ or $\kappa = 1/3$ correspond to isotropic distributions.

Two loading configurations, typically used to characterize connective tissues, have been selected for comparison: uniaxial and biaxial tension. The tissue is considered as an incompressible material; therefore, the condition $\det \mathbf{C} = 1$ holds. The comparison between the angular integration and structure tensor formulations

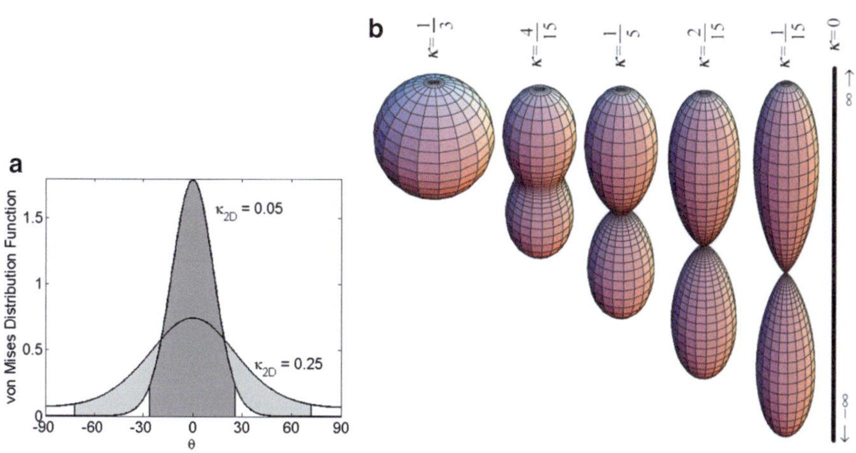

Fig. 2.6 Graphical representation of the fiber distribution of (**a**) planar distribution for several values of $\kappa_{2\mathrm{D}}$ and (**b**) a transversely isotropic distribution for different values of κ. (Adapted from Gasser et al. (2006) and Cortes et al. (2010))

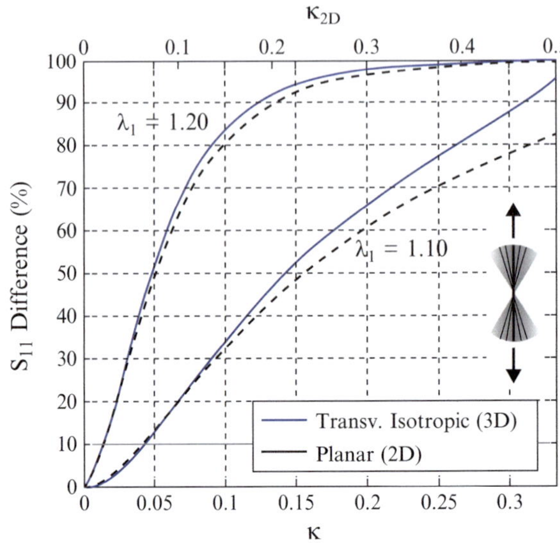

Fig. 2.7 Difference of the second Piola-Kirchhoff stresses in the direction x_1 (S_{11}) between angular integration and structure tensor formulations increases with the applied stretch (λ_1) and decreases when the fiber distribution (κ for transversely isotropic (3D) and κ2D for planar (2D) distributions) is small. Difference defined as $(S_{AI} - S_{ST})/S_{AI} \times 100$. Notice that distributions go from perfectly aligned ($\kappa = \kappa 2D = 0$) to isotropic ($\kappa = 1/3$, $\kappa 2D = 1/2$). Adapted from Cortes et al. (2010)

for planar (2D) and transversely isotropic (3D) distributions under uniaxial tension is shown in Fig. 2.7. It can be observed that the difference between these formulations tends to zero when the fibers are aligned ($\kappa \to 0$). This can be attributed to the reduced number of buckled fibers and that the average fiber stretch is closer to the actual stretch of the fibers. A difference of 10 % in the longitudinal stress is obtained for $\kappa = 0.015$ ($\kappa_{2D} = 0.022$) when a stretch of 1.2 is applied. These values of the distribution parameters κ and κ_{2D} represent distributions where 95 % of the fibers are oriented within 17° of the mean fiber direction in the reference configuration. This is in agreement with the analysis of Federico and Herzog (2008) who asserted that approximations as those presented in (2.17) and (2.18) are a good approximation of the general case ((2.8) and (2.9)) when there is a weak directional dispersion around the mean direction.

For a transversely isotropic distribution, in the equi-biaxial case ($\lambda_1 = \lambda_2$), a fraction of the fibers buckle due to the out-of-plane contraction and therefore do not contribute to total stresses in the tissue. However, in the GST formulation, those fibers are considered in the calculation of the average stretch. To illustrate this, a single fiber of families with transversely isotropic distribution (2.5) under equal-biaxial stretch is analyzed. Figure 2.8 shows the stress difference as a function of fiber distribution for several values of the applied stretch. A 10 % difference is obtained for $\kappa = 0.014$ and a stretch equal to 1.2. This value is very close to that obtained for the case of uniaxial tension (Fig. 2.7). The results shown in Figs. 2.8 and 2.9 were confirmed by a subsequent study (Pandolfi and Vasta 2012).

To decrease the difference between these formulations, a method that considers second order terms of the pseudo-invariant $I_{\langle 4 \rangle}$ was proposed (Pandolfi and Vasta 2012). The inclusion of the dependence of the second order parameters improves the

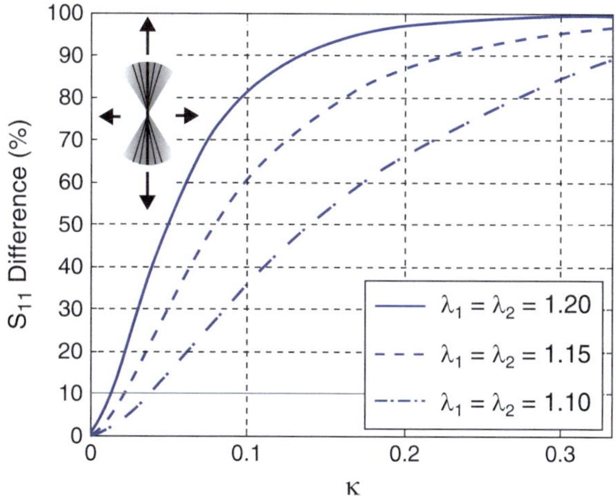

Fig. 2.8 Transversely isotropic (3D) family of fibers shows differences in the second Piola–Kirchhoff stresses (S_{11}) even for the case equal-biaxial testing. Difference defined as $(S_{\mathrm{AI}} - S_{\mathrm{GST}})/S_{\mathrm{AI}} \times 100$

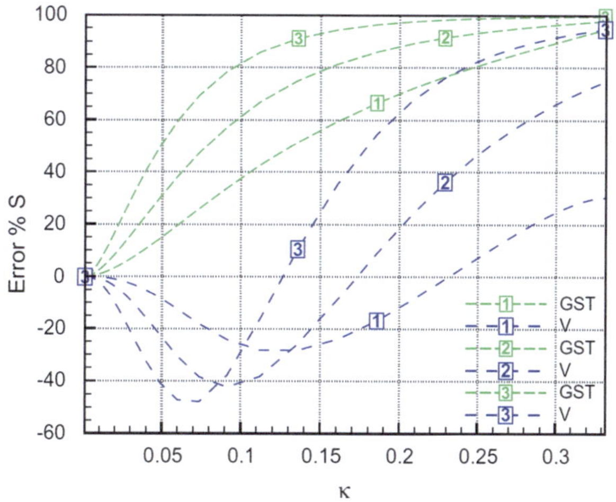

Fig. 2.9 Error percentage with respect to the response of the angular integration approach for the generalized structure tensor model (GST) (Gasser et al. 2006) and the fourth order structure tensor model (V, Pandolfi and Vasta 2012) for equi-biaxial deformation. *Curves* refer to the stretches $\lambda = 1.1$ (label 1), $\lambda = 1.15$ (label 2), and $\lambda = 1.2$ (label 3). Adapted from Pandolfi and Vasta (2012)

approximation, still conserving a rather simple formulation that avoids the explicit integration of the stretch in the spatial direction. A Taylor expansion of the $\Psi(I_4)$ around it mean argument $\bar{I}_4$

$$\Psi\left(I_4\right) \approx \Psi\left(\bar{I}_4\right) + \left.\frac{\partial \Psi}{\partial I_4}\right|_{I_4=\bar{I}_4} \left\langle I_4 - \bar{I}_4 \right\rangle + \frac{1}{2} \left.\frac{\partial^2 \Psi}{\partial I_4^2}\right|_{I_4=\bar{I}_4} \left\langle \left(I_4 - \bar{I}_4\right)^2 \right\rangle, \tag{2.22}$$

where the operator $\langle\cdot\rangle$ is defined as $\langle\cdot\rangle = \int_0^{2\pi} \int_0^{\pi} R\left(\theta, \varphi\right)(\cdot) \sin\theta d\theta d\varphi$. The first term on the right hand side of (2.22) is equivalent to that proposed by Gasser et al.

(2006); the second term is equal to zero since $\langle I_4 \rangle = \bar{I}_4$; and the third term is a correction term proposed by Pandolfi and Vasta (2012), which can be rewritten as:

$$\left(2c_2\Psi\left(\bar{I}_4\right) + c_1\right)\left[2c_2(\bar{I}_4 - 1)^2 + 1\right]\left[\mathbf{C} : \mathbb{H} : \mathbf{C} - (\mathbf{H} : \mathbf{C})^2\right]$$

for the constitutive equation for the fiber shown in (2.19). Here $\mathbb{H} = \mathbf{a}_0 \otimes \mathbf{a}_0 \otimes \mathbf{a}_0 \otimes \mathbf{a}_0$ is a fourth order structure tensor. A comparison between angular integration (Lanir's formulation) and structure tensor formulations (Gasser et al. 2006; Pandolfi and Vasta 2012) for the biaxial case shows a decrease in the difference with respect to angular integration (Fig. 2.9).

2.3.4 *Remarks on Modeling Approaches*

In this section, the two major approaches, angular integration and structure tensors, used to analyze the mechanical behavior of tissues with distributed fibers were described in detail. The angular integration method (Lanir's formulation) is considered as the gold standard formulation for the analysis of the mechanics of distributed fibers. It has been successfully used for many applications. However, it requires a numerical integration every time a value of stress is required. Consequently, when this formulation is implemented in numerical tools, such as finite elements, the computation time increases considerably. The icosahedron method used to discretize the unit sphere increases the numerical integration efficiency reducing computational time. The generalized structure tensor formulation was proposed as an alternative to reduce the number of integrals required to calculate fiber stress. However, differences have been reported for some loading cases and orientation distributions when compared to the angular integration approach. Consequently, choosing one method over the other depends on the application at hand. In the following section, some applications of both formulations are described.

2.4 Applications

Several fibrous tissues have been modeled using fiber orientation distributions. Although this approach to model the fibers was first applied by Lanir to tendons and skin (Lanir 1983; Lanir et al. 1996), a great number of studies have analyzed cardiovascular tissues using fiber orientation distributions (Vidal Bde and Mello 2009; Stoller et al. 2002; Freed et al. 2005; Thomopoulos et al. 2006; Holzapfel et al. 2004; Soltz and Ateshian 2000; Lanir et al. 1996). Recently, a few studies have applied orientation distributions to the modeling of other fibrous tissues such as articular cartilage, cornea, and annulus fibrosus (Ateshian et al. 2009; Pandolfi and Holzapfel 2008; Caner et al. 2007). In general, a better prediction of the mechanical behavior of these tissues has been obtained when fiber orientation distributions are considered. Showing each of these cases in detail is out of the scope of this chapter. Instead, we summarize and discuss only those studies where the effect of fiber distribution has been quantified by comparing angular integration, generalized structure tensor, and models without fiber dispersion.

2.4.1 Arteries

The arterial wall is grossly divided in three layers: intima, media, and adventitia. The intimal layer is the thin innermost layer of the arterial wall. It is composed of a layer of endothelial cells, a sub-endothelial layer of loose connective tissue with a large angular deviation, and an elastic layer that separates the intima and the media. The media is the layer of the arterial and is composed of smooth muscle cells, elastin fibers, and collagen fiber bundles. The collagen fibers in the medial layer are highly aligned in the circumferential direction. Finally, the adventitia is the outermost layer and composed mainly of fibroblasts and fibrocytes, collagen fibers and ground matrix. The collagen fibers in the adventitia are arranged in two helical fiber families with significant angular dispersion.

Gasser et al. (2006) modeled the mechanical behavior of the adventitial layer using a model that consisted of an incompressible neo-Hookean matrix reinforced by two families of distributed fibers. The fibers were modeled using the "κ" model with an exponential strain energy shown here in (2.18). The parameters used for the analysis were representative to the human iliac artery: shear modulus of the matrix $c = 7.64$ kPa, $k1 = 996.6$ kPa, $k2 = 524.6$, $\kappa = 0.226$, and the angle between the circumferential direction and each of the fiber families as $\gamma = 49.98°$. The model was used to describe the response of the adventitial layer for the inflation and uniaxial tests. To analyze the effect of the dispersion parameter κ and the mean fiber angle γ on the mechanical response of the adventitia, simulations with the following parameters were also calculated: $\gamma = 39.98°$ and $59.98°$ and $\kappa = 0$ and 0.333. A major effect of including the fiber orientation distribution is the reduction of the dependence of the response of the tube with the mean fiber angle γ. The simulations also show a stiffening effect on the adventitial tube as the axial and circumferential stretches are reduced when the orientation dispersion increases (Fig. 2.10a). On uniaxial tests in the circumferential and axial directions, an increase on the fiber orientation dispersion decreases the amount of lateral contraction, i.e., apparent Poisson's ratio (Fig. 2.10b). This indicates that, for aligned fibers, big rotations occur before fibers start taking load.

In summary, the inclusion of fiber distribution using the structure tensor model (Stoller et al. 2002) proposed by Gasser et al. (2006) revealed that characteristics of the mechanical behavior of the human iliac arteries are more consistent with experimental observations than simulations with aligned fibers. In particular, the stiffening effect in the axial direction and elastic properties in the uniaxial behavior of the adventitia layer were captured closely by considering the fiber orientation distribution.

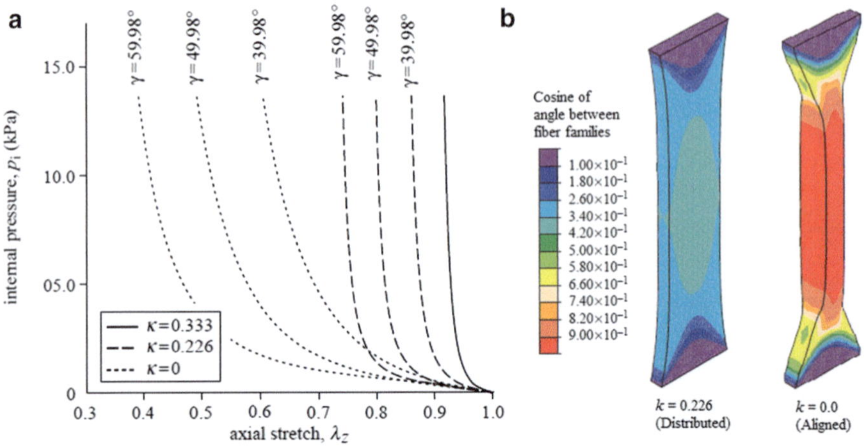

Fig. 2.10 Effect of fiber distribution on the mechanical behavior of arterial tissue. (**a**) Simulations of an inflation tests show that the axial stretch is lower for models with distributed fibers (stiffening effect). (**b**) The lateral contraction (Poisson's effect) is higher for samples with aligned fibers. Adapted from Gasser et al. (2006)

2.4.2 *Aortic Valves*

Another cardiovascular tissue that has been modeled using fiber angle dispersion is the aortic valve. This valve is located between the left ventricle and the aorta. It is composed of three leavelets that are planar tissues composed of proteoglycans, and fibers of collagen type I and III (Dainese et al. 2006; Eriksen et al. 2006). Billiar and Sacks (2000) measured the fiber orientation using SALS and used a bimodal distribution (Fig. 2.11). Although the fibers of both fiber families contribute to $R(\theta)$, Billiar and Sacks (2000) concluded that mechanical contribution of the broader distribution is negligible and the in-plane biaxial response can be modeled using the highly aligned family of fibers. Freed et al. (2005) used a structure tensor approach to improve the computational efficiency compared to the angular integration approach. Additionally, the quality of the curve fit of biaxial tension experiments was evaluated and compared to the angular integration approach.

The leaflets of the aortic valve were modeled using the distribution shown in (2.1) and a combination of the three different strain energy functions: dilatational, distortional isotropic, and distortional anisotropic. In this case, the dilatational and distortional isotropic components can be regarded as the ground matrix, and the distortional anisotropic as the fiber component. Angular integration approach and structure tensor formulations were used for the fiber term. It was found that both formulations provided a good fit to the experimental biaxial data (Fig. 2.12). The structure tensor approach slightly under-predicted in the fiber direction and over-predicted in the transverse direction the stiffness at the toe region. These results are in agreement with the comparison between angular integration and structure tensors

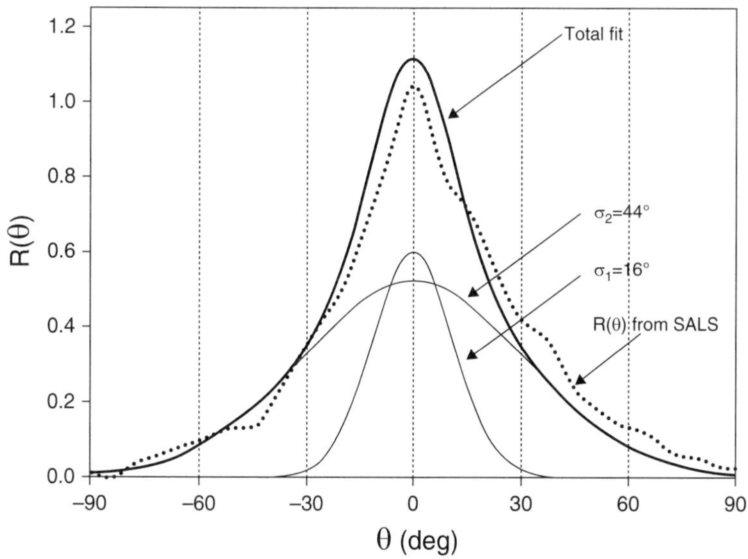

Fig. 2.11 Representative fiber orientation distribution for an aortic valve leaflet to a dual Gaussian distribution. The distribution was principally composed of a highly aligned population ($\sigma_1 = 16°$) and a broader distribution ($\sigma_1 = 44°$). Adapted from Billiar and Sacks (2000)

presented in the previous section. Since only the fiber family with highly aligned fibers contributes to the in-plane mechanics (Billiar and Sacks 2000), the difference between both the formulations should be small (Fig. 2.9).

2.4.3 *Articular Cartilage*

Articular cartilage is mostly composed of proteoglycans and collagen type II. Collagen fibers in articular cartilage have a characteristic orientation distribution that goes from radially aligned at the insertion in the subchondral bone, randomly oriented in the middle zone and aligned parallel to the surface. Articular cartilage has a complex mechanical behavior that includes tension compression nonlinearity (Soltz and Ateshian 2000); high Poisson's ratios in tension and the opposite in compression (Elliott et al. 2002); and stiffening of glycosaminoglycan digested samples (Schmidt et al. 1990). A model consisting of osmotic pressure (representing the matrix) and a distributed fiber family (as that of (2.6)) is able to replicate these experimental observations (Ateshian et al. 2009). Most of these could not be obtained from constitutive models which use a discrete number of aligned fiber populations.

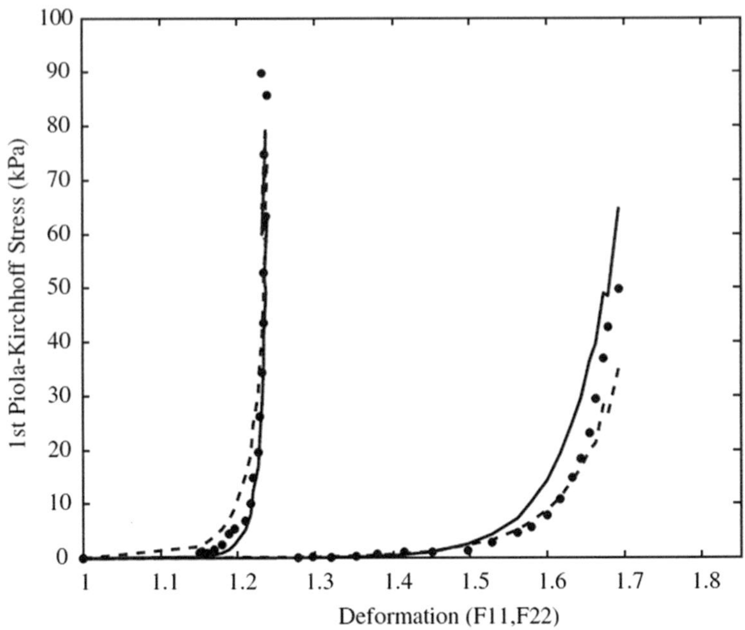

Fig. 2.12 Curve fitting of the biaxial tension tests. *Dots* correspond to experimental data, *dashed lines* correspond to the angular integration formulation, and *solid lines* to the structure tensor approach. Protocols 2–6 correspond to tension ratio Circumferential:Radial = 30:60 (N/m). Adapted from Freed et al. (2005)

2.4.4 Annulus Fibrosus

The use of fiber orientation distributions in the annulus fibrosus has not been as common as for other tissues. However, two studies show the importance of the orientation dispersion in the mechanical behavior of the annulus fibrosus (Caner et al. 2007; Guo et al. 2012). Annulus fibrosus is a tissue which is composed of concentric lamellae with alternating mean fiber angles. It has been typically modeled as an isotropic matrix reinforced with two aligned fiber families. However, a material model composed of matrix and fibers is not enough to describe the multiaxial mechanical behavior of the annulus fibrosus (Guerin and Elliott 2006; Guo et al. 2012; Wagner and Lotz 2004; O'Connell et al. 2009). Specifically, uniaxial and biaxial experiments demonstrate that the stiffness of the matrix increases with fiber stretch (Guo et al. 2012). Therefore, several studies have included strain energy terms representing fiber–matrix interactions to improve model predictions. However, these fiber–matrix interactions may be artificial concepts to quantify the real strain energy from homogeneous macro-scale deformations (Guo et al. 2012).

Caner et al. (2007) proposed a model which included a fiber orientation distribution and proposed that the stiffening effect associated with fiber distributions could describe the mechanics of annulus fibrous without explicitly including fiber–matrix interactions. The transversely isotropic fiber orientation distribution was described by the function:

$$R(\theta) = c_1 e^{c_2 \theta^2}, \tag{2.23}$$

where c_1 and c_2 were assumed as 18.5 and −60, respectively. The model parameters were obtained by curve fitting of uniaxial tests of Acaroglu et al. (1995), Skaggs et al. (1994), Wu and Yao (1976), and Elliott and Setton (2001). The predictions of the model with fiber distribution were also compared to models with aligned fibers and aligned fibers with explicit fiber–matrix interactions.

All models were able to describe the uniaxial tension behavior in the fiber direction (Fig. 2.13a). For the aligned without fiber–matrix interaction, all model parameters were obtained from this experiment. Tension in the circumferential direction was used to obtain the remainder of the model parameters for the model with aligned fibers and fiber–matrix interaction and the fiber distribution model. Prediction of the change in fiber angle and lateral stretch in uniaxial tension in the circumferential direction show the good agreement between experimental measurements and the models with fiber–matrix interaction and fiber distribution. Additionally, the model with distributed fibers was able to predict the lateral stretch in the uniaxial test in the circumferential direction (Fig. 2.13b). This comparison shows that the fiber distribution can explain the discrepancies observed in the predictions of models that only include aligned fibers and matrix. Additionally, the response of the models with fiber–matrix interaction and fiber distribution was very similar suggesting that the "apparent" shear fiber–matrix interaction can be

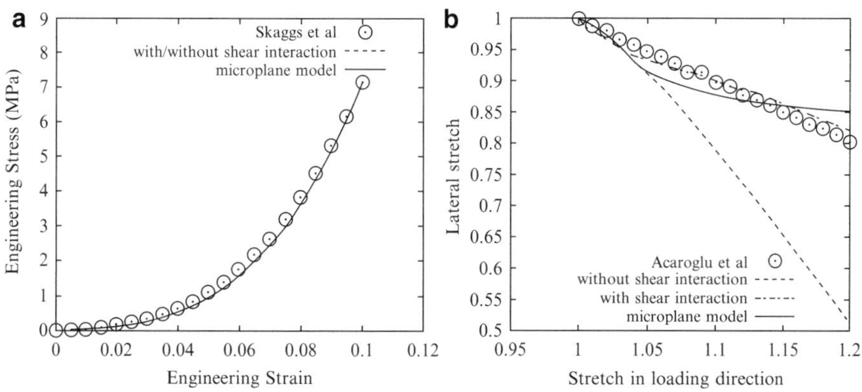

Fig. 2.13 Comparison of the ability of predicting the uniaxial mechanical behavior of annulus fibrosus of three models: aligned fibers without shear fiber–matrix interaction, aligned fibers with shear fiber–matrix interaction, and including fiber distribution (microplane). Adapted from Caner et al. (2007)

explained by including the fiber orientation distribution in the model. Whether or not the annulus fibrosus fiber distribution as suggested by the model is physically present within the tissue structure has not yet been determined.

2.5 Conclusions

In this chapter, several methods to measure fiber orientation distribution were described. All the experimental methods presented here used different approaches; however, they all have a common output: a fiber distribution histogram. The choice of a particular method depends on the application: real-time acquisition, micro/macroscopic scale, and shape of the sample, just to mention few. For simplicity, a continuous distribution function can be curve-fitted to the experimental distribution, but this is not a requirement, the formulation can use both types of distributions since there are no restrictions on their shape.

Two approaches to model the mechanical behavior of distributed fibers were presented in this chapter: Angular Integration and Structure Tensors. The angular integration approach, proposed by Lanir in 1983, adds the contribution of fiber at different orientations, excluding fibers undergoing compression. This approach is considered the gold standard formulation to describe the mechanics of distributed fibers. An alternative approach, generalized structure tensors, simplifies the angular integration formulation by defining a tensor that represents the spatial organization of the fibers. Once the structure tensor is defined, stress can be calculated just using algebraic operation. Consequently, the number of integrals required in a numerical method such as finite elements is drastically reduced. However, some differences compared to the angular integration approach have been reported for specific distributions and loading cases. Applications of both approaches show the importance of including the angular distribution in the modeling of collagenous tissues. Models that include distributed fiber orientations were able to describe several key experimental observations that models considering aligned fibers cannot predict.

References

Acaroglu ER, Iatridis JC, Setton LA, et al. Degeneration and aging affect the tensile behavior of human lumbar anulus fibrosus. Spine. 1995;20:2690–701.

Ateshian GA, Rajan V, Chahine NO, et al. Modeling the matrix of articular cartilage using a continuous fiber angular distribution predicts many observed phenomena. J Biomech Eng. 2009;131:061003. doi:10.1115/1.3118773.

Ayres C, Bowlin GL, Henderson SC, et al. Modulation of anisotropy in electrospun tissue-engineering scaffolds: analysis of fiber alignment by the fast Fourier transform. Biomaterials. 2006;27:5524–34. doi:10.1016/j.biomaterials.2006.06.014.

Ayres CE, Jha BS, Meredith H, et al. Measuring fiber alignment in electrospun scaffolds: a user's guide to the 2D fast Fourier transform approach. J Biomater Sci Polym Ed. 2008;19:603–21. doi:10.1163/156856208784089643.

Berendsen H. Nuclear magnetic resonance study of collagen hydration. J Chem Phys. 1962;36:3297–305. doi:10.1063/1.1732460.

Billiar K, Sacks M. Biaxial mechanical properties of the native and glutaraldehyde-treated aortic valve cusp: part II—a structural constitutive model. J Biomech Eng. 2000;122:327–35. doi:10.1115/1.1287158.

Bischoff J. Continuous versus discrete (invariant) representations of fibrous structure for modeling non-linear anisotropic soft tissue behavior. Int J Non-Linear Mech. 2006;41:167–79. doi:10.1016/j.ijnonlinmec.2005.06.008.

Caner FC, Guo Z, Moran B, et al. Hyperelastic anisotropic microplane constitutive model for annulus fibrosus. J Biomech Eng. 2007;129:632–41. doi:10.1115/1.2768378.

Chang Y, Deng X. Characterization of excitation beam on second-harmonic generation in fibrillous type I collagen. J Biol Phys. 2010;36:365–83. doi:10.1007/s10867-010-9190-8.

Comninou M, Yannas IV. Dependence of stress-strain nonlinearity of connective tissues on the geometry of collagen fibers. J Biomech. 1976;9:427–33.

Cortes DH, Lake SP, Kadlowec JA, et al. Characterizing the mechanical contribution of fiber angular distribution in connective tissue: comparison of two modeling approaches. Biomech Model Mechanobiol. 2010;9:651–8. doi:10.1007/s10237-010-0194-x.

Dainese L, Barili F, Topkara VK, et al. Effect of cryopreservation techniques on aortic valve glycosaminoglycans. Artif Organs. 2006;30:259–64. doi:10.1111/j.1525-1594.2006.00213.x.

de Visser SK, Bowden JC, Wentrup-Byrne E, et al. Anisotropy of collagen fibre alignment in bovine cartilage: comparison of polarised light microscopy and spatially resolved diffusion-tensor measurements. Osteoarthritis Cartilage. 2008;16:689–97. doi:10.1016/j.joca.2007.09.015.

Deng X, Farley M, Nieminen MT, et al. Diffusion tensor imaging of native and degenerated human articular cartilage. Magn Reson Imaging. 2007;25:168–71. doi:10.1016/j.mri.2006.10.015.

Diamant J, Keller A, Baer E, et al. Collagen; ultrastructure and its relation to mechanical properties as a function of ageing. Proc R Soc Lond B Biol Sci. 1972;180:293–315.

Elliott DM, Setton LA. Anisotropic and inhomogeneous tensile behavior of the human anulus fibrosus: experimental measurement and material model predictions. J Biomech Eng. 2001;123:256–63.

Elliott DM, Narmoneva DA, Setton LA. Direct measurement of the Poisson's ratio of human patella cartilage in tension. J Biomech Eng. 2002;124:223–8.

Eriksen HA, Satta J, Risteli J, et al. Type I and type III collagen synthesis and composition in the valve matrix in aortic valve stenosis. Atherosclerosis. 2006;189:91–8. doi:10.1016/j.atherosclerosis.2005.11.034.

Fechete R, Demco D, Blumich B. Order parameters of the orientation distribution of collagen fibers in Achilles tendon by H-1 NMR of multipolar spin states RID C-3671-2011. NMR Biomed. 2003;16:479–83. doi:10.1002/nbm.854.

Federico S, Herzog W. Towards an analytical model of soft biological tissues. J Biomech. 2008;41:3309–13. doi:10.1016/j.jbiomech.2008.05.039.

Fine S, Hansen W. Optical second harmonic generation in biological systems. Appl Opt. 1971;10:23503. doi:10.1364/AO.10.002350.

Franken P, Weinreich G, Peters C, Hill A. Generation of optical harmonics. Phys Rev Lett. 1961;7:118. doi:10.1103/PhysRevLett.7.118.

Fratzl P. Collagen: structure and mechanics, an introduction. In: Fratzl P, editor. Collagen: structure and mechanics. New York: Springer; 2008.

Freed AD, Einstein DR, Vesely I. Invariant formulation for dispersed transverse isotropy in aortic heart valves: an efficient means for modeling fiber splay. Biomech Model Mechanobiol. 2005;4:100–17. doi:10.1007/s10237-005-0069-8.

Gasser TC, Ogden RW, Holzapfel GA. Hyperelastic modelling of arterial layers with distributed collagen fibre orientations. J R Soc Interface. 2006;3:15–35. doi:10.1098/rsif.2005.0073.

Gilbert TW, Wognum S, Joyce EM, et al. Collagen fiber alignment and biaxial mechanical behavior of porcine urinary bladder derived extracellular matrix. Biomaterials. 2008;29: 4775–82. doi:10.1016/j.biomaterials.2008.08.022.

Girard MJA, Downs JC, Burgoyne CF, Suh J-KF. Peripapillary and posterior scleral mechanics—part I: development of an anisotropic hyperelastic constitutive model. J Biomech Eng. 2009;131(5):051011.

Guerin HAL, Elliott DM. Degeneration affects the fiber reorientation of human annulus fibrosus under tensile load. J Biomech. 2006;39:1410–8. doi:10.1016/j.jbiomech.2005.04.007.

Guo Z, Shi X, Peng X, Caner F. Fibre-matrix interaction in the human annulus fibrosus. J Mech Behav Biomed Mater. 2012;5:193–205. doi:10.1016/j.jmbbm.2011.05.041.

Holzapfel GA, Ogden RW. Constitutive modelling of arteries. Proc R Soc A Math Phys Eng Sci. 2010;466:1551–96. doi:10.1098/rspa.2010.0058.

Holzapfel G, Gasser T, Ogden R. A new constitutive framework for arterial wall mechanics and a comparative study of material models RID B-3906-2008. J Elast. 2000;61:1–48. doi:10.1023/A:1010835316564.

Holzapfel G, Gasser T, Ogden R. Comparison of a multi-layer structural model for arterial walls with a fung-type model, and issues of material stability RID B-3906-2008. J Biomech Eng. 2004;126:264–75. doi:10.1115/1.1695572.

Hulmes D. Collagen diversity, synthesis and assembly (Chapter 2). In: Fratzl P, editor. Collagen: structure and mechanics. New York: Springer; 2008.

Kadlowec JA. A hyperelastic model with distributed fibers to describe the human supraspinatus tendon tensile mechanics; 2009.

Lake SP, Barocas VH. Mechanical and structural contribution of non-fibrillar matrix in uniaxial tension: a collagen-agarose co-gel model. Ann Biomed Eng. 2011;39:1891–903. doi:10.1007/s10439-011-0298-1.

Lake SP, Miller KS, Elliott DM, Soslowsky LJ. Effect of fiber distribution and realignment on the nonlinear and inhomogeneous mechanical properties of human supraspinatus tendon under longitudinal tensile loading. J Orthop Res. 2009;27:1596–602. doi:10.1002/jor.20938.

Lake SP, Hald ES, Barocas VH. Collagen-agarose co-gels as a model for collagen-matrix interaction in soft tissues subjected to indentation. J Biomed Mater Res A. 2011;99A:507–15. doi:10.1002/jbm.a.33183.

Lanir Y. Structure-strength relations in mammalian tendon. Biophys J. 1978;24:541–54. doi:10.1016/S0006-3495(78)85400-9.

Lanir Y. Constitutive equations for fibrous connective tissues. J Biomech. 1983;16:1–12.

Lanir Y, Lichtenstein O, Imanuel O. Optimal design of biaxial tests for structural material characterization of flat tissues. J Biomech Eng. 1996;118:41–7. doi:10.1115/1.2795944.

Merboldt K-D, Hanicke W, Frahm J. Self-diffusion NMR imaging using stimulated echoes. J Magn Reson. 1985;64:479–86. doi:10.1016/0022-2364(85)90111-8.

Nagel T, Kelly DJ. Apparent behaviour of charged and neutral materials with ellipsoidal fibre distributions and cross-validation of finite element implementations. J Mech Behav Biomed Mater. 2012;9:122–9. doi:10.1016/j.jmbbm.2012.01.006.

Nguyen TD, Jones RE, Boyce BL. A nonlinear anisotropic viscoelastic model for the tensile behavior of the corneal stroma. J Biomech Eng. 2008;130(4):041020. doi:10.1115/1.2947399.

O'Connell GD, Guerin HL, Elliott DM. Theoretical and uniaxial experimental evaluation of human annulus fibrosus degeneration. J Biomech Eng. 2009;131:111007. doi:10.1115/1.3212104.

Pandolfi A, Holzapfel GA. Three-dimensional modeling and computational analysis of the human cornea considering distributed collagen fibril orientations. J Biomech Eng. 2008;130:061006. doi:10.1115/1.2982251.

Pandolfi A, Vasta M. Fiber distributed hyperelastic modeling of biological tissues. Mech Mater. 2012;44:151–62. doi:10.1016/j.mechmat.2011.06.004.

Pierce DM, Trobin W, Raya JG, et al. DT-MRI based computation of collagen fiber deformation in human articular cartilage: a feasibility study. Ann Biomed Eng. 2010;38:2447–63. doi:10.1007/s10439-010-9990-9.

Quinn KP, Winkelstein BA. Preconditioning is correlated with altered collagen fiber alignment in ligament. J Biomech Eng. 2011;133(6):064506. doi:10.1115/1.4004205.
Quinn KP, Bauman JA, Crosby ND, Winkelstein BA. Anomalous fiber realignment during tensile loading of the rat facet capsular ligament identifies mechanically induced damage and physiological dysfunction. J Biomech. 2010;43:1870–5. doi:10.1016/j.jbiomech.2010.03.032.
Raghupathy R, Barocas VH. A closed-form structural model of planar fibrous tissue mechanics. J Biomech. 2009;42:1424–8. doi:10.1016/j.jbiomech.2009.04.005.
Raghupathy R, Witzenburg C, Lake SP, et al. Identification of regional mechanical anisotropy in soft tissue analogs. J Biomech Eng. 2011;133(9):091011. doi:10.1115/1.4005170.
Raya JG, Melkus G, Adam-Neumair S, et al. Change of diffusion tensor imaging parameters in articular cartilage with progressive proteoglycan extraction. Invest Radiol. 2011;46:401–9. doi:10.1097/RLI.0b013e3182145aa8.
Sacks M. Incorporation of experimentally-derived fiber orientation into a structural constitutive model for planar-collagenous tissues. J Biomech Eng. 2003;125:280–7. doi:10.1115/1.1544508.
Sacks MS, Chuong CJ. Characterization of collagen fiber architecture in the canine diaphragmatic central tendon. J Biomech Eng. 1992;114:183–90.
Schmidt MB, Mow VC, Chun LE, Eyre DR. Effects of proteoglycan extraction on the tensile behavior of articular cartilage. J Orthop Res. 1990;8:353–63. doi:10.1002/jor.1100080307.
Skaggs DL, Weidenbaum M, Iatridis JC, et al. Regional variation in tensile properties and biochemical composition of the human lumbar anulus fibrosus. Spine. 1994;19:1310–9.
Soltz MA, Ateshian GA. A conewise linear elasticity mixture model for the analysis of tension-compression nonlinearity in articular cartilage. J Biomech Eng. 2000;122:576–86.
Spencer A. Continuum theory of the mechanics of fibre-reinforced composites. New York: Springer; 1984.
Stoller P, Reiser KM, Celliers PM, Rubenchik AM. Polarization-modulated second harmonic generation in collagen. Biophys J. 2002;82:3330–42. doi:10.1016/S0006-3495(02)75673-7.
Thomopoulos S, Marquez JP, Weinberger B, et al. Collagen fiber orientation at the tendon to bone insertion and its influence on stress concentrations. J Biomech. 2006;39:1842–51. doi:10.1016/j.jbiomech.2005.05.021.
Tower T, Neidert M, Tranquillo R. Fiber alignment imaging during mechanical testing of soft tissues. Ann Biomed Eng. 2002;30:1221–33. doi:10.1114/1.1527047.
Vidal Bde C, Mello MLS. Structural organization of collagen fibers in chordae tendineae as assessed by optical anisotropic properties and Fast Fourier transform. J Struct Biol. 2009;167:166–75. doi:10.1016/j.jsb.2009.05.004.
Wagner DR, Lotz JC. Theoretical model and experimental results for the nonlinear elastic behavior of human annulus fibrosus. J Orthop Res. 2004;22:901–9. doi:10.1016/j.orthres.2003.12.012.
Waldman SD, Sacks MS, Lee JM. Imposed state of deformation determines local collagen fibre orientation but not apparent mechanical properties. Biomed Sci Instrum. 1999;35:51–6.
Wu HC, Yao RF. Mechanical behavior of the human annulus fibrosus. J Biomech. 1976;9:1–7.
Yang C-F, Crosby CM, Eusufzai ARK, Mark RE. Determination of paper sheet fiber orientation distributions by a laser optical diffraction method. J Appl Polym Sci. 1987;34:1145–57. doi:10.1002/app.1987.070340323.

Chapter 3
Emergent Behaviors in Cell Mechanics

Robert L. Steward Jr., Sonia R. Rosner, and Jeffrey J. Fredberg

Abstract Cells and therefore tissues ordinarily experience some form of mechanical stimulation as they are often in mechanically diverse and dynamic environments. As a result we have learned that cells have the astounding ability to sense and respond to their environment. This seemingly innate behavior of cells has intrigued many researchers in the field of cell mechanics for decades and compelled efforts aimed at characterizing its behaviors and underlying mechanisms. While many techniques exist, in the context of this chapter, novel techniques we have developed and implemented will be examined as well as new emergent behaviors we have discovered. The behaviors that will be discussed have relevance in various areas of pathology and physiology including collective cell migration and cancer metastasis.

3.1 Introduction

Professor Yoram Lanir has performed pioneering works in the field of biomechanics. His efforts helped lay the groundwork for theoretical and experimental studies aimed at elucidating the basic functions of cells, tissues, and organs and the constitutive mechanical laws that govern them. More importantly, he has contributed significantly to our understanding of how lessons learned from structures at the cellular level influence physiologic and pathophysiologic function at the tissue level. Complementing his work, here we present new methodologies in cell mechanics developed by our group and describe the emergent cellular behaviors we have discovered using those approaches.

Mechanics are part of our everyday life. For the father teaching his son to catch a ball as for the commuter running to catch a train, cells cooperate to allow bones, tissues, and other structures to exert and experience mechanical force. Throughout their lifetime, cells and tissues experience mechanical stimulation, and depending on the cell's location and physiology such stimuli may be in the form of tension,

R.L. Steward Jr. • S.R. Rosner • J.J. Fredberg (✉)
Program in Molecular and Integrative Physiological Sciences, School of Public Health, Harvard University, Boston, MA 02115, USA
e-mail: jfredber@hsph.harvard.edu

G.S. Kassab, M.S. Sacks (eds.), *Structure-Based Mechanics of Tissues and Organs*,
DOI 10.1007/978-1-4899-7630-7_3

compression, or shear, and may be static or cyclic (Garanich et al. 2007; Butcher et al. 2004; Cheng et al. 2009; Richard et al. 2007; Owan et al. 1997). As a result, cells have evolved the ability to sense and respond to their local microenvironment (Kung 2005; Eastwood et al. 1998; Hu et al. 2004; Smith et al. 2003; Thompson 1961; Steward et al. 2009). Such behavior has intrigued scientists from the time of D'Arcy Thompson (Thompson 1961) and inspired efforts aimed at characterizing these behaviors and their underlying mechanisms. While a plethora of approaches exists (Cheng et al. 2009; Steward et al. 2009; Bao and Suresh 2003; Zhu et al. 2000; Chien 2007; Puig-De-Morales et al. 2001; Bellin et al. 2009), we will examine techniques used to define the physical forces a cell exerts on its substrate and upon neighboring cells, and the techniques used to characterize cellular material properties.

3.2 Contractile Forces and Traction Microscopy

Since the pioneering work of Cyril Harris, cellular contractility has been of interest to many groups (Harris 1988; Harris et al. 1980; Oliver et al. 1995, 1998, 1999; Dembo and Wang 1999; Dembo et al. 1996). For the adherent cell in isolation, cellular contractility provides insights into the mechanism of force transmission between the cell and its underlying substrate. Force transmission involves a feedback loop in which integrins connected to the extracellular matrix (ECM) sense and transmit mechanical information through focal adhesions, which contain proteins including vinculin, talin, paxillin, tensin, and zyxin (Zhu et al. 2000; Ridley et al. 2003; Wang et al. 1993; Yoshigi et al. 2005; Vogel and Sheetz 2006). Focal adhesions transmit these mechanical signals to the cytoskeleton (Ridley et al. 2003; Yoshigi et al. 2005; Revenu et al. 2004; DeMali et al. 2003; Cao et al. 1993). Drawing from this information, the cytoskeleton adjusts its physical or chemical activity until intracellular homeostasis is achieved (Ridley et al. 2003; Vogel and Sheetz 2006; DeMali et al. 2003; Norman et al. 1998; Insall and Machesky 2009; Rodriguez et al. 2003). Force transmission and therefore contractility is known to differ between cell types as well as between single cells and integrated monolayers of the same cell type. Contractility is also important in numerous cellular processes including migration, embryogenesis, morphogenesis, metastasis, and wound healing (Ridley et al. 2003; Larsen et al. 2006; Poukkula et al. 2011; Tambe et al. 2011; Friedl et al. 2004; Rorth 2007, 2011; Inaki et al. 2012; Trepat and Fredberg 2011; Trepat et al. 2009). To measure contractile forces or tractions at the cellular level we use traction force microscopy (Dembo and Wang 1999; Dembo et al. 1996; Trepat et al. 2007).

Dembo and Wang were the first to quantify the spatial distributions of the tractions exerted by single cells (Dembo and Wang 1999). Tractions were obtained by measuring displacement fields generated by a single cell adherent upon a flexible, polyacrylamide gel. Embedded within that gel were fluorescent markers located beneath the gel surface. Although the approach described by Dembo et al. is computationally intensive, this method provides detailed maps of the traction forces exerted by a single cell. An alternative to this approach consists of using microfabricated, elastomeric microposts (Maruthamuthu et al. 2011; Sniadecki and Chen 2007). Using this method, the deflection caused by an adherent cell on vertically aligned microposts can be directly measured and used to deduce tractions (Maruthamuthu et al. 2011; Sniadecki and Chen 2007).

We extended the approach of Dembo by developing Fourier Transform Traction Microscopy (FTTM) (Trepat et al. 2009; Butler et al. 2002). Compared with Dembo's method, FTTM is computationally efficient. Efficiency arises from its mathematical foundation in Fourier analysis, and can be applied in either of two subcases, constrained and unconstrained FTTM. In the former tractions are deduced without prior specification of cellular boundaries, while in the latter tractions are deduced including prespecified cellular boundaries.

Experimentation and implementation are as follows. A single cell is cultured on a ligand-coated, flexible substrate and allowed to spread. Phase contrast images of cells on top of the gel and fluorescent images of beads embedded within the gel are obtained. In addition, another image is taken after the cells have been detached from the gel, representing an unstressed traction-free state referred to as the reference image (Butler et al. 2002). After image acquisition, post-processing begins by initially compensating for artifactual translational shifts between the two pairs of fluorescent images, which are ordinarily attributable to drift of the microscope stage. A fast Fourier transform algorithm is then used to compute a two-dimensional cross-correlation function between the two images. Cross-correlation is used to generate a uniform local displacement from which one image is translated with respect to another, yielding shift-corrected images. The shift-corrected images are divided into smaller windows or pixel areas from which the correlation function is once again used to calculate the local displacement between the reference window and its corresponding window in the experimental image. The displacement, which is calculated for each pixel within the image, produces a discretized gel displacement field. One advantage of FTTM worth noting is its insensitivity to explicit bead identification and bead density due to the utilization of the cross-correlation approach. Alternatively, a potential disadvantage is that this method provides an estimated displacement field while others provide a direct measurement (Dembo and Wang 1999; Butler et al. 2002). The estimated displacement field represents the "raw data" from which unconstrained and constrained tractions are calculated; these tradeoffs were analyzed by the group of Gardel (Maruthamuthu et al. 2011). The Boussinesq solution in Fourier space provides the direct solution of the tractions as a function of the two-dimensional inverse Fourier transform (Butler et al. 2002). This method yields unconstrained tractions as it does not impose traction boundary conditions upon the solution, contrary to the Dembo approach (Butler et al. 2002). Calculating constrained tractions is slightly more complex as this presents a mixed boundary value problem, and is subject to serious potential artifacts caused by improperly identifying the cell boundary. Computing constrained tractions is an iterative process that requires the initial calculation of the traction field as an unconstrained traction and calculation of new traction field by specifying all tractions outside of a user-specified boundary to zero. We initially developed this method for single cells and later expanded this to monolayers (Tambe et al. 2011; Trepat et al. 2009), as shown in Fig. 3.1.

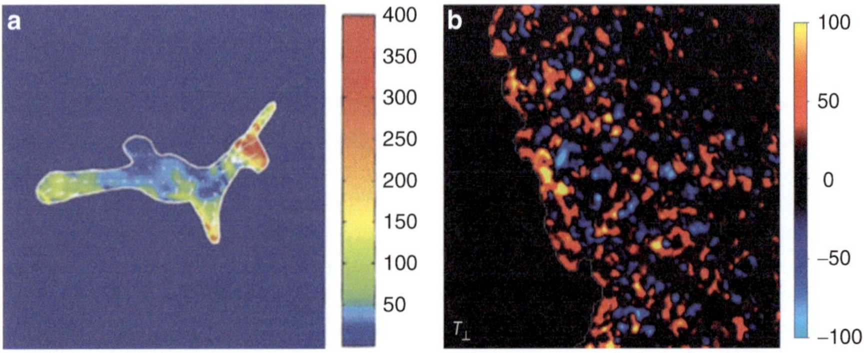

Fig. 3.1 Tractions (in Pascals) generated by a single cell (**a**) and monolayer (**b**). Adapted from Trepat et al. (2009) and Butler et al. (2002)

3.3 Monolayer Stress Microscopy and Intercellular Stresses

Having introduced FTTM we now extend the discussion to a new technique we recently developed, monolayer stress microscopy (MSM) (Tambe et al. 2011; Trepat and Fredberg 2011). The traction forces that the individual cell exerts on its substrate have been well established and useful in elucidating various cellular processes (Oliver et al. 1995, 1998; Ridley et al. 2003). However, extending our analysis to a monolayer of cells, we know that there are also intercellular forces and therefore intercellular stresses between each cell that they exert on their nearest neighbors. Newton's Laws demand that the traction forces produced at the cell–substrate interface across the monolayer be balanced by intercellular stresses between the cells (Tambe et al. 2011; Trepat and Fredberg 2011). Therefore combining FTTM with MSM we are now able to measure previously elusive intercellular stresses within a monolayer. The intercellular stress within a monolayer is defined as the local intercellular force per unit area of cell–cell contact (Tambe et al. 2011) and is composed of two components: a normal stress defined as the stress acting perpendicular to the local intercellular junction, and shear stress defined as the stress acting parallel to the local intercellular junction (Fig. 3.2). By assuming the monolayer to be a thin elastic sheet, a computationally less complex, yet rigorous two-dimensional force balance can be done (Tambe et al. 2011). The two-dimensional force balance from Newton's laws yields the distribution of line forces (force per unit length) everywhere within the monolayer, which are converted to stresses (force per unit area) using the average monolayer height (Tambe et al. 2011). This method is almost model-independent and is built upon the key assumption that the monolayer be treated as a continuum in which all forces are in balance.

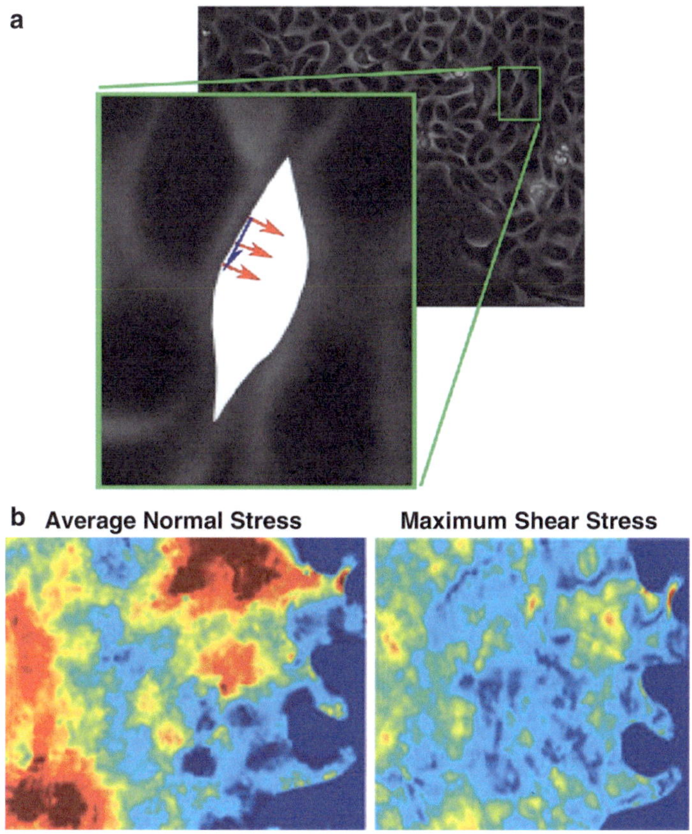

Fig. 3.2 *Arrows* depicting intercellular stresses at the cell junction (**a**) and intercellular stress maps of migrating monolayers of RPMEC (**b**). Adapted from Tambe et al. (2011)

3.4 Optical Magnetic Twisting Cytometry and Cell Material Properties

Optical magnetic twisting cytometry (OMTC) is used to measure cell material properties including cell stiffness and has helped elucidate the physical mechanisms of force transmission across the cell membrane (Wang et al. 1993). OMTC was first introduced by Crick (Crick and Hughes 1950; Crick 1950), describing a novel method that used phagocytosed small, magnetic particles to be manipulated by three principal motions; twisting, dragging, or prodding. Other groups, including our own (Hu et al. 2004; Puig-De-Morales et al. 2001; Maksym et al. 2000; Fabry et al. 2001a), later improved upon this technique. While slightly more complex, the basic operation remains similar to what was introduced some time ago. In general an OMTC system contains the following components: high-voltage generator for generating current in coils to magnetize magnetic particles, separate current sources

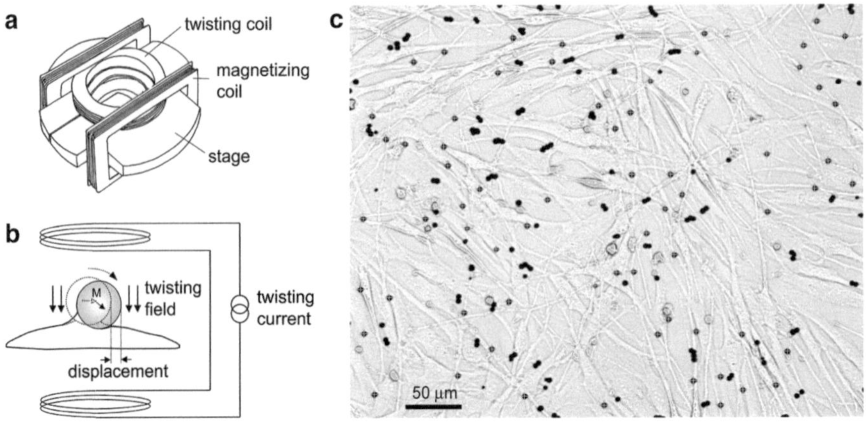

Fig. 3.3 Top (**a**) and side (**b**) schematic of OMTC device. Image of magnetic beads on single cells marked with bead-tracking program (**c**). Adapted from Fabry et al. (2001a)

and computer to manipulate and control the particles, and a microscope and camera with complementary image acquisition software to observe and acquire images of particle movement (Hu et al. 2004; Maksym et al. 2000). Cell preparation and OMTC measurement protocols used by us as well as others begin by incubating sparsely plated or confluent cells with ligand-coated ferromagnetic beads for a period of 15–20 min in an incubator. In contrast with the Crick method, this brief period allows beads to bind to integrin receptors on the cell surface, linking them to the cytoskeleton through focal adhesions, instead of being phagocytosed (Wang et al. 1993; Crick and Hughes 1950; Crick 1950). After incubation, beads are initially magnetized with a strong magnetic field. Subsequently, a set of twisting coils create a twisting field perpendicular to the initial magnetization and provide a twisting torque at frequencies ranging from 0.5 Hz to frequencies ranging in excess of 5 Hz, Fig. 3.3. During twisting, image acquisition occurs at a specific temporal resolution synchronized with the twisting cycle. A bead-tracking program is then used to compute the bead motion signal. This software is designed to search for beads with specific size, contrast, and shape properties (Hu et al. 2004) (Fig. 3.3). Once, identified, an intensity-based weighted center of mass of the bead is performed in which the respective x and y coordinates of the bead center is calculated, defining the bead motion signal. A Fourier transformation of the bead motion signal ($x\sim$) and torque (T) allow the calculation of the complex modulus (G) (in dimensions of Pascals per nanometer), which can be related to the stiffness (G'), friction (G''), and hysteresivity (η) (3.1). Therefore OMTC can be a useful tool in determining the effects of various agonist and antagonist on cell material properties.

$$G = T/x \sim = G' + jG'' = G'\left(1 + j\eta\right), \quad \text{where } j = \sqrt{-1} \tag{3.1}$$

3.5 Three Emergent Behaviors

Many cellular behaviors have been elegantly explained by detailed biochemical mechanisms, while explanation from a physical perspective remains limited and poorly understood (Richard et al. 2007; Cao et al. 1993; Norman et al. 1998; Huang et al. 2004; Lauffenburger 2000; Zheng et al. 1995; Hall and Nobes 2000). This lack of explanation of cellular phenomena from a physical perspective served as a motivation for our group, as a complete physical and chemical explanation is needed to better understand cellular behavior. Here, we present three emergent cellular behaviors that raise exciting questions and controversies.

3.5.1 Fluidization/Reinforcement

Cells have the ability to sense and respond to mechanical forces. For example, pulmonary vascular smooth muscle cells exhibit stretch-induced VEGF and FGF-2 expression (Quinn et al. 2001) and endothelial cells exhibit fluid shear stress-induced polarization (Chien 2007). Stretch in particular is important in various cellular processes including proliferation, differentiation, and gene expression (Eastwood et al. 1998; Wang and Thampatty 2006). An important response to stretch is reinforcement (Choquet et al. 1997; Matthews et al. 2006; von Wichert et al. 2003). Reinforcement describes a phenomenon in which the cytoskeleton increases stiffness and recruits actin stress fibers as stretch is experienced by the cell, exhibiting an active strain-stiffening behavior along with associated structural changes (Krishnan et al. 2009; Chen et al. 2010). Contrary to stretch-induced reinforcement, earlier studies suggested the cytoskeleton fluidized in response to stretch (Choquet et al. 1997; Matthews et al. 2006). These paradoxical findings were baffling as no agreed-upon consensus existed on a simple question: How does the cell respond to stretch? To address this paradox we developed a novel system that allows us to stretch cells while simultaneously using OMTC (Trepat et al. 2007). We used our system on multiple cell types including human airway smooth muscle (HASM) cells, human lung fibroblasts, Madin–Darby canine kidney epithelial (MDCK) cells, and human bronchial epithelial cells to (1) determine if the cell fluidizes or resolidifies in response to stretch and (2) determine if this response is shared among mammalian cell types of diverse mesenchymal lineages. Following stretch, cell stiffness immediately decreased significantly and then gradually returned to baseline over time, suggesting that in response to stretch, cells initially fluidize and gradually resolidify (Trepat et al. 2007). Fluidization in response to stretch was also observed to occur despite chemical perturbation, but with slight change in magnitude and temporal scales (Trepat et al. 2007). Contrary to fluidization, chemical perturbation revealed resolidification to be ATP-dependent (Trepat et al. 2007). Our findings revealed that among mammalian cells resolidification and fluidization are universal responses (Trepat et al. 2007).

While fluidization remains relatively new within the field it raises many new and exciting questions: What are the molecular components involved in fluidization and resolidification? How do these two responses stay in balance? The latter is exactly what we next sought out to answer.

Fluidization was further probed as a function of cell contractility using a method we developed called Cell Mapping Rheometry (CMR) (Krishnan et al. 2009). CMR uses a novel punch-indentation system to induce biaxial or uniaxial homogeneous deformations on cells cultured on a flexible substrate (Krishnan et al. 2009). The displacements of the substrate are calculated by tracking fluorescent markers embedded within the substrate using FTTM. HASM cells were subjected to biaxial and uniaxial stretch to probe whether fluidization was dependent on stretch isotropy. In response to stretch, cells immediately decreased contractility followed by a gradual recovery. This stretch-induced fluidization was independent of stress isotropy and agreed quite well with results found previously using OMTC (Trepat et al. 2007; Krishnan et al. 2009). However the question of whether fluidization or resolidification dominated in a mechanical physiological loading condition remained unanswered. To address this, we extended our system to induce non-homogeneous stretch on cells by applying force at a localized area on the substrate (Krishnan et al. 2009). Cells that underwent non-homogeneous stretch responded by reinforcement, while cells experiencing homogenous stretch fluidized. Being that we as well as others believe that homogenous stretch is more physiologically relevant, we concluded that mammalian cells primarily fluidize in response to stretch (Pirentis et al. 2011). While fluidization is believed to be a critical determinate of cell response to mechanical stimulation we should not disregard the fact that in certain physiological conditions reinforcement does prevail. This raises the question of how, if at all, does the interplay between fluidization and reinforcement change in response to other forces such as fluid shear, compression, or even a combination of both? All of this has yet to be elucidated.

3.5.2 *Plithotaxis*

Shifting our attention from single cells to cell monolayers, we now focus on a recent emergent phenomena describing collective cellular migration. Cellular migration is an essential step for many physiological processes including morphogenesis, wound healing, and regeneration (Ridley et al. 2003; Friedl et al. 2004). For example, organs including the kidney, lung, breast, and salivary glands form through branching morphogenesis that requires the coordinated collective migration of mesenchymal and epithelial cells to form sprouting vessels and ducts (Rorth 2007; Ewald et al. 2008; Vasilyev et al. 2009). Collective migration has also recently been shown to be important in cancer, suggesting cellular migration to be ubiquitous in not only physiology, but pathology as well (Friedl et al. 2004; Scotton et al. 2001). A suggested and widely accepted mechanism proposes this process to be solely dependent on biochemical signaling (Ridley et al. 2003; Rorth 2007).

For example the EGF receptor has been shown to guide oogenesis (Duchek and Rorth 2001) and FGF has been shown to guide tracheal branching morphogenesis of the Drosophila (Ribeiro et al. 2002). Other studies report cell migration to be driven by internal biochemical mechanisms in which cells within a monolayer relay their chemical state to each other through intercellular connections (Rorth 2007), further illustrating a common theme whereby cells moving as a collective do so by sensing and responding to endogenous and exogenous chemical cues. While the role of biochemistry in cell migration is clearly evident, the role of physics is less so. Cells within a monolayer are physically linked to their underlying substrate through integrins and other transmembrane proteins including syndecans (Bellin et al. 2009; Poukkula et al. 2011). These transmembrane proteins are linked to focal adhesions, which are coupled to the cytoskeleton, but this is not the full story. In addition to being physically linked to their underlying substrate, the cells themselves are physically linked to each other through cell–cell junctions (Gomez et al. 2011; Weber et al. 2012; Borghi et al. 2010). While the mechanics of cell–cell junctions remain poorly understood, it is agreed upon that they sense and exert forces of their own.

Liu et al. initially examined intercellular forces between a pair of geometrically constrained cells and reported intercellular force magnitude to regulate cell–cell contact size (Liu et al. 2010). We later expanded this finding to geometrically constrained endothelial monolayers and determined intercellular forces to be dependent on substrate stiffness and Rho kinase activity (Krishnan et al. 2011). Complementary to our work the Gardel group later concluded intercellular forces to be independent of cell size and morphology (Maruthamuthu et al. 2011). These studies demonstrated that groups of cells exert intercellular forces at their cell–cell junctions, opening the door for many more exciting questions to be answered. For example: How do these forces differ for larger groups of cells (>1000) as would be seen physiologically? What if any are the spatial and temporal fluctuations observed by these forces? How, if at all, does this affect collective cellular migration?

To answer these questions we used traction microscopy and MSM to observe the physical dynamics of migrating cell monolayers (Tambe et al. 2011). We cultured monolayers of rat pulmonary microvascular endothelial cells and MDCK cells on elastic gels and observed their migration. Using MSM we reported for the first time high-resolution normal and shear intercellular stress maps, revealing stress distributions that were extremely heterogeneous. Also the velocity vector of each migrating cell was found to correlate with the maximum normal stress (or minimal shear stress), implying that cells collectively migrating do so such as to minimize shear stress on its junctions, a phenomenon we have termed plithotaxis (Tambe et al. 2011; Trepat and Fredberg 2011). Plithotaxis represents the first physical explanation of collective cellular migration. This mechanically guided behavior was inhibited by calcium chelation and treatment of anti-cadherin antibodies in MCF10A breast cancer cells, suggesting cells must be physically linked to each other in order to migrate via plithotaxis (Tambe et al. 2011). Whether plithotaxis works cooperatively with other well-known cell guidance mechanisms such as chemotaxis, mechanotaxis, and durotaxis or if plithotaxis exists in vivo remains unanswered. There is also the possibility that plithotaxis is integral in cellular processes beyond collective cell migration where the boundary conditions of cells are different.

3.5.3 *Viewing the Cell as a Soft Glassy Material*

Stretch-induced fluidization, reinforcement, and plithotaxis all imply an emerging yet surprising concept of cells being comparable to a class of materials in physics known as soft glassy materials (SGMs). SGM are quite ubiquitous in nature and include colloidal suspensions, pastes, foams, and slurries (Sollich 1998; Sollich et al. 1997). While materials classified as SGM may vary in chemical properties, they share similar mechanical properties. All materials classified as SGM share these three common mechanical characteristics: (1) they are soft (young's modulus = <1 kPA), (2) their dynamics are "scale free," and (3) the frictional stress is proportional to the elastic stress with a constant of proportionality known as the hysteresivity, η (where η is on the order of 0.1) (Sollich 1998; Sollich et al. 1997; Krishnan et al. 2008). In accordance, SGM also have the ability to selectively phase transition between solid-like and liquid-like states (Sollich 1998). An initial suggestion that cells are analogous to SGM was provided by Fabry (Fabry et al. 2001b) when OMTC was used to demonstrate that multiple cell types exhibited a scaling law behavior that governed their elastic and frictional properties over a wide range of temporal scales and biological conditions. Trepat later combined OMTC with stretch on single cells, revealing the cell's elastic and frictional properties to not only exhibit a scaling law behavior, but to be scale-free as well (Trepat et al. 2007). Complementary to our stretch studies, single cells subjected to osmotically induced compressive stress were observed to become much more solid-like (Zhou et al. 2009). Cell material properties and their similarities to SGM have been documented for single cells, but the question of whether this behavior remains true for groups of cells such as monolayers needed to be investigated.

To address this we used FTTM and MSM to observe cooperativity of intercellular stresses between cells within monolayers over relatively long distances (10–15 cell diameters) (Tambe et al. 2011). Intercellular stress cooperativity was observed to become enhanced over greater distances when comparing intercellular stress transmission to increasing cell density over time (Tambe et al. 2011), reflecting an increase in the dynamic heterogeneity of intercellular stress as cell density increased. Complementary to cooperativity of intercellular forces, Angelini (Angelini et al. 2011) recently reported cell migration velocity to decrease as cell density increased, a behavior remarkably similar to a glassy system transitioning from a liquid to solid phase (Krishnan et al. 2008).

This glassy-like behavior has led us as well as others to propose that cellular migration be viewed in terms of a glassy system. It should be cautioned that while these cellular systems have been observed to possess similar qualities of glassy systems, exhaustive studies still need to be done to prove beyond a doubt that these living, complex systems are in fact comparable to the inert glassy systems we see and use everyday.

3.6 Concluding Remarks

We have presented emergent behaviors on the cellular level that illustrate physical mechanisms which dictate biological responses and have forced researchers to rethink how they view the cell and its processes. We now know that mammalian cells primarily fluidize in response to stretch and epithelial and endothelial monolayers are mechanically guided as they collectively migrate through their environment. Finally, the ability of cells to fluidize, reinforce, and migrate by plithotaxis are all properties similar to SGMs. Supplementary to our findings is evidence linking these phenomena to systems beyond our fields of research, including cancer and morphogenesis. As there still exist many unknowns, these newly defined concepts represent only the tip of the iceberg of the underlying physical phenomena that influence cellular behavior.

References

Angelini TE, Hannezo E, Trepat X, Marquez M, Fredberg JJ, Weitz DA. Glass-like dynamics of collective cell migration. Proc Natl Acad Sci U S A. 2011;108(12):4714–9. doi:1010059108 [pii] 10.1073/pnas.1010059108.

Bao G, Suresh S. Cell and molecular mechanics of biological materials. Nat Mater. 2003;2(11):715–25. doi:10.1038/nmat1001 nmat1001 [pii].

Bellin RM, Kubicek JD, Frigault MJ, Kamien AJ, Steward Jr RL, Barnes HM, Digiacomo MB, Duncan LJ, Edgerly CK, Morse EM, Park CY, Fredberg JJ, Cheng CM, LeDuc PR. Defining the role of syndecan-4 in mechanotransduction using surface-modification approaches. Proc Natl Acad Sci U S A. 2009;106(52):22102–7. doi:0902639106 [pii] 10.1073/pnas.0902639106.

Borghi N, Lowndes M, Maruthamuthu V, Gardel ML, Nelson WJ. Regulation of cell motile behavior by crosstalk between cadherin- and integrin-mediated adhesions. Proc Natl Acad Sci U S A. 2010;107(30):13324–9. doi:1002662107 [pii] 10.1073/pnas.1002662107.

Butcher JT, Penrod AM, Garcia AJ, Nerem RM. Unique morphology and focal adhesion development of valvular endothelial cells in static and fluid flow environments. Arterioscler Thromb Vasc Biol. 2004;24(8):1429–34. doi:10.1161/01.ATV.0000130462.50769.5a01.ATV.0000130462.50769.5a [pii].

Butler JP, Tolic-Norrelykke IM, Fabry B, Fredberg JJ. Traction fields, moments, and strain energy that cells exert on their surroundings. Am J Physiol Cell Physiol. 2002;282(3):C595–605. doi:10.1152/ajpcell.00270.2001.

Cao LG, Fishkind DJ, Wang YL. Localization and dynamics of nonfilamentous actin in cultured cells. J Cell Biol. 1993;123(1):173–81.

Chen C, Krishnan R, Zhou E, Ramachandran A, Tambe D, Rajendran K, Adam RM, Deng L, Fredberg JJ. Fluidization and resolidification of the human bladder smooth muscle cell in response to transient stretch. PLoS One. 2010;5(8):e12035. doi:10.1371/journal.pone.0012035.

Cheng CM, Steward Jr RL, LeDuc PR. Probing cell structure by controlling the mechanical environment with cell-substrate interactions. J Biomech. 2009;42(2):187–92. doi:S0021-9290(08)00520-4 [pii]10.1016/j.jbiomech.2008.10.014.
Chien S. Mechanotransduction and endothelial cell homeostasis: the wisdom of the cell. Am J Physiol Heart Circ Physiol. 2007;292(3):H1209–24. doi:01047.2006 [pii] 10.1152/ajpheart.01047.2006.
Choquet D, Felsenfeld DP, Sheetz MP. Extracellular matrix rigidity causes strengthening of integrin-cytoskeleton linkages. Cell. 1997;88(1):39–48. doi:S0092-8674(00)81856-5 [pii].
Crick FC. The physical properties of the cytoplasm. A study by means of the magnetic particle method. Part 2. Theoretical treatment. Exp Cell Res. 1950;1:505–33.
Crick FC, Hughes AW. The physical properties of the cytoplasm. A study by means of the magnetic particle method. Part 1. Exp Cell Res. 1950;1:37–80.
DeMali KA, Wennerberg K, Burridge K. Integrin signaling to the actin cytoskeleton. Curr Opin Cell Biol. 2003;15(5):572–82. doi:S0955067403001091 [pii].
Dembo M, Wang YL. Stresses at the cell-to-substrate interface during locomotion of fibroblasts. Biophys J. 1999;76(4):2307–16. doi:S0006-3495(99)77386-8 [pii] 10.1016/S0006-3495(99)77386-8.
Dembo M, Oliver T, Ishihara A, Jacobson K. Imaging the traction stresses exerted by locomoting cells with the elastic substratum method. Biophys J. 1996;70(4):2008–22. doi:S0006-3495(96)79767-9 [pii] 10.1016/S0006-3495(96)79767-9.
Duchek P, Rorth P. Guidance of cell migration by EGF receptor signaling during Drosophila oogenesis. Science. 2001;291(5501):131–3. doi:10.1126/science.291.5501.131291/5501/131 [pii].
Eastwood M, McGrouther DA, Brown RA. Fibroblast responses to mechanical forces. Proc Inst Mech Eng H. 1998;212(2):85–92.
Ewald AJ, Brenot A, Duong M, Chan BS, Werb Z. Collective epithelial migration and cell rearrangements drive mammary branching morphogenesis. Dev Cell. 2008;14(4):570–81. doi:S1534-5807(08)00111-1 [pii] 10.1016/j.devcel.2008.03.003.
Fabry B, Maksym GN, Shore SA, Moore PE, Panettieri Jr RA, Butler JP, Fredberg JJ. Selected contribution: time course and heterogeneity of contractile responses in cultured human airway smooth muscle cells. J Appl Physiol. 2001a;91(2):986–94.
Fabry B, Maksym GN, Butler JP, Glogauer M, Navajas D, Fredberg JJ. Scaling the microrheology of living cells. Phys Rev Lett. 2001b;87(14):148102.
Friedl P, Hegerfeldt Y, Tusch M. Collective cell migration in morphogenesis and cancer. Int J Dev Biol. 2004;48(5–6):441–9. doi:10.1387/ijdb.041821 041821 [pii].
Garanich JS, Mathura RA, Shi ZD, Tarbell JM. Effects of fluid shear stress on adventitial fibroblast migration: implications for flow-mediated mechanisms of arterialization and intimal hyperplasia. Am J Physiol Heart Circ Physiol. 2007;292(6):H3128–3135. doi:00578.2006 [pii]10.1152/ajpheart.00578.2006.
Gomez GA, McLachlan RW, Yap AS. Productive tension: force-sensing and homeostasis of cell-cell junctions. Trends Cell Biol. 2011;21(9):499–505. doi:S0962-8924(11)00105-X [pii] 10.1016/j.tcb.2011.05.006.
Hall A, Nobes CD. Rho GTPases: molecular switches that control the organization and dynamics of the actin cytoskeleton. Philos Trans R Soc Lond B Biol Sci. 2000;355(1399):965–70. doi:10.1098/rstb.2000.0632.
Harris AK. Fibroblasts and myofibroblasts. Methods Enzymol. 1988;163:623–42.
Harris AK, Wild P, Stopak D. Silicone rubber substrata: a new wrinkle in the study of cell locomotion. Science. 1980;208(4440):177–9.
Hu S, Eberhard L, Chen J, Love JC, Butler JP, Fredberg JJ, Whitesides GM, Wang N. Mechanical anisotropy of adherent cells probed by a three-dimensional magnetic twisting device. Am J Physiol Cell Physiol. 2004;287(5):C1184–91. doi:10.1152/ajpcell.00224.2004 00224.2004 [pii].

Huang H, Kamm RD, Lee RT. Cell mechanics and mechanotransduction: pathways, probes, and physiology. Am J Physiol Cell Physiol. 2004;287(1):C1–11. doi:10.1152/ajpcell.00559.2003 287/1/C1 [pii].

Inaki M, Vishnu S, Cliffe A, Rorth P. Effective guidance of collective migration based on differences in cell states. Proc Natl Acad Sci U S A. 2012;109(6):2027–32. doi:1115260109 [pii] 10.1073/pnas.1115260109.

Insall RH, Machesky LM. Actin dynamics at the leading edge: from simple machinery to complex networks. Dev Cell. 2009;17(3):310–22. doi:S1534-5807(09)00348-7 [pii] 10.1016/j.devcel.2009.08.012.

Krishnan R, Trepat X, Nguyen TT, Lenormand G, Oliver M, Fredberg JJ. Airway smooth muscle and bronchospasm: fluctuating, fluidizing, freezing. Respir Physiol Neurobiol. 2008; 163(1–3):17–24. doi:S1569-9048(08)00099-2 [pii] 10.1016/j.resp.2008.04.006.

Krishnan R, Park CY, Lin YC, Mead J, Jaspers RT, Trepat X, Lenormand G, Tambe D, Smolensky AV, Knoll AH, Butler JP, Fredberg JJ. Reinforcement versus fluidization in cytoskeletal mechanoresponsiveness. PLoS One. 2009;4(5):e5486. doi:10.1371/journal.pone.0005486.

Krishnan R, Klumpers DD, Park CY, Rajendran K, Trepat X, van Bezu J, van Hinsbergh VW, Carman CV, Brain JD, Fredberg JJ, Butler JP, van Nieuw Amerongen GP. Substrate stiffening promotes endothelial monolayer disruption through enhanced physical forces. Am J Physiol Cell Physiol. 2011;300(1):C146–54. doi:ajpcell.00195.2010 [pii] 10.1152/ajpcell.00195.2010.

Kung C. A possible unifying principle for mechanosensation. Nature. 2005;436(7051):647–54. doi:nature03896 [pii]10.1038/nature03896.

Larsen M, Wei C, Yamada KM. Cell and fibronectin dynamics during branching morphogenesis. J Cell Sci. 2006;119(Pt 16):3376–84. doi:jcs.03079 [pii] 10.1242/jcs.03079.

Lauffenburger DA. Cell signaling pathways as control modules: complexity for simplicity? Proc Natl Acad Sci U S A. 2000;97(10):5031–3. doi:97/10/5031 [pii].

Liu Z, Tan JL, Cohen DM, Yang MT, Sniadecki NJ, Ruiz SA, Nelson CM, Chen CS. Mechanical tugging force regulates the size of cell-cell junctions. Proc Natl Acad Sci U S A. 2010;107(22):9944–9. doi:0914547107 [pii] 10.1073/pnas.0914547107.

Maksym GN, Fabry B, Butler JP, Navajas D, Tschumperlin DJ, Laporte JD, Fredberg JJ. Mechanical properties of cultured human airway smooth muscle cells from 0.05 to 0.4 Hz. J Appl Physiol. 2000;89(4):1619–32.

Maruthamuthu V, Sabass B, Schwarz US, Gardel ML. Cell-ECM traction force modulates endogenous tension at cell-cell contacts. Proc Natl Acad Sci U S A. 2011;108(12):4708–13. doi:1011123108 [pii]10.1073/pnas.1011123108.

Matthews BD, Overby DR, Mannix R, Ingber DE. Cellular adaptation to mechanical stress: role of integrins, Rho, cytoskeletal tension and mechanosensitive ion channels. J Cell Sci. 2006;119(Pt 3):508–18. doi:119/3/508 [pii] 10.1242/jcs.02760.

Norman JC, Jones D, Barry ST, Holt MR, Cockcroft S, Critchley DR. ARF1 mediates paxillin recruitment to focal adhesions and potentiates Rho-stimulated stress fiber formation in intact and permeabilized Swiss 3T3 fibroblasts. J Cell Biol. 1998;143(7):1981–95.

Oliver T, Dembo M, Jacobson K. Traction forces in locomoting cells. Cell Motil Cytoskeleton. 1995;31(3):225–40. doi:10.1002/cm.970310306.

Oliver T, Jacobson K, Dembo M. Design and use of substrata to measure traction forces exerted by cultured cells. Methods Enzymol. 1998;298:497–521.

Oliver T, Dembo M, Jacobson K. Separation of propulsive and adhesive traction stresses in locomoting keratocytes. J Cell Biol. 1999;145(3):589–604.

Owan I, Burr DB, Turner CH, Qiu J, Tu Y, Onyia JE, Duncan RL. Mechanotransduction in bone: osteoblasts are more responsive to fluid forces than mechanical strain. Am J Physiol. 1997;273(3 Pt 1):C810–5.

Pirentis AP, Peruski E, Iordan AL, Stamenovic D. A model for stress fiber realignment caused by cytoskeletal fluidization during cyclic stretching. Cell Mol Bioeng. 2011;4(1):67–80. doi:10.1007/s12195-010-0152-9.

Poukkula M, Cliffe A, Changede R, Rorth P. Cell behaviors regulated by guidance cues in collective migration of border cells. J Cell Biol. 2011;192(3):513–24. doi:jcb.201010003 [pii] 10.1083/jcb.201010003.

Puig-De-Morales M, Grabulosa M, Alcaraz J, Mullol J, Maksym GN, Fredberg JJ, Navajas D. Measurement of cell microrheology by magnetic twisting cytometry with frequency domain demodulation. J Appl Physiol. 2001;91(3):1152–9.

Quinn TP, Schlueter M, Soifer SJ, Gutierrez JA. Cyclic mechanical stretch induces VEGF and FGF-2 expression in pulmonary vascular smooth muscle cells. Am J Physiol Lung Cell Mol Physiol. 2001;282(5):L897–903.

Revenu C, Athman R, Robine S, Louvard D. The co-workers of actin filaments: from cell structures to signals. Nat Rev Mol Cell Biol. 2004;5(8):635–46. doi:10.1038/nrm1437.

Ribeiro C, Ebner A, Affolter M. In vivo imaging reveals different cellular functions for FGF and Dpp signaling in tracheal branching morphogenesis. Dev Cell. 2002;2(5):677–83. doi:S1534580702001715 [pii].

Richard MN, Deniset JF, Kneesh AL, Blackwood D, Pierce GN. Mechanical stretching stimulates smooth muscle cell growth, nuclear protein import, and nuclear pore expression through mitogen-activated protein kinase activation. J Biol Chem. 2007;282(32):23081–8. doi:M703602200 [pii]10.1074/jbc.M703602200.

Ridley AJ, Schwartz MA, Burridge K, Firtel RA, Ginsberg MH, Borisy G, Parsons JT, Horwitz AR. Cell migration: integrating signals from front to back. Science. 2003;302(5651):1704–9. doi:10.1126/science.1092053302/5651/1704 [pii].

Rodriguez OC, Schaefer AW, Mandato CA, Forscher P, Bement WM, Waterman-Storer CM. Conserved microtubule-actin interactions in cell movement and morphogenesis. Nat Cell Biol. 2003;5(7):599–609. doi:10.1038/ncb0703-599 ncb0703-599 [pii].

Rorth P. Collective guidance of collective cell migration. Trends Cell Biol. 2007;17(12):575–9. doi:S0962-8924(07)00246-2 [pii] 10.1016/j.tcb.2007.09.007.

Rorth P. Whence directionality: guidance mechanisms in solitary and collective cell migration. Dev Cell. 2011;20(1):9–18. doi:S1534-5807(10)00595-2 [pii] 10.1016/j.devcel.2010.12.014.

Scotton CJ, Wilson JL, Milliken D, Stamp G, Balkwill FR. Epithelial cancer cell migration: a role for chemokine receptors? Cancer Res. 2001;61(13):4961–5.

Smith PG, Deng L, Fredberg JJ, Maksym GN. Mechanical strain increases cell stiffness through cytoskeletal filament reorganization. Am J Physiol Lung Cell Mol Physiol. 2003;285(2): L456–63. doi:10.1152/ajplung.00329.2002 00329.2002 [pii].

Sniadecki NJ, Chen CS. Microfabricated silicone elastomeric post arrays for measuring traction forces of adherent cells. Methods Cell Biol. 2007;83:313–28. doi:S0091-679X(07)83013-5 [pii]10.1016/S0091-679X(07)83013-5.

Sollich P. Rheological constitutive equation for a model of soft glassy materials. Phys Rev E. 1998;58:738–59.

Sollich P, Lequeneux F, Hebraud P, Cates ME. Rheology of soft glassy materials. Phys Rev Lett. 1997;78:2020–3.

Steward Jr RL, Cheng CM, Wang DL, Leduc PR. Probing cell structure responses through a shear and stretching mechanical stimulation technique. Cell Biochem Biophys. 2009;56(2–3): 115–24. doi:10.1007/s12013-009-9075-2.

Tambe DT, Hardin CC, Angelini TE, Rajendran K, Park CY, Serra-Picamal X, Zhou EH, Zaman MH, Butler JP, Weitz DA, Fredberg JJ, Trepat X. Collective cell guidance by cooperative intercellular forces. Nat Mater. 2011;10(6):469–75. doi:nmat3025 [pii] 10.1038/nmat3025.

Thompson DAW. On growth and form. Abridged ed. Cambridge: Cambridge University Press; 1961.

Trepat X, Fredberg JJ. Plithotaxis and emergent dynamics in collective cellular migration. Trends Cell Biol. 2011;21(11):638–46. doi:S0962-8924(11)00127-9 [pii] 10.1016/j.tcb.2011.06.006.

Trepat X, Deng L, An SS, Navajas D, Tschumperlin DJ, Gerthoffer WT, Butler JP, Fredberg JJ. Universal physical responses to stretch in the living cell. Nature. 2007;447(7144):592–5. doi:nature05824 [pii]10.1038/nature05824.

Trepat X, Wasserman MR, Angelini TE, Millet E, Weitz DA, Butler JP, Fredberg JJ. Physical forces during collective migration. Nat Phys. 2009;5:426–30.

Vasilyev A, Liu Y, Mudumana S, Mangos S, Lam PY, Majumdar A, Zhao J, Poon KL, Kondrychyn I, Korzh V, Drummond IA. Collective cell migration drives morphogenesis of the kidney nephron. PLoS Biol. 2009;7(1):e9. doi:08-PLBI-RA-2396 [pii] 10.1371/journal.pbio.1000009.

Vogel V, Sheetz M. Local force and geometry sensing regulate cell functions. Nat Rev Mol Cell Biol. 2006;7(4):265–75. doi:nrm1890 [pii]10.1038/nrm1890.

von Wichert G, Haimovich B, Feng GS, Sheetz MP. Force-dependent integrin-cytoskeleton linkage formation requires downregulation of focal complex dynamics by Shp2. EMBO J. 2003;22(19):5023–35. doi:10.1093/emboj/cdg492.

Wang JH, Thampatty BP. An introductory review of cell mechanobiology. Biomech Model Mechanobiol. 2006;5(1):1–16. doi:10.1007/s10237-005-0012-z.

Wang N, Butler JP, Ingber DE. Mechanotransduction across the cell surface and through the cytoskeleton. Science. 1993;260(5111):1124–7.

Weber GF, Bjerke MA, DeSimone DW. A mechanoresponsive cadherin-keratin complex directs polarized protrusive behavior and collective cell migration. Dev Cell. 2012;22(1):104–15. doi:S1534-5807(11)00465-5 [pii] 10.1016/j.devcel.2011.10.013.

Yoshigi M, Hoffman LM, Jensen CC, Yost HJ, Beckerle MC. Mechanical force mobilizes zyxin from focal adhesions to actin filaments and regulates cytoskeletal reinforcement. J Cell Biol. 2005;171(2):209–15. doi:jcb.200505018 [pii]10.1083/jcb.200505018.

Zheng Y, Olson MF, Hall A, Cerione RA, Toksoz D. Direct involvement of the small GTP-binding protein Rho in lbc oncogene function. J Biol Chem. 1995;270(16):9031–4.

Zhou EH, Trepat X, Park CY, Lenormand G, Oliver MN, Mijailovich SM, Hardin C, Weitz DA, Butler JP, Fredberg JJ. Universal behavior of the osmotically compressed cell and its analogy to the colloidal glass transition. Proc Natl Acad Sci U S A. 2009;106(26):10632–7. doi:0901462106 [pii] 10.1073/pnas.0901462106.

Zhu C, Bao G, Wang N. Cell mechanics: mechanical response, cell adhesion, and molecular deformation. Annu Rev Biomed Eng. 2000;2:189–226. doi:2/1/189 [pii] 10.1146/annurev.bioeng.2.1.189.

Chapter 4
Histomechanical Modeling of the Wall of Abdominal Aortic Aneurysm

T. Christian Gasser

Abstract Vascular diseases are already the leading cause of death in the industrialized countries and many of the associated risk factors are increasing. A multidisciplinary approach including biomechanics is needed to better understand and more effectively treat these diseases. Specifically, constitutive modeling is critical in understanding the biomechanics of the vascular wall and to uncover pathologies like Abdominal Aortic Aneurysms (AAAs), i.e. local dilatations of the infrarenal aorta. Aneurysms are formed through irreversible pathological remodeling of the vascular wall and integrating this biological process in the constitutive description could improve our current understanding of aneurysm disease. It might also increase the predictability of biomechanical simulations towards augmenting clinical decisions. The present chapter develops histomechanical constitutive models for the AAA wall according to Lanir's pioneering approach. Consequently, macroscopic properties were derived through an integration of distributed fibers, where collagen was regarded as the most important protein of the aneurysmatic Extra Cellular Matrix (ECM). Collagen organization was quantified through Polarized Light Microscopy (PLM) of picrosirius red stained histological slices from tissue samples harvested during elective open AAA repair. This histological information was either directly integrated in the constitutive description or used to qualitatively validate the predicted remodeling of the AAA wall. Specifically, two descriptions for the AAA wall were used, where collagen was regarded either as a purely passive entity of the ECM or as an active entity. The suggested constitutive models were able to successfully capture salient features of the AAA wall, but a rigorous validation against detailed experimental data was beyond the scope of this chapter.

T.C. Gasser (✉)
Department of Solid Mechanics, School of Engineering Sciences,
KTH Royal Institute of Technology, Stockholm, Sweden
e-mail: gasser@kth.se

G.S. Kassab, M.S. Sacks (eds.), *Structure-Based Mechanics of Tissues and Organs*,
DOI 10.1007/978-1-4899-7630-7_4

4.1 Introduction

Vascular diseases are the leading cause of death in the industrialized countries and associated risk factors like obesity, diabetes, and life expectancy are increasing. Current clinical options are somewhat limited, and clearly there is an urgent need for multi-disciplinary approaches to improve our current understanding of vascular diseases. Biomechanical conditions play a key role in the genesis and development of vascular diseases (Bäck et al. 2013) and the identification of the specific causative links between biomechanics and biochemistry may help advance our current view of physiology and pathology.

Continued advances in computer technology and computational methods allow us nowadays to model patient-specific problems, where the nonlinear Finite Element Method (FEM) effectively solves the 3D (coupled) mechanical problems that arise. This allows us to incorporate the nonlinear character of the related problems and combines synergetically with medical imaging. Consequently, computer simulations of biomechanical phenomena have become potentially significant in order to explore, for instance, loads experienced by cells and extra-cellular components, the interaction between medical devices and biological material, drug delivery pathways, the interplay between structure and function of tissues, mechanotransduction, and the like. Although, to some extent, traditional mechanics concepts are directly applicable to solve biomechanical problems, they remain a modeling challenge due to complex spatial domains, constitutive nonlinearities, and coupling among structural, fluid, chemical, and electrical fields to mention just a few. Specifically, the inherent property of biological tissue to adapt to mechanical and chemical environments remains a challenging modeling task.

The present chapter focuses on Abdominal Aortic Aneurysms (AAAs), i.e. local dilations of the infrarenal aorta caused by pathological remodeling of structural proteins in the aortic wall (Choke et al. 2005; Tsamis et al. 2013). An AAA ruptures if the mechanical stress exceeds the local wall strength, and is a frequent cause of death in elderly male population (The UK Small Aneurysm Trial Participants 1998). Consequently, a rupture risk assessment is central to the management of AAA patients, and an accurate assessment could reduce the related mortality without unnecessarily increasing the rate of AAA repair interventions. Clinically, the risk of rupture is correlated with the aneurysm's maximum diameter, and surgical or endovascular AAA repair is indicated if the diameter exceeds 55 mm (The UK Small Aneurysm Trial Participants 1998). However, the diameter criterion has clear limitations and mechanical parameters, such as Peak Wall Stress (PWS) (Fillinger et al. 2002; Venkatasubramaniam et al. 2004; Heng et al. 2008) and Peak Wall Rupture Risk (PWRR) (Gasser et al. 2010; Maier et al. 2010), have been found to be more reliable indices to assess AAA rupture risk. The computation of PWS and PWRR requires an accurate patient-specific reconstruction of AAAs from medical images (Gasser 2012) and appropriate modeling of the aneurysmatic tissue's constitution (Martufi and Gasser 2013).

Throughout this chapter the pioneering constitutive approach by Lanir (1983) was followed, such that the macroscopic mechanical properties were governed by the arrangement of fibrous tissue components like collagen. According to Lanir's approach, the spatial orientation and undulation of collagen are the most influential micro-histological parameters that together with the fibers' constitution determine the macroscopic mechanical properties. Consequently, the macroscopic mechanical tissue properties are derived through two numerical integrations, i.e. (1) over the undulation and (2) over the fibers' spatial orientation. This requires extensive computations and makes the application of such an approach somewhat limited. However, both the use of phenomenological collagen fiber models and specific assumptions regarding the constitution and undulation of collagen can avoid the numerical integration over the undulation of collagen (Martufi and Gasser 2011). Consequently, the biomechanics of an entire AAA can be analyzed by such a two-scale approach within reasonable computational times. For specific problems, i.e. under certain constitutive and deformation conditions, even both integrations can be avoided (Gasser et al. 2006; Federico and Herzog 2008). It is also worth noting that independent from the applied constitutive formulation an efficient and implicit numerical implementation (Gasser and Holzapfel 2002; Gasser et al. 2006; Gasser 2011; Gasser and Forsell 2011) is beneficial for analyzing clinically relevant problems.

Constitutive modeling requires a sound histological understanding, such that Sect. 4.2 reviews the composition of the aneurysmatic Extracellular Matrix (ECM) from a structural mechanical point of view. Specifically, the role of collagen is discussed, and its three-dimensional organization in the AAA wall is studied from histological slices. In Sect. 4.3 two constitutive descriptions for the AAA wall are derived that regard collagen either as a purely passive entity or as an active entity of the ECM.

4.2 The Extracellular Matrix of the AAA Wall

The ECM provides an essential supporting scaffold for the structural and functional properties of vessel walls. The ECM mainly contains elastin, collagen, and Proteoglycans (PGs) (Carey 1991) and their three-dimensional organization is vital to accomplish proper physiological functions. The ECM, therefore, rather than being merely a system of scaffolding for the surrounding cells, is an active mechanical structure that controls the micro-mechanical and macro-mechanical environments to which vascular tissue is exposed. Consequently, a proper understanding of the mechanical properties of the ECM is critically important to estimate and quantify the amount of stress and/or strain transmitted from the macroscopic to the cellular levels of vascular tissue.

Collagen is one of the most dominant structural proteins in the ECM. Collagen fibrils, with diameters ranging from 50 to a few hundreds of nanometers are the basic building blocks of fibrous collagenous tissues (Fratzl 2008), and their organization into suprafibrilar structures has a large impact on the tissue's macroscopic mechanical properties. Already 60 years ago Roach and Burton (1957) reported that collagen mainly determined the mechanical properties of arterial tissue at high strain levels. Since that time a direct correlation between the collagen content and the stiffness and strength has become generally accepted. Earlier observations indicated that the collagen-rich abdominal aorta was stiffer than the collagen-poor thoracic aorta (Bergel 1961; Langewouters et al. 1984) and later regional variations of aortic properties were specifically documented, see Sokolis (2007) for example. Numerous further references were provided by the seminal works of Fung (1993) and Humphrey (2002). Apart from the amount of collagen in the wall its spatial orientation (Fratzl 2008), including the spread in orientations (Gasser et al. 2006) is the most influential micro-structural parameter that significantly affects the macroscopic mechanical properties.

While elastin in the ECM is a stable protein having half-life times of tens of years (Alberts et al. 1994), collagen is normally in a continuous state of deposition and degradation (Humphrey 1999) and has a normal half-life time of 60–70 days (Nissen et al. 1978). Collagen fibrils are locally secreted by fibroblast and understanding the formation of hierarchical collagen structures is crucial to assessing the mechanical properties of the vascular wall. Physiological maintenance of the collagen structure relies on a delicate (coupled) balance between degradation and synthesis. Fibroblasts perceive changes in the mechanical strains/stresses and adjust their expression and synthesis of collagen molecules in order to account for the changes in their micro-mechanical environment.

Collagen is also critically involved in the gradual remodeling and weakening of the aneurysmal wall (Choke et al. 2005). It is well understood that through aneurysm disease the elastin in the wall degrades (in larger AAAs it almost vanishes) while the collagen content increases significantly (Rizzo et al. 1989). However, at later stages of aneurysm disease, the collagen synthesis is insufficient to counteract the increased mechanical wall stress, i.e. the stress required to carry the blood pressure (Choke et al. 2005). Consequently, the structural integrity of the wall is not ensured and wall strength decreases, which in turn quickly increases risk for AAA rupture.

In conclusion, from a biomechanics perspective, collagen in the AAA wall is by far the most important structural protein that largely defines the wall's stiffness, strength and toughness.

4.2.1 Proteoglycan

Proteoglycan (PG) bridges have been suggested (Scott 2003, 2008) as providing interfibrillar load transition, a necessity for a load-carrying collagen fiber structures. Specifically, small proteoglycans such as decorin bind noncovalently but specifically to collagen fibrils and cross-link adjacent collagen fibrils at about 60 nm intervals (Scott 2003). Reversible deformability of the PG bridges is crucial to serve as shape-maintaining modules (Scott 2003) and fast and slow deformation mechanisms have been identified. The fast (elastic) deformation is supported by the sudden extension of about 10 % of the L-iduronate (an elastic sugar) at a critical load of about 200 pN (Haverkamp et al. 2005). The slow (viscous) deformation is based on a sliding filament mechanism of the twofold helix of the glycan (Scott 2003). In addition, the close packing and cross-linking of collagen molecules in fibrils defines a virtually inextensible fiber, such that the strain of collagen fibrils is always much smaller than the macroscopic strain in collagenous tissues. This also points towards the existence of gliding processes occurring at the interfibrilar and/or the interfiber levels (Gupta et al. 2010).

PG-based cross-linking is supported by numerous experimental studies showing that PGs play a direct role in inter-fibril load sharing (Liao and Vesely 2007; Robinson et al. 2005; Scott 2003; Sasaki and Odajima 1996). This has also been verified through theoretical investigations (Fessel and Snedeker 2011; Redaelli et al. 2003; Vesentini et al. 2005). However, it should also be noted that the biomechanical role of PGs is somewhat controversial, and some data indicates minimal, if any, PG contribution to the tensile properties of the tissue (Fessel and Snedeker 2011; Rigozzi et al. 2009, 2010).

4.2.2 Imaging the Collagen in the AAA Wall

Collagen is intrinsically birefringent and Polarized Light Microscopy (PLM) provides an ideal method for its detection and analysis (Vidal et al. 1982; Lindeman et al. 2009; Weber et al. 1990). PLM provides a clear qualitative image of the organization of collagen and, when combined with a Universal Rotary Stage (URS), it allows a quantitative representation of the collagen orientation (Canham et al. 1989; Canham and Finlay 2004). To analyze the collagen organization in the AAA wall, anterior tissue samples were selected during elective open surgical repair, see Fig. 4.1. Wall specimens from 24 aneurysm patients were fixed, embedded in paraffin, and sliced at a thickness of 7 μm. Finally, picrosirius red staining was used to enhance the birefringent properties of collagen, which improved the precision of measurement while preserving the optical axis of the unstained collagen fiber (Smith et al. 1981). Further details of specimen preparation are given elsewhere (Gasser et al. 2012), and two representative samples are shown in Fig. 4.2, which already allows a qualitative assessment of the collagen organization in the AAA wall. The images illustrate a large mix bag of azimuthal alignment and the diameter of the collagen fibers varied significantly between the two wall samples. Extinctions within the larger collagen fibers (see Fig. 4.2c) arose from the planer zig-zag structure of collagen fibrils, which was first observed in tail tendon (Diamant et al. 1972) and later verified by electron microscopy (Gathercole et al. 1974).

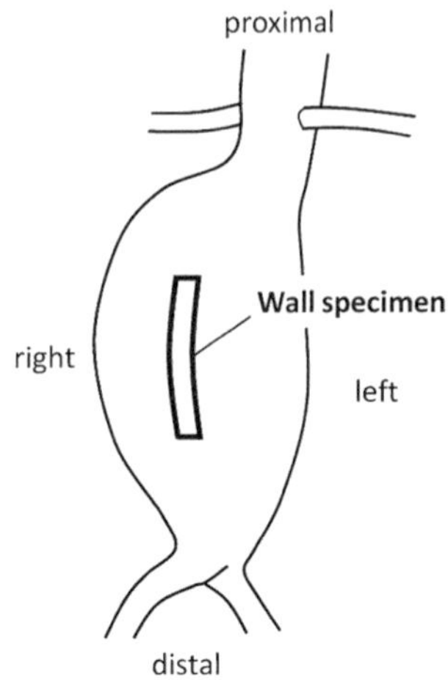

Fig. 4.1 Schematic drawing of an Abdominal Aortic Aneurysm (AAAs) that indicates the orientation and site of wall specimens that were harvested during open elective repair. Image taken from Gasser et al. (2012)

4.2.3 Three-dimensional Collagen Orientation

The orientation of a collagen fiber, or a coherent bundle of fibers, in the three-dimensional space is uniquely defined by its azimuthal angle θ and its elevation angle ϕ that are measured by (in-plane) rotating and (out-of-plane) tilting the URS, respectively. See Fig. 4.3 for the definition of these angles. An artifact-free part of the slice (as shown in Fig. 4.2a, b) was considered and the specimen edge was used as a reference for the collagen orientations. The black regions correspond to collagen fibers perpendicular to the linear polarized light ray, and hence the appearance of the image changes during rotating and tilting the stage. Note that the intensity of the non-extinguished (red) areas does not only depend on the angle of the collagen fibers with respect to the light ray but also on the amount of collagen in the tissue and stain used in specimen preparation.

To provide a statistically random selection, a 6 by 6 grid of measurement points was used to measure the collagen orientation from the histological slices. Azimuthal θ and elevation ϕ angles were identified by sequentially rotating and tilting the specimen until the particular measurement point of at least 7 by 7 μm extinguished in the PLM (Smith et al. 1981). Measurements from all slices were pooled, grouped into frequency plots, and fitted to a Bingham distribution (Bingham 1974; Alastrué et al. 2010)

$$\rho(\theta,\phi) = c^{-1}\exp[\kappa_1(\cos\theta\cos\phi)^2 + \kappa_2(\cos\phi\sin\theta)^2]. \tag{4.1}$$

The normalization parameter c was used to normalize the collagen fiber density function, such that $\int_{\phi=-\pi/2}^{\pi/2}\int_{\theta=-\pi/2}^{\pi/2}\rho\cos\phi \mathrm{d}\phi \mathrm{d}\theta = 1$ held. This condition was satisfied for

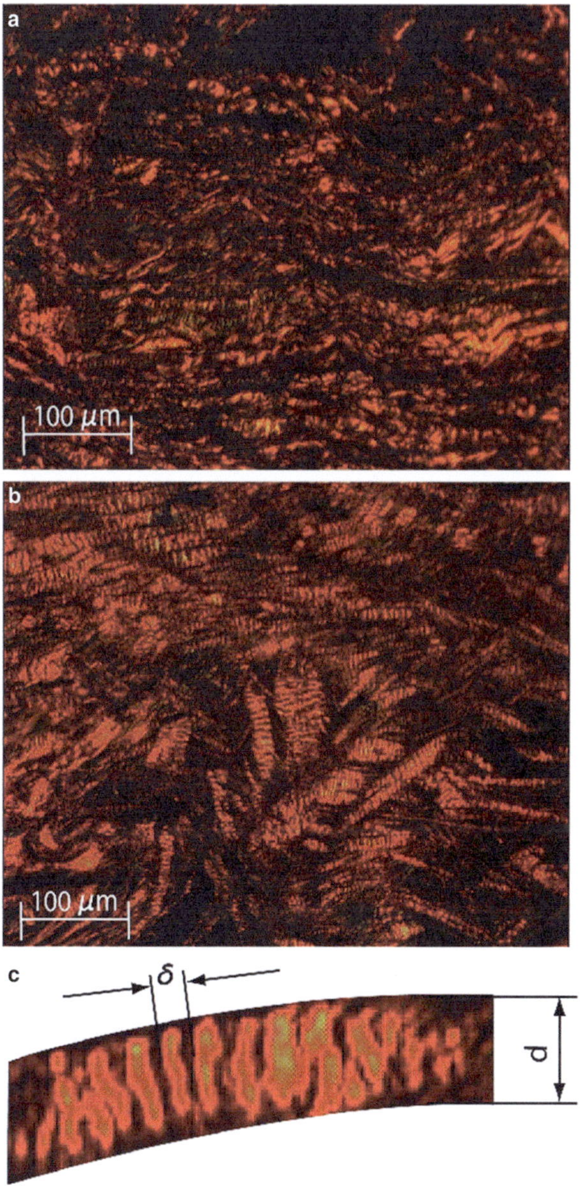

Fig. 4.2 Polarized Light Microscopy (PLM) images taken from the Abdominal Aortic Aneurysm (AAA) wall. The horizontal denotes the circumferential vessel direction. The collagen that is oriented perpendicular to the linear polarized light defines the extinctions (black) seen in the image. Picrosirius red was used as a birefringent enhancement stain and the images were taken at crossed polars on the microscope. (**a, b**) Typically observed collagen organizations in the AAA wall, showing a large mix bag of azimuthal alignment. (**c**) Segmented portion of a single collagen fiber of diameter d that is formed by a bundle of collagen fibrils. Extinctions at distances of δ denote the wavelength of the collagen fibrils that form the collagen fiber

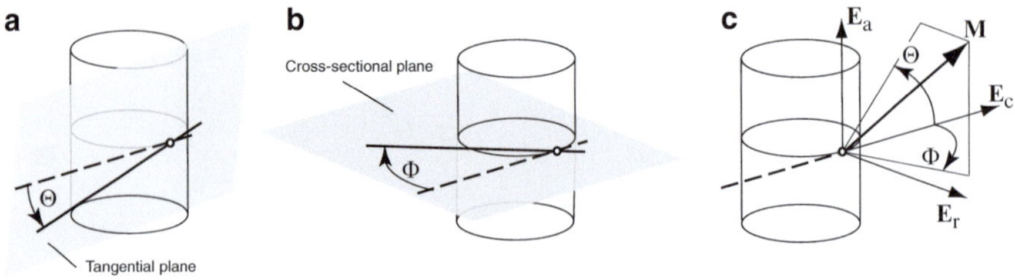

Fig. 4.3 Definition of the azimuthal angle θ (**a**) and the elevation angle ϕ (**b**) and their relation to the unit orientation vector **M** of a collagen fiber or a coherent bundle of fibers (**c**). The cylinder schematically represents the vessel with the *dashed line* denoting the circumferential direction. The azimuthal angle θ and the elevation angle ϕ are measured in planes in parallel (tangential plane) and perpendicular (cross-sectional plane) to the vessel axis, respectively. Both angles together define the three-dimensional orientation vector **M** of a collagen fiber within a Cartesian coordinate system $\mathbf{E}_c$, $\mathbf{E}_a$, $\mathbf{E}_r$, where the indices c, a, and r denote the local circumferential, axial, and radial directions, respectively

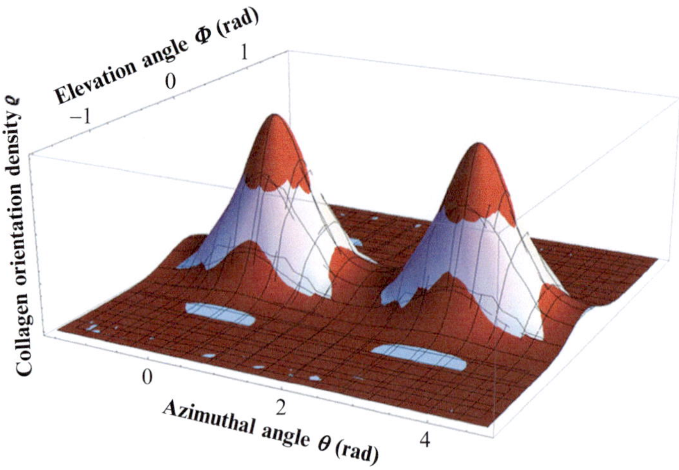

Fig. 4.4 Bingham distribution function (*red*) fitted to the experimentally measured fiber orientation distribution (*light-blue*) in the Abdominal Aortic Aneurysm (AAA) wall

$$c(\kappa_1, \kappa_2) = \sqrt{\pi} \sum_{i,j=0}^{\infty} \frac{\Gamma(i+\frac{1}{2})\Gamma(j+\frac{1}{2})\kappa_1^i \kappa_2^j}{\Gamma(i+j+\frac{3}{2})i!j!} \tag{4.2}$$

where Γ and $i!$ denoted the Euler gamma function and the factorial of i, respectively. The identified collagen orientation in the AAA wall is illustrated in Fig. 4.4, where the red surface denotes the Bingham distribution with parameters $\kappa_1 = 11.6$ and $\kappa_2 = 9.7$. The light-blue surface denotes the experimentally measured orientation distribution. Details regarding the applied optimization method for parameter estimation are given elsewhere (Gasser et al. 2012).

Note that the Bingham distribution is symmetric, i.e. $\rho(\phi, \theta) = \rho(\phi + \pi, \theta + \pi)$ and able to capture a large spectrum of distributions. Specifically, it accounts for different fiber dispersions in the tangential and cross-sectional planes, such that it provides more flexibility than a transversely isotropic distribution, i.e. as was earlier suggested to model the collagen fiber organization in the arterial wall (Gasser et al. 2006).

4.3 Constitutive Modeling of AAA Tissue

Constitutive modeling of vascular tissue is an active field of research and numerous descriptions have been reported. However, the phenomenological approaches (Vaishnav et al. 1972; Fung et al. 1979; Chuong and Fung 1983; Takamizawa and Hayashi 1987; Humphrey et al. 1990) that have been successfully used to fit experimental data cannot allocate stress or strain to the different histological constituents in the vascular wall. In contrast, structural constitutive descriptions (Lanir 1983; Wuyts et al. 1995; Holzapfel et al. 2000; Zulliger et al. 2004; Gasser et al. 2006; Gasser 2011; Pena et al. 2011; Martufi and Gasser 2011) overcome this limitation and integrate histological and mechanical information of the arterial wall.

Several descriptions, histomechanical constitutive models say, aim at integrating collagen fiber density and orientation according to Lanir's pioneering work (Lanir 1983). We followed this powerful approach and assume that the macroscopic Cauchy stress was defined by a superposition of individual collagen fiber contributions, i.e.

$$\boldsymbol{\sigma} = \frac{2}{\pi} \int_{\phi=0}^{\pi/2} \int_{\theta=0}^{\pi/2} \rho(\phi, \theta)\sigma(\lambda)\mathrm{dev}(\mathbf{m} \otimes \mathbf{m}) \cos\phi \mathrm{d}\phi \mathrm{d}\theta + p\mathbf{I}, \tag{4.3}$$

where $\mathbf{m} = \mathbf{FM}/|\mathbf{FM}|$ denoted the spatial orientation vector of the collagen fiber. In Eq. (4.3) the constitution of the collagen fiber is incorporated through its Cauchy stress $\sigma(\lambda)$. Assuming the collagen fiber is incompressible, Cauchy stress $\sigma(\lambda)$ and First Piola-Kirchhoff stress $P(\lambda)$ are linked through $\sigma(\lambda) = \lambda P(\lambda)$. In Eq. (4.3) the term $p\mathbf{I}$ denoted the hydrostatic stress with the Lagrange parameter p that is independent from the tissue's constitution and defined by the boundary conditions of the problem of interest.

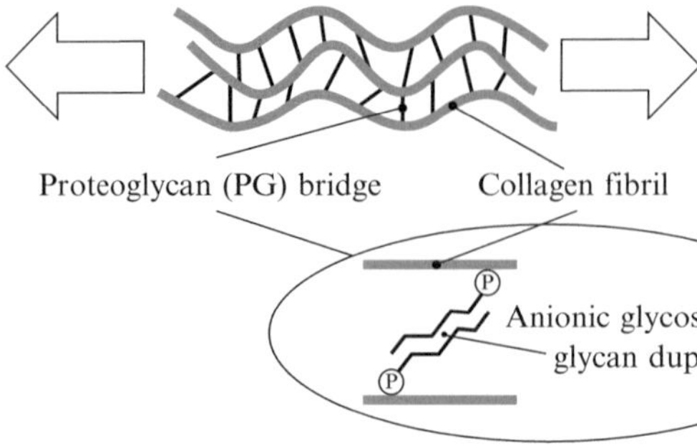

Fig. 4.5 Hierarchical structure of a collagen fiber, i.e. a fiber built-up by undulated collagen fibrils that are interlinked through proteoglycan (PG) bridges. Image taken from Martufi and Gasser (2011)

4.3.1 A Passive Histomechanical AAA Wall Model

AAA wall tissue was regarded as a fibrous collagenous composite, where fibers of collagen reinforced an otherwise isotropic matrix material. Each collagen fiber was assembled by a bundle of collagen fibrils mutually interconnected by PG bridges (Scott 2003, 2008) that provided interfibrillar load transition, see Sect. 4.2.1, for a micro-histological justification. In addition Fig. 4.2c provides a kind of visual imagination of the collagen fiber assembly. This structural view defines a basic load carrying unit, Collagen Fibril PG-complex (CFPG-complex) say, as it is illustrated in Fig. 4.5. Stretching a collagen fiber involves continuous recruitment of collagen fibrils, such that they gradually start carrying a load. A straightening stretch λ_{st} defines the stretch beyond which the collagen fibril is stretched elastically, i.e. elastic energy is stored in the CFPG-complex, i.e. in the collagen fibril itself and in the PG-rich matrix between the fibrils.

Finite Strain Kinematics

A fibrous tissue at finite deformations was considered, where the unit direction vector $\mathbf{M}$ denoted the local collagen fiber direction in the reference configuration Ω_0, see Fig. 4.6. The deformation $\mathbf{F}_{\text{st}\,i}$ straightened the i-th collagen fibril, i.e. it mapped its crimped referential configuration into a straight but still unstressed intermediate configuration $\Omega_{\text{st}\,i}$. In contrast the deformation $\mathbf{F}_{\text{c}\,i}$ recorded deformation relative to $\Omega_{\text{st}\,i}$ and mapped the fibril to its spatial configuration Ω. Here, the intermediate configuration served as a local reference configuration, with fibril stretch $\lambda_{\text{c}\,i}$ and fibril tension $T_{\text{c}\,i} = 0$, relative to which the fibril deformed elastically. Consequently, multiplicative kinematics related the continuum deformation to the introduced sub-deformations through $\mathbf{F} = \mathbf{F}_{\text{c}\,i}\mathbf{F}_{\text{st}\,i}$.

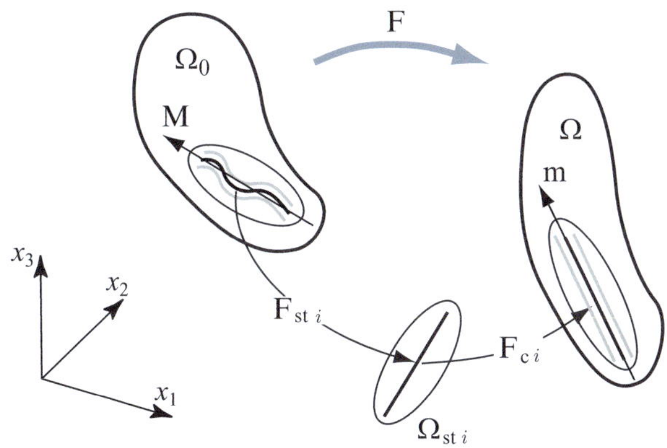

Fig. 4.6 Multi-scale kinematics of the collagen-fiber-reinforced tissue. Configurational map, where the intermediate configuration $\Omega_{\mathrm{st}\,i}$ separates the straightening and stretching of the i-th collagen fibril. A collagen fiber is thought to be built-up of a number of undulated collagen fibrils that are interlinked through proteoglycan (PG) bridges

For simplicity and due to the lack of micro-structural data we assumed that collagen fibrils that formed a collagen fiber straightened according to a symmetric triangular probability distribution (Kotz and vanDorp 2004). Specifically, the first and last fibrils within a collagen fiber straightened at fiber stretches $\lambda_{\mathrm{min}} = 1$ and λ_{max}, respectively. Note that elastin in AAA tissue is decreased (and fragmented) and collagen may determine the tissues unloaded configuration, which in turn justifies setting λ_{min} to one. Finally, affine deformation between the continuum and the collagen fiber, i.e. $\lambda = |\mathbf{FM}| = |\mathbf{m}|$, and incompressible macro-deformation ($\det\mathbf{F} = J = 1$), was considered.

Constitutive Description of the CFPG-Complex

Collagen fibrils have an approximately linear stress-stretch property (Miyazaki and Hayashi 1999; Shen et al. 2008), which is nicely captured by the First Piola-Kirchhoff stress $T_{\mathrm{c}\,i} = k\lambda_{\mathrm{c}\,i}\log\lambda_{\mathrm{c}\,i}$. It is assumed that the same constitutive relation also describes the i-th CFPG-complex sufficiently accurately. Considering the continuous recruitment of collagen fibrils (see Fig. 4.5), the First Piola-Kirchhoff stress of a collagen fiber is

$$T(\lambda) = k\int_0^{\lambda} \mathrm{CDF}(\overline{\lambda})\mathrm{d}\overline{\lambda}, \tag{4.4}$$

where $\mathrm{CDF}(\overline{\lambda})$ denotes the Cumulative Density Function of the triangular probability distribution (Martufi and Gasser 2011). Integrating $\mathrm{CDF}(\overline{\lambda})$ and using the abbreviations $a = \lambda_{\max} - 1$ and $b = (\lambda_{\max} + 1)/2$ yields the piecewise analytical expressions for the First Piola-Kirchhoff stress of a collagen fiber

$$T(\lambda) = \begin{cases} 0, & 0 < \lambda \leq 1, \\ k\dfrac{2}{3a^2}(\lambda - 1)^3, & 1 < \lambda \leq b, \\ k[\lambda - \dfrac{2}{3a^2}(\lambda - \lambda_{\max})^3 - b], & b < \lambda \leq \lambda_{\max}, \\ k(\lambda - b), & \lambda_{\max} < \lambda \leq \infty. \end{cases} \tag{4.5}$$

This set of equations exhibits the typically non-linear property of soft biological tissues and its numerical implementation is detailed elsewhere (Martufi and Gasser 2011).

Parameter Identification and Model Validation

Constitutive parameters were identified from macroscopic planar biaxial testing, where λ_c and λ_a denoted the circumferential and axial stretches, respectively. Introducing the kinematics of the biaxial deformation of an incompressible solid, i.e. $\mathbf{F} = \text{diag}[\lambda_c, \lambda_a, (\lambda_c\lambda_a)^{-1}]$, and expressing the orientation of the collagen fiber through the Eulerian angles $\mathbf{M} = [\cos\phi\cos\theta \;\; \sin\theta \;\; \sin\phi\cos\theta]^{\mathrm{T}}$ (see Fig. 4.3), the circumferential σ_c and axial σ_a Cauchy stresses were

$$\sigma_i = \frac{2}{\pi}\int\limits_{\phi=0}^{\pi/2}\int\limits_{\theta=0}^{\pi/2} \rho(\phi,\theta)\sigma(\lambda)(a_{ii} - a_{\mathrm{rr}})\cos\phi \mathrm{d}\phi \mathrm{d}\theta \;\; ; \;\; i = \mathrm{c}, \mathrm{a}. \tag{4.6}$$

Here, a_{cc}, a_{aa} and a_{rr} were the diagonal coefficients of the spatial tensor $\mathbf{a} = \mathbf{m} \otimes \mathbf{m} = (\mathbf{FM}) \otimes (\mathbf{FM})$. In addition incompressibility of the collagen fiber was assumed, such that $\sigma(\lambda) = \lambda T(\lambda)$ held. Knowing the collagen orientation density function $\rho(\phi, \theta)$ in Eq. (4.6) made it possible to estimate the constitutive parameters from the macroscopic biaxial experimental data.

Comprehensive experimental data from the AAA wall was recorded from tension-based biaxial loading protocols (Vande Geest et al. 2006). However, such data could not be compared directly to predictions through Eq. (4.6), i.e. where the stretch rather than tension is prescribed. Consequently, patient-specific model parameters that have been reported elsewhere (Ferruzzi et al. 2010) were used to generate synthetic experimental data instead. The synthetic data was then used to estimate constitutive parameters based on Eq. (4.6) (Mathematica, Wolfram Research). In order to achieve a reasonable fit with the experimental data, it was essential that collagen fibers changed their constitution with respect to their azimuthal alignment, and $\lambda_{\max} = \boldsymbol{\lambda}_{\max}\mathbf{M}$ with $\boldsymbol{\lambda}_{\max} = \text{diag}[\lambda_{\max\,c}, \lambda_{\max\,a}, (\lambda_{\max\,c}\lambda_{\max\,a})^{-1}]$ was used to set the upper undulation limit. Further details are given elsewhere (Gasser et al. 2012).

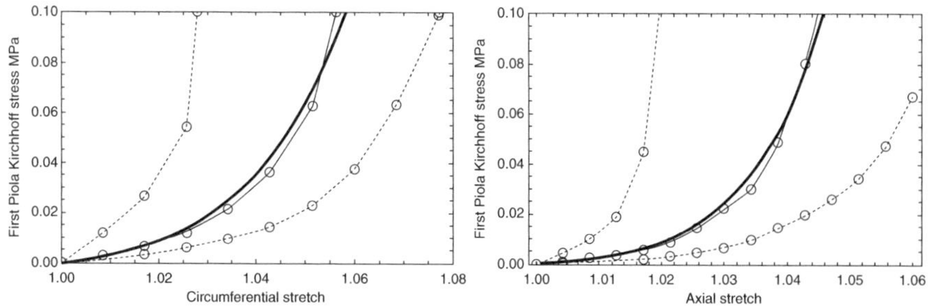

Fig. 4.7 Comparison between (synthetic) experimental and analytical stress-stretch response of the Abdominal Aortic Aneurysm (AAA) wall. Results were derived from a stretch-based biaxial loading protocol with a stretch ratio of $\lambda_c/\lambda_a = 2$. *Thick solid lines* show the analytical results. The median and the 50 % quartile of the experimental data are illustrated by the *thin solid* and *dotted lines*, respectively

Table 4.1 Model parameters used to predict the mean-population response of Abdominal Aortic Aneurysm (AAA) wall tissue (sr denotes steradian)

Bingham distribution parameters	$\kappa_1 = 11.6$, $\kappa_1 = 9.7$
Collagen undulation parameters	$\lambda_{\text{max c}} = 1.44$, $\lambda_{\text{max a}} = 1.2$
CFPG-complex stiffness	$k = 2.0$ GPa

Model predictions in relation to experimental data are illustrated in Fig. 4.7. To this end the parameters shown in Table 4.1 and a stretch ratio of $\lambda_c/\lambda_a = 2$ were used.

Discussion

The constitutive model discussed in this section was based on a sound histomechanical framework that integrated the collagen's fibril and fiber levels with the tissue's continuum level. Specifically, cross-linked collagen fibrils were thought to form collagen fibers, which in turn were integrated over the unit sphere to define the tissue's macroscopic properties (Lanir 1983; Federico and Gasser 2010).

The applied constitutive model was based on the collagen organization measured with PLM and was able to replicate anisotropic properties of the AAA wall as reported from planar biaxial testing (Vande Geest et al. 2006). It was important to equip collagen fibers with mechanical properties that changed with respect to their azimuthal orientation. Specifically, the upper undulation limit of a circumferentially aligned collagen fiber was about twice that of an axially aligned fiber, see Table 4.1. The undulation of collagen fibrils could be determined by the continuous collagen turnover that is superimposed on the pulsating wall tissue. Consequently, the higher undulation limit of collagen fibrils aligned with the circumferential direction might be the direct consequence of the higher pulsating strains in the circumferential direction.

It is well understood that inter-fiber and inter-fibril sliding plays a significant role in tendon deformation (Gupta et al. 2010). While the present collagen fiber model clearly accounted for inter-fibril sliding any inter-fiber sliding was suppressed through the prescribed affine kinematics between collagen fiber and tissue levels. If inter-fiber sliding also relates to vascular tissue, the applied affine transformation should be relaxed, by introducing an interface model for example. Unlike the affine fiber-continuum kinematics used in the present work, fiber-reinforced tissues were modeled by interlinked network structures (Chandran and Barocas 2006). Such an approach is clearly justified for hydrated collagen networks, i.e. where the absence of inter-fibrillar (solid) material allows a largely unconstraint motion of collagen fibrils. In contrast, the inter-fibrillar material in the AAA wall was expected to define a rather affine transformation between the collagen fibers and the continuum. Likewise, in the media of the normal aorta, bundles of collagen are not woven together (O'Connell et al. 2008), i.e. cross-linking among fibers is missing, which further reinforced the affine kinematics approach. However, a sound experimental validation of affine kinematics for vascular tissue has not yet been reported in the literature.

4.3.2 An Adaptation Model for the AAA Wall

Vascular tissue responds to mechanical stimuli, a mechanism necessary to optimize cardio-vascular function under defined boundary conditions. However, malfunction of vascular adaption can lead to pathologies like aneurysms. Aneurysms grow at different rates, and in addition to a diameter that exceeds 55 mm, a growth rate that reaches 10 mm per year is typically regarded as an AAA repair indication. The growth rate and wall strength of AAAs are thought to be linked (Wilson et al. 2003), such that a reliable growth predictor could reinforce an AAA rupture risk assessment. Describing the growth of aneurysms is an active field of research and several models have been reported, see among others (Volokh and Vorp 2008; Watton and Hill 2009; Kroon and Holzapfel 2009; Zeinali-Davarani and Baek 2012; Wilson et al. 2012). However, these models require significant further development to augment clinical decisions. This section extends the passive AAA wall model from Sect. 4.3.1 by adding a description for collagen turnover, i.e. the synthesis and degradation of collagen.

Collagen Turnover Model

Collagen turnover in the vascular wall is accomplished by cells like fibroblasts that are spread throughout the collagen network. Specifically, the present model assumes that fibroblast senses the state of strain and pre-stretches collagen fibrils prior to their deposition. This is formulated through three distinct sub-models, denoted as *the sensing model*, *the collagen turnover model*, and *the structure update model*, respectively. The sensing model defines the physical quantity that stimulates collagen turnover, the collagen turnover model quantifies the relation between the sensed stimulus and the change of collagen mass, and the structure update model details the way collagen is integrated/disintegrated into/from the existing collagen structure.

Sensing Model

It is assumed that at homeostasis the stretch of a collagen fiber in the vascular wall tends to λ_{ph}, such that

$$\xi(\mathbf{M}) = \lambda(\mathbf{M})/\lambda_{\text{ph}} \to 1 \tag{4.7}$$

was used as optimality condition. Here, ξ serves as a stretch-based mechanical stimulus. For $\xi > 1$, the existing collagen is stretched too much, such that in total more collagen is required to reach homeostasis. In this case the collagen turnover needs to be amplified in order to increase the total collagen density in the tissue. Equivalently, for $\xi < 1$, the collagen turnover needs to slow down to reach homeostasis through a net loss of collagen. It is emphasized that ξ depends on the orientation through the unit direction vector $\mathbf{M}$, that defines the orientation of the particular collagen fiber, see Fig. 4.6.

Collagen Turnover Model

It is assumed that collagen degrades isotropically according to

$$\dot{\rho}^{-} = -\eta\rho, \tag{4.8}$$

where η defines the time-scale of the degradation process and ρ is the total collagen density. This relation is independent from the orientation $\mathbf{M}$ and the local strain, such that degrading collagen is purely time-based, and matured collagen continuously resolves.

In contrast, the production of collagen is based on the idea that fibroblast produces collagen in an effort to maintain a homeostatic environment. Consequently, for a particular direction $\mathbf{M}$, the production of collagen fibrils is related to the stimulus $\xi(\mathbf{M})$ and reads

$$\dot{\rho}(\mathbf{M})^{+} = \min[\eta\rho\xi(\mathbf{M}), \dot{\rho}^{+}_{\max}] \tag{4.9}$$

Here, $\dot{\rho}^{+}_{\max}$ denotes the maximum collagen production rate that reflects that fibroblasts can only produce a certain amount of collagen in time. A similar limit has been reported for stretch-induced collagen synthesis of cultured vascular smooth muscle cells (Li et al. 1998).

Structure Update Model

It is assumed that collagen fibrils are disintegrated from the existing collagen structure without changing their undulation characteristics, i.e. without changing the triangular PDF that describes the continuous recruitment of fibrils when stretching a collagen fiber, see Sect. 4.3.1.

In contrast to the process that resolves collagen, newly formed collagen fibrils are integrated at a certain distribution of pre-stretches. We again applied multiplicative kinematics and considered Fig. 4.6 to analyze the deformation of a newly formed collagen fibril. Here, the i-th collagen fibril is straight but does not carry any load, i.e. it is stress-free, at its local intermediate configuration $\Omega_{\text{st}\,i}$. This individual collagen fibril is thought to be integrated (cross-linked) in the current (deformed) configuration at a certain stretch λ^{new} that is measured relative to $\Omega_{\text{st}\,i}$. Consequently, the straightening stretch $\lambda_{\text{st}} = \lambda/\lambda^{\text{new}}$ is required to straighten this collagen fibril relative to the reference (undeformed) configuration Ω_0, where λ denotes the total macroscopic stretch in direction of the fibril. Similar to the passive model, collagen fibrils were considered in a statistical sense, and the undulation of the newly formed collagen fibrils followed a triangular PDF having the limits $\lambda^{\text{new}}_{\text{min}}$ and $\lambda^{\text{new}}_{\text{max}}$, respectively. Finally, and as demonstrated in Martufi and Gasser (2012), adding the newly formed collagen to the already existing collagen structure can be regarded as a continuous update of the PDF that represents the collagen fibril distribution.

Although the outlined structure update model assumed a pre-stretched deposition of collagen fibrils, one could also consider collagen fibrils being stretched by fibroblasts after their deposition. Note that it is widely accepted that fibroblast adds tension to a collagen network (Alberts et al. 1994), but Fig. 5 in Silver et al. (2003) (taken from McBridge 1984) nicely shows that collagen fibrils might already be under tension when synthesized.

Results

The outlined adaptation model was implemented in an FE environment (FEAP, University of California at Berkeley) (Taylor 2007) and a patient-specific AAA was reconstructed from standard Computer Tomography-Angiography images (A4clinics Research Edition, VASCOPS GmbH). To this end deformable segmentation models were used that provided an artifact-free segmentation with minimal user interaction (Auer and Gasser 2010; Gasser 2012). Although the intra-luminal thrombus is known to be an important solid structure (Gasser et al. 2008) that increases the predictability of biomechanical AAA models (Gasser et al. 2010), it was not considered in the present study. Based on the reconstructed geometry an FE model was developed and exposed to a constant blood pressure of 100 mmHg. The wall was allowed to remodel over time and Fig. 4.8 shows the development of the aneurysms. Here, the parameters that determined collagen turnover were set such that a growth of 3.2 mm per year was reached, which matched available data from patient follow-up study (Martufi et al. 2013).

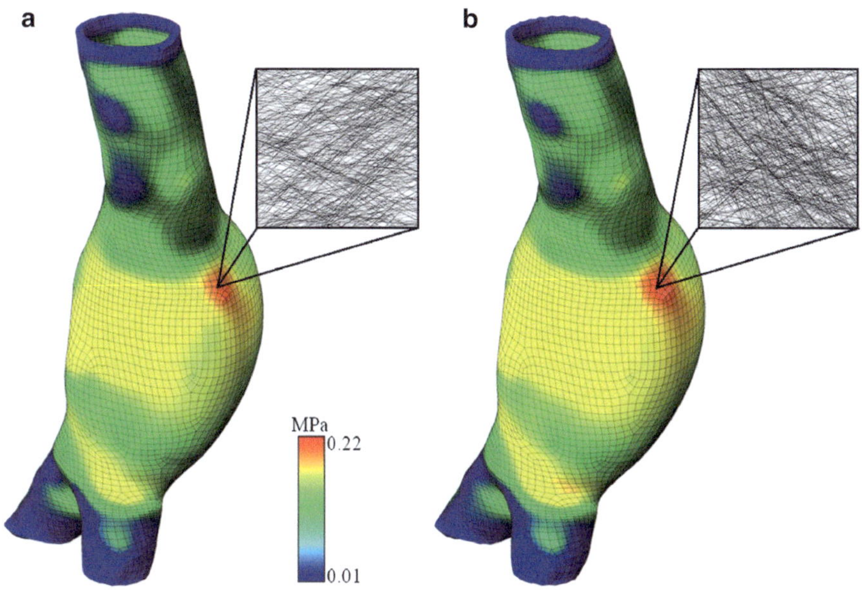

Fig. 4.8 Development of a patient-specific Abdominal Aortic Aneurysm (AAA) over a time period of 1 year. The *color code* denotes the maximum principal Cauchy stress and the zoom-ups show the predicted collagen organization at the site of Peak Wall Stress (PWS). (**a**) Baseline prediction at homeostatic conditions, i.e. such that the aneurysm remains stable over time. (**b**) Prediction after 1 year with parameters for newly formed collagen ($\eta = 1/60$ [day]$^{-1}$, $\lambda_{\min}^{\text{new}} = 0.98$, $\lambda_{\max}^{\text{new}} = 1.22$, $\dot{\rho}_{\max}^{+} = 5.3 \cdot 10^{-3}$ [sr day]$^{-1}$) that matched the growth of small AAAs

Discussion

The suggested adaptation model was able to predict the growth of small aneurysms, where the undulation limits $\lambda_{\min}^{\text{new}}$ and $\lambda_{\max}^{\text{new}}$ of the newly formed collagen as well as the maximum collagen production rate $\dot{\rho}_{\max}^{+}$ were the most influential parameters. The experimental identification of these parameters is not straightforward, and, when only considering macroscopic observations, not unique either. While the present model assumed that the collagen production rate depends on mechanical factors (see Eq. (4.9)) and its degradation rate not (see Eq. (4.8)), other models assumed precisely the opposite (Loerakker et al. 2014).

Compared to passive constitutive models, predictions that considered collagen turnover led to a smoother wall stress distribution. Specifically, the adaptation model avoided high stress gradients across the vessel (Martufi and Gasser 2013), which are thought to be non-physiological. Apart from these macroscopic consequences of collagen turnover, it shifted the initially isotropic collagen orientation density into a locally orthotropic distribution. This agrees well with an histological study of the AAA wall, see Sect. 4.2.3.

By limiting the collagen production in our model, i.e. enforcing being lower or equal to $\dot{\rho}^{+}_{\max}$, the mechanical stimulus was deactivated and the collagen synthesis could no longer support a net increase of collagen. Considering the rapidity of collagen remodeling, this circumstance could over time lead to aneurysm rupture.

4.3.3 Conclusions

Vascular biomechanics is critical in order to define new diagnostic and therapeutic methods that could have a significant influence on our medical understanding and even on the lifestyle of human beings. Nowadays, continued advances in computer technology and computational methods allow us to model patient-specific vascular problems. Apart from other challenges, such simulations critically depend on an accurate constitutive description of vascular tissue. Like other biological tissues the vascular wall responds to its mechanical environment and predictions based on passive constitutive models, i.e. suppressing tissue remodeling and growth, can only cover a limited time period. Vascular tissue develops at a loaded in-vivo configuration, which induces residual strains in its (hypothetical) load-free configuration, i.e. in the setting that typically serves as a reference for FE computations. Predicting realistic physiological stress states with passive constitutive models requires residual strains in the load-free configuration, which, for complex geometries are unfortunately unknown. Consequently, the key for improving biomechanical models is to understand the tissue's inherent properties to adapt to mechanical environments, and as demonstrated in Sect. 4.3.2 computational simulations can be potentially helpful.

The present chapter focused on structural aspects and it is known that aneurysm disease also alters the blood flow through the aorta (Taylor and Humphrey 2009). The enlarged lumen lowers the wall shear stress (Biasetti et al. 2009) and vortical structure dynamics differ remarkably between normal and aneurysmatic aortas (Biasetti et al. 2011). These hemodynamic alterations dictate the distribution of chemical species like thrombin (Biasetti et al. 2011), and their associated biological consequences are unknown.

Finally, although the constitutive models presented were able to successfully capture some features of an AAA wall, a rigorous validation against experimental data is crucial to evaluate its descriptive and predictive capabilities. In that respect, a mixed experimental numerical approach that accounts for tissue growth and remodeling seems to be most appropriate.

Acknowledgements This work has been financially supported by the Project Grant No. 2010-4446 provided by the Swedish Research Council, and the EC Seventh Framework Programme, Fighting Aneurysmal Disease (FAD-200647).

References

Alastrué V, Saez P, Martínez MA, Doblaré M. On the use of the bingham statistical distribution in microsphere-based constitutive models for arterial tissue. Mech Res Commun. 2010;37:700–6.

Alberts B, Bray D, Lewis J, Raff M, Roberts K, Watson JD. Molecular biology of the cell. New York: Garland Publishing; 1994.

Auer M, Gasser TC. Reconstruction and finite element mesh generation of Abdominal Aortic Aneurysms from computerized tomography angiography data with minimal user interaction. IEEE Trans Med Imaging. 2010;29:1022–8.

Bäck M, Gasser TC, Michel J-B, Caligiuri G. Review. biomechanical factors in the biology of aortic wall and aortic valve diseases. Cardiovascular Research doi: 10.1093/cvr/cvt040, 2013.

Bergel DH. The static elastic properties of the arterial wall. J Physiol. 1961;156:445–57.

Biasetti J, Gasser TC, Auer M, Hedin U, Labruto F. Hemodynamics conditions of the normal aorta compared to fusiform and saccular abdominal aortic aneurysms with emphasize on thrombus formation. Ann Biomed Eng. 2009;38:380–90.

Biasetti J, Hussain F, Gasser TC. Blood flow and coherent vortices in the normal and aneurysmatic aortas. A fluid dynamical approach to Intra-Luminal Thrombus formation. J R Soc Interface. 2011;8:1449–61.

Biasetti J, Spazzini PG, Gasser TC. An integrated fluido-chemical model towards modeling the formation of intra-luminal thrombus in abdominal aortic aneurysms. Front Physiol. 2011;3:article 266.

Bingham C. An antipodally symmetric distribution on the sphere. Ann Stat. 1974;2:1201–25.

Canham PB, Finlay HM. Morphometry of medial gaps of human brain artery branches. Stroke 2004;35:1153–7.

Canham PB, Finlay HM, Dixon JG, Boughner DR, Chen A. Measurements from light and polarised light microscopy of human coronary arteries fixed at distending pressure. Cardiovasc Res. 1989;23:973–82.

Carey D. Control of growth and differentiation of vascular cells by extracellular matrix proteins. Annu Rev Physiol. 1991;53:161–77.

Chandran PL, Barocas VH. Affine versus non-affine fibril kinematics in collagen networks: theoretical studies of network behavior. J Biomech Eng. 2006;128:259–70.

Choke E, Cockerill G, Wilson WR, Sayed S, Dawson J, Loftus I, Thompson MM. A review of biological factors implicated in Abdominal Aortic Aneurysm rupture. Eur J Vasc Endovasc Surg. 2005;30:227–44.

Chuong CJ, Fung YC. Three-dimensional stress distribution in arteries. J Biomech Eng. 1983;105:268–74.

Diamant J, Keller A, Baer E, Litt M, Arridge RGC. Collagen: ultrastructure and its relation to mechanical properties as a function of ageing. Proc R Soc Lond B. 1972;180:293–315.

Federico S, Gasser TC. Non-linear elasticity of biological tissues with statistical fibre orientation. J R Soc Interface. 2010;7:955–66.

Federico S, Herzog W. Towards an analytical model of soft tissues. J Biomech. 2008;41:3309–13.

Ferruzzi J, Vorp DA, Humphrey JD. On constitutive descriptors of the biaxial mechanical behaviour of human abdominal aorta and aneurysms. J R Soc Interface. 2010;8:435–540.

Fessel G, Snedeker JG. Equivalent stiffness after glycosaminoglycan depletion in tendon–an ultra-structural finite element model and corresponding experiments. J Theor Biol. 2011;268:77–83.

Fillinger MF, Raghavanand ML, Marra SP, Cronenwett J-L, Kennedy FE. In vivo analysis of mechanical wall stress and Abdominal Aortic Aneurysm rupture risk. J Vasc Surg. 2002;36:589–97.

Fratzl P, editor. Collagen: structure and mechanics. New York: Springer; 2008.

Fung YC. Biomechanics. Mechanical properties of living tissues. 2nd ed. New York: Springer; 1993.

Fung YC, Fronek K, Patitucci P. Pseudoelasticity of arteries and the choice of its mathematical expression. Am J Physiol. 1979;237:H620–31.

Gasser TC. An irreversible constitutive model for fibrous soft biological tissue: a 3rd microfiber approach with demonstrative application to Abdominal Aortic Aneurysms. Acta Biomater. 2011;7:2457–66.
Gasser TC, Auer M, Labruto F, Swedenborg J, Roy J. Biomechanical rupture risk assessment of Abdominal Aortic Aneurysms. Model complexity versus predictability of finite element simulations. Eur J Vasc Endovasc Surg. 2010;40:176–85.
Gasser TC, Forsell C. The numerical implementation of invariant-based viscoelastic formulations at finite strains. an anisotropic model for the passive myocardium. Comput Meth Appl Mech Eng. 2011;200:3637–45.
Gasser TC, Gallinetti S, Xing X, Forsell C, Swedenborg J, Roy J. Spatial orientation of collagen fibers in the Abdominal Aortic Aneurysm wall and its relation to wall mechanics. Acta Biomater. 2012;8:3091–103.
Gasser TC, Görgülü G, Folkesson M, Swedenborg J. Failure properties of intra-luminal thrombus in Abdominal Aortic Aneurysm under static and pulsating mechanical loads. J Vasc Surg. 2008;48:179–88.
Gasser TC, Holzapfel GA. A rate-independent elastoplastic constitutive model for (biological) fiber-reinforced composites at finite strains: continuum basis, algorithmic formulation and finite element implementation. Comput Mech. 2002;29:340–60.
Gasser TC, Ogden RW, Holzapfel GA. Hyperelastic modelling of arterial layers with distributed collagen fibre orientations. J R Soc Interface. 2006;3:15–35.
Gasser TC. Patient-specific computational modeling. Lecture notes in computational vision and biomechanics, chapter Bringing vascular biomechanics into clinical practice. Simulation-based decisions for elective abdominal aortic aneurysms repair. Dordrecht: Springer; 2012. p. 1–37.
Gathercole LGJ, Keller A, Shah JS. Periodic wave pattern in native tendon collagen: correlation of polarising with scanning electron microscopy. J Microsc. 1974;10:95–105.
Vande Geest JP, Sacks MS, Vorp DA. The effects of aneurysm on the biaxial mechanical behavior of human abdominal aorta. J Biomech. 2006;39:1324–34.
Gupta HS, Seto J, Krauss S, Boesecke P, Screen HRC. In situ multi-level analysis of viscoelastic deformation mechanisms in tendon collagen. J Struct Biol. 2010;169:1183–91.
Haverkamp R, Williams MW, Scott JE. Stretching single molecules of connective tissue glycans to characterize their shape-maintaining elasticity. Biomacromols 2005;6:1816–8.
Heng MS, Fagan MJ, Collier JW, Desai G, McCollum PT, Chetter IC. Peak wall stress measurement in elective and acute Abdominal Aortic Aneurysms. J Vasc Surg. 2008;47:17–22.
Holzapfel GA, Gasser TC, Ogden RW. A new constitutive framework for arterial wall mechanics and a comparative study of material models. J Elast. 2000;61:1–48.
Humphrey JD. Remodeling of a collagenous tissue at fixed lengths. J Biomech Eng. 1999;121: 591–7.
Humphrey JD. Cardiovascular solid mechanics. Cells, tissues, and organs. New York: Springer; 2002.
Humphrey JD, Strumpf RK, Yin FCP. Determination of constitutive relation for passive myocardium—Part I and II. J Biomech Eng. 1990;112:333–46.
Kotz S, vanDorp JR. Beyond beta: other continuous families of distributions with bounded support. Singapore: World Scientific; 2004.
Kroon M, Holzapfel GA. A theoretical model for fibroblast-controlled growth of saccular cerebral aneurysms. J Theor Biol. 2009;257:73–83.
Langewouters GJ, Wesseling KH, Goedhard WJA. The static elastic properties of 45 human thoracic and 20 abdominal aortas in vitro and the parameters of a new model. J Biomech. 1984;17:425–35.
Lanir Y. Constitutive equations for fibrous connective tissues. J Biomech. 1983;16:1–12.
Li Q, Muragaki Y, Hatamura I, Ueno H, Ooshima A. Stretch-induced collagen synthesis in cultured smooth muscle cells from rabbit aortic media and a possible involvement of angiotensin ii and transforming growth factor-beta. J Vasc Res. 1998 Mar-Apr;35(2):93–103
Liao J, Vesely I. Skewness angle of interfibrillar proteoglycans increases with applied load on mitral valve chordae tendineae. J Biomech. 2007;40:390–8.

Lindeman JHN, Ashcroft BA, Beenakker J-WM, van Es M, Koekkoek NBR, Prins FA, Tielemans JF, Abdul-Hussien H, Bank RA, Oosterkamp TH. Distinct defects in collagen microarchitecture underlie vessel-wall failure in advanced abdominal aneurysms and aneurysms in marfan syndrome. Proc Natl Acad Sci USA. 2009;107:862–5.

Loerakker, S., Obbink - Huizer, C. & Baaijens, F.P.T. (2014). A physically motivated constitutive model for cell-mediated compaction and collagen remodeling in soft tissues. Biomechanics and Modeling in Mechanobiology, 13(5), 985–1001

Maier A, Gee MW, Reeps C, Pongratz J, Eckstein HH, Wall WA. A comparison of diameter, wall stress, and rupture potential index for abdominal aortic aneurysm rupture risk prediction. Ann Biomed Eng. 2010;38:3124–34.

Martufi G, Gasser TC. A constitutive model for vascular tissue that integrates fibril, fiber and continuum levels. J Biomech. 2011;44:2544–50.

Martufi G, Gasser TC. Turnover of fibrillar collagen in soft biological tissue with application to the expansion of abdominal aortic aneurysms. J. Vasc. Surg. 58, 748–55. 2013.

Martufi G, Gasser TC. Review: the role of biomechanical modeling in the rupture risk assessment for abdominal aortic aneurysms. J Biomed Eng. 2013;135:021010.

Martufi G, Roy J, Swedenborg J, Sakalihasan N, Panuccio G, Gasser TC. Multidimensional growth measurements of abdominal aortic aneurysms. J. Vasc. Surg. 2013;58:748–55.

McBridge DJ. Hind limb extensor tendon development in the chick: a light and transmission electron microscopic study. Master's Thesis. Newark, NJ: Rutgers University; 1984.

Miyazaki H, Hayashi K. Tensile tests of collagen fibers obtained from the rabbit patellar tendon. Biomed Microdevices. 1999;2:151–7.

Nissen R, Cardinale GJ, Udenfriend S. Increased turnover of arterial collagen in hypertensive rats. Proc Natl Acad Sci USA. 1978;75:451–3.

O'Connell MK, Murthy S, Phan S, Xu C, Buchanan J, Spilker R, Dalman RL, Zarins CK, Denk W, Taylor CA. The three-dimensional micro- and nanostructure of the aortic medial lamellar unit measured using 3rd confocal and electron microscopy imaging. Matrix Biol. 2008;27:171–81.

Pena J, Martínez MA, Pena E. A formulation to model the nonlinear viscoelastic properties of the vascular tissue. Acta Mech. 2011;217:63–74.

Redaelli A, Vesentini S, Soncini M, Vena P, Mantero S, Montevecchi FM. Possible role of decorin glycosaminoglycans in fibril to fibril force transfer in relative mature tendons—a computational study from molecular to microstructural level. J Biomech. 2003;36:1555–69.

Rigozzi S, Mueller R, Snedeker JG. Local strain measurement reveals a varied regional dependence of tensile tendon mechanics on glycosaminoglycan content. J Biomech. 2009;42:1547–52.

Rigozzi S, Mueller R, Snedeker JG. Collagen fibril morphology and mechanical properties of the achilles tendon in two inbred mouse strains. J Anat. 2010;216:724–31.

Rizzo RJ, McCarthy WJ, Dixit SN, Lilly MP, Shively VP, Flinn WR, Yao JST. Collagen types and matrix protein content in human abdominal aortic aneurysms. J Vasc Surg. 1989;10:365–73.

Roach MR, Burton AC. The reason for the shape of the distensibility curve of arteries. Can J Biochem Physiol. 1957;35:681–90.

Robinson PS, Huang TF, Kazam E, Iozzo RV, Birk DE, Soslowsky LJ. Influence of decorin and biglycan on mechanical properties of multiple tendons in knockout mice. J Biomech Eng. 2005;127:181–5.

Sasaki N, Odajima S. Elongation mechanism of collagen fibrils and force-strain relations of tendon at each level of the structural hierarchy. J Biomech. 1996;29:1131–6.

Scott JE. Elasticity in extracellular matrix 'shape modules' of tendon, cartilage, etc. A sliding proteoglycan-filament model. J Physiol. 2003;553(2):335–43.

Scott JE. Cartilage is held together by elastic glycan strings. Physiological and pathological implications. Biorheology 2008;45:209–17.

Shen ZL, Dodge MR, Kahn H, Ballarini R, Eppell SJ. Stress-strain experiments on individual collagen fibrils. Biophys J. 2008;95:3956–63.

Silver FH, Freeman JW, Seehra GP. Collagen self-assembly and the development of tendon mechanical properties. J Biomech. 2003;36:1529–53.

Smith JF, Canham PB, Starkey J. Orientation of collagen in the tunica adventitia of the human cerebral artery measured with polarized light and the universal stage. J Ultrastruct Res 1981;77:133–45.

Sokolis DP. Passive mechanical properties and structure of the aorta: segmental analysis. Acta Physiol. 2007;190:277–89.

Takamizawa K, Hayashi K. Strain energy density function and uniform strain hypothesis for arterial mechanics. J Biomech. 1987;20:7–17.

Taylor CA, Humphrey JD. Open problems in computational vascular biomechanics: Hemodynamics and arterial wall mechanics. Comput Meth Appl Mech Eng. 2009;198: 3514–23.

Taylor RL. FEAP: a finite element analysis program, Version 8.2 user manual. Berkeley, CA: University of California at Berkeley; 2007.

The UK Small Aneurysm Trial Participants. Mortality results for randomised controlled trial of early elective surgery or ultrasonographic surveillance for small abdominal aortic aneurysms. Lancet 1998;352:1649–55.

Tsamis A, Krawiec JT, Vorp DA. Elastin and collagen fibre microstructure of the human aorta in ageing and disease: a review. J R Soc Interface. 2013;10:20121004. http://dx.doi.org/10.1098/rsif.2012.1004.

Vaishnav RN, Young JT, Janicki JS, Patel DJ. Nonlinear anisotropic elastic properties of the canine aorta. Biophys J. 1972;12:1008–27.

Venkatasubramaniam AK, Fagan MJ, Mehta T, Mylankal KJ, Ray B, Kuhan G, Chetter IC, McCollum PT. A comparative study of aortic wall stress using finite element analysis for ruptured and non-ruptured Abdominal Aortic Aneurysms. Eur J Vasc Surg. 2004;28:168–76.

Vesentini S, Redaelli A, Montevecchi FM. Estimation of the binding force of the collagen molecule-decorin core protein complex in collagen fibril. J Biomech. 2005 Mar;38(3):433–43.

Vidal BC, Mello MLS, Pimentel ER. Polarization microscopy and microspectrophotometry of sirius red, picrosirius and chlorantine fast red aggregates and of their complexes with collagen. Histochem J. 1982;14:857–78.

Volokh KY, Vorp DA. A model of growth and rupture of abdominal aortic aneurysm. J Biomech. 2008;41:1015–21.

Watton PN, Hill NA. Evolving mechanical properties of a model of Abdominal Aortic Aneurysm. Biomech Model Mechanobiol. 2009;8:25–42.

Weber KT, Pick R, Silver MA, Moe GW, Janicki JS, Zucker IH, Armstrong PW. Fibrillar collagen and remodeling of dilated canine left ventricle. Circulation 1990;82:1387–401.

Wilson JS, Baek S, Humphrey JD. Importance of initial aortic properties on the evolving regional anisotropy, stiffness and wall thickness of human abdominal aortic aneurysms. J R Soc Interface. 2012 Sep 7;9(74):2047–58. doi: 10.1098/rsif.2012.0097.

Wilson KA, Lee AJ, Hoskins PR, Fowkes FG, Ruckley CV, Bradbury A. The relationship between aortic wall distensibility and rupture of infrarenal abdominal aortic aneurysm. J Vasc Surg. 2003;37:112–7.

Wuyts FL, Vanhuyse VJ, Langewouters GJ, Decraemer WF, Raman ER, Buyle S. Elastic properties of human aortas in relation to age and atherosclerosis: a structural model. Phys Med Biol. 1995;40:1577–97.

Zeinali-Davarani S, Baek S. Medical image-based simulation of abdominal aortic aneurysm growth. Mech Res Commun. 2012;42:107–117.

Zulliger MA, Fridez P, Hayashi K, Stergiopulos N. A strain energy function for arteries accounting for wall composition and structure. J Biomech. 2004;37:989–1000.

Chapter 5
The Biomechanics of Fat: From Tissue to a Cell Scale

Naama Shoham and Amit Gefen

Abstract Adipose tissues are weight-bearing biological structures that are involved in central medical problems such as obesity, diabetes and their comorbidities. Their biomechanical properties are relevant in several fields such as cosmetics, acute and chronic wound treatments, implantable drug delivery systems, and plastic surgery. In this chapter, we relate the mechanical behavior of the adipose tissue continuum to the biological activities of adipocytes. We demonstrate that first, at a macroscopic scale, the mechanical behavior of adipose tissues depends on the anatomical site and hence on physiological function. At a microscopic scale, mechanical function such as cell stiffness properties depends on the triglyceride contents that in turn depend on the level of differentiation, which has recently been shown to be regulated by mechanical loading. Hence, based on the empirical data, we propose a novel hypothesis regarding structure–function–adaptation processes and relationships in fat, which might open new research paths for studying adipose-related diseases from a biomechanical point of view, involving the mechanotransduction and structure–function–adaptation concepts that are well known to exist for other tissues but were so far very poorly studied in fat.

5.1 The Adipose Tissue

The adipose tissue is a specialized connective tissue, located in three major anatomical depots: subcutaneous, dermal, and intraperitoneal (Shoham and Gefen 2012a). In human adults, the main component of adipose tissue is the white adipocyte cell which contains primarily triglycerides, and the remaining components are water (5–30 % weight) and proteins (2–3 % weight) (Geerligs et al. 2010). Adipose tissue is highly vascularized with each adipocyte being in close proximity to at least one capillary vessel that supports active metabolism (Christiaens and Lijnen 2010). The main function of adipose tissue is to store excess energy in the form of lipids,

N. Shoham • A. Gefen (✉)
Faculty of Engineering, Department of Biomedical Engineering, Tel Aviv University, Tel Aviv 69978, Israel
e-mail: gefen@eng.tau.ac.il

G.S. Kassab, M.S. Sacks (eds.), *Structure-Based Mechanics of Tissues and Organs*,
DOI 10.1007/978-1-4899-7630-7_5

which can later be mobilized to other tissues in response to metabolic needs (Avram et al. 2005). Though poorly studied from a biomechanical perspective with respect to say musculoskeletal or other connective tissues, adipose tissue is a weight-bearing biological structure (e.g., it supports the bodyweight during sitting or lying). Fat tissues are also involved in central medical problems, such as obesity, diabetes, and their comorbidities, and its biomechanical behavior and properties are also relevant in the fields of cosmetics, acute and chronic wounds, implantable drug delivery systems, and plastic surgery (Shoham and Gefen 2012a; Gefen and Haberman 2007; Geerligs et al. 2010). Hence it is very important to characterize and formulate structure–function relationships for adipose tissue, and, in particular, to understand how changes that occur at a cellular level eventually lead to macroscopic changes in the contents, structure and mechanical properties of the fatty tissues.

5.2 Tissue-Scale Studies

Mechanical properties of adipose tissues extracted from different species and/or different anatomical sites have been reported in the literature (Table 5.1) (Geerligs et al. 2008, 2010; Gefen and Haberman 2007; Gefen et al. 2001; Iatridis et al. 2003; Krouskop et al. 1998; Samani et al. 2007; Sinkus et al. 2005; Weaver et al. 2005). The biomechanical behavior of the fat tissues in the buttocks is relevant, for example, when developing mathematical or computational models of deep pressure ulcers, where adipose tissue damage occurs due to prolonged compression by bony prominences (Gefen and Haberman 2007). In vitro viscoelastic mechanical properties of the adipose tissue covering the gluteus muscles of sheep were therefore measured by Gefen and Haberman (2007) using confined compression and swift indentations techniques. They found that the short-term aggregated modulus in confined compression (which is defined as the ratio of the measured reaction force divided by the cross-sectional area of the upper platen) was 28.9 ± 14.9 and 18.1 ± 6.9 kPa for nonpreconditioned and preconditioned specimens, respectively. The corresponding short-term elastic moduli were 0.85 ± 0.4 and 0.53 ± 0.2 kPa for the nonpreconditioned and preconditioned specimens, respectively. The long-term aggregated modulus property, however, was not affected by the preconditioning; being 10.3 ± 4.2 kPa constantly, and accordingly the long-term elastic modulus of the tissue was calculated to be 0.3 ± 0.12 kPa. The plateau phase in the transient aggregate modulus function was always reached within 2 min. The short-term elastic modulus strongly depended on the deformation rate, since in the high indentation rate experiments the short-term elastic moduli were 22.6 ± 10 and 15.8 ± 9.4 kPa for the nonpreconditioned and preconditioned specimens, respectively.

Mechanical properties of subcutaneous adipose tissue, which influence, for example, on the mechanical environment in plastic reconstructive surgeries or in transdermal drug-delivery systems were examined in the studies of Iatridis and colleagues (2003) and later by Geerligs and co-authors (2008, 2010). Iatridis et al. (2003) studied mechanical properties of rat subcutaneous adipose tissues under

Table 5.1 Summary of reported mechanical properties for adipose tissue

Tissue type		Experiment	Parameter	Value (kPa)	Reference
Adipose tissue covering the gluteus muscle of sheep		Confined compression	Short-term elastic modulus	0.53–0.85	Gefen and Haberman (2007)
			Long-term elastic modulus	0.3	
		Swift indentation	Short-term elastic modulus	15.8–22.6	
Subcutaneous adipose tissue	Porcine	Shear experiments on a rotational rheometer	Shear modulus	7.5	Geerligs et al. (2008)
			Long-term shear modulus	1.5–15	Geerligs et al. (2010)
	Rat	Uniaxial tension with incremental stress relaxation experiments	Instantaneous tensile modulus	4.77	Iatridis et al. (2003)
			Long-term tensile modulus	2.75	
Human breast adipose tissue		Compression loading	Elastic modulus	18–24	Krouskop et al. (1998)
		Magnetic resonance elastography	Shear modulus	1	Sinkus et al. (2005)
		Indentation technique	Young's modulus	3.25	Samani et al. (2007)
Human heel fat pad		Magnetic resonance elastography	Shear modulus	8–12	Weaver et al. (2005)
		Digital radiographic fluoroscopy/contact pressure display method	Compression modulus	105–306	Gefen et al. (2001)

uniaxial tension while applying incremental stress relaxation protocols. They reported a linear elastic response of the tissue, with instantaneous and long-term tensile moduli of 4.77 and 2.75 kPa, respectively. Given the linearity in their data, they were able to describe the constitutive tissue behavior using a 5-parameter Maxwell solid model and a continuous relaxation function with a constant amplitude relaxation spectrum. Geerligs and colleagues (2008) performed shear experiments on porcine subcutaneous adipose tissue using a rotational rheometer. They found that within the linear viscoelastic regime, which was up to 0.1 % strain, the storage and loss moduli showed a frequency- and temperature-dependent behavior. Conversely, the ratio between the two moduli, i.e., the phase angle, did not show any dependence on temperature and frequency. The shear modulus of the porcine adipose tissue was 7.5 kPa at 10 rad/s and 37 °C. Finally, a power-law function model was used to describe both the frequency-dependent behavior at constant temperature and the stress relaxation behavior. In a complementary study, Geerligs and colleagues (2010) further examined the long-term behavior of adipose viscoelastic properties under small strains (tissue deformations were applied for at least 45 min), as well as the tissue responses to various large strain profiles which were delivered again through rheological methods. They found that the shear modulus dramatically increased (from 1.5 to 15 kPa) after a loading period that lasted between 250 and 1250 s, but then the modulus returned to its initial value within 3 h of recovery from the loading. Additionally, the strain–stress responses for various large strain history sequences were reproducible up to strains of 0.15. For larger strains, the stress decreased in subsequent loading cycles, and, above strains of 0.3, the structure of the tissue changed such that the stress became independent of the applied strain. Given that an anti-thixotropic behavior is defined as a time-dependent increase in the viscosity or stiffness properties of the tissue which is induced by deformation loading, but is reversible when the deformation ceases, Geerligs et al. concluded that, based on their experimental results, subcutaneous adipose tissues likely behave as an anti-thixotropic material. In addition, they suggested that a Mooney–Rivlin model would be appropriate for representing the constitutive behavior of this tissue at physiologically relevant large strains.

Mechanical properties of breast adipose tissues are of interest considering that palpation for detecting local stiffening of the tissue is commonly used in clinical practice as a means to screen for breast cancer. Krouskop et al. (1998) investigated the viscoelastic behavior of human breast fat tissue when subjected to compression loading at three strain rates (0.1, 1, and 4 Hz) and found that the tissue had an approximately constant elastic modulus. Additionally, the modulus of the fat tissues did not change significantly in their studies when the pre-compression strain level was increased. Specifically, the elastic moduli of the fat were 18 ± 7, 19 ± 7, and 22 ± 12 kPa for the 0.1, 1, and 4 Hz loading frequencies, respectively, under 5 % pre-compression strain. When a 20 % preconditioning strain was applied, the elastic moduli were 20 ± 8, 20 ± 6, and 24 ± 6 kPa for the corresponding loading frequencies, which was statistically indistinguishable from the previous dataset. Since the elastic moduli of the tissue did not change significantly with the frequency of the applied deformations, it was concluded that breast fat tissues behave as an elastic material (i.e., the viscous component can be considered negligible). Mechanical properties of fat breast tissue were examined by means of other methods as well (Sinkus et al. 2005; Samani et al. 2007). Specifically, using magnetic resonance (MR) elastography, Sinkus et al. (2005) found that the shear modulus of this tissue was 1 ± 0.3 kPa. In brief, MR elastography quantifies viscoelastic properties of tissues indirectly, by recording mechanical low-frequency shear waves while they spread in the tissue (Sinkus et al. 2005). Another attempt to determine mechanical properties of fat breast tissues was made by Samani and colleagues (2007), who used an indentation technique and found that the Young's modulus of the tissue (for small deformations) was 3.25 ± 0.91 kPa.

Mechanical properties of the heel fat pad are relevant, for example, when studying how to biomechanically protect tissues in diabetic feet. Gefen et al. (2001) measured deformations and contact stresses in heel pad tissues simultaneously, in vivo, during the stance phase of gait, by combining fluoroscopic and plantar pressure measurement techniques. These researchers found that the stress–strain relationship for heel fat pads is highly nonlinear, with a compression modulus of 105 ± 11 kPa initially and 306 ± 16 kPa at 30 % strain. Additionally, the energy dissipation in this tissue during heel strike was evaluated to be 17.8 ± 0.8 %. Using MR elastography, Weaver et al. (2005) later reported that the shear modulus of the heel fat pad increased from 8 to 12 kPa with increasing pressures, again indicating a strong nonlinearity of the constitutive law.

To summarize this part, the stiffness of adipose tissues appears to depend on the anatomical site (Table 5.1), and hence on the physiological function (e.g., the magnitudes of the mechanical loads that are normally transferred through the tissue), which is consistent with biomechanical knowledge regarding other tissue types such as cancellous bone. However, it is rather difficult to compare tissue mechanical properties across literature reports, considering the diversity in measurement techniques and protocols, inter-specie differences, and the different constitutive parameters that were determined in each study. Still, and as could be well expected based on the structure–function concept in tissue biomechanics, the intensively loaded heel fat pad is stiffer by orders of magnitudes under compression with respect to the fat tissues of the breast (which are much more lightly loaded, by gravity or perhaps during breastfeeding). Such substantial differences in mechanical properties must relate to the fatty acid composition and turnover rates in adipose tissues. For example, weight-bearing buttocks fat depots generally have a lower proportion of saturated fatty acids, elevated proportions of monosaturated and polysaturated fat, and lower lipolysis rates with respect to abdominal depots that are mainly loaded by gravity (apart from the time when a person spends sleeping prone) (Gefen and Haberman 2007). Understanding the factors causing these differences in mechanical performance of fat across anatomical sites requires that the biomechanics of fat be studied at a cellular scale.

5.3 Cell Scale Studies

It is very likely that the differences across anatomical sites which were documented in the literature with respect to mechanical properties of fat at the tissue scale (Table 5.1) would relate to composition of cells (and/or extracellular matrix composition) at the microscopic scale. Using atomic force microscopy (AFM), Darling and colleagues (2008) studied mechanical properties of adipocytes that were harvested from human fat pads in the joints as well as properties of adipose-derived adult stem cells which were obtained from human subcutaneous abdominal adipose tissues. The adipose-derived adult stem cells were tested in two configurations: short seeding times (which produce nearly spherical cell morphology) and long seeding times (that allow cells to spread over the culturing substrata). These researchers found that the elastic moduli (calculated by processing the AFM data by means of the Hertz theory) of the spherical adipocytes from the fat pads, and the spherical and spread adipose-derived adult stem cells were 0.9 ± 0.8, 2.6 ± 1.6, and 2.5 ± 1.2 kPa, respectively. The long-term elastic moduli of these cells (at 30 s) were 0.71 ± 0.74, 0.37 ± 0.31, and 1.7 ± 1.1 kPa, respectively. The relaxed moduli (i.e., the stresses at a given deformation value, after a certain time of relaxation) were 0.61 ± 0.54, 0.37 ± 0.26, and 1.7 ± 1 kPa, for the spherical adipocytes from the fat pads, and the spherical and spread adipose-derived adult stem cells, respectively. Corresponding times for reaching plateaus of the creep response were 55.5 ± 129, 31 ± 41.5, and 21.5 ± 78.5 s, respectively, and times for plateau of the stress relaxation curves were 31.1 ± 63.8, 7.3 ± 4.3, and 9.6 ± 16 s. Though the stiffness properties and creep/relaxation times reported by the Darling group were around the same order of magnitude for the different cells and seeding conditions, their data suggest that mechanical properties of adipocytes change during cellular maturation, and perhaps also depend on the interaction of the cells with the substrate (for in vitro studies).

Other than studying the effective mechanical properties of adipocytes, it is also important to characterize their biomechanical behavior as a (micro) structure, since it is the cellular structure which supports and transfers loads at the microscopic scale, but also given the potential influence of mechanical loads on function of cellular organelles which could, in turn, imply on how cells would respond biologically to the loading—which is generally termed mechanotransduction. Or-Tzadikario and Gefen (2011) developed cell-specific finite element models of a fibroblast-like pre-adipocyte cell versus a mature adipocyte. In brief, they observed both cell types under confocal microscopy, acquiring z-stack cellular images that were then assembled to create a three-dimensional solid model of each cell. In their modeling, the cytoplasm, plasma membrane, nucleus, and lipid droplets (in the adipocyte) were all incorporated and were assumed to behave as isotropic compressible materials that obey a neo-Hookean strain energy density function. They simulated quasi-static cell compression tests by lowering a rigid platen onto the cell surface until reaching a global cell deformation (GCD = difference between the undeformed and compressed cell heights over the undeformed cell height, in %) of 35 % per each cell type. Or-Tzadikario and Gefen then found that the %-plasma membrane that was subjected to large tensional strains (>5 %) increased nonlinearly with the GCD level for both cells, and large tensional strains started to appear in their plasma membrane for GCD > ~15 %. In the adipocyte model, means of peak tension, compression, and shear strains in cytoplasmic material around lipid droplets increased in a nearly linear relationship with GCD ≥ 15 %, reaching 17.9 %, 13.4 %, and 25.6 %, respectively, for GCD = 35 %. Additionally, strains were highly inhomogeneous across the individual lipid droplets, and this inhomogeneity increased with the rise in GCD. The Or-Tzadikario et al. (2010) paper again provides support to the hypothesis that cell stiffness, intracellular strains, and intracellular stresses which develop in response to a certain external loading condition depend on the maturity of the cells, as cells progress along the process of differentiation and produce more, and larger lipid droplets. Differentiation in these cells is influenced in turn by the mechanical environment and applied mechanical loads, as recently shown by our group, where statically stretching pre-adipocyte cultures up to large strains for 2 weeks or more accelerated adipogenesis and development of larger intracytoplasmic lipid droplets (Levy et al. 2011; Shoham et al. 2012b). Therefore, it appears that there is ongoing feedback between mechanical loads that are sensed by adipocytes, their differentiation response, and the effective stiffness properties of the cells—which depend on the stage of differentiation. These interactions, which are schematically illustrated in Fig. 5.1, were studied by several research groups in vitro, as described in the next section.

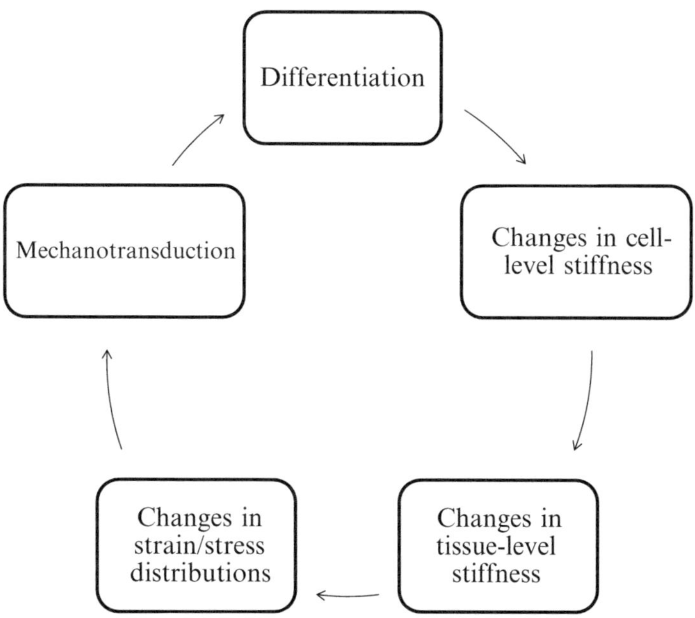

Fig. 5.1 The proposed interactions between mechanotransduction at the cell scale and tissue- or organ-level biomechanics in weight-bearing fat tissues

5.4 Cell-Level Structure–Function–Adaptation Relationships in Fat

Recent studies employing different in vitro model systems point to some very interesting structure–function–adaptation relationships in adipose tissues, which are probably analogue to those documented in other connective tissues such as bone. First, at a cellular scale, adipocytes are becoming increasingly recognized as being mechanosensitive and mechanoresponsive cells, much like osteoblasts, myocytes, or tenocytes (Levy et al. 2011; Shoham et al. 2012b). Second, dynamic loading regimes such as cyclic stretching or vibration were generally found to suppress adipogenesis in cultured adipocytes harvested from various sources, e.g., pre-adipocytes, mesenchymal stem cells, and adipose tissue stromal cells. From a molecular biology perspective, different signaling pathways are activated by the dynamic mechanical conditioning (Case et al. 2010; David et al. 2007; Huang et al. 2010; Sen et al. 2008, 2009, 2011; Tanabe et al. 2004, 2008; Tirkkonen et al. 2011; Turner et al. 2008) and readers are referred to our recent review paper in the *Journal of Biomechanics* in this regard (Shoham and Gefen 2012a). Contrarily to the suppressing effect that dynamic loading has on adipogenesis, static loads that are delivered chronically, i.e., for periods of days to weeks, appear to have a dual influence on adipogenesis, where static stretching accelerates differentiation and consequent formation of new and larger lipid droplets (Hara et al. 2011; Levy et al. 2011; Shoham et al. 2012b) but static compression impedes it (Hossain et al. 2010). Here, again, multiple signaling pathways are involved, one of which being the ERK/MEK pathway (Shoham et al. 2012b). These findings in cell cultures in vitro are supported by several animal studies and even a few human subject trials where dynamic loads delivered to adipose tissues have generally been found to suppress adipogenesis and reduce body fat (Luu et al. 2009; Maddalozzo et al. 2008; Marques et al. 2011; Monteux and Lafontan 2008; Ozcivici et al. 2010; Rubin et al. 2007; Vissers et al. 2010), whereas static stretching promoted the adipogenesis (Kato et al. 2010). The different nature of responses of adipose tissues to dynamic versus static loads resembles the nature of response of bone tissue to loading in the sense that bone, as well, is known to respond differently to dynamic versus static loading. However, in bone, chronic static loads would generally have a catabolic effect and dynamic loading would have an anabolic effect, whereas adipose tissues respond the other way around (Shoham and Gefen 2012a).

The role of mechanotransduction in regulating lipid contents in adipocytes and adipose tissues becomes extremely interesting when realizing that cells in human adipose tissues are physiologically exposed to compound mechanical loading: tensile, compressive, and shear strains/stresses, which are associated with bodyweight loads and weight-bearing and whose loading profile depends on the lifestyle, age, injury, and disease in the individual (Linder-Ganz et al. 2007). For example, in seated healthy individuals the adipose tissues at the buttocks are subjected to peak tensile, compressive, and shear strains of ~30 %, ~45 %, and ~40 %, respectively. A lying posture induces peak strains that are approximately half these magnitudes, but in case that there are anatomical changes that are associated with an injury or disease, such as muscle atrophy and/or weight-gain post spinal cord injury, these loads may increase considerably above the normative levels in sitting or lying (Linder-Ganz et al. 2007, 2008). During dynamic loading such as when performing a physical exercise, strains in adipose tissues are also large.

Taken together, the studies reviewed above lead to a novel and perhaps out-of-the-box hypothesis regarding structure–function–adaptation processes and relationships in fat, as follows. Given that during their differentiation and growth process, adipocytes accumulate increasing amounts of triglycerides in intracytoplasmic lipid droplets, the structure and composition of the individual cells change over the timecourse of differentiation (Levy et al. 2011; Shoham et al. 2012b) (Fig. 5.2). Hence, it is very likely that the effective stiffness of the cells, as well as the distribution of intracellular stiffnesses between, e.g., the lipid dropletsand cytosol in each cell are differentiation-process-dependent, and so, adipocyte stiffnesses would change over a time course (Or-Tzadikario and Gefen 2011) depending on the biochemical and mechanical environments of the cells (Or-Tzadikario et al. 2010; Shoham and Gefen 2012b). This immediately implies that over time, tissue-level (or continuum scale) changes in stiffnesses would result in, which would then affect the strain/stress distributions in weight-bearing fat, reflect back on the mechanical environment of the cells—which would trigger a mechanotransduction response and vice versa (Fig. 5.1). Massive research work is needed to look into the components of this hypothesis in detail, but the pieces of information that we already have indicate that it is very likely that these interactions exist in vivo, and moreover, that they could play a fundamental role in common diseases such as obesity. Our current work in this regard focuses on characterizing how the effective stiffness of adipocytes is influenced by the progress of the differentiation process. For this purpose, we currently image individual adipocytes using wide field digital interferometric microscopy (Park et al. 2010; Shaked et al. 2011). By means of this state-of-the-art microscopy method, we were able to evaluate experimentally, though in a small number of experiments so far, how the effective adipocyte stiffness depends on the contents of intracytoplasmic lipid droplets (Shoham et al. 2012a) (Fig. 5.3). Obtaining quantitative formulations for how adipocytes change structure and mechanical properties in response to biochemical stimuli such as insulin (Or-Tzadikario et al. 2010) or biomechanical stimuli such as static stretching (Shoham et al. 2012b; Shoham and Gefen 2012b), which is our nearest goal, would clarify whether there exists a positive or negative feedback in the structure–function–adaption loop for fat, that is, whether formation or growth of intracytoplasmic lipid droplets leads to softening or stiffening of the cells and tissues. If this process ultimately results in continuum-scale stiffening, the extent of tissue deformations, particularly their tensional components would eventually plateau, hence the mechanical driving force for the differentiation would also stabilize, which would lead to a physiological equilibrium in the tissue. If on the other hand adipocytes (and fat tissues) soften as they mature—that would increase deformability at the tissue scale, would deliver more stretching to pre-adipocytes and promote further differentiation, in a positive feedback loop which

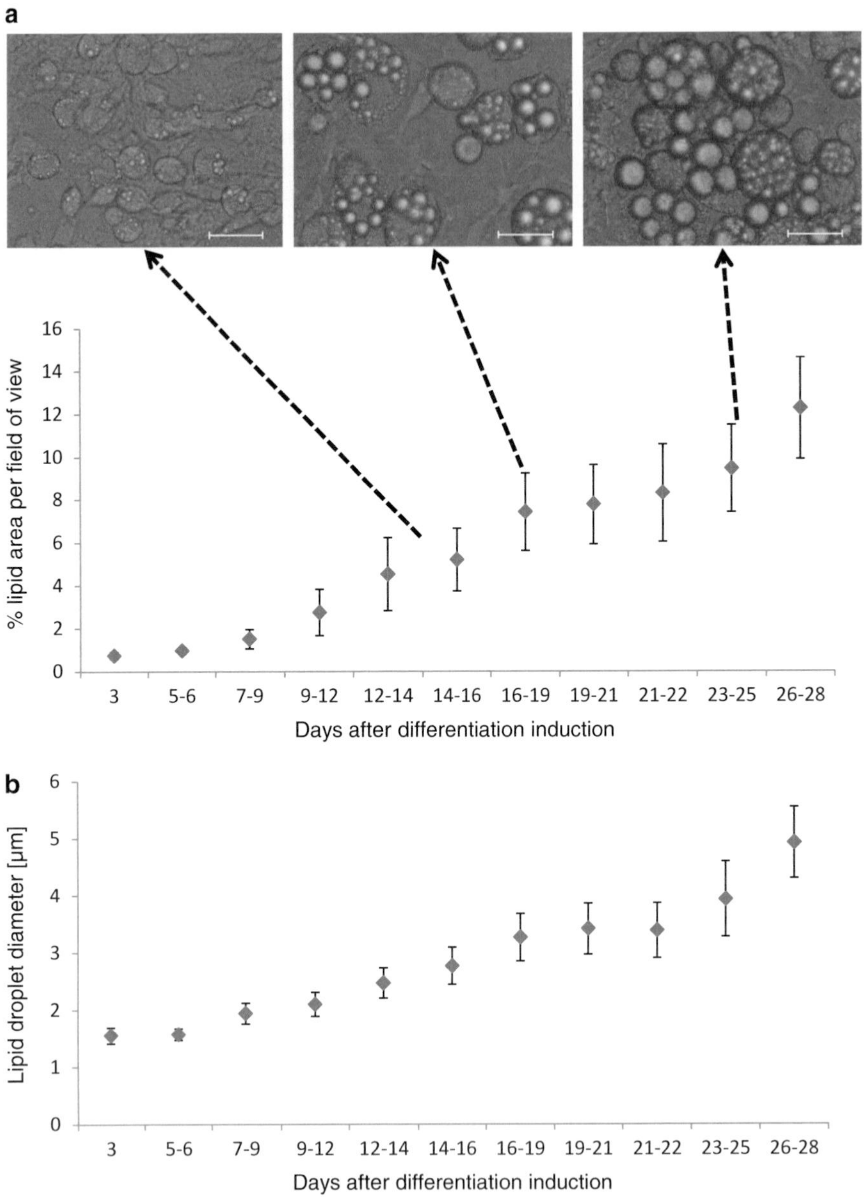

Fig. 5.2 Differentiation of 3T3-L1 adipocytes in culture: (**a**) percentage lipid area per field of view and (**b**) mean diameters of lipid droplets over time. Example micrographs for three selected time points are shown at the *top frames*. The *scale bars* indicate a 50 μm length

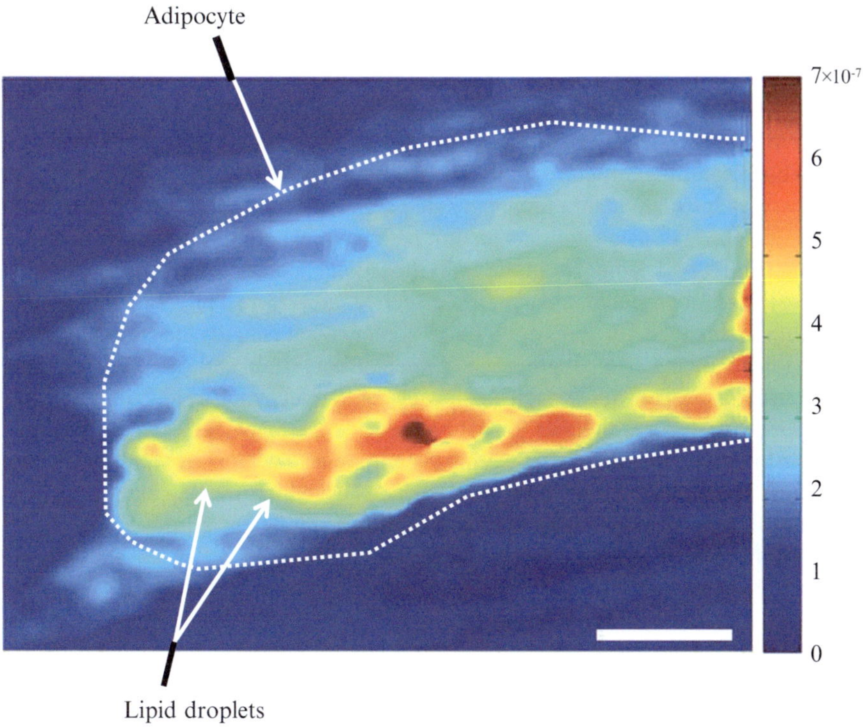

Fig. 5.3 The optical path delay of a 3T3-L1 adipocyte, which was used in order to evaluate the mechanical stiffnesses of the subcellular components. The units of the scale are meters. The *scale bar* indicates a 10 μm length. (Adapted from Shoham et al. 2012a)

could contribute to obesity or weight gaining, and eventually to hyperlipidemia as well. These exciting ideas open completely new research paths for studying obesity, diabetes, and related diseases from a biomechanical point of view, involving the mechanotransduction and structure–function–adaptation concepts that are well known to exist for other tissues but were so far very poorly studied in fat.

5.5 Summary and Concluding Comments

In this chapter, we described the up-to-date research work concerning the mechanical properties of adipose tissues, as well as the most recent experimental work concerning mechanical behavior of, and mechanotransduction in the microscopic building block of adipose tissues—the adipocytes. We proposed to relate the mechanical behavior of the adipose tissue continuum to the biological activities of adipocytes, which follows the classic structure-based mechanics theory for tissues but have not been previously addressed for fat. We demonstrate that first, at a macroscopic scale, the mechanical behavior of adipose tissues depends on the anatomical site and hence on physiological function. At a microscopic scale, mechanical function such as cell stiffness properties depends on the triglyceride contents that in turn depend on the level of differentiation, which has recently been shown to be regulated by mechanical loading. This have led us to develop a new hypothesis regarding structure–function–adaptation processes and relationships in fat, where bodyweight loads chronically shape the mass and composition of weight-bearing fat through mechanotransduction pathways, where the influence is on the levels of triglyceride contents in the individual cells and rate of cell differentiation in mass. The consequent changes in effective adipocyte cell stiffness—which is highly influenced by the triglyceride contents—eventually cause changes in continuum-scale mechanical properties, which affects strain and stress distributions in the weight-bearing fat, which again acts as a mechanotransduction stimulus on adipocytes. Hence, there is probably ongoing feedback between macro-scale tissue loads and synthesis of triglycerides in adipocytes, which—if validated further by future research—may revolutionize the way by which we perceive fat-related diseases such as obesity, diabetes, and hyperlipidemia, as in all these conditions, biomechanics, particularly at the cell level, may play a key role.

References

Avram AS, Avram MM, James WD. Subcutaneous fat in normal and diseased states: 2. Anatomy and physiology of white and brown adipose tissue. J Am Acad Dermatol. 2005;53:671–83.

Case N, Xie Z, Sen B, Styner M, Zou M, O'Conor C, Horowitz M, Rubin J. Mechanical activation of β-catenin regulates phenotype in adult murine marrow-derived mesenchymal stem cells. J Orthop Res. 2010;28:1531–8.

Christiaens V, Lijnen HR. Angiogenesis and development of adipose tissue. Mol Cell Endocrinol. 2010;318:2–9.

Darling EM, Topel M, Zauscher S, Vail TP, Guilak F. Viscoelastic properties of human mesenchymally-derived stem cells and primary osteoblasts, chondrocytes, and adipocytes. J Biomech. 2008;41:454–64.

David V, Martin A, Lafage-Proust MH, Malaval L, Peyroche S, Jones DB, Vico L, Guignandon A. Endocrinology. 2007;148:2553–62.

Geerligs M, Peters GW, Ackermans PA, Oomens CW, Baaijens FP. Linear viscoelastic behavior of subcutaneous adipose tissue. Biorheology. 2008;45:677–88.

Geerligs M, Peters GW, Ackermans PA, Oomens CW, Baaijens FP. Does subcutaneous adipose tissue behave as an (anti-)thixotropic material? J Biomech. 2010;43:1153–9.

Gefen A, Haberman E. Viscoelastic properties of ovine adipose tissue covering the gluteus muscles. J Biomech Eng. 2007;129:924–30.

Gefen A, Megido-Ravid M, Itzchak Y. In vivo biomechanical behavior of the human heel pad during the stance phase of gait. J Biomech. 2001;34:1661.

Hara Y, Wakino S, Tanabe Y, Saito M, Tokuyama H, Washida N, Tatematsu S, Yoshioka K, Homma K, Hasegawa K, Minakuchi H, Fujimura K, Hosoya K, Hayashi K, Nakayama K, Itoh H. Rho and Rho-kinase activity in adipocytes contributes to a vicious cycle in obesity that may involve mechanical stretch. Sci Signal. 2011;4:ra3.

Hossain MG, Iwata T, Mizusawa N, Shima SW, Okutsu T, Ishimoto K, Yoshimoto K. Compressive force inhibits adipogenesis through COX-2-mediated down-regulation of PPARgamma2 and C/EBPalpha. J Biosci Bioeng. 2010;109:297–303.

Huang SC, Wu TC, Yu HC, Chen MR, Liu CM, Chiang WS, Lin KM. Mechanical strain modulates age-related changes in the proliferation and differentiation of mouse adipose-derived stromal cells. BMC Cell Biol. 2010;10:11–8.

Iatridis JC, Wu J, Yandow JA, Langevin HM. Subcutaneous tissue mechanical behavior is linear and viscoelastic under uniaxial tension. Connect Tissue Res. 2003;44:208–17.

Kato H, Suga H, Eto H, Araki J, Aoi N, Doi K, Iida T, Tabata Y, Yoshimura K. Reversible adipose tissue enlargement induced by external tissue suspension: possible contribution of basic fibroblast growth factor in the preservation of enlarged tissue. Tissue Eng Part A. 2010;16:2029–40.

Krouskop TA, Wheeler TM, Kallel F, Garra BS, Hall T. Elastic moduli of breast and prostate tissues under compression. Ultrason Imaging. 1998;20:260–74.

Levy A, Enzer S, Shoham N, Zaretsky U, Gefen A. Large, but not small sustained tensile strains stimulate adipogenesis in culture. Ann Biomed Eng. 2011;40(5):1052–60. doi:10.1007/s10439-011-0496-x.

Linder-Ganz E, Shabshin N, Itzchak Y, Gefen A. Assessment of mechanical conditions in sub-dermal tissues during sitting: a combined experimental-MRI and finite element approach. J Biomech. 2007;40:1443–54.

Linder-Ganz E, Shabshin N, Itzchak Y, Yizhar Z, Siev-Ner I, Gefen A. Strains and stresses in sub-dermal tissues of the buttocks are greater in paraplegics than in healthy during sitting. J Biomech. 2008;41:567–80.

Luu YK, Capilla E, Rosen CJ, Gilsanz V, Pessin JE, Judex S, Rubin CT. Mechanical stimulation of mesenchymal stem cell proliferation and differentiation promotes osteogenesis while preventing dietary-induced obesity. J Bone Miner Res. 2009;24:50–61.

Maddalozzo GF, Iwaniec UT, Turner RT, Rosen CJ, Widrick JJ. Whole-body vibration slows the acquisition of fat in mature female rats. Int J Obes (Lond). 2008;32:1348–54.

Marques MA, Combes M, Roussel B, Vidal-Dupont L, Thalamas C, Lafontan M, Viguerie N. Impact of a mechanical massage on gene expression profile and lipid mobilization in female gluteofemoral adipose tissue. Obes Facts. 2011;4:121–9.

Monteux C, Lafontan M. Use of the microdialysis technique to assess lipolytic responsiveness of femoral adipose tissue after 12 sessions of mechanical massage technique. J Eur Acad Dermatol Venereol. 2008;22:1465–70.

Or-Tzadikario S, Gefen A. Confocal-based cell-specific finite element modeling extended to study variable cell shapes and intracellular structures: the example of the adipocyte. J Biomech. 2011;44:567–73.

Or-Tzadikario S, Sopher R, Gefen A. Quantitative monitoring of lipid accumulation over time in cultured adipocytes as function of culture conditions: toward controlled adipose tissue engineering. Tissue Eng Part C Methods. 2010;16:1167–81.

Ozcivici E, Luu YK, Rubin CT, Judex S. Low-level vibrations retain bone marrow's osteogenic potential and augment recovery of trabecular bone during reambulation. PLoS One. 2010;5:e11178.

Park Y, Best CA, Badizadegan K, Dasari RR, Feld MS, Kuriabova T, Henle ML, Levine AJ, Popescu G. Measurement of red blood cell mechanics during morphological changes. Proc Natl Acad Sci U S A. 2010;107:6731–6.

Rubin CT, Capilla E, Luu YK, Busa B, Crawford H, Nolan DJ, Mittal V, Rosen CJ, Pessin JE, Judex S. Adipogenesis is inhibited by brief, daily exposure to high-frequency, extremely low-magnitude mechanical signals. Proc Natl Acad Sci U S A. 2007;104:17879–84.

Samani A, Zubovits J, Plewes D. Elastic moduli of normal and pathological human breast tissues: an inversion-technique-based investigation of 169 samples. Phys Med Biol. 2007;52:1565–76.

Sen B, Xie Z, Case N, Ma M, Rubin C, Rubin J. Mechanical strain inhibits adipogenesis in mesenchymal stem cells by stimulating a durable beta-catenin signal. Endocrinology. 2008;149:6065–75.

Sen B, Styner M, Xie Z, Case N, Rubin CT, Rubin J. Mechanical loading regulates NFATc1 and beta-catenin signaling through a GSK3beta control node. J Biol Chem. 2009;284:34607–17.
Sen B, Xie Z, Case N, Styner M, Rubin CT, Rubin J. Mechanical signal influence on mesenchymal stem cell fate is enhanced by incorporation of refractory periods into the loading regimen. J Biomech. 2011;44:593–9.
Shaked NT, Satterwhite LL, Rinehart MT, Wax A. Quantitative analysis of biological cells using digital holographic microscopy. In: Rosen J, editor. Holography, research and technologies. Rijeka: InTech; 2011.
Shoham N, Gefen A. Mechanotransduction in adipocytes. J Biomech. 2012a;45:1–8.
Shoham N, Gefen A. The influence of mechanical stretching on mitosis, growth, and adipose conversion in adipocyte cultures. Biomech Model Mechanobiol. 2012b;11(7):1029–45. doi:10.1007/s10237-011-0371-6.
Shoham N, Girshovitz P, Shaked NT, Gefen A (2012a) Evaluating stiffness distributions of differentiating adipocytes using a wide-field digital interferometric method. In: Proceedings of the 10th international symposium on computer methods in biomechanics and biomedical engineering, Berlin, Germany, 11–14 April 2012.
Shoham N, Gottlieb R, Sharabani-Yosef O, Zaretsky U, Benayahu D, Gefen A. Static mechanical stretching accelerates lipid production in 3T3-L1 adipocytes by activating the MEK signaling pathway. Am J Physiol Cell Physiol. 2012b;302:C429–41.
Sinkus R, Tanter M, Xydeas T, Catheline S, Bercoff J, Fink M. Viscoelastic shear properties of in vivo breast lesions measured by MR elastography. Magn Reson Imaging. 2005;23:159–65.
Tanabe Y, Koga M, Saito M, Matsunaga Y, Nakayama K. Inhibition of adipocyte differentiation by mechanical stretching through ERK-mediated downregulation of PPARgamma2. J Cell Sci. 2004;117:3605–14.
Tanabe Y, Matsunaga Y, Saito M, Nakayama K. Involvement of cyclooxygenase-2 in synergistic effect of cyclic stretching and eicosapentaenoic acid on adipocyte differentiation. J Pharmacol Sci. 2008;106:478–84.
Tirkkonen L, Halonen H, Hyttinen J, Kuokkanen H, Sievänen H, Koivisto AM, Mannerström B, Sándor GK, Suuronen R, Miettinen S, Haimi S. The effects of vibration loading on adipose stem cell number, viability and differentiation towards bone-forming cells. J R Soc Interface. 2011;8:1736–47.
Turner NJ, Jones HS, Davies JE, Canfield AE. Cyclic stretch-induced TGFbeta1/Smad signaling inhibits adipogenesis in umbilical cord progenitor cells. Biochem Biophys Res Commun. 2008;377:1147–51.
Vissers D, Verrijken A, Mertens I, Van Gils C, Van de Sompel A, Truijen S, Van Gaal L. Effect of long-term whole body vibration training on visceral adipose tissue: a preliminary report. Obes Facts. 2010;3:93–100.
Weaver JB, Doyley M, Cheung Y, Kennedy F, Madsen EL, Van Houten EE, Paulsen K. Imaging the shear modulus of the heel fat pads. Clin Biomech (Bristol, Avon). 2005;20:312–9.

Chapter 6
Glaucoma and Structure-Based Mechanics of the Lamina Cribrosa at Multiple Scales

Rafael Grytz, Günther Meschke, Jost B. Jonas, and J. Crawford Downs

Abstract Glaucoma is among the leading causes of blindness worldwide. The disease involves damage to the retinal ganglion cell axons that transmit visual information from the eye to the brain. Experimental evidence indicates that biomechanical mechanisms at different length scales are involved in pathophysiology of glaucoma, where chronic intraocular pressure (IOP) elevation at the organ level initiates axonal insult at the level of the lamina cribrosa. The lamina cribrosa consists of a porous collagen structure through which the axons of retinal ganglion cells (RGCs) pass on their path from the retina to the brain. The extent to which the structural mechanics of the lamina cribrosa contribute to the axonal insult remains unclear. In this book chapter, we give a short review of the present understanding of the structural mechanics of the lamina cribrosa and its role in glaucoma. The main aim is to present a first computationally coupled two-scale analysis of the lamina cribrosa that translates the IOP load at the macroscale to the mechanical insult of the axons within the mesostructure of the lamina cribrosa. The numerical results of two-scale analysis suggest that the collagen structures of the lamina cribrosa and its surrounding peripapillary sclera effectively provide mechanical support to the axons by protecting them from high tensile stresses even at elevated IOP levels. However, in-plane shear stresses in the axonal tissue may increase with increasing IOP at the posterior lamina insertion region and contribute to a mechanical insult of the RGC axons in glaucoma.

R. Grytz (✉) • J.C. Downs
Department of Ophthalmology, University of Alabama at Birmingham, Birmingham, Alabama
e-mail: grytz@uab.edu

G. Meschke
Institute for Structural Mechanics, Ruhr-University Bochum, Bochum, Germany

J.B. Jonas
Department of Opthalmology, Medical Faculty Mannheim, Ruprecht-Karls-University Heidelberg, Mannheim, Germany

G.S. Kassab, M.S. Sacks (eds.), *Structure-Based Mechanics of Tissues and Organs*,
DOI 10.1007/978-1-4899-7630-7_6

6.1 Introduction

The collagen structures of corneoscleral shell provide the structural integrity necessary to resist the intraocular pressure (IOP) load and also give the eye the stable shape necessary for focused vision (see Fig. 6.1a for ocular anatomy). At the posterior side of the scleral shell the collagen structure is interrupted to form the scleral canal through which the retinal ganglion cell (RGC) axons pass on their way from the retina to the brain. The tissues within in the scleral canal can be histologically divided into three regions (Fig. 6.1b). In the prelaminar region, the RGC axons converge and turn 90° to form the optic nerve. Posterior to the prelaminar region is the lamina cribrosa. The lamina cribrosa is characterized by a porous collagenous structure (Fig. 6.1c). The RGC axons pass as aggregated bundles through the pores of the lamina cribrosa. The collagen architecture of the lamina cribrosa provides mechanical support to the RGC axons as they pass from the high pressure environment in the eye to the lower pressure environment in cerebrospinal fluid space. Posterior to the lamina cribrosa is the retrolaminar region. Here the RGC axons become myelinated, which results in an increased diameter of the optic nerve.

Glaucoma is characterized by irreversible loss of vision, and is often associated with chronically elevated IOP. There are varying levels of individual susceptibility to IOP, yet IOP lowering is currently the only clinical treatment that has been proven to arrest or delay the onset and progression of the disease. There is considerable body of evidence to indicate that glaucomatous damage to the RGC is due to RGC axonal insult at the level of the lamina cribrosa (Lampert et al. 1968; Gaasterland et al. 1978; Anderson and Hendrickson 1974; Quigley and Addicks 1980; Quigley et al. 1981). The porous collagen structure of the lamina cribrosa represents a relatively vulnerable structure in the otherwise robust corneoscleral pressure vessel. Recent experimental studies have shown that the connective tissues of the lamina

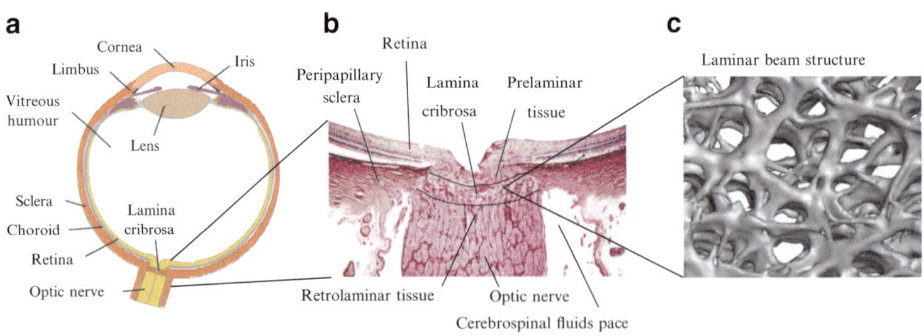

Fig. 6.1 (**a**) The anatomy of the human eye. (**b**) Histologic section through the optic nerve head region illustrating the prelaminar, laminar cribrosa, and retrolaminar tissues. (**c**) Three-dimensional reconstruction of the porous laminar beam structure

cribrosa and its surrounding tissues undergo extensive structural changes during the development and progression of glaucoma (Roberts et al. 2009; Yang et al. 2007, 2011).

The mechanisms that lead from IOP elevation to RGC axonal insult and, ultimately, to RGC death are still unclear. It has been shown that RGCs have an increased vulnerability to IOP-induced blockade of axonal transport at the level of the lamina cribrosa (Quigley and Anderson 1976; Minckler et al. 1977; Radius and Anderson 1979a, b, 1981; Minckler 1980), which may lead to RGC death through apoptosis. It was suggested that axonal transport blockade might be caused by a direct IOP-dependent mechanical insult of the RGC axons (Grytz et al. 2011a) or indirectly through the remodeling of the lamina cribrosa structure or other mechanisms. The blood supply for the lamina cribrosa and its contained RGC axons is delivered through capillaries within the load-bearing lamina cribrosa beams themselves, which directly couples the mechanical and vascular responses of these tissues (Downs et al. 2008). Hence, a reduction in blood flow or nutrient delivery might contribute to the IOP-related axonal insult. Whether the axonal damage results form direct mechanical insult or indirectly through remodeling or reduced blood flow, the structural mechanics of the lamina cribrosa seems to play major role in the pathophysiology of glaucoma.

In this book chapter, we give a short review of the current understanding of the structural mechanics of the lamina cribrosa and its relevance in glaucoma research. The main aim of this chapter is to present the first computationally coupled two-scale analysis to bridge structural mechanics between the macro- and mesostructure of the lamina cribrosa and to investigate the potential mechanical insult of RGC axons due to IOP elevation.

6.2 Structural Mechanics of the Lamina Cribrosa at Multiple Scales

Most of the research on the biomechanics of the lamina cribrosa is based on the eye macrostructure and its mechanical response (Dongqi and Zeqin 1999; Sigal et al. 2005, 2007, 2009a, b, 2011a, b; Bellezza et al. 2003). Sigal et al. showed in a series of publications that many macroscopic factors influence the mechanics of the lamina cribrosa and these factors act both independently and in interaction (Sigal et al. 2005, 2007, 2009a, b, 2011a, b). Major factors influencing the lamina cribrosa biomechanics are peripapillary scleral thickness and stiffness, and laminar position, thickness, and stiffness. These studies were performed using simplified parametric, axisymmetric models with homogeneous, isotropic, elastic material properties. Roberts et al. (2009, 2010a, b) showed in several studies of eye-specific models incorporating inhomogeneous orthotropic material properties of the lamina cribrosa based on local connective tissue volume fraction and predominant lamina beam orientation, that the stress and strain distribution in the lamina cribrosa are correlated with local laminar density.

The structural mechanics of the human eye are strongly related to the collagen structures within its tissues. At the mesoscale, organized collagen structures form networks that introduce anisotropic and nonlinear elastic properties to eye tissues (Girard et al. 2009a; Grytz et al. 2011a; Grytz and Meschke 2010; Petsche et al. 2012; Pinsky and Datye 1991; Pinsky et al. 2005). The lamina cribrosa represents a weak spot in an otherwise stiff scleral shell. Consequently, its mechanical response to IOP is not only influenced by its own collagen structures but also by the structures of the surrounding peripapillary sclera. It was experimentally observed that the scleral canal is surrounded by a circumferentially oriented ring of collagen fibrils in the peripapillary sclera (Goldbaum et al. 1989; Morrison et al. 1989; Winkler et al. 2010). Grytz et al. (2011a) predicted the same collagenous ring-like structure in the peripapillary sclera using a computational model based on a remodeling rule for collagen fibrils. The numerical results revealed that the collagen fibril ring of peripapillary sclera shields the lamina cribrosa from high tensile forces.

The porous architecture of the laminar beams provides mechanical support to the RGC axons passing through the lamina cribrosa. The overall organization of laminar beams seems rather random and mainly designed to provide enough pore space for the RGC axon bundles. However, Roberts et al. (2009) qualitatively analyzed the laminar beam structure within three-dimensional reconstructions, and showed that laminar beams of the monkey lamina cribrosa have a preferred radial orientation in the periphery of the lamina cribrosa. Grytz et al. (2011a) predicted the same radial structure using a computational remodeling approach for collagen fibrils. The numerical results also suggested that these radial laminar beams reinforce the laminar cribrosa against high transverse shear forces and reduce posterior deformation.

Several numerical studies incorporated the collagen structures of the peripapillary sclera and lamina cribrosa through constitutive models and investigated their impact on the biomechanics of the lamina cribrosa (Roberts et al. 2009, 2010a, b; Grytz et al. 2011a; Girard et al. 2009b). While microstructurally motivated constitutive models are very useful to study the impact of micro- and mesostructures on the biomechanical response of the tissue at the macroscale, these model are not suited to localize biomechanical effects at lower scales such as the potential mechanical insult of RGC axons in the glaucomatous lamina cribrosa. Downs et al. (2009) generated finite element models of the porous laminar mesostructure from three-dimensional reconstructions. The mesostructure was subjected to a displacement field, which was previously computed at the macroscale. Consequently, macroscopic displacements were transfered to the mesoscale to investigate the biomechanical environment of the laminar beam structure. This numerical study revealed the complex mechanical environment of the laminar beam structure. Strains were four to five times higher locally in the laminar beam structure at the mesoscale compared with the strains in the homogenized parent macroscale models (Fig. 6.2). While macroscopic

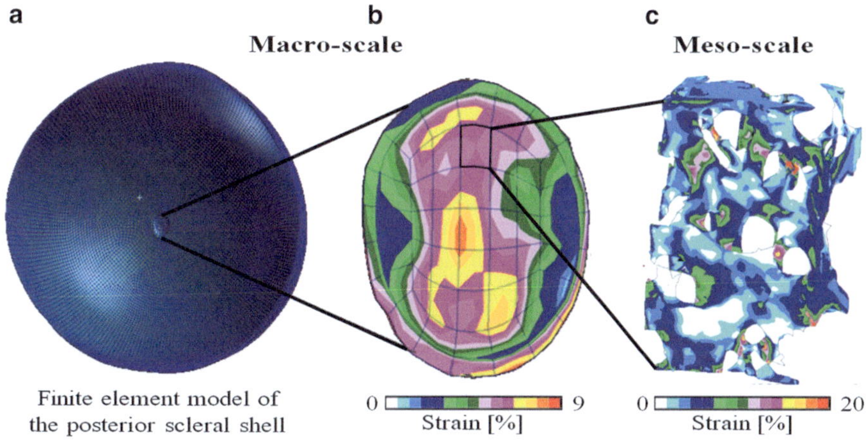

Fig. 6.2 (**a**) The finite element mesh of the posterior scleral shell of a monkey eye. (**b**) The maximum principal strain of the lamina cribrosa computed at the macroscale for IOP elevation from 10 to 45 mmHg. (**c**) For one parent finite element of the macrostructure the maximum principal strain of the underlying laminar beam mesostructure. The displacements calculated for each finite element of the macrostructure were applied as displacement boundary conditions to the underlying laminar beam structure at the mesoscale. The meso- and macrostructures were obtained from three-dimensional reconstructions of the connective tissues (Downs et al. 2009)

deformations were transfered to the mesoscale, no information from the mesoscale was transfered back to the macroscale. Consequently, the mesostructure and its mechanical properties did not impact the structural mechanics of the macrostructure. As soft tissues are subjected to large deformations, nonlinear effects may lead to unexpected results when the feedback of the mesoscale to the macroscale is neglected.

At the present time, most existing numerical models investigate the structural mechanics of the lamina cribrosa at the macroscale only. In the recent past, substantial effort was put forward to incorporate the collagen structures of the eye into computational models by homogenizing meso- and microstructural information and using microstructurally motivated constitutive formulations. On the other hand, only very limited work was done on localizing the mechanical environment from the macro- to the meso- or microscale of the lamina cribrosa. To the best knowledge of the authors, no scale-bridging study has been proposed that couples the meso- and macrostructure of the lamina cribrosa in both ways: (1) from the meso- to the macroscale through homogenization and (2) from the macro- to the mesoscale through localization. The first computationally coupled two-scale analysis of the lamina cribrosa is presented here based on the generalized computational homogenization scheme proposed in Grytz and Meschke (2008).

6.3 Computationally Coupled Two-Scale Analysis of the Lamina Cribrosa

To investigate interactions between the meso- and the macrostructure of the lamina cribrosa, the first computationally coupled two-scale model of the human eye is presented in this section (Fig. 6.3). At the macroscale, the model consists of the corneoscleral shell and the lamina cribrosa. Only the lamina cribrosa has an underlying mesostructure, which includes the porous laminar beam structure and RGC axon bundles. The generalized computational homogenization scheme is applied to computationally couple the meso- and macrostructures of the lamina cribrosa. Macroscopic deformations are transfered to the mesoscale through the macroscopic deformation gradient and the stress response of the mesostructure is homogenized to link the mesostructure back to the macroscale (Grytz and Meschke 2008). To gain insight into the multi-scale phenomena related to glaucomatous optic neuropathy the two-scale analysis is used to investigate the nonlinear relationship between the IOP at the macro-level and the stress and strain environment of the RGC axon bundles at the meso-level of the lamina cribrosa.

6.3.1 *Generalized Computational Homogenization*

If the length scale of the mesostructure is much smaller than the scale of the macroscopic boundary value problem, the argument of scale separation is applicable.

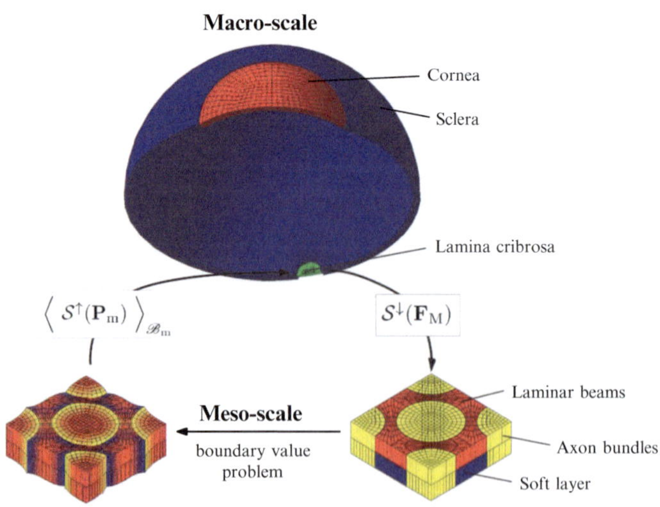

Fig. 6.3 Numerical two-scale model of the human eye illustrating the numerical coupling between macro- and mesoscale using the generalized computational homogenization scheme

In this case, the macroscopic overall properties at a typical point of a body $\mathscr{B}_{\mathrm{M}}$ can be defined by the homogenized response of a representative volume of its mesostructure $\mathscr{B}_{\mathrm{m}}$. The scale transitions in existing computational homogenization frameworks are usually formulated in one physical space for both scales (Miehe et al. 1999a, b; Löhnert and Wriggers 2003; Terada and Kikuchi 2001; Feyel and Chaboche 2000; Miehe 1996, 2003; Kouznetsova et al. 2004; Miehe and Bayreuther 2007; Balzani et al. 2010). This approach is sufficient for problems where the physical material directions follow the same Cartesian coordinate lines at both scales. For problems with physical directions following curvilinear paths, a more general formulation using different physical spaces at different scales may ease solution considerably. In this subsection, the basic equations of the generalized computational homogenization scheme are presented (Grytz and Meschke 2008). A detailed derivation of the computational homogenization scheme can be found in Miehe et al. (1999a). To account for different physical spaces at different scales Grytz and Meschke (2008) introduced the so-called *scale-up* $\mathcal{S}^{\uparrow}(\cdot)$ and *scale-down* $\mathcal{S}^{\downarrow}(\cdot)$ operations, which are used to transform tensor variables between different physical spaces at different scales. The mesoscopic deformation gradient $\mathbf{F}_{\mathrm{m}}$

$$\mathbf{F}_{\mathrm{m}} = \mathrm{GRAD}_{\mathrm{m}}\mathbf{x}_{\mathrm{m}} = \frac{\partial \mathbf{x}_{\mathrm{m}}}{\partial \mathbf{X}_{\mathrm{m}}} \tag{6.1}$$

is related with its macroscopic counterpart $\mathbf{F}_{\mathrm{M}}$ via the volume average under consideration of associated physical reference directions through

$$\mathbf{F}_{\mathrm{M}} := \left\langle \mathcal{S}^{\uparrow}\left(\mathbf{F}_{\mathrm{M}}\right)\right\rangle_{\mathscr{B}_{\mathrm{m}}} \text{ in } \mathscr{B}_{\mathrm{M}}, \tag{6.2}$$

where $\langle\cdot\rangle_{\mathscr{B}} = \frac{1}{V}\int_{\mathscr{B}} \cdot \mathrm{d}V$. The same relation holds for the work-conjugate stress measure of first *Piola-Kirchhoff* type

$$\mathbf{P}_{\mathrm{M}} := \left\langle \mathcal{S}^{\uparrow}\left(\mathbf{P}_{\mathrm{M}}\left(\mathbf{F}_{\mathrm{M}}\right)\right)\right\rangle_{\mathscr{B}_{\mathrm{m}}} \text{ in } \mathscr{B}_{\mathrm{M}}, \tag{6.3}$$

where $\mathbf{P}_{\mathrm{M}}$ and $\mathbf{P}_{\mathrm{m}}$ denote the first *Piola-Kirchhoff* stress tensors the at macro- and mesoscale, respectively. The stress at the mesoscale $\mathbf{P}_{\mathrm{m}}$ is assumed to be related to $\mathbf{F}_{\mathrm{m}}$ by a constitutive relation that governs the local response of the constituents of the mesostructure. The deformation of the mesostructure is assumed to be linked to the local deformations of the macroscopic point via

$$\mathbf{x}_{\mathrm{m}} = \mathcal{S}^{\downarrow}\left(\mathbf{F}_{\mathrm{M}}\right)\mathbf{X}_{\mathrm{m}} + \tilde{\mathbf{w}}_{\mathrm{m}} \text{ in } \mathscr{B}_{\mathrm{m}}. \tag{6.4}$$

The deformation consists of a homogeneous part $\mathcal{S}^{\downarrow}\left(\mathbf{F}_{\mathrm{M}}\right)\mathbf{X}_{\mathrm{m}}$ and an inhomogeneous superimposed field $\tilde{\mathbf{w}}_{\mathrm{m}}\left(\Theta^{i}_{\mathrm{m}}, t\right)$ usually referred to as the fluctuation field.

By inserting (6.4) into (6.1) the definition of the mesoscopic deformation gradient can be rewritten into

$$\mathbf{F}_{\mathrm{m}} = \mathcal{S}^{\downarrow}\left(\mathbf{F}_{\mathrm{M}}\right) + \tilde{\mathbf{F}}_{\mathrm{m}} \quad \text{with} \quad \tilde{\mathbf{F}}_{\mathrm{m}} = \mathrm{GRAD}_{\mathrm{m}}\tilde{\mathbf{w}}_{\mathrm{m}}. \tag{6.5}$$

Inserting (6.5) into (6.2) yields the constraint $\left\langle \tilde{\mathbf{F}}_{\mathrm{m}} \right\rangle_{\mathscr{B}_{\mathrm{m}}} = \mathbf{0}$ for the fluctuation field $\tilde{\mathbf{w}}_{\mathrm{m}}$. This constraint and the so-called averaging theorem may be directly enforced by alternative conditions (see, e.g., Kouznetsova et al. 2004) for the superimposed deformation field. Here, we assume periodic fluctuations at the boundaries of the mesostructure $\tilde{\mathbf{w}}_{\mathrm{m}}^{+} = \tilde{\mathbf{w}}_{\mathrm{m}}^{-}$ on $\partial\mathscr{B}_{\mathrm{m}}$. Note that boundary problem at the mesoscale becomes a nonlinear problem when considering the existence of a fluctuation field $\tilde{\mathbf{w}}_{\mathrm{m}} \neq \mathbf{0}$.

For a given macroscopic deformation gradient $\mathbf{F}_{\mathrm{M}}$, the boundary value problem of the mesostructure is governed by the equilibrium condition, the constitutive equation for the mesoscopic stresses and the boundary condition for the unknown fluctuation field $\tilde{\mathbf{w}}_{\mathrm{m}}$ up to a constant translation, which does not affect the stress state. To avoid a singular stiffness matrix when solving the boundary value problem of the mesostructure within the finite element method, the fluctuation $\tilde{\mathbf{w}}_{\mathrm{m}} = \mathbf{0}$ is held fixed at one corner node in addition to the periodic boundary condition.

Assuming a static equilibrium state of the meso-continuum governed by the field equation

$$\mathrm{DIV}_{\mathrm{m}}\mathbf{P}_{\mathrm{m}} = \mathbf{0} \quad \text{in} \quad \mathscr{B}_{\mathrm{m}} \tag{6.6}$$

the standard *Galerkin* procedure yields the weak formulation of the meso-boundary value problem

$$\delta W_{\mathrm{m}} = \int_{\mathscr{B}_{\mathrm{m}}} \mathbf{P}_{\mathrm{m}}\left(\mathbf{F}_{\mathrm{m}}\right) : \delta\tilde{\mathbf{F}}_{\mathrm{m}} \mathrm{d}V_{\mathrm{m}}. \tag{6.7}$$

The solution of the weak formulation results in the localized stress and strain response at the mesoscale driven by the deformation gradient of the macroscale. The homogenization of the localized stress field according to (6.3) serves as the stress response at the macroscale. The consistent linearization of the two-scale problem also requires for the homogenization of the elasticity tensor, which can be found in Grytz and Meschke (2008).

6.3.2 *The Human Eye Model at the Macro-Level*

We use our previous generic model (Grytz and Meschke 2010; Grytz et al. 2011a) to approximate the macrostructure of the human eye. The geometry of the human eye globe is modeled by means of two spherical shells representing the corneoscleral

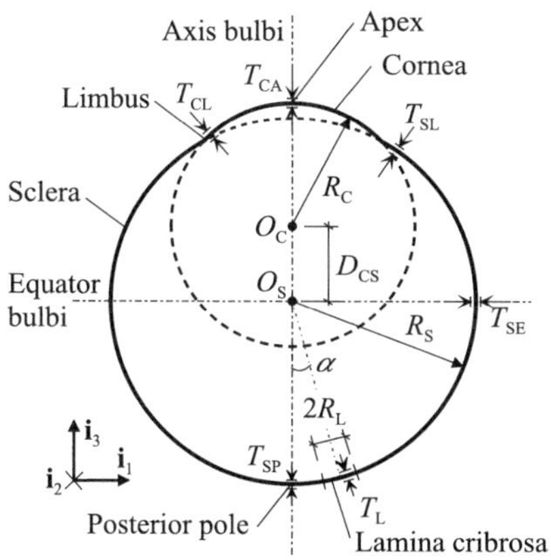

Cornea geometry:
R_C = 7.56 mm
T_{CA} = 0.55 mm
T_{CL} = 0.65 mm
D_{CS} = 5.0 mm
Sclera geometry:
R_S = 11.5 mm
T_{SL} = 0.8 mm
T_{SE} = 0.55 mm
T_{SP} = 1.0 mm
Lamina cribrosa geometry:
R_L = 1.0 mm
T_L = 0.3 mm
α = 15°

Fig. 6.4 Geometry of the human eye model at the macro-level

shell (Fig. 6.4). We also use our previous estimations of the constitutive parameters of the sclera and cornea (Grytz and Meschke 2010; Grytz et al. 2011a), which were fitted to inflation tests performed by Woo et al. (1972). The anisotropic collagen structures were adopted from our previous numerical remodeling predictions and include the circumcorneal ring of collagen fibrils at the limbus (Grytz and Meschke 2010) and the circumferentially aligned collagen fibrils around the scleral canal (Grytz et al. 2011a). The two predicted ring-like structures of collagen fibrils are in good agreement with experimental observations (Aghamohammadzadeh et al. 2004; Winkler et al. 2010). Due to the symmetry of the model, only one half of the eye was modeled using 1372 finite shell elements. The bilinear finite shell elements are based on a quadratic kinematic assumption in thickness direction and were designed to model incompressible shell structures at large strains and finite rotations (Başar and Grytz 2004).

6.3.3 *The Lamina Cribrosa Model at the Meso-Level*

The lamina cribrosa is composed of collagenous beams which form a porous connective tissue architecture (see Fig. 6.1c). RGC axon bundles exit the eye through the porous structure of the lamina cribrosa. The laminar beams are composed of a core of extracellular matrix consisting of collagen fibrils and are lined by astrocytes (see the schematic illustrations in Fig. 6.5). The astrocytes are separated from the core of the laminar beams by a continuous basement membrane (Anderson 1969). In young adults, the core of the laminar beams contains abundant fibers of elastin, and a network of fibrillar collagen of type I and III (Goldbaum et al. 1989; Hernandez 1992; Morrison et al. 1989; Rehnberg et al. 1987).

We designed a generic representative volume element of the lamina cribrosa mesostructure inspired by the schematic illustrations in Fig. 6.5a, b proposed by

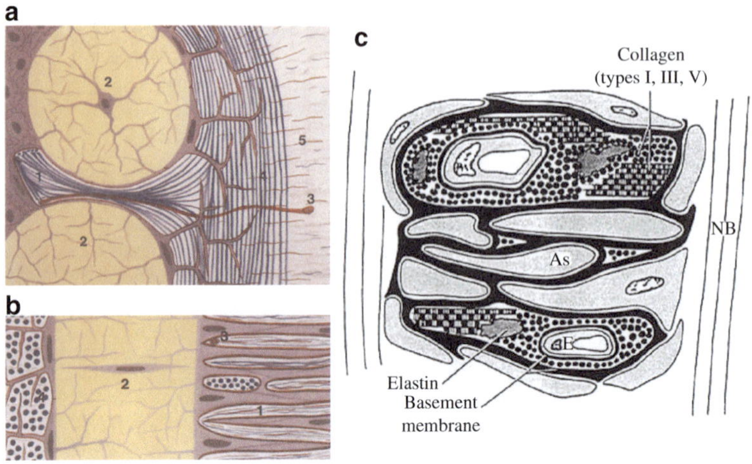

Fig. 6.5 Schematic drawing of the (**a**) cross section and (**b**) sagittal section of the lamina cribrosa showing different extracellular matrix components and their relationships. *1* laminar beams, *2* axon bundles, *3* blood vessels, *4* insertion region, *5* sclera. Color code: *red* basement membranes, *lavender* astrocytes, *blue* collagen and elastin fibers (reproduced from Hernandez et al. 1987). (**c**) Schematic cross section of two laminar beams illustrating the components–collagen types I, III and V and elastin. The capillary endothelial cell (E) basement membranes collagen type IV and V, laminin, and heparan sulfate. Astrocytes (As) separate the extracellular matrix from axonal bundles (NB) (reproduced from Morrison et al. 1989)

Hernandez et al. (1987) and in Fig. 6.5c proposed by Morrison et al. (1989). We assume that the mesostructure is composed of three constituents: the porous laminar structure, the axon bundles, and a soft layer representing a matrix mainly composed of astrocytes and the basement membrane (Fig. 6.6). The mesostructure of the lamina cribrosa is assumed to be periodic in all three dimensions. The laminar beams are assumed to contain aligned collagen fibrils, where collagen fibril orientations are illustrated in Fig. 6.7a. The soft layer is assumed to be isotropic. The axon bundles passing through the laminar beam structure are modeled as circular cylinders, where Fig. 6.7b illustrates the orientation of the RGC axons within the axon bundles. We assumed no local variation in mesostructure and the proposed structure in Fig. 6.6 was used as the underlying mesostructure of every macroscopic point of the lamina cribrosa model. Note that the proposed homogenization technique does account for the volume fraction, distribution, and morphology of the constituents, but not the absolute size of the mesostructure. Thus the absolute dimensions of the mesostructure L_{m} and H_{m} are irrelevant. The relative dimensions of the meso-model are selected such that the volume fraction of the axon bundles correlates with the average value of 58 % experimentally obtained in rhesus monkeys by Minckler et al. (1976).

The meso-model of the lamina is discretized into 320 trilinear brick elements. Preliminary convergence studies showed small changes of the homogenized stress tensor for finer discretizations. The proposed model is a highly simplified

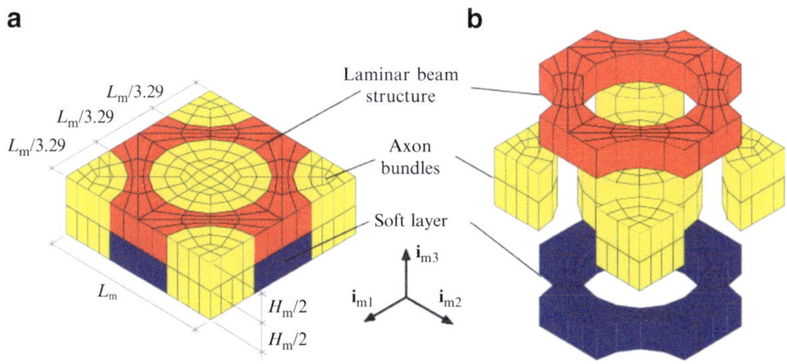

Fig. 6.6 Lamina cribrosa model at the mesoscale. (**a**) Finite element model. (**b**) Exploded view

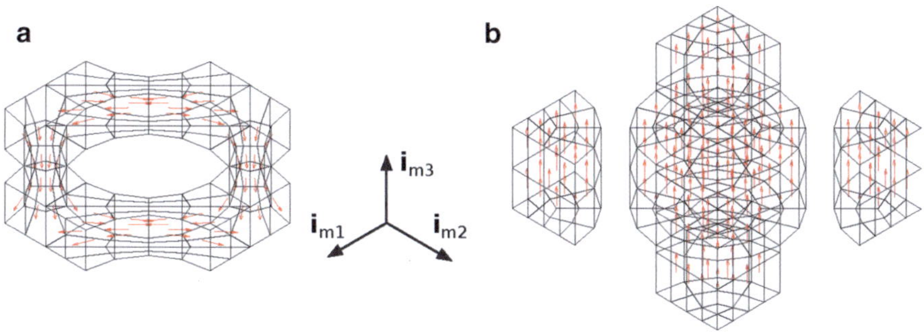

Fig. 6.7 Lamina cribrosa model at the mesoscale containing three constituents: the porous laminar beam structure, the axon roundels, and a soft layer. (**a**) Collagen fibril directions of the laminar beams. (**b**) Axon fiber directions of the axon bundles

representation of the true mesostructure (Roberts et al. 2009). A qualitatively better representation of the laminar mesostructure would also require for a much higher number of finite elements. However, increasing the unknown degrees of freedom at the meso-level leads to an exponential increase in computational cost for the coupled multi-scale analysis. Thus our current computational resources limited the complexity of the meso-model.

6.3.4 Collagen Fibrils and RGC Axons at the Micro-Level

The constitutive response of the laminar mesostructure is based on the microstructure-motivated constitutive model presented by Grytz and Meschke (2009). The strain energy density W_m of the mesostructural constituents is decomposed into three parts

$$W_{\mathrm{m}} = U\left(J_{\mathrm{m}}\right) + \breve{W}_{\mathrm{mat}}\left(I_{\breve{\mathbf{C}}_{\mathrm{m}}}\right) + W_{\mathrm{fib}}\left(\sqrt{IV_{\mathbf{C}_{\mathrm{m}}}}\right) = \frac{1}{2}\kappa(\ln J_{\mathrm{m}})^2 + \frac{1}{2}\mu\left(I_{\breve{\mathbf{C}}_{\mathrm{m}}} - 3\right) + W_{\mathrm{fib}}\left(\sqrt{IV_{\mathbf{C}_{\mathrm{m}}}}\right). \tag{6.8}$$

The following invariant quantities were used in (6.8)

$$J_{\mathrm{m}} = \det \mathbf{F}_{\mathrm{m}}, \quad I_{\breve{\mathbf{C}}_{\mathrm{m}}} = J_{\mathrm{m}}^{-2/3}\mathrm{tr}\mathbf{C}_{\mathrm{m}}, \quad IV_{\mathbf{C}_{\mathrm{m}}} = \mathbf{e}_0\mathbf{C}_{\mathrm{m}}\mathbf{e}_0,$$

where $\mathbf{C}_{\mathrm{m}} = \mathbf{F}_{\mathrm{m}}^{\mathrm{T}}\mathbf{F}_{\mathrm{m}}$ is the Cauchy-Green tensor at the mesoscale and $\mathbf{e}_0$ is a unit vector representing the collagen fibril and axon direction of the laminar beams and RGC axon bundles, respectively. In (6.8), $\breve{W}_{\mathrm{mat}}$ represents a deviatoric Neo-Hookean energy contribution with one material parameter: the shear modulus μ. U represents a pure volumetric part of the energy function with the bulk modulus κ, which was assumed to be sufficiently higher than the shear modulus to account for near incompressibility $\kappa = 1000\mu$. W_{fib} represents the one-dimensional energy contribution of the microstructure-motivated model proposed by Grytz and Meschke (2009), which was originally derived to model the constitutive response of crimped collagen fibrils. W_{fib} is used here to model the fibrillar components of the laminar beams and the axon bundles. The helix model contains one material parameter (the elastic model of the fibril E_{fib}) and two microstructural parameters (the crimp angle of the fibril θ_0 and the ratio between the amplitude of the helix and the radius of the fibril's cross section R_0/r_0). The energy function W_{fib} cannot be presented in a closed form. We, therefore, refer to the original paper for the detailed derivation of the model (Grytz and Meschke 2009).

Parameter Identification of the Axon Bundles

For a realistic approximation of the model parameters, isolated tests of each constituent would be most favorable. However, to extract each constituent from the mesostructure of the lamina cribrosa first and to perform mechanical tests thereafter, is not possible. Here, we use uniaxial extension tests performed on the optic nerve (Meaney 2003) to estimate the constitutive parameters for the axon bundles and a previous fit of the lamina cribrosa to estimate the properties of the laminar beams.

Bain and Meaney (2000) as well as Bain et al. (2003) observed axons in the unstretched optic nerve to have a wavy (undulated) structure. Meaney (2003) demonstrated that the constitutive response of white matter under tension can be nicely captured by microstructurally motivated models considering the crimp of nerve axons rather than classical approaches of *Ogden* or *Mooney–Rivlin* type. Inspired by these observations, the microstructure-motivated constitutive model for crimped collagen fibrils (Grytz and Meschke 2009) was used here to predict the stress–strain relationship of the RGC axon bundles. Figure 6.8 presents the fit of the crimped fibril model to experimental data of simple elongation tests on

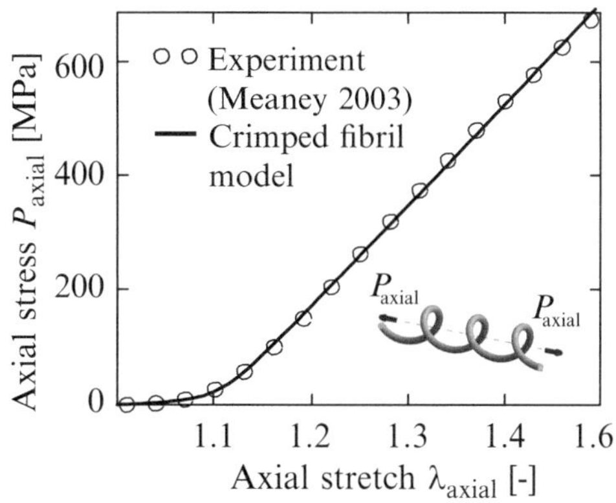

Fig. 6.8 A fit of the crimped fibril model (Grytz and Meschke 2009) to data taken from simple elongation tests on optic nerves (Meaney 2003)

optic nerves reported by Meaney (2003). The crimped fibril model was capable to replicate the stress-stretch response in excellent agreement with the experimental data. The model parameters have been identified as $E_{fib} = 1766$ Pa, $\theta_0 = 24.81°$, and $R_0/r_0 = 2.51$. The so-called locking stretch $\lambda_{lock} = 1/\cos(\theta_0)$ of the crimped fibril model represents the initial undulation (true length over end-to-end distance) of the crimped axons. Bain et al. (2003) measured a mean undulation of 1.13 for axons existing in unstretched optic nerves. The locking stretch estimated from the fitting of the crimped fibril model $\lambda_{lock} = 1.10$ is very close to the experimentally obtained undulation value, which supports the physical relevance of the constitutive model for modeling optic nerve tissue.

Note that the experimental data used to fit our model (Fig. 6.8) are obtained from experiments on complete optic nerves which are composed of myelinated axons. The axon bundles existing in the lamina cribrosa are not myelinated. Also, the experimental data is based on uniaxial extension test along the optic nerve. The axon bundles within the lamina cribrosa are expected to be subjected to very different loading conditions. Accordingly, the optimized model parameters can only be considered as a rough approximation of the constitutive response of the axon bundles within the lamina cribrosa.

Parameter Identification of the Laminar Beam Structure and the Soft Layer

The constitutive parameters of two remaining constituents—the laminar beam structure and the soft layer—are indirectly identified using the experimental inflation data presented by Woo et al. (1972). The remaining unknown parameters are optimized towards the equibiaxial stress-stretch response of our single-scale lamina model (Fig. 4 in Grytz et al. 2011a) previously fitted to the inflation experiments performed by Woo et al. (1972). Therefore, a computational two-scale analysis was performed using one finite element at the macroscale and the meso-model of the lamina cribrosa (Fig. 6.6) attached to each macroscopic *Gauss* point. The constitutive

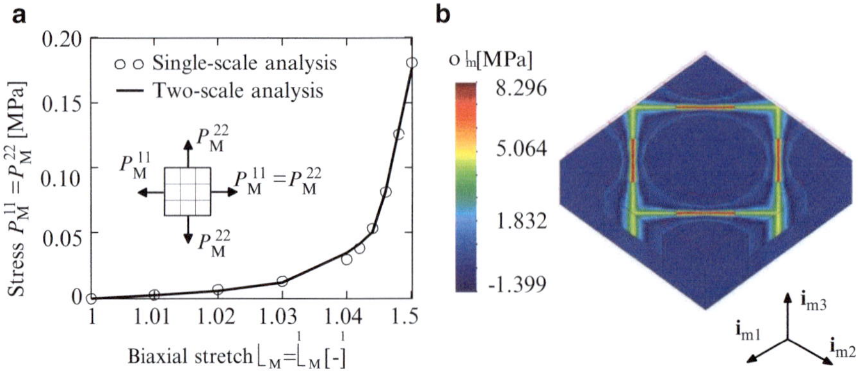

Fig. 6.9 (**a**) The indirect fit of the coupled two-scale model of the lamina cribrosa to the equibiaxial numerical data from Grytz et al. (2011a) (Fig. 6.4), which was initially fitted to inflation experiments on eye segments by Woo et al. (1972). (**b**) Contour plot of the maximum principal *Cauchy* stress σ_m^1 within the mesostructure subjected to a macroscopic stretch level $\lambda_{M1} = \lambda_{M2} = 1.05$

Table 6.1 The constitutive parameters used for the three constituents of the lamina cribrosa mesostructure

	μ (MPa)	E_{fib} (MPa)	θ_0 (°)	R_0/r_0 (–)
Laminar beams	0.2	1421.8	16.7	29.9
Soft layer	0.2	0	–	–
Axon bundles	10^{-5}	0.0018	24.81	2.51

parameters of the axon bundles were set according to the fitted parameters obtained ("Parameter Identification of the Axon Bundles"). We assume that the isotropic strain energy response $\breve{W}_{mat}$ is identical for the laminar beam structure and the soft layer. In accordance with the assumption that solely the laminar beams contain fibrillar collagen, the transversely isotropic part $\breve{W}_{fib}$ of the strain energy function (6.8) was neglected for the soft layer.

To identify the remaining unknown parameters, the macroelement was subjected to an equibiaxial stretch state $\lambda_M^1 = \lambda_M^2$ and the homogenized macroscopic stress components of first *Piola-Kirchhoff* type $P_M^{11} = P_M^{22}$ were used to compute the sum of square residuals with respect to the numerical data of the single-scale analysis from our previous publication (Fig. 4 in Grytz et al. 2011a). The nonlinear *Levenberg-Marquardt* algorithm was applied to optimize the unknown parameters of the mesostructure such that the sum of square residuals was minimized. In Fig. 6.9a, the good agreement between the optimized stress-stretch response of the homogenized mesostructure and the single-scale analysis can be seen. The complete set of optimized model parameters that belong to the three constituents of the meso-model are presented in Table 6.1.

Figure 6.9b shows a contour plot of the maximum principal *Cauchy* stress values of the mesostructure for an equibiaxial stretch level of $\lambda_{M1} = \lambda_{M2} = 1.05$ at the macroscale. High stress concentrations can be seen in the narrow parts of the laminar beams.

6.3.5 Computational Coupled Two-Scale Analysis of the Human Eye

Figure 6.3 illustrates the numerical coupling between the meso- and macroscale, where the deformation of the mesoscale is driven by the deformation gradient of the macroscale and the constitutive response of the macrostructure is computed from the homogenized response of the mesostructure after solving the boundary value problem (6.7) at the mesoscale. We use the finite element method to solve both, the boundary value problem at the meso- and macroscale. Each macroelement contains eight *Gauss* points according to the finite element formulation presented in Başar and Grytz (2004). One representative volume element of the underlying mesostructure is attached to each macroscopic *Gauss* point. In accordance with the definition of the *scale-up* and *scale-down* operations (Grytz and Meschke 2009) the physical reference base vectors of the mesoscale relate to the macroscale such that $\mathbf{i}_{m3}$ (Fig. 6.6) coincides with the thickness direction of the lamina cribrosa macrostructure. According to the computational homogenization scheme, the non-linear boundary value problem of the mesostructure has to be solved at each iteration step of each load increment for each *Gauss* point of the lamina macroelements of the boundary value problem at the macroscale. To reduce computation time, the eight mesostructure boundary value problems of each macro-finite element were solved simultaneously.

The two-scale analysis presented here was used to investigate the structural mechanics of lamina cribrosa at the two scales of the model. At both scales, we investigated maximum principal stretch λ^1 and *Cauchy* stress values σ^1. Note that the principal stretch values can be easily computed from the principal values of the *Cauchy-Green* tensor $C^i = (\lambda^i)^2$. At the mesoscale, we investigated in addition the axial stretch of collagen fibrils in the laminar beams $\lambda^{\text{axial}}_{\text{m,LB}}$ and of axons in the axon bundles $\lambda^{\text{axial}}_{\text{m,AB}}$. Furthermore, to investigate a potential direct mechanical insult of the axon bundles, two alternative measures previously introduced by Grytz et al. (2011a) were used in addition: the *in-plane change in area*

$$dA_{\text{m}}/dA_{0\text{m}} = J_{\text{m}}\left(\mathbf{i}_{\text{m}3}\mathbf{C}_{\text{m}}^{-1}\mathbf{i}_{\text{m}3}\right)^{1/2} \tag{6.10}$$

and the *in-plane shear stress*

$$\sigma_{\text{m}}^{\text{IPS}} = \left(|\mathbf{i}^*_{\text{m}3}\boldsymbol{\sigma}_{\text{m}}|^2 - \left(\mathbf{i}^*_{\text{m}3}\boldsymbol{\sigma}_{\text{m}}\mathbf{i}^*_{\text{m}3}\right)^2\right)^{1/2} \quad \text{with } \mathbf{i}^*_{\text{m}3} = \mathbf{i}_{\text{m}3}\mathbf{F}_{\text{m}}^{-1}/\left|\mathbf{i}_{\text{m}3}\mathbf{F}_{\text{m}}^{-1}\right|. \tag{6.11}$$

The first measure is used to investigate the potential squeezing of the axon bundles perpendicular to their axis, while the second measure is used to investigate the potential shearing of axon bundles. Both mechanisms were proposed as potential indicators for mechanical insult to axonal transport within RGC axons (Grytz et al. 2011a).

6.3.6 Numerical Results

In what follows, numerical results are presented and discussed for two IOP levels: a normal IOP level $p_{\mathrm{IOP}} = 16$ mmHg and an elevated IOP level $p_{\mathrm{IOP}} = 40$ mmHg. In Fig. 6.10, the evolution of peak values of strain- and stress-like quantities defined at the macro- and mesoscale are plotted for IOP levels between 0 and 45 mmHg. Selected macro- and mesoscale results are presented as contour plots at the macroscale of the lamina cribrosa in Fig. 6.11.

Collagen fibrils existing in the laminar beams which are stretched beyond their locking stretch (here $\lambda_{\mathrm{lock}} = 1/\cos(\theta_0) = 1.044$) are mainly responsible for bearing the IOP-dependent load in the lamina cribrosa. Figure 6.10b presents the IOP-dependent development of the absolute peak value of the axial stretch of collagen fibrils within the laminar beams. This value increased as the eye model was loaded until the normal IOP level was reached. Further IOP elevation did not change (<0.2 %) the peak stretch of collagen max $\lambda^{\mathrm{axial}}_{\mathrm{m,LB}}$. Instead, collagen fibrils in a wider region of the lamina cribrosa were stretched beyond their locking stretch (see Fig. 6.11b). This result suggests that the lamina cribrosa may resist elevated IOP levels by recruiting additional collagen fibrils to bear a significant part of the load instead of stretching already locked fibrils further.

In Fig. 6.11a, the plots of the through-the-thickness stretch suggest that the lamina cribrosa macrostructure was subjected to compression strains in the thickness direction of maximal $\overset{0}{\lambda}_{\mathrm{M}} = 0.991$ and $\overset{0}{\lambda}_{\mathrm{M}} = 0.979$ at normal and elevated IOP, respectively. Interestingly, the fact that overall the macrostructure of the lamina cribrosa was subjected to compressive strains in the thickness direction at both IOP levels (see the thickness stretch $\overset{0}{\lambda}_{\mathrm{M}}$ distribution in Fig. 6.11a), but some axons within the attached mesostructure were under tension, especially at the high IOP level (see the peak values of the axon bundle axial stretch $\lambda^{\mathrm{axial}}_{\mathrm{m,AB}} > 1$ at the bottom surface in the periphery and at the top surface in the center of the macrostructure Fig. 6.11c). This finding suggests that axons which are unstressed under physiological pressure conditions may suffer tension when eye is exposed to an elevated IOP level even though IOP-dependent thinning of the lamina cribrosa is observed. This phenomena was mainly caused by the nonlinear constitutive response of the ring-like architecture of collagen fibrils existing in the peripapillary sclera of our eye model. As IOP was increasing, the scleral canal was expanding until the collagen fibrils that form this ring around the scleral canal started to lock. After this point, posterior (bending) deformations dominated the deformation response of the lamina

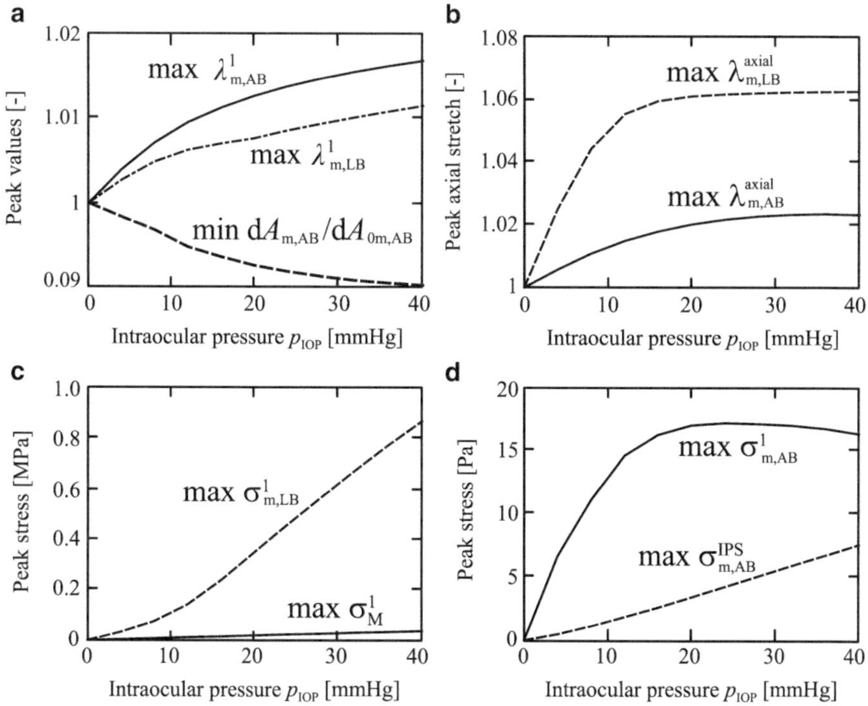

Fig. 6.10 Peak values within the meso- or macrostructure of the lamina cribrosa for varying levels of IOP. (**a**) Upper peak values of the maximum principal stretch within the laminar beam structure max $\lambda^1_{m,LB}$ and the axon bundles max $\lambda^1_{m,AB}$, lower peak values of the change of area normal to the axon direction min $dA_{m,AB}/dA_{0m,AB}$. (**b**) Upper peak values of the axial stretch of collagen fibrils max $\lambda^{axial}_{m,LB}$ and axons max $\lambda^{axial}_{m,AB}$. (**c**) Upper peak values of the macroscopic maximum principal *Cauchy* stress σ^1_M and the maximum principal *Cauchy* stress within the laminar beams max $\sigma^1_{m,LB}$. (**d**) Upper peak values of the maximum principal stress max $\sigma^1_{m,AB}$ and the in-plane shear stress $\sigma^{IPS}_{m,AB}$ within the axon bundles

cribrosa in favor of in-plane (membrane) deformations. In the present example, the local in-plane compressive deformations caused by the posterior deformation of the lamina cribrosa (bending mode) outweigh the local in-plane tensile deformations related to the membrane mode at the top surface of central region of the lamina cribrosa. For nearly incompressible tissues, a local biaxial compression leads to tensile deformations in the perpendicular direction, which, in our case, represents the thickness direction of the lamina cribrosa. The axonal tensile strain seen at the lower surface in the periphery of the lamina was caused by high transverse shear forces, which are also related to the bending dominated loading situation. However, axon fibers remain undulated even at high IOP levels as their axial stretch remains below their critical locking stretch $\lambda_{lock} = 1.1$ (Fig. 6.10b).

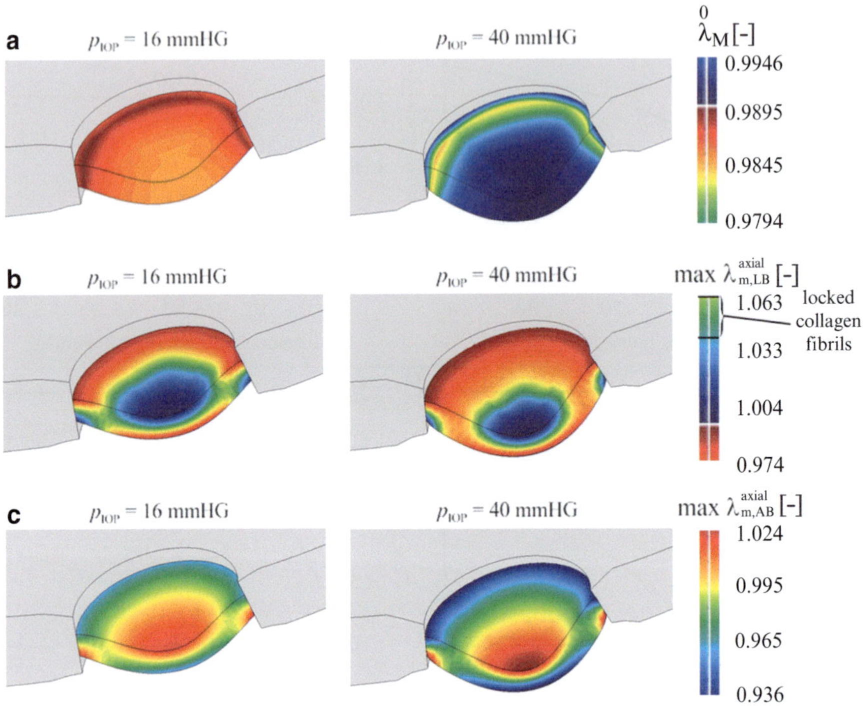

Fig. 6.11 Contour plots of the lamina cribrosa at low $p_{IOP} = 16$ mmHg and high $p_{IOP} = 40$ mmHg IOP levels at the macroscale (deformations fivefold magnified). (**a**) Through-the-thickness stretch $\overset{0}{\lambda}_M$ of the macrostructure. (**b**) Upper peak values of the axial stretch of collagen fibrils within the laminar beams of the attached mesostructures max $\lambda^{axial}_{m,LB}$. (**c**) Upper peak values of the axial stretch within the axon bundles of the attached mesostructures max $\lambda^{axial}_{m,AB}$

The maximum principal stress distribution of the human eye macro-model is presented in Fig. 6.12a identifying high stress concentrations at the limbus and at the peripapillary sclera around the scleral canal. These two regions were characterized by circumferentially aligned collagen fibrils forming ring-like structures. We have shown that the circumcorneal ring of collagen fibrils allows the cornea to maintain its curvature while the eye is subjected to IOP (Grytz and Meschke 2010). The ring of collagen fibrils in the peripapillary sclera was identified to shield the lamina cribrosa from high tensile forces (Grytz et al. 2011a).

Contour plots of the mesoscopic maximum principal stress $\widehat{\sigma}^1_m$ and maximum principal stretch λ^1_m of the laminar beam structure and axon bundles are presented in Figs. 6.13, 6.14, 6.15, and 6.16 for four selected mesostructures, two at the center region (point P_{MB}) and two at the periphery of the lamina cribrosa (point P_{MA}). The peak values of the mesostructure are used to draw the contour plots at the macroscale. Results are presented for the normal and the elevated IOP level. In Fig. 6.13, high strain levels can be observed at the narrow parts of the

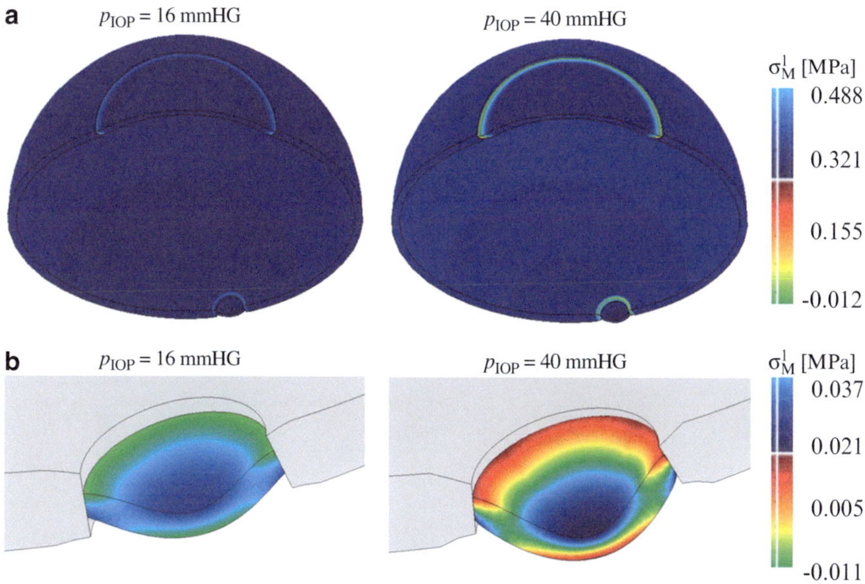

Fig. 6.12 Contour plots of the maximum principal *Cauchy* stress $\hat{\sigma}^1_M$ of the macrostructure at low $p_{IOP} = 16$ mmHg and high $p_{IOP} = 40$ mmHg IOP levels (deformations fivefold magnified). (**a**) The complete human eye model. (**b**) The macrostructure of the lamina cribrosa

laminar beams existing in the meso-model that is attached to the anterior *Gauss* point of macro-point P_{MA} (Fig. 6.13) while moderate strain levels can be found at the mesostructure attached to the posterior *Gauss* point at P_{MA}. Note that stress and strain tensors are usually not *coaxial* when using anisotropic constitutive formulations. Consequently, the distribution of maximum principal stress and strain values can differ substantially in anisotropic tissues. Stress concentrations evolved at the inside of the laminar beam pore structure of the meso-model attached to the anterior *Gauss* point at P_{MA} and posterior *Gauss* point at P_{MB} but not at the posterior *Gauss* point at P_{MA} (Fig. 6.14). Both, the absolute peak value of the maximum principal stretch max $\lambda^1_{m,LB}$ and stress max $\sigma^1_{m,LB}$ of the laminar beams were continuously increasing for increasing IOP levels (Fig. 6.10a, c).

By comparing the numerical results for the elevated IOP level in Fig. 6.12b with Figs. 6.10c and 6.14, one can observe that the maximum principal stress values can be up to 25-fold higher at the mesoscale than at the macroscale of the lamina cribrosa. As the laminar beams are the main load-bearing constituent of the mesostructure, the distribution of σ^1_M and max $\sigma^1_{m,LB}$ were qualitatively identical at the macroscale.

The strain environment of the axon bundles presented in Fig. 6.15 was very similar to strain environment of the laminar beams (Fig. 6.13), with strain concentrations at the top surface in the periphery of the lamina and continuously increasing peak values for increasing IOP levels (Fig. 6.10a). In contrast to the laminar beams,

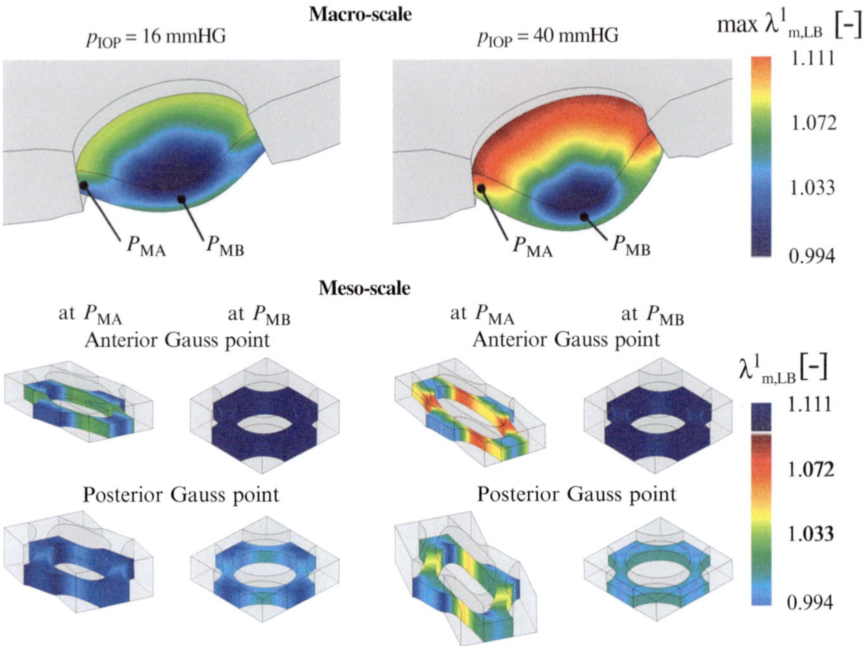

Fig. 6.13 Contour plots of the lamina cribrosa at low $p_{IOP} = 16$ mmHg and high $p_{IOP} = 40$ mmHg IOP levels (deformations fivefold magnified). The maximum principal stretch values $\lambda^1_{m,LB}$ within the laminar beams of selected mesostructures at the mesoscale and upper peak values of attached mesostructures max $\lambda^1_{m,LB}$ at the macroscale

only small changes in the distribution of the maximum principal stress values were observed in the axon bundles for IOP elevation from 16 to 40 mmHg (Fig. 6.16). According to Fig. 6.10d, the absolute peak value of the maximum principal stress within the axon bundles of all mesostructures increased until normal IOP was reached and remained almost constant for higher IOP levels. This numerical finding indicates that the meso-architecture of the lamina cribrosa is effectively protecting the embedded axon bundles from high tensile stresses even at elevated IOP levels.

In (6.10), the changes of area normal to the RGC axons was introduced as a potential indicator for axonal insult. Only small changes in axonal area were observed in the distribution of the scalar quantity when IOP was increased from 16 to 40 mmHg IOP (Fig. 6.17). However, the lower peak value min $dA_{m,N}/dA_{0m,N}$ decreased with increasing IOP (Fig. 6.10a). In accordance with our single-scale analysis (Grytz et al. 2011a), lower peak values representing the contraction of the axon bundle cross sections were found at the top surface in the center of the lamina cribrosa and at the bottom surface in the periphery of the lamina (Fig. 6.17).

The in-plane shear stress was introduced in (6.11) to indicate regions of axonal insult due to axon shearing. The numerical results of our single-scale analysis at the macroscale (Grytz et al. 2011a) showed the highest in-plane shear stress values at

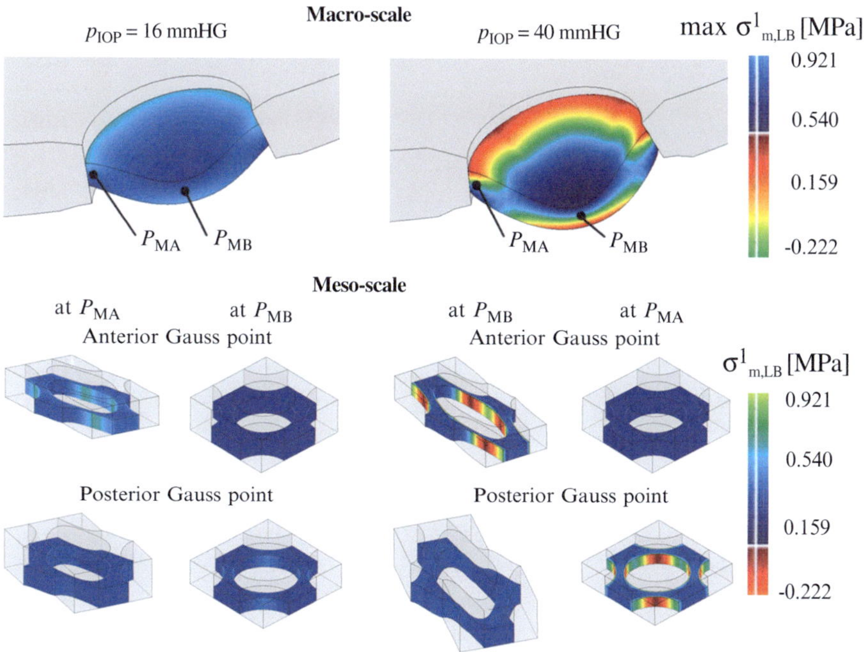

Fig. 6.14 Contour plots of the lamina cribrosa at low $p_{IOP} = 16$ mmHg and high $p_{IOP} = 40$ mmHg IOP levels (deformations fivefold magnified). The maximum principal *Cauchy* stress $\sigma^1_{m,LB}$ within the laminar beams of selected mesostructures at the mesoscale and upper peak values of attached mesostructures max $\sigma^1_{m,LB}$ at the macroscale

the anterior periphery of the lamina cribrosa. In contrast, the two-scale analysis used here revealed the highest in-plane shear stress σ^{IPS}_m to exist within the axon bundles at the posterior periphery of the lamina cribrosa (Fig. 6.18). Obviously, numerical results obtained from single-scale analyses may not be representative for lower-scale results and may lead to wrong interpretations. At the elevated IOP level, finite deformations and fluctuations can be observed in the representative volume element attached to the posterior *Gauss* point at P_{MA}. In contrast to the maximum principal stress, the absolute peak value of the in-plane stress existing in the axon bundles max σ^{IPS}_m continuously increased with increasing IOP (Fig. 6.10d).

6.4 Discussion

The two-scale analysis has revealed that the lamina cribrosa seems to resist elevated IOP levels by recruiting additional collagen fibrils to bear a significant part of the IOP-related load instead of stretching already locked fibrils further. These numerical findings demonstrate the important role of the nonlinear constitutive

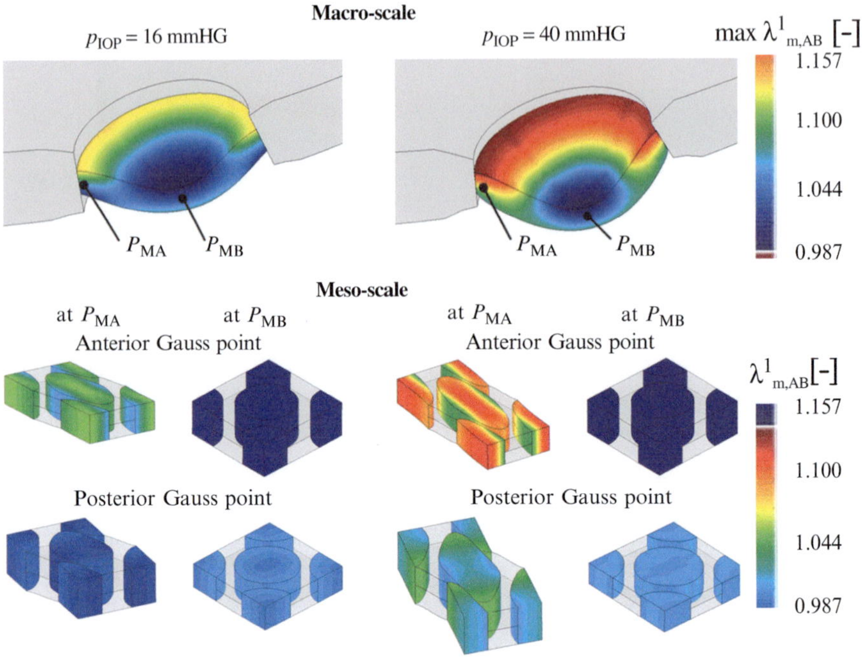

Fig. 6.15 Contour plots of the lamina cribrosa at low $p_{IOP} = 16$ mmHg and high $p_{IOP} = 40$ mmHg IOP levels (deformations fivefold magnified). The maximum principal stretch λ^1_m within the axon bundles of selected mesostructures at the mesoscale and upper peak values of attached mesostructures max $\lambda^1_{m,AB}$ at the macroscale

response of collagen fibrils on the biomechanics of the laminar beams that linear elastic constitutive models cannot account for.

Furthermore, the findings of the two-scale analysis have suggested that RGC axons that are unstressed under physiological pressure conditions may suffer tension when exposed to an elevated IOP level. However, axons were predicted to remain undulated at high IOPs. Accordingly, tearing of RGC axons seems unlikely to occur in ocular hypertension. Single-scale analyses performed solely at the macroscale could not provide such information.

The numerical investigation has shown that the absolute peak value of the maximum principal stress within the axon bundles remains almost constant at elevated IOPs. This numerical finding indicates that the meso-architecture of the lamina cribrosa together with the ring of collagen fibrils in the peripapillary sclera are effectively protecting the axon bundles within the lamina cribrosa from high tensile stresses even at elevated IOPs. In contrast, the in-plane shear stresses within the axon bundles seem to increase with increasing IOP at the posterior periphery of the lamina cribrosa. This observation suggests that IOP-induced shearing of RGC axons may be a potential mechanism for axonal insult in ocular hypertension.

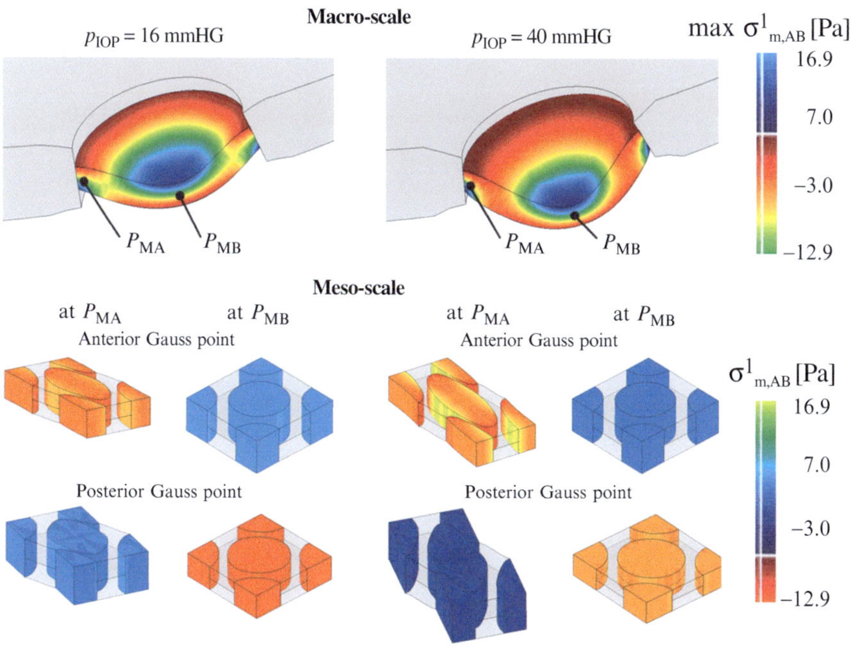

Fig. 6.16 Contour plots of the lamina cribrosa at low $p_{\text{IOP}} = 16$ mmHg and high $p_{\text{IOP}} = 40$ mmHg IOP levels (deformations fivefold magnified). The maximum principal *Cauchy* stress $\sigma^1_{\text{m,AB}}$ within the axon bundles of selected mesostructures at the mesoscale and upper peak values of attached mesostructures max $\sigma^1_{\text{m,AB}}$ at the macroscale

Note that the physical and biological relevance of the numerical results presented in this chapter are limited by the approximation of the geometry and constitutive parameters assumed for the meso- and macrostructures. Both structures are characterized by highly simplified geometries. On the macroscale, the quality of the contour plots are strongly affected by considering only one element with two *Gauss* points through the shell thickness. Accordingly, interpretations of the numerical results should be carefully drawn. For a higher resolution of the numerical results at the macroscale, higher order element formulations or multi-layer shell element could be applied.

The geometry of the proposed mesostructure represents solely a schematic representation of the structure existing in the living tissue. It is well known that the mesostructure of the lamina beams varies significantly in different regions of the lamina cribrosa (Quigley and Addicks 1981) and is characterized by a much more complex geometry (Burgoyne et al. 2004) than assumed in the present work. The physical and biological relevance of the two-scale analysis could be greatly improved by considering three-dimensional reconstructions of experimentally obtained geometries of the meso- (Burgoyne et al. 2004) and macrostructure.

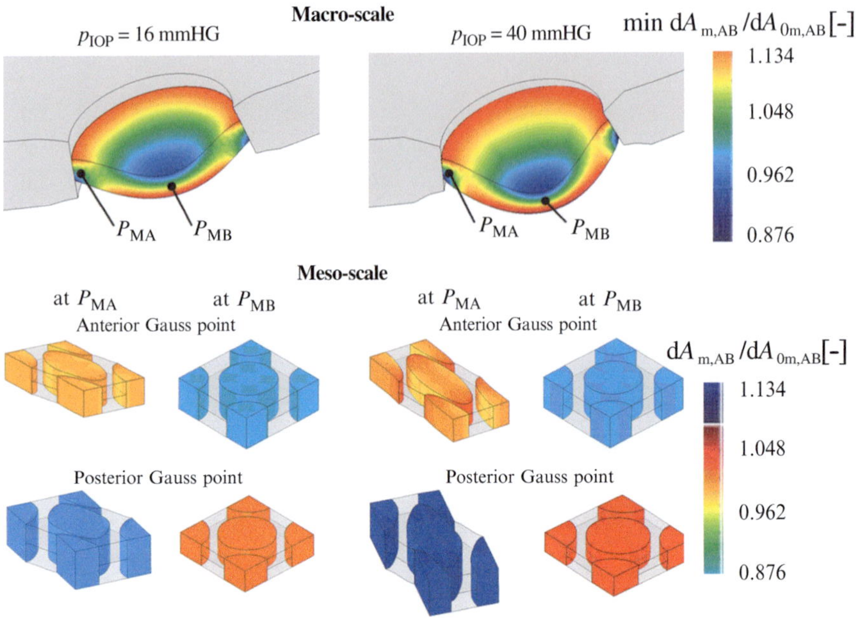

Fig. 6.17 Contour plots of the lamina cribrosa at low $p_{IOP} = 16$ mmHg and high $p_{IOP} = 40$ mmHg IOP levels (deformations fivefold magnified). The change of area $dA_{m,AB}/dA_{0m,AB}$ normal to the RGC axons within the axon bundles of selected mesostructures at the mesoscale and lower peak values of attached mesostructures min $dA_{m,AB}/dA_{0m,AB}$ at the macroscale

The proposed scale-bridging technique between the meso- and the macrostructure does account for the volume fraction, distribution, and morphology of the constituents, but it is completely insensitive to the absolute size of the mesostructure. As a consequence, effects related to the absolute size at the mesoscale cannot be investigated. Furthermore, the presented homogenization method is based on the principle of separation of scales. The argument of scale separation can be applied if the ratio between the characteristic length scale of the macrostructure d_M and the size of the mesostructural constituents d_m is very small $d_m/d_M \ll 1$. The characteristic length scales of our two-scale lamina cribrosa model can be approximated by means of the shell thickness ($d_M \approx 0.3$ mm) at the macroscale and by the typical diameter of the axon bundles ($d_m \approx 58$ μm) at the mesoscale. Unfortunately, the argument of scale separation $d_m/d_M \approx 0.16$ does not hold for the given multi-scale problem. Consequently, the results presented here can only be considered as a first approximation of the multi-scale structural mechanics of the lamina cribrosa. The requirement for scale separation can be overcome by using alternative scale-bridging algorithms based on a higher-order continuum formulation (Kouznetsova et al. 2004), multi-grid solver (Miehe and Bayreuther 2007), or locally constraint models (Hund and Ramm 2007).

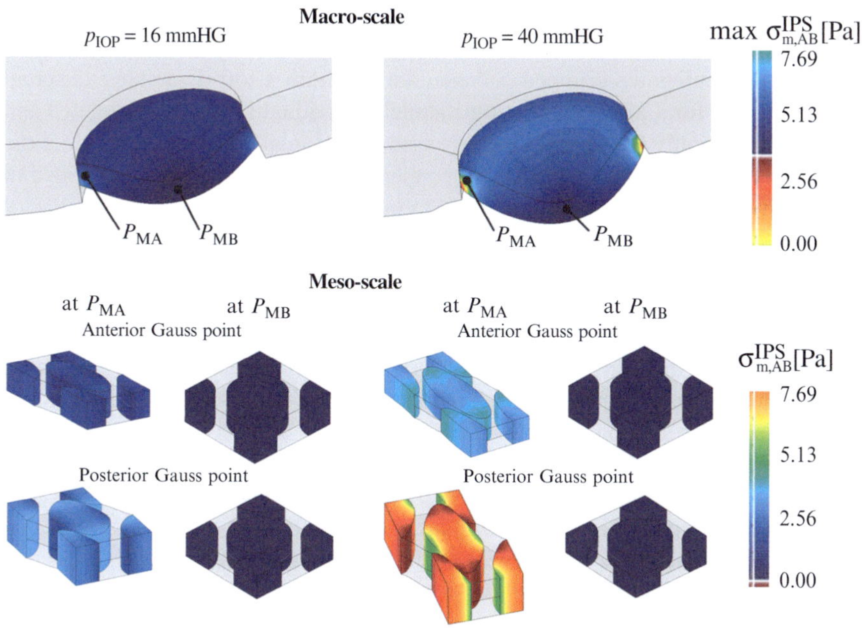

Fig. 6.18 Contour plots of the lamina cribrosa at low $p_{IOP} = 16$ mmHg and high $p_{IOP} = 40$ mmHg IOP levels (deformations fivefold magnified). The in-plane *Cauchy* shear stress $\sigma^{IPS}_{m,AB}$ within the laminar beams of selected mesostructures at the mesoscale and upper peak values of attached mesostructures max $\sigma^{IPS}_{m,AB}$ at the macroscale

Considering glaucoma, further deficiencies of the proposed models are the assumed static loading conditions and the time-independent constitutive properties. It is well known that glaucomatous eyes are usually subjected not only to an increased average level of IOP but also to increased circadian fluctuations. Downs et al. (2011) has shown that IOP fluctuates considerably on the second, minute, hour, and day timescales. It has been shown that chronic IOP elevation in monkey eyes alter the viscoelastic (Downs et al. 2005) and hyperelastic material properties (Girard et al. 2011) of the peripapillary sclera. Furthermore, the mesostructure of the lamina cribrosa undergoes significant structural changes during the development and progression of glaucoma, which may manifest as thickening, thinning, and migration at the macroscale of the lamina cribrosa (Grytz et al. 2011b; Roberts et al. 2009; Yang et al. 2007, 2011).

Our current understanding of soft tissue mechanics is strongly coupled with the advances in structure-based mechanics at different length. Computational and analytical model of tissue structures at lower length scales changed our current understanding of hyperelasticity, viscoelasticity, residual stresses, preconditioning effects, osmotic swelling, remodeling, growth, and rheology of soft tissues (Lanir 1979, 1983, 2009, 2012; Hollander et al. 2011; Chen et al. 2011; Lokshin and Lanir 2009; Raz and Lanir 2009; Sverdlik and Lanir 2002; Taber and Humphrey 2001;

Ambrosi et al. 2011; Grytz et al. 2011b). Pioneering work in this class of models was performed throughout his career by Lanir. In the case of hyperelasticity, Lanir's idea was that individual crimped collagen fibrils within a soft tissue have different lengths so that for a given macroscopic stretch individual fibrils are undulated and stretched differently (Lanir 1979). Lanir used a statistical distribution in either the stretch of the fibrils or their length. Since then Lanir's idea has been duplicated in many constitutive formulations and used to investigate the structural mechanics of many different tissues. Also the present study was highly impacted by Lanir's work, as the constitutive response of collagen fibrils within the lamina beams and crimped axons within the axon bundles were modeled based on a microstructurally motivated constitutive model for crimped fibrils (Grytz and Meschke 2009).

The structural mechanics of the lamina cribrosa are complex and highly impacted by the complex mesostructure of the laminar beams. To investigate the structural mechanics of the lamina cribrosa at multiple scales and its potential impact in glaucoma remains a challenge for computational and experimental researchers. However, computational methods and experimental methods are evolving to gain insight into the multi-scale mechanisms that are involved in the development and progression of glaucoma. The numerical results presented in this chapter show that deeper understanding of potential mechanical insults of RGC axons in glaucoma can be gained from computational models that couple the structural mechanics of the lamina cribrosa at different length scales.

The proposed two-scale model of the lamina cribrosa bridges biomechanical mechanisms between the macro- and mesoscale. To investigate mechanisms down to the meso-level might be insufficient to investigate the phenomena that lead to axonal transport blockade. Balaratnasingam et al. (2007) reported that elevated IOP induced changes to the cytoskeleton existing in the optic nerve axons. Changes to the cytoskeleton may contribute to the axonal transport abnormalities that occur in glaucoma. To investigate this hypothesis using computational multi-scale simulations would require the development of biomechanical models down to the nano-scale of living cells and incorporation of the nano-structure of the cytoskeleton into such models.

References

Aghamohammadzadeh H, Newton R, Meek K. X-ray scattering used to map the preferred collagen orientation in the human cornea and limbus. Structure. 2004;12:249–56.

Ambrosi D, Ateshian GA, Arruda EM, Cowin SC, Dumais J, Goriely A, Holzapfel GA, Humphrey JD, Kemkemer R, Kuhl E, Olberding JE, Taber LA, Garikipati K. Perspectives on biological growth and remodeling. J Mech Phys Solids. 2011;59:863–83.

Anderson DR. Ultrastructure of human and monkey lamina cribrosa and optic nerve head. Arch Ophthalmol. 1969;82:800–14.

Anderson DR, Hendrickson A. Effect of intraocular pressure on rapid axoplasmic transport in monkey optic nerve. Invest Ophthalmol. 1974;13:771–83.

Bain A, Meaney D. Tissue-level thresholds for axonal damage in an experimental model of central nervous system white matter injury. J Biomech Eng. 2000;122:615–22.
Bain A, Shreiber D, Meaney D. Modeling of microstructural kinematics during simple elongation of central nervous system tissue. J Biomech Eng. 2003;125:798–804.
Balaratnasingam C, Morgan WH, Bass L, Matich G, Cringle SJ, Yu DY. Axonal transport and cytoskeletal changes in the laminar regions after elevated intraocular pressure. Invest Ophthalmol Vis Sci. 2007;48:3632–44.
Balzani D, Schröder J, Brands D. FE 2-simulation of microheterogeneous steels based on statistically similar RVEs. In: IUTAM symposium on variational concepts with applications to the mechanics of materials. Springer; 2010. pp. 15–28.
Başar Y, Grytz R. Incompressibility at large strains and finite-element implementation. Acta Mechanica. 2004;168:75–101.
Bellezza A, Rintalan C, Thompson H, Downs J, Hart R, Burgoyne C. Deformation of the lamina cribrosa and anterior scleral canal wall in early experimental glaucoma. Invest Ophthalmol Vis Sci. 2003;44:623–37.
Burgoyne CF, Downs JC, Bellezza AJ, Hart RT. Three-dimensional reconstruction of normal and early glaucoma monkey optic nerve head connective tissues. Invest Ophthalmol Vis Sci. 2004;45:4388–99.
Chen H, Liu Y, Zhao X, Lanir Y, Kassab G. A micromechanics finite-strain constitutive model of fibrous tissue. J Mech Phys Solids. 2011;59(9):1823–37.
Dongqi H, Zeqin R. A biomathematical model for pressure-dependent lamina cribrosa behavior. J Biomech. 1999;32:579–84.
Downs JC, Suh JK, Thomas KA, Bellezza AJ, Hart RT, Burgoyne CF. Viscoelastic material properties of the peripapillary sclera in normal and early-glaucoma monkey eyes. Invest Ophthalmol Vis Sci. 2005;46:540–6.
Downs JC, Roberts MD, Burgoyne CF. Mechanical environment of the optic nerve head in glaucoma. Optom Vis Sci. 2008;85:425–35.
Downs J, Roberts M, Burgoyne C, Hart R. Multiscale finite element modeling of the lamina cribrosa microarchitecture in the eye. In: Engineering in Medicine and Biology Society, IEEE; 2009. p. 4277–80.
Downs JC, Burgoyne CF, Seigfreid WP, Reynaud JF, Strouthidis NG, Sallee V. 24-hour IOP telemetry in the nonhuman primate: implant system performance and initial characterization of IOP at multiple timescales. Invest Ophthalmol Vis Sci. 2011;52:7365–75.
Feyel F, Chaboche JL. FE^2 multiscale approach for modeling the elastoviscoplastic behavior of long fiber SiC/Ti composite material. Comput Meth Appl Mech Eng. 2000;183:309–30.
Gaasterland D, Tanishima T, Kuwabara T. Axoplasmic flow during chronic experimental glaucoma. 1. Light and electron microscopic studies of the monkey optic nervehead during development of glaucomatous cupping. Invest Ophthalmol Vis Sci. 1978;17:838–46.
Girard MJA, Downs JC, Bottlang M, Burgoyne CF, Suh JF. Peripapillary and posterior scleral mechanics—part II: experimental and inverse finite element characterization. J Biomech Eng. 2009a;131:051012.
Girard MJA, Downs JC, Burgoyne CF, Suh JF. Peripapillary and posterior scleral mechanics—part I: development of an anisotropic hyperelastic constitutive model. J Biomech Eng. 2009b;131:051011.
Girard MJA, Suh JKF, Bottlang M, Burgoyne CF, Downs JC. Biomechanical changes in the sclera of monkey eyes exposed to chronic IOP elevations. Invest Ophthalmol Vis Sci. 2011;52:5656–69.
Goldbaum MH, Jeng SY, Logemann R, Weinreb RN. The extracellular matrix of the human optic nerve. Arch Ophthalmol. 1989;107:1225–31.
Grytz R, Meschke G. Consistent micro-macro transitions at large objective strains in curvilinear convective coordinates. Int J Numer Meth Eng. 2008;73:805–24.
Grytz R, Meschke G. Constitutive modeling of crimped collagen fibrils in soft tissues. J Mech Behav Biomed Mater. 2009;2:522–33.

Grytz R, Meschke G. A computational remodeling approach to predict the physiological architecture of the collagen fibril network in corneo-scleral shells. Biomech Model Mechanobiol. 2010;9:225–35.
Grytz R, Meschke G, Jonas JB. The collagen fibril architecture in the lamina cribrosa and peripapillary sclera predicted by a computational remodeling approach. Biomech Model Mechanobiol. 2011a;10:371–82.
Grytz R, Sigal IA, Ruberti JW, Meschke G, Downs JC. Lamina cribrosa thickening in early glaucoma predicted by a microstructure motivated growth and remodeling approach. Mech Mater. 2011b;44:99–109.
Hernandez MR. Ultrastructural immunocytochemical analysis of elastin in the human lamina cribrosa. changes in elastic fibers in primary open-angle glaucoma. Invest Ophthalmol Vis Sci. 1992;33:2891–903.
Hernandez MR, Luo XX, Igoe F, Neufeld AH. Extracellular matrix of the human lamina cribrosa. Am J Ophthalmol. 1987;104:567–76.
Hollander Y, Durban D, Lu X, Kassab GS, Lanir Y. Experimentally validated microstructural 3D constitutive model of coronary arterial media. J Biomech Eng. 2011;133:031007.
Hund A, Ramm E. Locality constraints within multiscale model for non-linear material behaviour. Int J Numer Meth Eng. 2007;70:1613–32.
Kouznetsova V, Geers M, Brekelmans W. Multi-scale second-order computational homogenization of multi-phase materials: a nested finite element solution strategy. Comput Meth Appl Mech Eng. 2004;193:5525–50.
Lampert PW, Vogel MH, Zimmerman LE. Pathology of the optic nerve in experimental acute glaucoma. Invest Ophthalmol. 1968;7:199–213.
Lanir Y. A structural theory for the homogeneous biaxial stress-strain relationship in flat collagenous tissues. J Biomech. 1979;12:423–36.
Lanir Y. Constitutive equations for fibrous connective tissues. J Biomech. 1983;16:1–12.
Lanir Y. Mechanisms of residual stress in soft tissues. J Biomech Eng. 2009;131:044506.
Lanir Y. Osmotic swelling and residual stress in cardiovascular tissues. J Biomech. 2012;45:780–9.
Löhnert S, Wriggers P. Homogenisation of microheterogeneous materials considering interfacial delamination at finite strains. Tech Mechanik. 2003;23:167–77.
Lokshin O, Lanir Y. Micro and macro rheology of planar tissues. Biomaterials. 2009;30:3118–27.
Meaney DF. Relationship between structural modeling and hyperelastic material behavior: application to CNS white matter. Biomech Model Mechanobiol. 2003;1:279–93.
Miehe C. Numerical computation of algorithmic (consistent) tangent moduli in large-strain computational inelasticity. Comput Meth Appl Mech Eng. 1996;134:223–40.
Miehe C. Computational micro-to-macro transitions for discretized micro-structures of heterogeneous materials at finite stains based on the minimization of averaged incremental energy. Comput Meth Appl Mech Eng. 2003;192:559–91.
Miehe C, Bayreuther C. On multiscale FE analyses of heterogeneous structures: from homogenization to multigrid solvers. Int J Numer Meth Eng. 2007;71:1135–80.
Miehe C, Schotte J, Schröder J. Computational micro-macro transitions and overall moduli in the analysis of polycrystals at large strains. Comput Mater Sci. 1999a;16:372–82.
Miehe C, Schröder J, Schotte J. Computational homogenization analysis in finite plasticity simulation of texture development in polycrystalline materials. Comput Meth Appl Mech. 1999b;171:387–418.
Minckler DS. The organization of nerve fiber bundles in the primate optic nerve head. Arch Ophthalmol. 1980;98:1630–6.
Minckler DS, McLean IW, Tso MO. Distribution of axonal and glial elements in the rhesus optic nerve head studied by electron microscopy. Am J Ophthalmol. 1976;82:179–87.
Minckler DS, Bunt AH, Johanson GW. Orthograde and retrograde axoplasmic transport during acute ocular hypertension in the monkey. Invest Ophthalmol Vis Sci. 1977;16:426–41.
Morrison JC, L'Hernault NL, Jerdan JA, Quigley HA. Ultrastructural location of extracellular matrix components in the optic nerve head. Arch Ophthalmol. 1989;107:123–9.

Petsche SJ, Chernyak D, Martiz J, Levenston ME, Pinsky PM. Depth-dependent transverse shear properties of the human corneal stroma. Invest Ophthalmol Vis Sci. 2012;53:873–80.
Pinsky PM, Datye DV. A microstructurally-based finite element model of the incised human cornea. J Biomech. 1991;24:907–22.
Pinsky PM, van der Heide D, Chernyak D. Computational modeling of the mechanical anisotropy in the cornea and sclera. J Cataract Refract Surg. 2005;31:136–45.
Quigley HA, Addicks EM. Chronic experimental glaucoma in primates. II. Effect of extended intraocular pressure elevation on optic nerve head and axonal transport. Invest Ophthalmol Vis Sci. 1980;19:137–52.
Quigley HA, Addicks EM. Regional differences in the structure of the lamina cribrosa and their reaction to glaucomatous nerve damage. Arch Ophthalmol. 1981;99:137–43.
Quigley HA, Anderson DR. The dynamics and location of axonal transport blockade by acute intraocular pressure elevation in primate optic nerve. Invest Ophthalmol. 1976;15:606–16.
Quigley HA, Addicks EM, Green WR, Maumenee AE. Optic nerve damage in human glaucoma. II. The site of injury and susceptibility to damage. Arch Ophthalmol. 1981;99:635–49.
Radius RL, Anderson DR. The course of axons through the retina and optic nerve head. Arch Ophthalmol. 1979a;97:1154–8.
Radius RL, Anderson DR. The histology of retinal nerve fiber layer bundles and bundle defects. Arch Ophthalmol. 1979b;97:948–50.
Radius RL, Anderson DR. Rapid axonal transport in primate optic nerve. Distribution of pressure-induced interruption. Arch Ophthalmol. 1981;99:650–9.
Raz E, Lanir Y. Recruitment viscoelasticity of the tendon. J Biomech Eng. 2009;131:111008.
Rehnberg M, Ammitzböll T, Tengroth B. Collagen distribution in the lamina cribrosa and the trabecular meshwork of the human eye. Br J Ophthalmol. 1987;71:886–92.
Roberts MD, Grau V, Grimm J, Reynaud J, Bellezza AJ, Burgoyne CF, Downs JC. Remodeling of the connective tissue microarchitecture of the lamina cribrosa in early experimental glaucoma. Invest Ophthalmol Vis Sci. 2009;50:681–90.
Roberts MD, Liang Y, Sigal IA, Grimm J, Reynaud J, Bellezza A, Burgoyne CF, Downs JC. Correlation between local stress and strain and lamina cribrosa connective tissue volume fraction in normal monkey eyes. Invest Ophthalmol Vis Sci. 2010a;51:295–307.
Roberts MD, Sigal IA, Liang Y, Burgoyne CF, Downs JC. Changes in the biomechanical response of the optic nerve head in early experimental glaucoma. Invest Ophthalmol Vis Sci. 2010b;51:5675–84.
Sigal IA, Flanagan JG, Ethier CR. Factors influencing optic nerve head biomechanics. Invest Ophthalmol Vis Sci. 2005;46:4189–99.
Sigal IA, Flanagan JG, Tertinegg I, Ethier CR. Predicted extension, compression and shearing of optic nerve head tissues. Exp Eye Res. 2007;85:312–22.
Sigal IA, Flanagan JG, Tertinegg I, Ethier CR. Modeling individual-specific human optic nerve head biomechanics. Part I: IOP-induced deformations and influence of geometry. Biomech Model Mechanobiol. 2009a;8:85–98.
Sigal IA, Flanagan JG, Tertinegg I, Ethier CR. Modeling individual-specific human optic nerve head biomechanics. Part II: influence of material properties. Biomech Model Mechanobiol. 2009b;8:99–109.
Sigal IA, Yang H, Roberts MD, Burgoyne CF, Downs JC. IOP-induced lamina cribrosa displacement and scleral canal expansion: an analysis of factor interactions using parameterized eye-specific models. Invest Ophthalmol Vis Sci. 2011a;52:1896–907.
Sigal IA, Yang H, Roberts MD, Grimm JL, Burgoyne CF, Demirel S, Downs JC. IOP-induced lamina cribrosa deformation and scleral canal expansion: independent or related? Invest Ophthalmol Vis Sci. 2011b;52:9023–32.
Sverdlik A, Lanir Y. Time-dependent mechanical behavior of sheep digital tendons, including the effects of preconditioning. J Biomech Eng. 2002;124:78–84.
Taber LA, Humphrey JD. Stress-modulated growth, residual stress, and vascular heterogeneity. J Biomech Eng. 2001;123:528–35.

Terada K, Kikuchi N. A class of general algorithms for multi-scale analyses of heterogeneous media. Comput Meth Appl Mech Eng. 2001;190:5427–64.

Winkler M, Jester B, Nien-Shy C, Massei S, Minckler DS, Jester JV, Brown DJ. High resolution three dimensional reconstruction of the collagenous matrix of the human optic nerve head. Brain Res Bull. 2010;81:339–48.

Woo SL, Kobayashi AS, Schlegel WA, Lawrence C. Nonlinear material properties of intact cornea and sclera. Exp Eye Res. 1972;14:29–39.

Yang H, Downs JC, Girkin C, Sakata L, Bellezza A, Thompson H, Burgoyne CF. 3-D histomorphometry of the normal and early glaucomatous monkey optic nerve head: lamina cribrosa and peripapillary scleral position and thickness. Invest Ophthalmol Vis Sci. 2007;48:4597–607.

Yang H, Williams G, Downs JC, Sigal IA, Roberts MD, Thompson H, Burgoyne CF. Posterior (outward) migration of the lamina cribrosa and early cupping in monkey experimental glaucoma. Invest Ophthalmol Vis Sci. 2011;52:7109–21.

Chapter 7
From Stress–Strain Relations to Growth and Remodeling Theories: A Historical Reflection on Microstructurally Motivated Constitutive Relations

J.D. Humphrey

Abstract As noted early on by Y.C. Fung, one of the greatest needs in biomechanics is formulation of constitutive relations for tissues that experience multiaxial loading. Although most investigators today seek to glean ideas on constitutive formulations from the latest papers, there is often much to learn from the earliest papers wherein truly original ideas can be found. In this Chapter, I provide a brief historical reflection on the formulation of constitutive relations for cardiovascular tissues, with particular focus on contributions by Y. Lanir. In this way, we can recall seminal works upon which much of our field has been built as well as see how past work continues to influence constitutive formulations, even in frontier areas such as soft tissue growth and remodeling.

7.1 Introduction

Emergence of the modern field of biomechanics can be traced primarily to the mid-1960s. Amongst the many reasons for this timing, I have suggested before (Humphrey 2002) that biomechanics needed to await the renaissance in nonlinear continuum mechanics that occurred following WWII and that was summarized to large extent in the works of Truesdell and Toupin (1960), Green and Zerna (1960), Truesdell and Noll (1965), and Green and Adkins (1970). That is, it was only because of this rich period that modern biomechanics could be founded upon the strong theoretical framework that was needed due to the complex material behaviors exhibited by most tissues. Because of the complex geometries and loading conditions inherent to most physiologically and pathologically relevant situations, biomechanics also had to await the development of appropriate computational methods, as, for example, the finite element method that developed rapidly from

J.D. Humphrey (✉)
Department of Biomedical Engineering, Yale University, New Haven, CT 06520, USA
e-mail: jay.humphrey@yale.edu

G.S. Kassab, M.S. Sacks (eds.), *Structure-Based Mechanics of Tissues and Organs*,
DOI 10.1007/978-1-4899-7630-7_7

the late 1950s to the early 1970s and was found in suitable form for nonlinear problems in the important book of Oden (1972). Both the need for computational methods and computer-control of complex multiaxial experiments also necessitated that biomechanics await the development of the computer, which progressed rapidly following the design of transistor-based machines in the late 1950s and early 1960s. In addition, of course, one can also point to the important influence on modern biomechanics by the emergence of molecular and cell biology in the mid-1950s and beyond, which followed the discovery of the structure of DNA as well as important structural proteins such as collagen. Finally, the Apollo program in the USA also provided significant new motivation to advance biomechanics, in particular, to predict how humans might respond to high G-forces due to lift-off or to a sustained microgravity environment in outer space. Hence, based on enabling theoretical and technological developments, and motivated by many needs including space travel, biomechanics emerged as an important and exciting new field in need of ideas and approaches.

7.2 Stress–Strain Relations (1967–1990)

Many contributed to the early successes of modern biomechanics, but Y.C. Fung (UCSD) and R. Skalak (Columbia University) provided particularly important leadership and scientific contributions (cf. Figs. 1.3 and 1.4 in Humphrey 2002). In particular, in a seminal paper published in the late 1960s (which has been cited over 750 times to date, not including the many related citations to his first book on biomechanics—Fung 1981), Fung (1967) showed via a clever one-dimensional experiment on mesentery that soft tissues tend to exhibit an exponential stress–strain behavior. Given that this behavior is relatively insensitive to strain-rate and that the associated hysteresis (i.e., area between the loading and unloading curves) is typically small, Fung suggested that preconditioned soft tissue behavior (i.e., that which follows multiple cycles of testing) is "pseudoelastic" and thus can be treated as (hyper)elastic separately in loading and unloading (where hyperelastic simply means an elastic response describable using a strain energy function). From 1967 through 1983, Fung and colleagues continued to develop their exponential relations for soft tissue behavior, yielding, for example, forms that were quadratic in the Green strain and containing 10, 6, and 3 material parameters for orthotropic, transversely isotropic, and isotropic relations, respectively. These parameters could be determined via nonlinear regression of experimental data. Among other successes achieved using such relations, Choung and Fung (1986) showed that the experimentally observed opening of a radially cut, traction-free, arterial ring revealed the presence of residual stresses on the order of a few kPa, which due to the strong geometric and material nonlinearities helped offset the expected gradients in wall stress in a pressurized artery (note: it is interesting that this formulation of the residual stress problem was enabled by a solution found in the aforementioned work by Green and Adkins (1970), which did not consider biomechanical problems in particular). That is, Choung and Fung showed that transmural stresses tend to be uniform within the normal arterial wall, an observation that contributed significantly to the subsequent formulation of a fundamental mechanobiological hypothesis: mechanosensitive cells tend to produce, organize, and remodel the extracellular matrix to establish, maintain, or restore a mechanical homeostasis (Humphrey 2008).

Although the "phenomenological approach" of Fung, and others, has been and continues to be very useful in many applications in soft tissue biomechanics, the associated material parameters do not have physical meaning and thus cannot be related, except loosely, to histological characteristics of the tissue. Consequently, in parallel with similar attempts in rubber elasticity (cf. Treolar 1975), investigators began to seek microstructurally motivated constitutive relations for soft tissues. Y. Lanir was not the first to seek such relations, but two of his early papers (Lanir 1979, 1983) represented the first rigorous multiaxial formulations (note: it is interesting that these papers were similarly founded in the theoretical foundation of the aforementioned work by Green and Zerna (1960), which serves as a reminder that continued advances in biomechanics can and should be founded in rigorous mathematical foundations provided by nonlinear continuum mechanics). In particular, these two seminal papers of Lanir have been cited over 250 and 450 times, respectively, and continue to inspire many studies. The interested reader is strongly encouraged to read these papers, but I will focus on a few particularly important aspects of these papers, especially from the 1979 paper wherein most of the new ideas first appeared.

Consistent with basic histological evidence, Lanir correctly observed that "The mechanical behavior of flat soft collagenous tissue depends primarily on the response of its constituents and their structure. They consist primarily of collagen fibers, elastin fibers and amorphous ground substances of mucopolysaccharides." He thus assumed that constitutive relations for elastic behaviors by such tissues could be formulated via a total strain energy function W that represents the sum of the energy stored in each constituent. In the case of both elastin and collagen, Lanir assumed further that these fibers could be modeled as 1D structures, hence their stored energy depended only on the 1D strain (or stretch) experienced by the individual fiber, which could be determined easily from the overall multiaxial strain field given the assumption of affine motions. Given the energy stored in an individual fiber, he then noted the importance of knowing how the fibers are oriented within the tissue, which could be prescribed via continuous distribution functions and thus integral relations. Another novel assumption was that the highly nonlinear behavior of a collagen fiber could be modeled by prescribing both an undulation function and a linearly elastic behavior for the fiber. That is, nonlinear behaviors could arise from the gradual recruitment of initially undulated fibers that exhibit a linearly elastic behavior when straight; overall tissue nonlinearity thus results in part from differential recruitment of fibers that are initially undulated to different degrees. Of course, once the specific form of the strain energy function is known, formulation and solution of the initial-boundary value problem can proceed as for the phenomenological hyperelastic constitutive relations. In strong form, one must satisfy linear momentum balance and boundary conditions (Humphrey 2002); in weak form (e.g., finite element), solutions can similarly be found easily (Holzapfel 2000).

Of particular interest here, however, let us return to other constitutive assumptions offered by Lanir. For example, in his 1979 paper, Lanir considered two "conceptual models" (which we would now call competing hypotheses) for the origin of undulation in the collagen fibers. One model assumed that "undulations originate from the subfibrillar or macromolecular organization within the collagen fiber" whereas the other assumed that "undulations are induced by prestressed elastin (thin) fibers attached to each collagen (thick) fiber at numerous points." Although the latter model did not garner much detailed investigation, particularly from the perspective of matrix biology, it is interesting to note that over the years it has become apparent in arterial wall mechanics that prestressed elastin plays fundamental roles in arterial homeostasis and pathophysiology (Humphrey et al. 2009). In particular, arterial elastin is produced primarily during the perinatal period and its half-life is on the order of the life-span of the individual (e.g., over 50 years in humans). Hence, elastic fibers (which consist of ~90 % elastin and 10 % microfibrillar glycoproteins) are stretched during normal development and they retain this "prestretch/prestress" during much of maturity. It thus appears that part of the undulation of collagen in an unloaded, intact tissue does indeed result from the prestretched elastin trying to unload (cf. Ferruzzi et al. 2011b), consistent with the conceptual model of Lanir. Moreover, it now appears that this prestretch/prestress of elastin probably contributes significantly to the existence of residual stresses in arteries (Cardamone et al. 2009), the importance of which was nevertheless identified via the phenomenological constitutive relations of Fung.

Another important assumption found in the 1979 paper of Lanir is that "the fibers possess only extensional rigidity and their compressive and bending rigidities are negligible ... Any interaction between the matrix (ground substance) and the fibers is negligible. The only possible contribution of the matrix to the behavior of the tissue is in the form of hydrostatic pressure which may develop upon stretch." This assumption continues to be invoked by many today and enables the tensile mechanical behavior of planar soft tissues to be described well as originally intended by Lanir; indeed, these ideas also provide some structural insight into values of the Lagrange multiplier (having units of pressure) that is often used to enforce incompressibility in such problems. Nevertheless, we must remain mindful of the intended use of this, indeed all, constitutive assumptions. For example, continuing histological insight (cf. Dingemans et al. 2000) reveals that glycosaminoglycans/proteoglycans (or GAGs/PGs, the main components of the so-called ground substance matrix) tend to orient perpendicular to and interact directly with collagen fibrils in arteries (and probably other tissues). In this way, the GAGs/PGs likely confer some effective bending stiffness to the collagen fibers (not unlike a laterally supported Euler bucking column). Hence, there is a need in certain problems, such as the problem of residual stresses in arteries wherein compressive and tensile stresses must self-equilibrate (Cardamone et al. 2009), to model the compressive stiffness associated with GAG/PG interactions with fibrillar collagens.

Not long after publication of the 1983 paper, Lanir and colleagues employed their microstructurally motivated constitutive relations to describe the biaxial mechanical behavior of passive myocardium (Horowitz et al. 1988). Specifically, they modeled the data reported by Demer and Yin (1983) and Yin et al. (1987) based on biaxial stretching tests of thin excised slabs of canine myocardium. Among the basic assumptions, one reads: "The myocytes and the collagen fibers are initially assumed to be the main structural elements of the tissue, their main mechanical role being to carry tensile stress... The third element appearing in the formulation is the interstitial fluid... which is assumed to transmit only hydrostatic pressure." Similar to the 1979 and 1983 papers, the fibers constituting the myocardium were assumed to be undulated in the unstressed configuration, to not bear load until straight, and to support tension not compression. Finally, it was assumed that "the total strain energy of the tissue, W, is obtained by summing up the strain energies of the muscle fibers, w_1, and of the collagen fibers, w_2." Because of a lack of detailed histological information, the distribution functions for both collagen fiber direction and waviness were assumed to be Gaussian.

As noted in their introduction, one of the motivations of this 1988 paper was to overcome a limitation of the phenomenological (Fung-type) relation used by Yin et al. (1987), namely, the inability of the proposed strain energy function to describe equally well the behaviors observed in very different loading protocols. Although the microstructural relation was found to fit the data as well as or better than the prior phenomenological model, a subtle but very important advance was not emphasized. Prior quantification of biaxial testing data by Fung and many of his associates (e.g., Lanir and Fung 1974; Vawter et al. 1978; Yin et al. 1987; Humphrey et al. 1987) was based on separate nonlinear regressions of data from the multiple biaxial protocols. The obvious outcome of such an approach was a different set of "best-fit" material parameters for each protocol, which necessarily rendered each set of material parameters not predictive of behavior from other protocols. It appears that Horowitz et al. (1988) were the first to combine data from multiple protocols within a single nonlinear regression to find best-fit values, which thereby had a better predictive capability. Soon thereafter this use of "combined data sets" became standard (cf. Choi and Vito 1990; Humphrey et al. 1990b) and thus was a major advance. Within the context of comparison of the utility of the microstructural (Horowitz et al. 1988) versus phenomenological (Yin et al. 1987) descriptors of myocardial behavior, however, the use of different approaches to regression masked potential real differences.

More importantly, however, following from their assumption that "the values of the material constants are physically meaningful," Horowitz et al. (1988) concluded that "during the process of estimation ... the contribution of the muscle fibers to the stress–strain behavior observed in the experiments is insignificant." This finding was supported further in their Discussion via references to the histology and stiffness of collagen. Yet, it was soon discovered that the data (from Demer and Yin 1983;

Yin et al. 1987) upon which this conclusion was based inadvertently came from tests wherein the myocardium was partially in contracture and thus significantly stiffer than that of truly passive myocardium (cf. Humphrey et al. 1990a). That is, although the microstructurally based model suggested that the myocardial fibers did not contribute to load bearing, these contracted fibers actually supported most of the measured loads in the experiments that were analyzed. This finding was an important reminder that although motivated by gross histological features, the lack of complete microstructural information (including precise distributions of orientations and waviness as well as interactions amongst the many constituents) demands that interpretations be made with caution. Soon thereafter, Nevo and Lanir (1989) relaxed some of the inherent complexity of the microstructural formulation in modeling myocardial behavior as justified by the prior findings. The use of "effective fiber stress–strain relations" subsequently became common (cf. Sacks 2000).

At this point, it is useful to remember that there are, in principal, three basic approaches to formulate a constitutive relation: theoretically, based on a detailed understanding of the microstructure; directly from special classes of experiments that reveal the forms of the response functions; and finally, via trial-and-error and experience that lead to postulated forms. Although theoretical formulations should be the best in principle, they are notoriously difficult in practice. Notwithstanding the success of the ideal gas law, attempts in elasticity have been much less successful. We recall, for example, the controversy surrounding early attempts by Navier to formulate a stress–strain relation for linearly elastic behaviors from a molecular theory that was not resolved fully until the phenomenological approach of G. Green (Timoshenko 1983). Similarly, we recall the improvement of the phenomenological Mooney–Rivlin relation over the theoretically derived neo-Hookean relation in describing the nonlinearly elastic isotropic behavior of rubber (Treolar 1975). It should not be surprising, therefore, that there remains no truly validated microstructural model for soft tissues that exhibit nonlinearly elastic anisotropic behaviors due to the complex responses of and interactions amongst the many proteins, glycoproteins, and glycosaminoglycans that constitute the tissue. Indeed, nearly all such formulations are based on the tacit assumption that the total stored energy is the sum of the energies stored by each of the constituents, which necessarily ignores possible interaction energies. Comparison of structurally motivated (Humphrey and Yin 1987) and experimentally based (Humphrey et al. 1990a) formulations of constitutive relations for passive myocardium reveal, for example, that such interactions are generally not negligible.

Limitations notwithstanding, microstructurally motivated constitutive relations continue to contribute significantly to our ability to describe better the complex stress–strain behaviors exhibited by many tissues as well as to interpret implications of the associated best-fit values of the material parameters. For example, the so-called two-fiber family constitutive model proposed by Holzapfel et al. (2000) and its extension to "four-fiber families" by Baek et al. (2006) have each proven particularly useful in arterial mechanics. Via the simple assumption that passive arterial behavior results primarily from an amorphous elastin-dominated contribution, a circumferentially oriented smooth muscle contribution, and four families of oriented collagen fibers, constitutive results by Ferruzzi et al. (2011a, b) and Eberth et al. (2011) support the histologically observed loss of structural integrity of elastin in aging, aneurysms, genetic disorders, and hypertension. Clearly, much more will continue to be learned using microstructurally motivated constitutive relations as originally envisioned by Lanir, provided that we respect the inherent limitations that result from an imprecise understanding of interactions amongst the many different constituents.

7.3 Growth and Remodeling

Y.C. Fung could not have selected a better title for his seminal 1981 book (now cited over 7500 times): *Biomechanics: Mechanical Properties of Living Tissues*. That is, the primary characteristics that separate biomechanics from classical mechanics are those of living, that is, functions such as metabolism, locomotion, responses to stimuli, reproduction, growth, and adaptation (cf. Dorland's Illustrated Medical Dictionary and The American Heritage Dictionary). There has long been interest in using mechanics to describe growth or adaptation (cf. Thompson 1999), but interest within modern biomechanics dates back mainly to the important works of Cowin and Hegedus (1976) and Skalak (1981). In particular, the so-called theory of "kinematic growth" of Skalak provided one way for applying nonlinear mechanics to study soft tissue growth and eventually remodeling. Whereas important reviews can be found elsewhere (e.g., Taber 1995), here we focus on the emerging use of microstructurally motivated approaches.

Towards that end, note that one usually assumes in the theory of kinematic growth that a materially uniform, stress-free body can be divided into fictitious parts, each of which can then grow independently. Because such growth need not be kinematically compatible, one then assumes that elastic deformations can be imposed to return the body to a contiguous configuration (which typically introduces residual stresses) that is then susceptible to typical applied loads. This approach (cf. Rodriguez et al. 1994) has garnered many adherents, primarily because of the mathematical tractability, yet it was not motivated by the biological mechanisms underlying growth and remodeling. That is, growth (change in mass) and remodeling (change in structure) result from cell-mediated production and removal of diverse constituents, often within evolving thermomechanical states. Moreover, such constituents may have different natural (stress-free) configurations, material properties, and rates of turnover. For this reason, Humphrey and Rajagopal (2002) introduced a different approach, a so-called constrained mixture theory. Briefly, in this approach, full mixture relations are used to satisfy mass balance (cf. Truesdell and Noll 1965) while a simple rule-of-mixtures relation is used to define the overall stress response (or stored energy function); the latter permits one to satisfy linear momentum balance in the usual ways, either in strong or weak form. Clearly then, the approach for modeling the stress response (e.g., stress–strain relation) is very similar to that advocated by Lanir in his microstructurally motivated approach (it is also similar in concept to the rule-of-mixtures approach advocated by A. Rachev in the 1970s). For example, one can assume that the total strain energy is the sum of the energy stored in the primary structurally significant constituents, including elastic fibers and collagen fibers. The primary difference, however, is that each cohort of constituents can consist not only of fibers having different orientations and undulations, but also different stress-free configurations, material properties, and rates of turnover.

Similar to aforementioned microstructurally motivated stress–strain relations, constrained mixture models of tissue growth and remodeling have proven useful in capturing salient features of diverse tissue responses (e.g., Baek et al. 2006; Valentin et al. 2009). Indeed, a very simple but interesting outcome of growth and modeling simulations is that the summation of stress responses over long periods can give rise to "smoother" (more realistic) nonlinear stress–strain behaviors (e.g., see results in Fig. 7 in Wilson et al. 2012) than sometimes seen in standard microstructural formulations (e.g., see the nearly bilinear responses in Fig. 4 in Lanir 1979). Nevertheless, one must remember not to over-interpret the results just because they are motivated by the microstructure, which is allowed to evolve. That is, current growth and remodeling formulations also suffer from many of the same limitations that remain in microstructural models, including the lack of information needed to model constituent-to-constituent interactions resulting from weak bonds, physical entanglements, or covalent cross-links. As with microstructurally motivated stress–strain relations, however, there is considerable promise that microstructurally and biologically motivated growth and remodeling models can continue to improve and to provide significantly increased insight into many important areas of biomechanics, including development, aging, disease progression, wound healing, regenerative medicine, surgical planning, and so forth.

7.4 Summary

It is natural for us to evaluate past achievements through our current understanding. Yet, to truly appreciate advances in any field of investigation, one must first understand the "status quo" during the associated period of study. For example, it is hard for us today to appreciate the genius of L. Euler and A. Cauchy in developing the concept of stress (which we now take for granted), yet it took over a century to move beyond the simple ideas of R. Hooke (as the force, so the extension) to the modern concept of stress. Similarly, to appreciate major advances during the early period of modern biomechanics, one must understand the state of the field in the mid-1960s and remember that biomechanics was truly in its infancy.

Professor Yoram Lanir contributed significantly to many different areas of soft tissue biomechanics, including study of skin, cartilage, arteries, and myocardium. Albeit not the first computer-controlled study, Lanir performed the first careful biaxial tests on planar soft tissues (skin) while working with Professor Fung at UCSD, and this general method of testing remains preferred to this day. He advanced the first rigorous microstructurally motivated constitutive relations for multiaxially loaded soft tissues, a general approach that has found broader applicability and continues to offer significant promise to this day. He proposed clever new approaches for nonlinear parameter estimation, an essential part of any constitutive formation. For these and many other advances, as discussed in other chapters within this volume, we must both thank and congratulate Professor Yoram Lanir.

In conclusion, I wish to note two things. First, it should be evident that I neither attempted to provide an exhaustive review of the early historical developments of modern soft tissue biomechanics, nor did I review all of Professor Lanir's contributions. Rather, I attempted to provide some historical context, largely based on personal recollections and observations (the interested reader is also referred to Humphrey (2003) for additional review). Given wonderful search engines such as PubMed, which by default list the most recent papers first, there is concern that many of the older papers will be forgotten even though there is much to learn from appreciating the evolution of ideas and methodologies. Second, much has been achieved and yet much more remains to be accomplished in continuum biomechanics. Indeed, given the continued rapid advance of complementary areas such as medical genetics, molecular and cell biology, medical imaging, and computational methods, there is great promise for biomechanics to contribute increasingly more and more to our understanding of physiology and pathophysiology and our advancing health care delivery. Hence as Y.C. Fung closed his Foreword in the inaugural issue of the international journal *Biomechanics and Modeling in Mechanobiology*, "let us enjoy the work."

Acknowledgments It is my privilege to contribute to this volume celebrating the 70th birthday of Professor Yoram Lanir and his many important contributions to the field of biomechanics. I have not only learned much from his many papers and lectures, I was also most fortunate to work alongside him in the mid-1980s, while I was a postdoctoral fellow, in the laboratory of Professor Frank Yin at Johns Hopkins University during Professor Lanir's mini-sabbatical. I remember well performing biaxial stretching tests together on excised parietal pericardium and learning many "tricks of the trade" that Professor Lanir shared as we designed and performed new experiments. These ideas ranged from different practical means of attaching thin specimens to the biaxial stretching device and keeping them submerged in the physiologic bathing solution to novel theoretical ideas for changing the basic boundary value problem in ways to yield increased information. Professor Lanir had, with Professor Y.C. (Bert) Fung, reported the first careful biaxial testing experiments on soft tissues (skin) in the 1970s while he was a postdoctoral fellow, hence it was a great opportunity for me to learn from his experiences. Moreover, Professor Lanir's seminal papers on microstructurally motivated constitutive relations for planar soft tissues had appeared relatively recently, in the late 1970s and early 1980s, and he had novel ideas of how to extract information on the constitutive parameters by designing experiments based on the constitutive theory, not simply the ease of measurement. Such lessons in both practical and theoretical approaches continue to serve me well.

I also remember working with Professor Lanir during that period to study the stability of rubber-like materials under biaxial dead loading. To do this, we designed and built in short order a simple biaxial system for loading thin sheets of rubber. In hindsight, our haste to design the setup actually lead to a somewhat dangerous experiment to apply relatively large dead loads to the sample, which we expected to be (and was) unstable or at least metastable. Nevertheless, the real lesson for me was in the design process and again I remember this time fondly. I thus am delighted to offer my warmest congratulations to Professor Lanir on the occasion of his 70th birthday. I thus thank Professors Sacks and Kassab for the invitation to contribute.

Finally, my continuing ability to study the fascinating field of biomechanics has been made possible by generous funding, primarily from the National Institutes of Health and National Science Foundation, for which I am grateful. I also acknowledge the wonderful contributions by so many excellent graduate students, postdoctoral fellows, and colleagues who continue to be a blessing to me.

References

Baek S, Rajagopal KR, Humphrey JD. A theoretical model of enlarging intracranial fusiform aneurysms. J Biomech Eng. 2006;128:142–9.
Cardamone L, Valentin A, Eberth JF, Humphrey JD. Origin of axial prestress and residual stress in arteries. Biomech Model Mechanobiol. 2009;8:431–46.
Choi HS, Vito RP. Two-dimensional stress-strain relationship for canine pericardium. J Biomech Eng. 1990;112:153–9.
Choung CJ, Fung YC. On residual stress in arteries. J Biomech Eng. 1986;108:189–92.
Cowin SC, Hegedus DH. Bone remodeling I. Theory of adaptive elasticity. J Elast. 1976;6:313–26.
Demer LL, Yin FCP. Passive biaxial mechanical properties of isolated canine myocardium. J Physiol. 1983;339:615–30.
Dingemans KP, Teeling P, Lagendijk JH, Becker AE. Extracellular matrix of the human aortic media: an ultrastructural histochemical study of the adult aortic media. Anat Rec. 2000;258:1–14.
Eberth JF, Cardamone L, Humphrey JD. Altered mechanical properties of carotid arteries in hypertension. J Biomech. 2011;44:2532–7.
Ferruzzi J, Vorp DA, Humphrey JD. On constitutive descriptors for the biaxial mechanical behavior of human abdominal aorta and aneurysms. J R Soc Interface. 2011a;8:435–50.
Ferruzzi J, Collins MJ, Yeh AT, Humphrey JD. Mechanical assessment of elastin integrity in fibrillin-1 deficient carotid arteries: implications for Marfan syndrome. Cardiovasc Res. 2011b;92:287–95.
Fung YC. Elasticity of soft tissues in simple elongation. Am J Physiol. 1967;28:1532–44.
Fung YC. Biomechanics: mechanical properties of living tissues. New York: Springer; 1981.
Green AE, Adkins JE. Large elastic deformations and non-linear continuum mechanics. Oxford: Oxford University Press; 1970.
Green AE, Zerna W. Theoretical elasticity. Oxford: Oxford University Press; 1960.
Holzapfel GA. Nonlinear solid mechanics: a continuum approach for engineering. New York: Wiley; 2000.
Holzapfel GA, Gasser TC, Ogden RW. A new constitutive framework for arterial wall mechanics and a comparative study of material models. J Elast. 2000;61:1–48.
Horowitz A, Lanir Y, Yin FCP, Perl M, Sheinman I, Strumpf RK. Structural three-dimensional constitutive law for the passive myocardium. J Biomech Eng. 1988;110:200–7.
Humphrey JD. Cardiovascular solid mechanics: cells, tissues, and organs. New York: Springer; 2002.
Humphrey JD. Continuum biomechanics of soft biological tissues. Proc R Soc Lond A. 2003;459:3–46.
Humphrey JD. Vascular adaptation and mechanical homeostasis at tissue, cellular, and sub-cellular levels. Cell Biochem Biophys. 2008;50:53–78.
Humphrey JD, Rajagopal KR. A constrained mixture model for growth and remodeling of soft tissues. Math Models Methods Appl Sci. 2002;12:407–30.
Humphrey JD, Yin FCP. On constitutive relations and finite deformations of passive cardiac tissue. I. A pseudostrain-energy function. J Biomech Eng. 1987;109:298–304.
Humphrey JD, Vawter DL, Vito RP. Pseudoelasticity of excised visceral pleura. J Biomech Eng. 1987;109:115–20.

Humphrey JD, Strumpf RK, Yin FCP. Determination of a constitutive relation for passive myocardium: I. A new functional form. J Biomech Eng. 1990a;112:333–9.
Humphrey JD, Strumpf RK, Yin FCP. Determination of a constitutive relation for passive myocardium: II. Parameter estimation. J Biomech Eng. 1990b;112:340–6.
Humphrey JD, Eberth JF, Dye WW, Gleason RL. Fundamental role of axial stress in compensatory adaptations by arteries. J Biomech. 2009;42:1–8.
Lanir Y. A structural theory for the homogeneous biaxial stress-strain relationships in flat collagenous tissues. J Biomech. 1979;12:423–36.
Lanir Y. Constitutive equations for fibrous connective tissues. J Biomech. 1983;16:1–12.
Lanir Y, Fung YC. Two-dimensional mechanical properties of rabbit skin. II. Experimental results. J Biomech. 1974;7:171–82.
Nevo E, Lanir Y. Structural finite deformation model of the left ventricle during diastole and systole. J Biomech Eng. 1989;111:342–9.
Oden JT. Finite elements of nonlinear continua. New York: McGraw-Hill; 1972.
Rodriguez EK, McCulloch AD, Hoger A. Stress-dependent finite growth in soft elastic tissues. J Biomech. 1994;27:455–67.
Sacks MS. Biaxial mechanical evaluation of planar biological materials. J Elast. 2000;61:199–246.
Skalak R. Growth as a finite displacement field. In: Carlson DE, Shield RT, editors. Proceed IUTAM symposium finite elasticity. The Hague: Martinus Nijhoff; 1981. p. 347–55.
Taber LA. Biomechanics of growth, remodeling, and morphogenesis. Appl Mech Rev. 1995;48:487–545.
Thompson DW. On growth and form. New York: Cambridge University Press; 1999.
Timoshenko SP. History of strength of materials. New York: Dover Publications; 1983.
Treolar LRG. Physics of rubber elasticity. Oxford: Clarendon Press; 1975.
Truesdell C, Noll W. The nonlinear field theories of mechanics. In: Flugge S, editor. Handbuch der Physik, vol. III/3. Berlin: Springer; 1965.
Truesdell C, Toupin RA. The classical field theories. In: Flugge S, editor. Handbuch der Physik, vol. III/1. Berlin: Springer; 1960.
Valentin A, Cardamone L, Baek S, Humphrey JD. Complementary vasoactivity and matrix remodeling in arterial adaptations to altered flow and pressure. J Roy Soc Interface. 2009;6:293–306.
Vawter DL, Fung YC, West JB. Elasticity of excised dog lung parenchyma. J Appl Physiol. 1978;45:261–9.
Wilson JS, Baek S, Humphrey JD. Importance of initial aortic properties on the evolving regional anisotropy, stiffness, and wall thickness of human abdominal aortic aneurysms. J R Soc Interface. 2012;9:2047–58.
Yin FCP, Strumpf RK, Chew PH, Zeger SL. Quantification of the mechanical properties of noncontracting canine myocardium under simultaneous biaxial loading. J Biomech. 1987;20:577–89.

Chapter 8
Relationship Between Structure and Mechanics for Membranous Tissues

Jessica W.Y. Jor, Thiranja P. Babarenda Gamage, Poul M.F. Nielsen, Martyn P. Nash, and Peter J. Hunter

Abstract Professor Yoram Lanir has pioneered the development of structurally based constitutive relations to describe the stress–strain response of soft biological tissues. This approach relates the mechanical response of the tissue to the intrinsic micro-structural properties of its constituents, such as collagen. This article summarises some of the work by the Auckland Bioengineering Institute contributing towards the goal of understanding the structure–function relationship of soft membranous tissue. Key aspects of our work are to (1) develop constitutive relations based on quantitative information of tissue structure; and (2) use rich sets of experimental data to aid in accurate and reliable constitutive parameter identification. We first outline several common techniques to quantify tissue structure, such as collagen fibre orientations. A detailed description of an extended-volume imaging system, developed in our laboratory, is then provided along with a few application examples. The gathered imaging data is incorporated into structural constitutive models by means of fitting to mathematical distributions. Based upon the observations made from some imaging studies, a conceptual fibre distribution model is proposed for modelling the collagen network in skin. We then introduce a selection of constitutive models, which have been developed to characterise the mechanical behaviour of soft connective tissues (skin in particular), with particular emphasis on structurally based models. Finite element models, used with appropriate constitutive relations, provide a means of interpreting experimental results. Some of our recent efforts in developing instrumentation to measure the two-dimensional and three-dimensional response of soft tissues are described. This includes a biaxial tensile rig, which is capable of deforming membranes in up to 16 directions, and a force-sensitive micro-robot. We highlight some of the challenges often associated with constitutive parameter identification using commonly used

J.W.Y. Jor (✉) • T.P.B. Gamage • P.J. Hunter
Auckland Bioengineering Institute, The University of Auckland, Auckland, New Zealand
e-mail: j.jor@auckland.ac.nz; psam012@aucklanduni.ac.nz; p.hunter@auckland.ac.nz

P.M.F. Nielsen • M.P. Nash
Auckland Bioengineering Institute and Department of Engineering Science, The University of Auckland, Auckland, New Zealand
e-mail: p.nielsen@auckland.ac.nz; martyn.nash@auckland.ac.nz

G.S. Kassab, M.S. Sacks (eds.), *Structure-Based Mechanics of Tissues and Organs*, DOI 10.1007/978-1-4899-7630-7_8

model based fitting approaches. These issues were examined and illustrated in depth by performing controlled studies on silicon gel phantoms, which allowed us to focus our attention solely on the identification problem. Lastly, future directions of applying structurally based models to understanding the biomechanics of soft tissues are discussed.

8.1 Introduction

More than three decades ago, Professor Yoram Lanir proposed the use of structural constitutive relations to describe the stress–strain response of soft connective tissues, thereby highlighting the important relationship between tissue composition and behaviour. In essence, structural models aim to take into consideration the intrinsic micro-structural properties of fibres (including fibre undulation and orientation) when deriving constitutive equations. In the skin, networks of collagen (60–80 % of dry weight) and elastin (1–4 % of dry weight) fibres are embedded in a fluid-like ground substance (70–90 % of skin volume), comprised of water and proteoglycan molecules (Wilkes et al. 1973; Silver et al. 1992). Collagen fibres are non-uniformly undulated in their natural state and form a three-dimensional interwoven network throughout the depth of the dermis (Finlay 1969). Elastin fibres are straight in their natural state and are less stiff compared with collagen fibres. They are capable of reversibly withstanding stretches to more than 100 % and hence provide the skin with its ability to recoil to its original shape after being stretched (Carton et al. 1962). Experiments in which elastin and ground substance were digested suggested that elastin's contribution to the mechanical response of skin is only significant at low strains, while the overall contribution from the ground substance is small (Daly 1969; Harkness and Harkness 1959). It is widely accepted that collagen is the predominant stress-bearing component in skin at high strains (Brown 1973). It has been observed in tendons that undulated collagen fibres became gradually recruited upon stretch, giving rise to the tissue nonlinearity commonly observed during the toe region of the stress–strain curve (Abrahams 1967). This is due to an increasing number of fibres being gradually straightened, and then stretched, contributing to the overall macroscopic tissue stiffness. In addition, tissue anisotropy is predominantly a result of the non-uniform orientation of the fibres.

Because there is a close relationship between tissue structure and mechanical behaviour, structural constitutive relations are advantageous by providing a means to directly relate model parameters to the underlying tissue microstructure. Prior to the use of structural models, the most common approach to modelling the elastic response of soft biological tissue was to use phenomenological models. In phenomenological models, mathematical expressions are chosen to best fit experimental data or reflect material behaviour, such as the widely used power

(Kenedi et al. 1965) and exponential forms (Tong and Fung 1976). A major disadvantage is that the parameters obtained for phenomenological models cannot be directly related to specific biological constituents.

Although the skin is used as a representative tissue throughout this paper to demonstrate the fundamental concepts of structural models, the observed structure–function relationship is applicable across other soft connective tissues. To date, structural approaches to modelling soft tissues have been employed to investigate the mechanical properties of various organs and tissues, including skin (Lanir 1983; Lokshin and Lanir 2009a,b; Jor et al. 2011a), tendon (Hurschler et al. 1997), pericardium (Sacks 2003), arterial walls (Holzapfel et al. 2002), passive myocardium (Horowitz et al. 1988; Holzapfel and Ogden 2009), and heart valves (Billiar and Sacks 2000; Freed et al. 2005; Gasser et al. 2006). This paper summarises some of our published works on characterising the structural and mechanical properties of soft tissues by bringing together imaging, experimental and computational modelling techniques into a single framework.

In Sect. 8.2, we first outline several common techniques to quantify tissue structure, such as collagen fibre orientations. A detailed description of an extended-volume imaging system, developed in our laboratory, is then provided along with a few application examples. The gathered imaging data is incorporated into structural constitutive models by means of fitting to mathematical distributions. Based upon the observations made from some imaging studies, a conceptual fibre distribution model is proposed for modelling the collagen network in skin. We then introduce, in Sect. 8.3, a selection of constitutive models, which have been developed to characterise the mechanical behaviour of soft connective tissues (skin in particular), with particular emphasis on structurally based models. Finite element (FE) models, used with appropriate constitutive relations, provide a means of interpreting experimental results. In Sect. 8.4, we describe some of our recent efforts in developing instrumentation to measure the two-dimensional (2D) and three-dimensional (3D) response of soft tissues. This includes a biaxial tensile rig, which is capable of deforming membranes in up to 16 directions, and a force-sensitive micro-robot. Several experimental studies on skin tissues will be summarised in this section. In Sect. 8.5, we highlight some of the challenges often associated with constitutive parameter identification using model based fitting approaches. These issues were examined and illustrated in depth by performing controlled studies on silicon gel phantoms, which allowed us to focus our attention solely on the identification problem. Finally, Sect. 8.6 is devoted to discussing future directions of applying structurally based models to soft tissues.

8.2 Quantification of Tissue Structure

One of the major challenges in modelling soft connective tissues is the difficulty in reliable and meaningful determination of constitutive parameters. To address this issue, quantification of structural parameters is therefore useful in minimising the effects of structural variability in constitutive modelling. Recognising the importance of tissue microstructure and its relationship to the mechanical behaviour has led to an increasing number of studies looking to quantify the structural properties of soft connective tissues. The structure of collagen networks has been studied extensively in both human and animal issues (Craik and McNeil 1964; Marcarian and Calhoun 1966; Finlay 1969; Meyer et al. 1982). Of particular interest is the orientation of collage fibres, due to its important role in providing the mechanical stability of the tissue and its contribution to the overall tissue behaviour.

Collagen fibres are rendered visible by traditional histological stains, such as Hematoxylin-Eosin, Pontamine Sky Blue-Eosin, Elastica van Gieson and picrosirius red (PSR). Microscopic techniques have been developed to visualise collagen fibres, such as polarised light microscopy (Junquiera et al. 1982), fluorescence microscopy (Dolber and Spach 1993; MacKenna et al. 1996) and confocal laser scanning microscopy (CLSM) (Young et al. 1998). However, these conventional microscopic techniques are often limited to relatively small regions in the tissue, making it difficult to quantify fibre structure on a larger scale. Early microscopic studies have suggested that collagen fibres are arranged in an organised, interwoven structure in the skin. In order to quantify fibre orientations, several studies represent the degree of fibre anisotropy using a relative measure, often referred to as the "collagen alignment index". The techniques used to quantify the degree of collagen anisotropy included fitting ovals to binary images of skin samples viewed under polarised light and taking the major axes of fitted ovals as the alignment index parameter (Melis et al. 2002), performing Fourier analysis on cross-sectional images of skin samples, and estimating the alignment index as the width-height ratio of the power spectrum (van Zuijlen et al. 2003; Noorlander et al. 2002). Typically, an alignment index value of zero indicates an isotropic distribution of fibres while a value of one indicates perfectly aligned fibres. Despite being able to describe the extent of fibre orientation anisotropy in a quantitative manner, the above studies do not provide fibre orientations with respect to the tissue material axes (i.e. an absolute orientation measure). Therefore, such data cannot be used for fitting to statistical distribution functions to model fibre orientation.

Sacks et al. (1997) characterised collagen orientations in samples of pericardium and intestinal submucosa using a small angle light scattering (SALS) device. In brief, a Helium–Neon (HeNe) laser was passed through the thin connective tissues. HeNe was chosen because its wavelength ($\lambda = 632.8\,\mathrm{nm}$) is within an order of magnitude of the diameter of the collagen and elastin fibres. As the laser is passed through the tissue, a pattern is formed as a result of light being scattered by the fibrous structures. The angular distribution of collagen fibres $R(\theta)$ can then be directly correlated to the angular distribution of scattered light $I(\theta)$. The experimentally derived angular distribution of collagen fibres was directly incorporated into structural constitutive models to predict the biaxial mechanical response of bovine pericardium. Although the SALS technique is suitable for determining collagen fibre orientations in thin connective tissues, such as the pericardium and heart valves, it is not appropriate for skin tissues. Meijer (1997) found that the charge-coupled device (CCD) camera employed in the SALS technique was not sensitive enough to detect scattered light from stained collagen fibres in rat skin tissues, which were approximately 380 μm thick. Due to its projective nature, the SALS technique is also limited to the detection of collagen orientations in the in-plane direction only, i.e. it cannot provide structural information in the transverse direction nor can it be used to build a 3D dataset of images.

The development of structurally based computational models of biological tissue relies on the ability to quantify and visualise 3D tissue structure across a large range of resolutions and scales. For imaging of large tissue volumes, it is often challenging to address the problems associated with image registration, section distortion, and realignment. To this end, the motivation to acquire high resolution images in a high throughput manner has led to the development of a novel automated CLSM imaging system by our group (Sands et al. 2005).

8.2.1 Extended-Volume CLSM

The automated extended-volume imaging system consists of a high precision XYZ translation stage, to which the specimen is mounted and imaged using a modified CLSM (Leica Microsystems AG, Germany) (Fig. 8.1). The use of the translation stage enables the acquisition of images over larger regions than previously possible using conventional microscope systems. It avoids the need to remove the specimen from the microscope stage for manual sectioning, thus re-positioning and alignment of the specimen is not required at each step. The dimensions that can be imaged in the Z direction using conventional CLSM is often limited by absorption and scattering of light in the specimen. To overcome this limitation, extended imaging into the depth of the tissue is enabled by removing the upper surface of the specimen using a specialised ultramill (Leica Microsystems). The translation stage, image acquisition and tissue removal are controlled simultaneously by specialised hardware and software to enable automated acquisition of a grid of image stacks to form a 3D image dataset.

The CLSM system has been used to visualise the collagen fibre structure in various soft tissues, including myocardial tissue (Sands et al. 2005; Pope et al. 2008), rat cortex tissues (Sands et al. 2005), engineered heart valves tissues (Eckert et al. 2011) and cardiac trabeculae carneae (Sands et al. 2011). To give a representative example of the capabilities of the CLSM system, the image acquisition of ventricular tissue will be presented. Rat hearts were excised, mounted on a Langedorff apparatus, and perfused with oxygenated Tyrode's solution. In a relaxed state, the heart was arrested and perfused with Bouin's fixative solution. To render collagen fibres visible, the heart was then perfused with collagen-specific stain PSR (Sweat et al. 1964). A transmural 3 mm section from the LV free wall was cut and embedded in Agar 100 resin for imaging. A 16x/0.5 NA Plan Fluotar oil-immersion lens was used to obtain 512 pixel × 512 pixel slices with lateral resolution of 1.22 μm. To image PSR-stained samples, a rhodamine filter set (excitation wavelength at 568 nm; high pass detection filter at 590 nm) was used, with 8× averaging for the ventricular

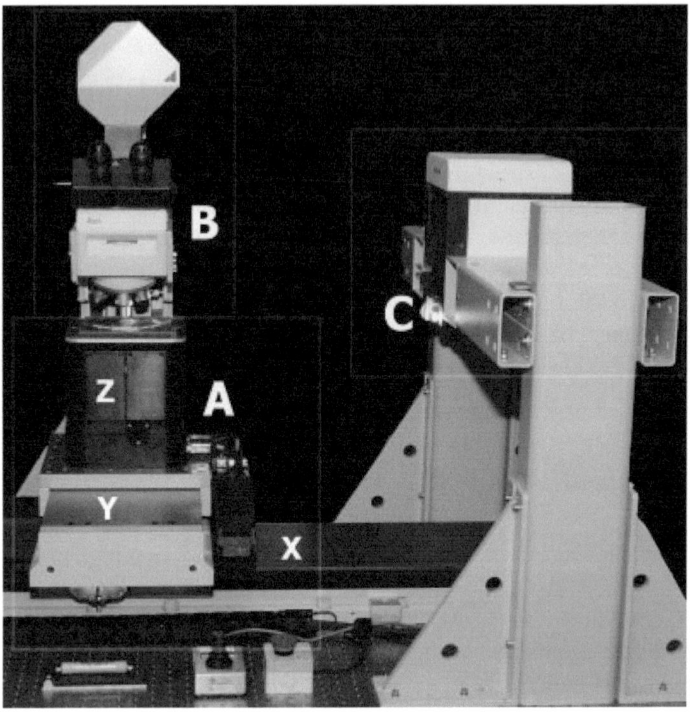

Fig. 8.1 The extended-volume confocal microscope imaging system consists of a high-precision three-axis translation stage (labelled as **A** with individual **X**,**Y** and **Z** translation stages), confocal microscope unit (**B**), and the ultramill (**C**). Reproduced with permission from Sands et al. (2005)

samples. A region of interest (5 × 3 mm) was specified to form a regular X-Y grid of overlapping slices. The depth to which the specimen may be imaged before the upper surface is milled was determined to be 35 μm, allowing 28 planes to be acquired at 1.22 μm spacing in the Z direction. At each milling step, 30 μm of tissue was removed. The volume rendering of the 3D reconstruction of the imaged specimen, covering 4.25 × 1.12 × 0.88 mm, is shown in Fig. 8.2. Brightly stained collagen fibres were shown to vary significantly in the transmural direction, and are densely arranged in sheets lining the epicardial and endocardial surfaces.

The imaging system has also been used to visualise collagen fibres in skin specimens (Jor et al. 2011b). Unlike cardiac tissues, stain penetration for the skin samples was achieved primarily via tissue diffusion (not via perfusion). However, because of the dense structure of the dermal tissue, stain penetration by means of diffusion was found to be limited in the direction of tissue depth. To overcome this problem, 30–60 μm thick transverse cryosections were obtained from the abdominal region of young pigs at four orientations (0°, 30°, 60°, 90°) with respect to the torso mid-line. Since the sample thickness was relatively thin, the milling procedure was omitted in this study. Skin sections were fixed in Bouin's solution, prior to staining in PSR. Images were acquired using the protocols described for myocardium tissues,

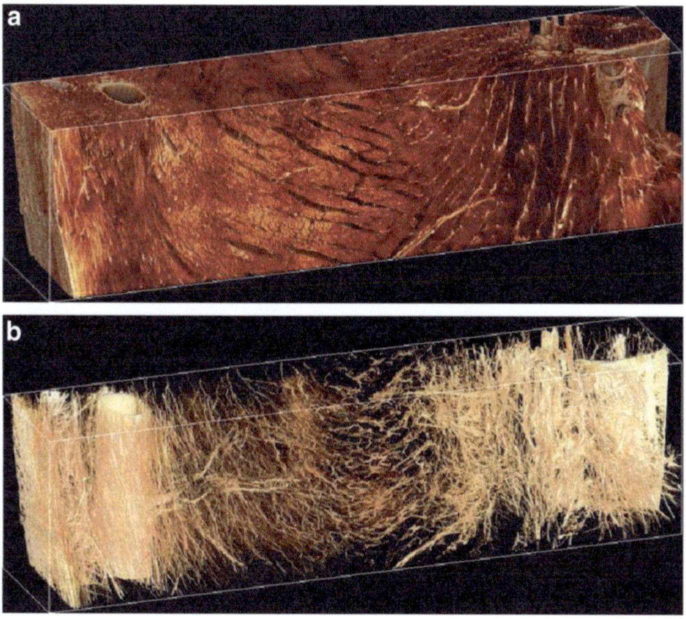

Fig. 8.2 (**a**) The volume rendering reconstruction of a transmural stack of myocardial tissue, and (**b**) with the collagen network revealed by adjusting for image transparency based upon intensity values. Reproduced with permission from Sands et al. (2005)

but using a 20× water immersion lens with 4× line averaging. The lateral resolution for the acquired images was 0.98 μm per pixel. Overlapping 512 pixel × 512 pixel images were acquired and montaged together to form a 3 × 2 mm slice mosaic at each z-depth. The imaging process was repeated at 1 μm steps throughout the entire tissue thickness, forming a stack of slice mosaics. For each tissue section, a maximum intensity projection was obtained from all slice mosaics through the entire z-depth.

Figure 8.3 shows a typical image of a PSR-stained sagittal section of porcine skin. CLSM images of pig skin in this study showed in the reticular dermis region thick collagen bundles crossing obliquely in two main directions, between epidermis and hypodermis. Smaller diameter fibres are interwoven between the main network. The oblique arrangement of collagen fibres is thought to be responsible for resisting in-plane shear deformation while allowing changes in thickness of the dermis as the skin is stretched or compressed. In contrast, thinner collagen bundles were found in a more parallel arrangement in the relatively thin papillary dermis layer as seen in Fig. 8.4. The collagen fibres were observed to form a dense and compact 3D meshwork in the dermis region. Such observations are consistent with early qualitative studies performed on pig skin using light microscopy and SEM (Mowafy and Cassens 1975; Meyer et al. 1982).

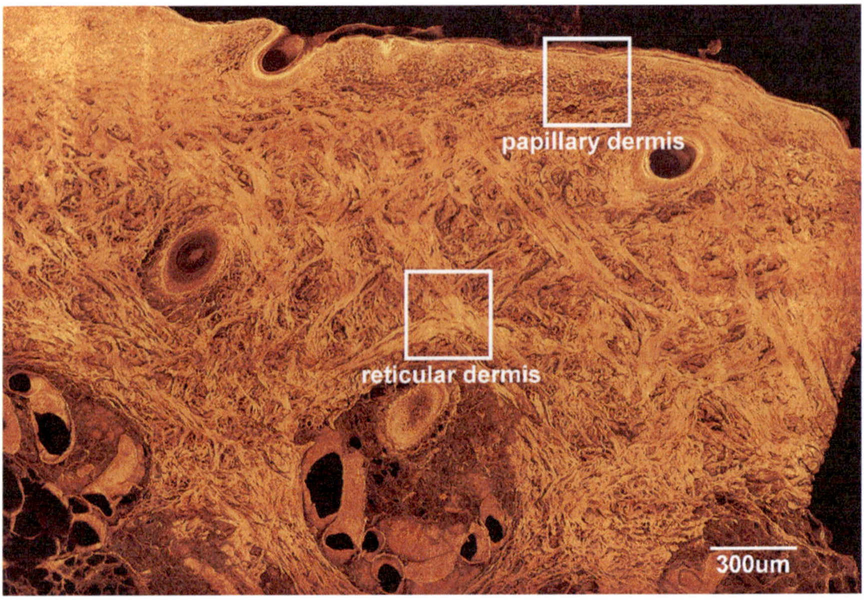

Fig. 8.3 Maximum intensity projection of a 60 μm sagittal section of picrosirius stained porcine skin, revealing the cross lattice arrangement of collagen fibres. Reproduced with permission from Jor et al. (2011b)

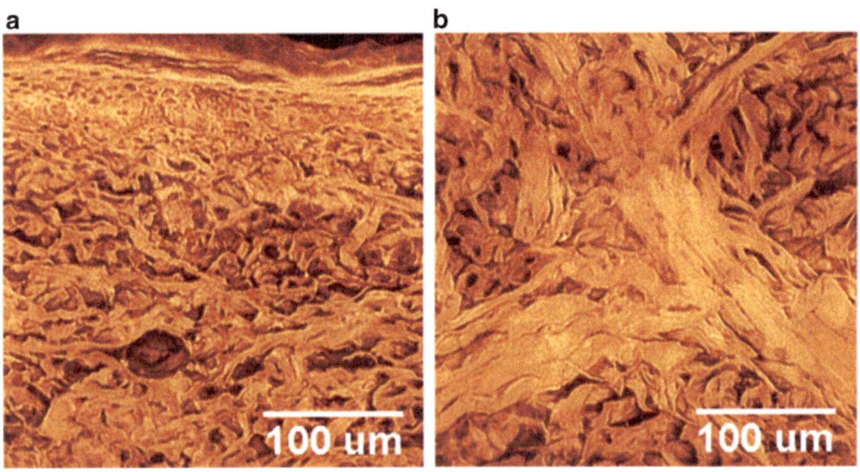

Fig. 8.4 300 × 300 μm cropped sections from Fig. 8.3: (**a**) thinner collagen fibre bundles are seen in the upper papillary dermis region; (**b**) much thicker collagen fibre bundles, arranged in a lattice structure, are found in the lower reticular dermis region. Reproduced with permission from Jor et al. (2011b)

8.2.2 *Structure Tensor Approach*

To analyse the structure of collagen of the acquired images of skin specimens, a structure-tensor approach was used to determine fibre orientations (Jahne 2004). The components of a structure tensor are the weighted sum of the first-order spatial derivatives of the image. The information within a local neighbourhood about the central point $\mathbf{x}$ is weighted by the window function $w(\mathbf{x} - \mathbf{x}')$

$$J_{pq}(\mathbf{x}) = \int_{-\infty}^{\infty} w(\mathbf{x} - \mathbf{x}') \left(\frac{\partial g(\mathbf{x}')}{\partial x_p} \frac{\partial g(\mathbf{x}')}{\partial x_q} \right) d\mathbf{x}' \tag{8.1}$$

where $g(\mathbf{x})$ represents the grey values and $\partial g(\mathbf{x}')/\partial x_p$ is the partial derivation along the p-axis direction.

The components of the structure tensor can be computed efficiently as a combination of linear convolution and nonlinear point operations. This involves convolving the image independently with partial derivative operators (D_p, D_q) associated with each coordinate (p, q), multiplying both images on a pixel basis prior to applying a smoothing operator (B) to the resultant image.

$$J_{pq} = B(D_p \times D_q) \tag{8.2}$$

After computing the structure tensor, an eigenvalue analysis is performed to extract local orientation vectors. For a 2D image, the form of the structure tensor is

$$\mathbf{J} = \begin{bmatrix} J_{xx} & J_{xy} \\ J_{xy} & J_{yy} \end{bmatrix} \tag{8.3}$$

The eigenvalue analysis involves only a single rotation from the image coordinate system to the principal axes coordinate system [Eq. (8.4)]. As a result, the two axes are aligned with the directions of the eigenvectors of the maximum and minimum eigenvalues, respectively.

$$\begin{bmatrix} J_x & 0 \\ 0 & J_y \end{bmatrix} = \begin{bmatrix} \cos\phi & -\sin\phi \\ \sin\phi & \cos\phi \end{bmatrix} \begin{bmatrix} J_{xx} & J_{xy} \\ J_{xy} & J_{yy} \end{bmatrix} \begin{bmatrix} \cos\phi & \sin\phi \\ -\sin\phi & \cos\phi \end{bmatrix} \tag{8.4}$$

By substitution of trigonometric identities, the orientation vector O in vectorial form can be written as

$$O = \begin{bmatrix} J_{yy} - J_{xx} \\ 2J_{xy} \end{bmatrix} \tag{8.5}$$

where the orientation angle is given by the phase of this vector (ϕ). As shown in Eq. (8.5), the components of the orientation vector are readily obtained from the components of the structure tensor and involve only one subtraction and one multiplication.

Fibre orientations in sagittal sections of porcine skin were defined with respect to the direction normal to the epidermis (ranging from $-\frac{\pi}{2}$ to $+\frac{\pi}{2}$), with clockwise angles from the Y-axis to X-axis in the imaging plane defined as positive. The structure-tensor algorithm was used to determine fibre orientations of a regular grid of 16 pixel × 16 pixel for each image projection, as shown in Fig. 8.5a. Areas containing non-collagenous features such as blood vessels, hair follicles and fat tissue were masked and excluded from the statistical analysis as shown in Fig. 8.5b.

8.2.3 Fitting Fibre Orientations

In order to incorporate micro-structural information into computational models, mathematical density distributions are employed in structural constitutive equations to represent the distributions of fibre undulation and orientations. One of the most common mathematical distributions used to represent circular data is the von Mises (VM) distribution, VM(μ, κ). The probability density distribution function of VM(μ, κ) is given by

$$R(\theta) = \frac{e^{\kappa \cos(\theta-\mu)}}{2\pi I_0(\kappa)} \quad 0 \leq \theta < 2\pi \quad 0 \leq \kappa < \infty \tag{8.6}$$

where

$$I_0(\kappa) = \frac{1}{2\pi}\int_0^{2\pi} \exp^{\kappa \cos(\phi-\mu)} d\phi \tag{8.7}$$

is the modified Bessel function of order zero. The parameter μ is the circular mean and κ is a measure of spread that is related to the inverse of the standard deviation in the conventional Gaussian distribution. As κ tends to zero, the distribution tends to a uniform distribution. As κ tends to large values, the distribution becomes concentrated about the mean orientation, μ.

Due to the bimodal nature of fibre orientations from porcine skin, a mixture of two symmetrical VM distributions was selected for fitting the acquired imaging data. The method of moments was used in this study to estimate parameters of the distribution (Spurr and Koutbeiy 1991). Figure 8.5c shows a representative histogram and fitted VM distributions to fibre data obtained from skin specimens aligned parallel to the torso mid-line. It was demonstrated that the two parameters of the distribution, the orientation mean and spread, may be directly determined using CLSM image analysis. An important advantage of this approach is that model parameters can be estimated directly from observable micro-structural features.

One important feature observed in the images was the presence of a distinct lattice pattern in the transverse samples. Based on fibre data obtained from a single sectioning orientation, there was insufficient information to deduce the collagen structure arrangement; thus, images from other sectioning orientations were required. Analysis of transverse sections excised at $0°, 30°, 60°$ and $90°$ relative to the torso mid-line revealed that the distributions of fibre orientations

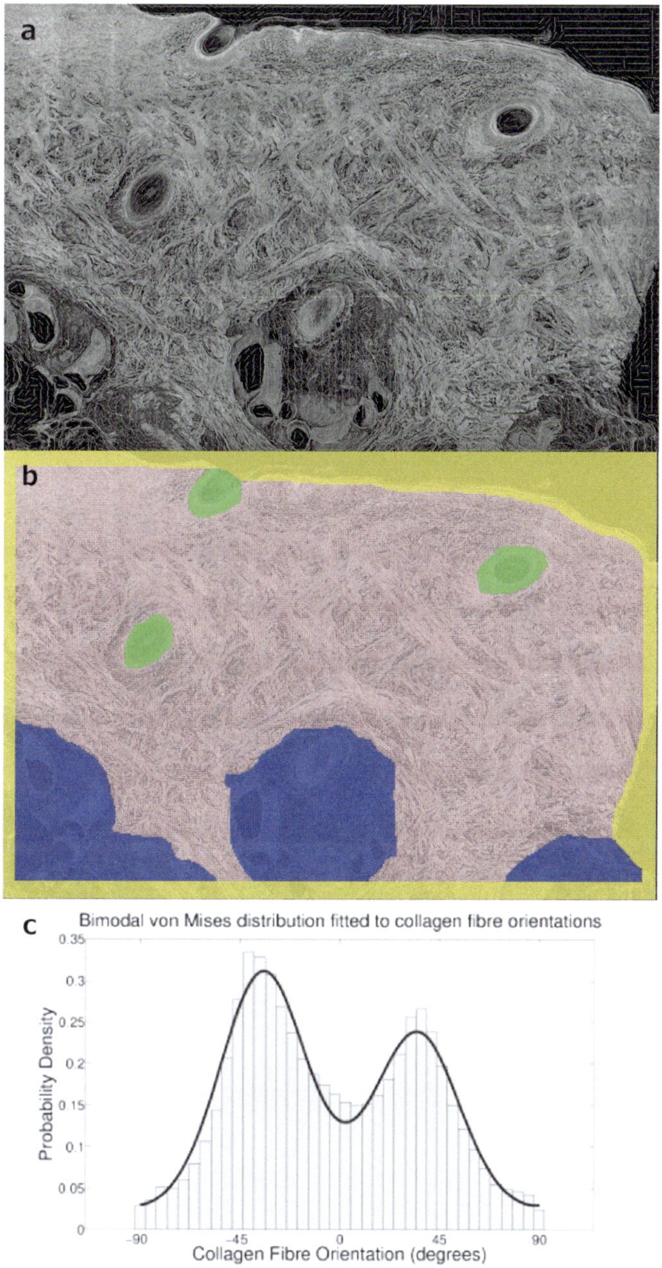

Fig. 8.5 Collagen orientation analysis: (**a**) A regular grid of computed fibre angles, using the structure tensor approach, shown as line segments; (**b**) Areas of the image containing non-collagenous structures, such as blood vessels, fat cells and glands, were masked out; (**c**) A mixture of two von Mises distributions fitted to collagen fibre orientations presented in the form of a histogram

were bi-modal for all sectioning orientations. To determine whether there was any difference in the collagen structure across the four different sectioning orientations, an analysis of variance (ANOVA) was performed on the fitted orientation means. It was shown that all p-values were greater than 0.05, indicating no significant differences in the fitted means across the four section orientation groups. The observations led to the proposal of a 3D conceptual model of collagen distribution for porcine skin, which will be described in the following section.

8.2.4 Conceptual Fibre Distribution Model

Based upon the observation of a lattice structure consistently observed in CLSM images of different sectioning orientations, it appears that porcine skin contains a rotationally symmetric 3D lattice structure about the normal direction to the epidermis. Thus, a 3D conceptual model to represent the observed collagen fibre distribution in porcine skin was proposed. Firstly, a material coordinate system (Fig. 8.6) was defined as follows: (1) N is normal to the epidermis; (2) L is parallel to the Langer's lines and (3) O is orthogonal to N and L.

Each fibre is defined in 3D space by its spherical coordinates $-\phi$ and θ (Fig. 8.6a). θ is the in-plane angle between L- and O-axes (with anti-clockwise defined as positive), whereas ϕ describes the fibre angle from the N-axis within the transverse section. It is assumed that the total probability of finding a fibre at

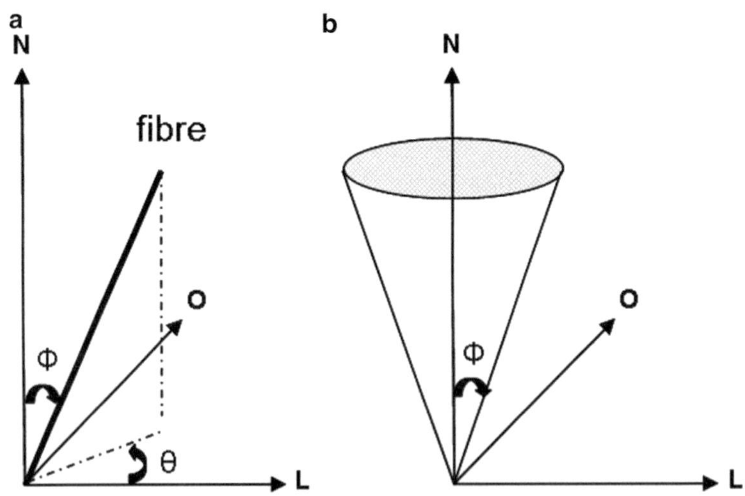

Fig. 8.6 3D conceptual model: (**a**) definition of a fibre with respect to the material coordinate system; (**b**) a circular distribution around the skin normal direction (N) is assumed based upon the similar collagen lattice network observed for the different sectioning orientations. L: Langer's line direction; O: direction orthogonal to N and L. Reproduced with permission from Jor et al. (2011b)

an orientation (ϕ,θ) can be determined by selecting appropriate density distribution functions $R_1(\phi)$ and $R_2(\theta)$ that independently represent ϕ and θ.

$$R(\phi, \theta) = R_1(\phi) \times R_2(\theta) \tag{8.8}$$

The in-plane angle θ describes structural information in the plane parallel to the epidermis. Although images acquired in this study do not provide any direct information, it is simplest to describe θ with a unimodal density distribution function, in which the fibre density in the direction of Langer's line has been observed to be greater than in any other direction. Cox (1941) showed that the preferred collagen orientation tends to be aligned along the Langer's lines. A unimodal π-periodic VM distribution, centred around the direction of Langer's lines, was chosen to describe the variation in θ.

For the idealised situation in which all the fibres are assumed to be perfectly aligned along two predominant orientations, the fibre distribution in the transverse planes may be represented by a rotationally symmetric distribution about the normal direction (Fig. 8.6b). This mathematical representation is consistent with the observations described in CLSM images of porcine skin collagen, i.e. that a similar lattice structure was apparent regardless of the sectioning orientation. Rather than being perfectly aligned, collagen fibres in biological tissues are likely to be dispersed about their mean directions within the lattice, thus requiring additional parameter(s) to describe the variation(s) about the predominant directions.

Based upon the bimodal nature of imaging data, a suitable density distribution function to describe ϕ is the mixture of two π-periodic VM distributions, defined as

$$R_1(\phi) = \frac{1}{2}\left(\frac{\exp\left[\kappa_{\phi+}\cos(2(\phi - \mu_{\phi+}))\right]}{\pi I_0(\kappa_{\phi+})} + \frac{\exp\left[\kappa_{\phi-}\cos(2(\phi + \mu_{\phi-}))\right]}{\pi I_0(\kappa_{\phi-})}\right) \tag{8.9}$$

where μ_ϕ and κ_ϕ are the mean and spread parameters, respectively, of the orientation distribution. As shown previously, both parameters can be directly estimated from analysis of CLSM images.

The total probability density function [Eq. (8.8)] is thus defined as the product of a unimodal VM distribution and a bimodal VM distribution to describe $R_1(\theta)$ and $R_2(\phi)$, respectively. By integrating the probability density function over the range $-\frac{\pi}{2} \leq \theta \leq \frac{\pi}{2}, -\frac{\pi}{2} \leq \phi \leq \frac{\pi}{2}$, all possible fibre orientations are covered by the unit hemisphere. Therefore, Eq. (8.8) must also satisfy the following normalisation constraint

$$\int_{\phi=-\frac{\pi}{2}}^{\frac{\pi}{2}} \int_{\theta=-\frac{\pi}{2}}^{\frac{\pi}{2}} R(\phi, \theta) d\theta d\phi = 1 \tag{8.10}$$

Consider a fibre in the reference configuration that is described by a vector $\mathbf{u}$. Since the stretch is identical in both the $+\mathbf{u}$ and $-\mathbf{u}$ directions, $R(\phi, \theta)$ must remain invariant under the symmetrical transformation ($\phi \rightarrow \pi - \phi, \theta \rightarrow \pi + \theta$) for all values of ϕ and θ such that

$$R(\phi, \theta) = R(\pi - \phi, \pi + \theta) \tag{8.11}$$

In examining some commonly used statistical distributions, it was found that not all proposed distributions satisfied this physical restriction. For example, the commonly used unimodal Gaussian distribution does not satisfy Eq. (8.11).

Lokshin and Lanir (2009a) proposed a bimodal distribution in the form of trigonometric functions, given by

$$R(\theta) = c_{11}(\cos^4(\theta - \mu_\theta)) + c_{12}(\sin^4(\theta - \mu_\theta)) + \frac{c_2}{\pi} \tag{8.12}$$

where c_{11}, c_{12}, c_2 and μ_θ are constants. It can be seen that Eq. (8.12) satisfies the physical restriction for the planar angle θ. However, if the same distribution was used to represent ϕ, i.e.

$$R(\phi) = c_{11}(\cos^4(\phi - \mu_\phi)) + c_{12}(\sin^4(\phi - \mu_\phi)) + \frac{c_2}{\pi} \tag{8.13}$$

Equation (8.11) would not be satisfied.

In the proposed conceptual model, it can be shown that by choosing unimodal and bimodal VM distributions to describe $R_1(\theta)$ and $R_2(\phi)$, respectively, Eq. (8.8) satisfies the physical symmetry restriction. This is due to a factor of two in the θ distribution and the appearance of both $(\phi - \mu_\phi)$ and $(\phi + \mu_\phi)$ in the $R_2(\phi)$ distribution.

In this section, we have described techniques to mathematically represent tissue structure by combining confocal imaging and statistical analyses. Structural data in this form act as important inputs to structurally based constitutive models, which will be introduced in the following section.

8.3 Constitutive Modelling of Connective Tissues

The model-based approach, which uses finite element (FE) models to simulate the mechanical testing of tissues, provides an effective means to analyse rich sets of experimental data. This iterative numerical-experimental technique allows the identification of constitutive parameters. A selection of the constitutive models used in characterising the response of soft membranes will be presented here.

8.3.1 Phenomenological Models

Phenomenological models are mathematical functions, such as the power or exponential forms, that are chosen to best fit experimental data or reflect material behaviour. Based upon the classical biaxial testing of rabbit skin, Tong and Fung developed the widely used Fung constitutive equations for soft tissues (Fung 1965, 1967). In order to model the J-shaped stress–strain curves observed from the experiment data (i.e. a low stiffness region followed by an abrupt transition to a high stiffness region), the following generalised form was proposed:

$$\begin{aligned} W = {} & \frac{1}{2}(\alpha_1 E_{11}^2 + \alpha_2 E_{22}^2 + \alpha_3 E_{12}^2 + 2\alpha_4 E_{11} E_{22}) \\ & + \frac{1}{2} c \exp(a_1 E_{11}^2 + a_2 E_{22}^2 + a_3 E_{12}^2 + 2a_4 E_{11} E_{22} \\ & + \gamma_1 E_{11}^3 + \gamma_2 E_{22}^3 + \gamma_4 E_{11}^2 E_{22} + \gamma_5 E_{11} E_{22}^2) \end{aligned} \tag{8.14}$$

where α_i, a_i, γ_i and c are material constants and E_{ij} is the Green strain tensor, resulting in a total of 13 constitutive parameters. It was found by Tong and co-workers that Eq. (8.14) can be simplified by omitting the α_i and γ_i terms without compromising the goodness of fit to a majority of stress–strain curves. In biaxial testing, this equation can be further simplified by assuming the shear strain E_{12} is zero, thus can be written as

$$W = \frac{c}{2} \exp(a_1 E_{11}^2 + a_2 E_{22}^2 + 2a_4 E_{11} E_{22}) \tag{8.15}$$

While phenomenological models often provide good fits to experimental data, a major limitation of phenomenological models is that the constitutive parameters do not bear any physical meaning. Driven by the need to relate tissue behaviour to structure, structural constitutive models have since been developed and employed in studies to investigate soft tissue mechanics.

8.3.2 Structural Models

Structural models aim to formulate the constitutive relations based upon the underlying tissue histology. Since constitutive parameters relate to specific biological features, structural models are capable of providing insights into the important relationship between tissue structure and behaviour. Lanir (1979, 1983) was one of the first researchers to consider the geometric arrangement of fibre networks in soft connective tissues. Following his pioneering work, structurally based constitutive models have since been developed over the years for a variety of tissues such as arterial walls (Holzapfel et al. 2002), pericardium (Sacks 2003), tendon/ligament

(Hurschler et al. 1997), passive myocardium (Horowitz et al. 1988; Holzapfel and Ogden 2009) and heart valves (Billiar and Sacks 2000).

Recently, Lokshin and Lanir (2009a) presented an extensive approach to model the in vitro uniaxial response of membranous tissues. The effects of tissue nonlinearity, anisotropy, viscoelasticity, orientation and recruitment of two fibre networks (collagen and elastin), and preconditioning adaptations have been considered in a single constitutive model. The model, with 31 constitutive parameters, provided good fits to biaxial data of rabbit skin. An important outcome from this study was the investigation and incorporation of preconditioning effects in the experimental setting and constitutive models, respectively. Since different test protocols recruit different sets of fibres, the loading history (and hence preconditioning adaptation) will alter the time-dependent response of the tissue. It has been suggested that the underlying mechanisms are different for the two types of fibres: strain- and time-dependent increases in the reference length of collagen fibres, and strain-dependent strain softening (i.e. Mullins effect) of elastin fibres. The study also confirms the notion that the ground matrix and elastin contribute predominantly in the low strain regions, while collagen fibres are important for higher strains.

Representing Fibre Orientations with Continuous Distributions

Although a comprehensive constitutive description of soft tissue mechanics is desirable, the bottleneck is often the lack of ability to accurately identify the unknown constitutive parameters. The identification of constitutive parameters is a challenging task, even in the case of simple, isotropic and heterogeneous materials (Babarenda Gamage et al. 2011). Therefore, to characterise the biaxial response of porcine skin (outlined in the following section), we considered a simplified version of the structural model proposed by Lanir. In this micro-structural model, a network of collagen fibres with varying orientations is embedded in a tissue block consisting of ground matrix. The tissue block was assumed to be very small compared to the whole tissue, such that the material can be assumed to be homogeneous within each block and that the deformation field varies linearly over the block. The following assumptions were made:

- each fibre is undulated and only resists load when completely straightened.
- fibres can only resist tensile loading and buckle under compressive loads, i.e. no load is required to fold or unfold fibres.
- each fibre undergoes a uniaxial strain, which is representative of the macroscopic tissue strain along the fibre direction.
- the fibres are assumed to be linearly elastic when stretched, and their mechanical properties are governed by a one-dimensional (1D) fibre load-stretch relation.
- viscoelastic effects are ignored.

The second Piola–Kirchhoff stress tensor, $\mathbf{S}$, sums the contributions from the ground matrix $\mathbf{S}_m$ and collagen fibres $\mathbf{S}_f$, respectively, by taking into consideration the fibre volume fraction V_f, such that

$$\mathbf{S} = (1 - V_f) \cdot \mathbf{S}_m + V_f \cdot \mathbf{S}_f \tag{8.16}$$

The ground matrix is assumed to be an isotropic neo-Hookean material; thus, the strain energy function of the ground matrix (W_m) is given by

$$W_m = K_m(I_1 - 3) \tag{8.17}$$

where K_m is the stiffness of the ground matrix and I_1 is the first principal strain invariant of the Cauchy–Green deformation tensor. For a 2D membrane, the second Piola–Kirchhoff stress tensor for the ground matrix can be expressed as

$$S_{ij}^m = \frac{\partial W_m}{\partial E_{ij}} = \frac{\partial W_m}{\partial I_1}\frac{\partial I_1}{\partial E_{ij}} = \begin{cases} 2K_m & i = j \\ 0 & i \neq j \end{cases} \tag{8.18}$$

where $\mathbf{E}$ is the Green–Lagrange strain tensor. A uniaxial fibre strain energy function (w_f) is defined as a function of the fibre stretch ratio λ

$$w_f = \frac{K_c}{2}(\lambda - 1)^2 \tag{8.19}$$

where K_c is the collagen fibre stiffness and λ is given by Eq. (8.20).

$$\begin{aligned} \lambda &= \sqrt{2\gamma_f' + 1} \\ &= \sqrt{2(E_{11}\cos^2\theta + E_{22}\sin^2\theta + 2E_{12}\cos\theta\sin\theta) + 1} \end{aligned} \tag{8.20}$$

where γ_f' represents the strain along the fibre in the deformed configuration and θ is the in-plane fibre angle.

The load per unit undeformed cross-sectional area in the fibre is

$$f(\lambda) = \frac{\partial w_f(\lambda)}{\partial \lambda} = \begin{cases} K_c(\lambda - 1) & \lambda > 1 \\ 0 & \lambda \leq 1 \end{cases} \tag{8.21}$$

The total strain energy of all fibres in an undeformed volume unit, W_f, is given by

$$W_f = \int_\theta R(\boldsymbol{\theta}) \cdot w_f(\lambda)\mathbf{d}\theta \tag{8.22}$$

where R is the probability density function for the fibre orientation. The summation can be replaced by integrals if there is a large enough number of fibres in each direction.

For a 2D membrane, the second Piola–Kirchhoff stress tensor for the fibres can be expressed as

$$S_{ij}^f = \int_\theta R(\theta)\frac{1}{\lambda}\frac{\partial \gamma_f'}{\partial E_{ij}} \cdot \int_{x=1}^{\lambda} D(x) \cdot f\left(\frac{\lambda}{x}\right)\mathbf{d}x\mathbf{d}\theta \tag{8.23}$$

where x is the stretch required to straighten an undulated fibre and $D(x)$ is the probability density function for fibre undulation. Continuous distribution functions are used to represent the orientation of collagen fibres, with the most commonly used being the Gaussian (Lanir 1983) and von Mises distributions. A unimodal π-periodic von Mises distribution was chosen to describe the in-plane fibre orientation θ. For fibre undulation, commonly used continuous distribution functions include the Gaussian, beta, Lorentz function and log-logistic distribution function. For present purposes, a Gaussian distribution was used to describe fibre undulation:

$$D(x) = \frac{1}{\sqrt{2\pi}\sigma_x}\exp\left\{-\frac{(x-\mu_x)^2}{2\sigma_x^2}\right\} \tag{8.24}$$

where μ_x and σ_x are the mean and standard deviation of the undulation distribution, respectively.

It is important to note that the above structural model has many simplifications, such as the omission of the elastic fibre response and time effects. However, we will demonstrate in Sect. 8.5 that, even with such a simplified model, there remains significant challenges with material parameter identification.

Discretised Fibre Model

Flynn et al. (2011a) proposed a discretised fibre structural model for soft tissues comprising of six fibre bundles. In this model, each fibre bundle was aligned parallel to lines that pass through opposing vertices of a regular icosahedron. Each fibre bundle was assumed to consist of a single elastin fibre in parallel with a distribution of undulated collagen fibres. The total strain energy function of the six fibre bundle ensemble was given by

$$W = f(J) + \sum_{i=1}^{6} w_i[W_e(\lambda_i) + W_c(\lambda_i)] \tag{8.25}$$

where $f(J)$ denotes the response to compression, and $W_e(\lambda_i)$ and $W_c(\lambda_i)$ is the strain energy of the elastin fibre and undulated bundle of collagen fibres, respectively. The total strain energy of equally weighted elastin fibres was represented by a modified neo-Hookean material.

To represent the strain energy of undulated collagen fibres, a novel approach involving analytical expressions was adopted. This approach does not require numerical integration of continuous distribution functions and is thus more computationally efficient. Two simple distribution functions (a step and triangular function) for collagen undulation that yield analytical stress–strain expressions were investigated. It was found that the fibre stresses closely matched those calculated using the normal distribution function, indicating that a two-parameter step function is sufficient to model collagen fibre undulation. The proposed discrete fibre model was used to fit experimental data from various biaxial tensile tests. Although providing relatively good fits, this model is limited by the fact that it does not guarantee a polyconvex formulation and it does not yield isotropic response when all of the weights w_i are equal.

In Sects. 8.2 and 8.3, we have established characterisation of tissue structure and introduced a selection of constitutive relations. The measurement of mechanical response relies on performing mechanical tests on the tissues of interest, which will be discussed in the next section.

8.4 Mechanical Experiments

Over the years, many research groups have attempted to characterise and quantify the mechanical functions of soft tissues by performing mechanical experiments. Two types of deformation are typically applied—loads parallel to the surface of the tissue (such as uniaxial, biaxial tension and torsion), or loads applied perpendicular to the surface (such as suction and indentation). In this section, we will present some of our recent instrumentation developments in measuring the mechanical response of skin tissues.

8.4.1 Biaxial Testing

Biaxial testing has been employed to determine the material properties of skin (Lanir and Fung 1974a,b), pericardium (Choi and Vito 1990; Billiar and Sacks 1997; Sacks 2003), lung parenchyma (Vawter et al. 1978) and thin sections of myocardium (Yin et al. 1987; Smaill and Hunter 1991; Humphrey et al. 1990). Prior to this, mechanical tensile tests were often restricted to uniaxial tests (Daly 1966; Manschot and Brakkee 1986) due to challenges in prescribing the appropriate boundary conditions. Biaxial testing presents many technical difficulties, including the need to control two boundary conditions, minimising gripping effects, applying forces uniformly along the specimen edge, and determining the optimal specimen size to promote homogeneity within the specimen while ensuring that the region of interest is located sufficiently far from the exterior edges (Sacks 2003). Lanir and Fung (1974a) were the first researchers to investigate the mechanical properties of planar soft tissue using biaxial testing. A rabbit skin specimen was mounted on the biaxial device in a trampoline-like manner at 68 attachment points using thin threads. The stress and strain states across the central region of interest were considered to be uniform. Tissue strain within this region was obtained by measuring the distance between pairs of lines marked on the specimen using video dimensional analysers. From the experimental data, it was observed that the skin tissues exhibited J-shaped directionally dependent stress–strain curves, highlighting nonlinearity and anisotropic mechanical behaviour. An important observation is that the stress–strain curves were independent of strain rate for both the loading and unloading cycles, within the range tested.

A major challenge in characterising and comparing the mechanical response of soft connective tissues, such as skin and pericardium, is the significant inter-specimen structural variability that exists, particularly relating to the anisotropy of the tissues. To minimise the underlying structural variability across specimens, Sacks (2003) developed biaxial testing protocols that incorporated the use of the SALS technique described in Sect. 8.2. Using the SALS imaging technique, tissue specimens with a high degree of structural uniformity were selected for biaxial testing. This resulted in consistent mechanical response across specimens and low variability in the identified constitutive parameters. Rectangular samples of bovine pericardial specimens and porcine aortic valve leaflets were attached to the biaxial rig in the trampoline-like manner with sutures. To track the strain field, four graphite markers (∅0.3 mm) were attached to the central region of the specimen using cyanoacrylate adhesive. The positions of these markers were continuously tracked during deformation using a CCD camera. It was demonstrated that the use of SALS with biaxial testing provided additional insight into the relationship between fibre alignment and mechanical anisotropic behaviour.

In order to identify anisotropic, nonlinear, and inhomogeneous properties of soft membranes, it is necessary to subject the sample to a rich set of deformations, as will be described in Sect. 8.5.2. In addition, the use of a model-based approach provides a framework within which to interpret the acquired experimental data. To this end, our group has developed a novel biaxial tensile device that is capable of imposing rich sets of complex deformation fields.

Biaxial Rig Hardware

The biaxial tensile rig (Fig. 8.7a) consists of up to 16 displacement actuators (Physik Instrument DC-Mike, Germany) arranged in a circular array to stretch the tissue specimen (Nielsen et al. 2002). Since the movement of each motor axis is controlled independently, the circular arrangement permits a rich set of strains to be applied to the membrane. Each actuator, controlled by a Hewlett Packard HCTL100 motor controller, has a displacement range of 50 mm and position accuracy of 0.2 μm. The minimum specimen size is restricted by the radius at which transducers can converge without colliding. For the 16 axis set-up, the minimum diameter is 50 mm. A smaller specimen size of 20 mm can be accommodated by using eight transducers.

Mounted at the tips of each displacement actuator is a 2D force transducer that measures the forces applied to the membrane during testing. Each force transducer is made up of two pairs of strain gauges, arranged as two half-Wheatstone bridges, bonded to the four necked surfaces of a cantilever. They measure the strain applied to the surface of the cantilever when force is applied at the attachment pin. The membrane is attached to the transducers via a sharp needle at the end of a 30 mm long pin. This provides a quick and easy tethering method, compared with

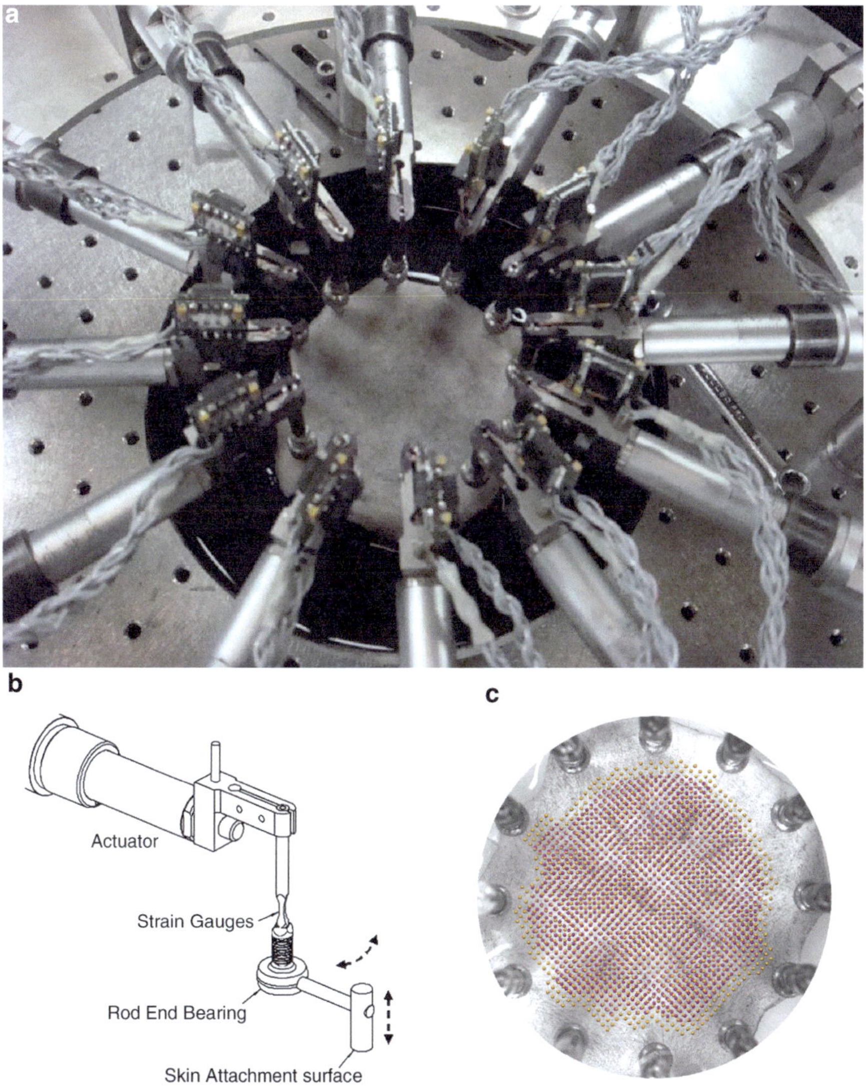

Fig. 8.7 Biaxial rig set-up: (**a**) Tensile tests were performed on in vitro samples of porcine skin bathed in phosphate buffered solution; (**b**) Alternative method of attaching in vivo samples to the biaxial rig using a rod end bearing approach; (**c**) Tracking of material points from the reference configuration to the deformed configuration. The CCD image of the deformed sample is shown in the background. Reproduced with permissions from Jor et al. (2011a) and Kvistedal and Nielsen (2009)

thin threads commonly used in biaxial devices. The deformed geometry and the associated deformed speckle pattern applied to the membrane surface are captured using a CCD camera (Atmel Corporation, France, Camelia 4 M). These images are used to compute the deformation field of the membrane using nonlinear cross-correlation techniques, which are described in the following section. The system uses a 12-bit monochrome camera with a 2048 pixel × 2048 pixel resolution. It is equipped with a bellows and an El-Nikkor (Tokyo, Japan) 80 mm enlarger lens, providing a field-of-view of 20–200 mm. The actuator control, force measurements, and image acquisition are controlled using an integrated software system developed with LabView (National Instruments).

Cross-Correlation Technique for Tracking Deformation

The measurement of strain fields poses a significant challenge in biaxial testing of soft biological tissues. Ideally, one would measure strain optically to avoid any mechanical interference. Malcolm et al. (2002) proposed a novel technique to measure the displacement of material points over the specimen using a phase-based cross-correlation technique, and subsequently compute the strain field via finite element modelling. The mathematics of this procedure have been previously described in detail by Malcolm (2000) and thus will only be briefly summarised here.

At each successive loading step, forces at the membrane boundary are recorded and an image of the speckled surface of the deformed specimen is acquired. In order to compute the displacement of a material point initially positioned at (x, y) in the undeformed configuration and later moved to $(x + \delta x, y + \delta y)$ using a phase-based Fourier transform cross-correlation (FTCC) method, signals with a wide spatial frequency bandwidth are required. In the biaxial testing set-up, this is achieved by applying a random speckle pattern on the surface of the membrane being tested. The FTCC method computes the 2D displacement vector of a region on the specimen when it is stretched from one state to another.

The total displacement is calculated by adding displacement obtained from the phase data to the displacement obtained by the amplitude data. This technique is capable of resolving displacements to a sub-pixel resolution (as small as 0.008 pixel). It is important to note that the cross-correlation gives the average displacement of the region within the subimage, and only works under the following conditions: (1) the visual pattern has a wide frequency bandwidth; (2) the displacement of the two subimages is relatively small compared with the size of the subimages, and; (3) there is little distortion or rotation associated with the deformation (i.e. deformation is homogeneous) within the subimage, although this can be accounted for using the underlying FE model.

To determine the strain field, a geometric FE geometric model of the undeformed membrane geometry is created and deformed to the measured displacements using least-squares fitting to obtain the geometric model of the deformed state. The deformation of the geometric model is used to compute the associated strain field.

Applications to Skin Tissues

The biaxial testing device was used to characterise the biaxial mechanical response of porcine skin (Jor et al. 2011a). For this experiment, a 12-transducer configuration was used to deform in vitro square samples (70×70 mm) of porcine skin excised along the abdominal mid-line of the animal. To determine the pretension, each sample was allowed to relax in phosphate buffered solution (PBS) over a period of 3–4 h (defined as the *unloaded* state). The changes in length along each of the 12 directions were estimated, by comparing an image of the sample in its unloaded and preloaded (i.e. in vivo or intact) states. In general, the skin samples retracted 40 % in the direction parallel to the torso mid-line (also the Langer's lines) and 20 % in the normal direction. Pretension forces were defined as the forces required to stretch a specimen from its unloaded to preloaded configuration. From the preloaded configuration, equi-axial deformations were imposed by stretching circular skin specimens uniformly along 12 directions using the biaxial rig, with the resultant loads at the membrane attachment points being measured. Displacement fields at each deformation step were tracked using the image 2D cross-correlation technique described above (Fig. 8.7c).

The biaxial rig has also been used to characterise the mechanical response of human forearm skin in vivo. Five healthy male subjects, between 29 years and 35 years of age, participated in this study (Kvistedal and Nielsen 2009). To avoid the use of fixed attachment points, which would lead to undesired vertical and torsional stresses in the sample, a miniature rod end bearing was used to attach the skin surface to the force transducers (Fig. 8.7b). Equi-axial deformations were imposed in increments of 0.2 mm along and across the Langer's lines of the forearm. Cross-correlation techniques were used to track the displacements between each successive loading state. It was observed that the stress–strain curves were different between individuals, further demonstrating variations in the mechanical properties of skin that is dependent on age, body locations and gender.

8.4.2 Three-Dimensional Testing

There are important limitations with biaxial testing of soft tissues. Researchers have commented on the inability to infer any in-plane shear information due to difficulties in imposing and controlling shear stresses in planar biaxial tests (Humphrey et al. 1990; Sacks 2000). Simple shear devices have been developed by Arbogast et al. (1997) and Dokos et al. (2000) for testing myocardial tissues, but these studies did not combine shear with biaxial stretching. Extension and inflation tests on tubular specimens, combined with torsion, have similar effects to planar biaxial tests with simple shear. The first experiment of this kind was performed by Humphrey et al. on cylindrical blood vessels (Humphrey et al. 1993). However, in order to completely characterise the 3D anisotropic mechanical response of soft tissues, an additional independent, out-of-plane deformation is required (Holzapfel and Ogden 2009).

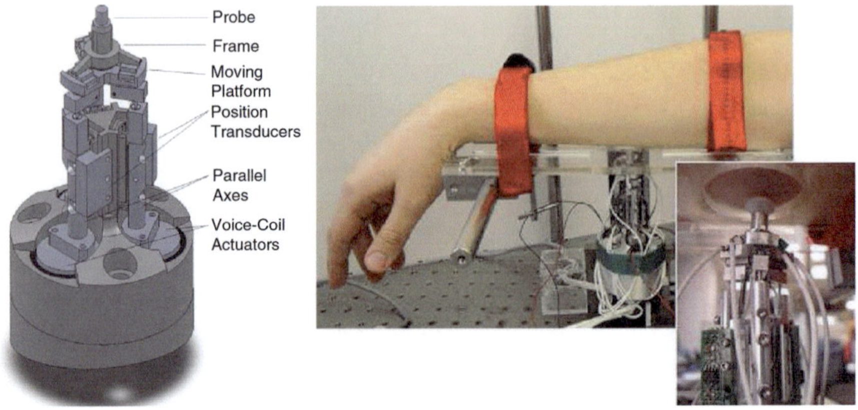

Fig. 8.8 Three-dimensional measurements using a force-sensitive micro-robot. Reproduced with permission from Flynn et al. (2011a)

To date, there is no biaxial device that will independently control and measure three components of stress or strain. In an attempt to impose a rich set of three-dimensional deformations on soft tissues, such as the skin, we have developed a novel force-sensitive micro-robot (Fig. 8.8).

Technical details of the micro-robot are provided in Flynn et al. (2011a) and will only be summarised here. The micro-robot consists of a 4-mm-cylindrical probe attached to three parallel axes, each driven by a voice-coil actuator to allow controlled movement of the probe tip within a working volume. Three linear position transducers are used to measure the displacement of each actuator. The displacement resolution is 50 μm with a repeatability of 60 μm. The 3D force vector at the probe tip can be computed from the force measured by three force transducers placed between each actuator and the probe, with a resolution of 6 mN with a repeatability of 8 mN. In a study to investigate the in vivo mechanical properties of human skin, 21 volunteers between the age of 21 and 52 years were recruited (Flynn et al. 2011a). To position the arm relative to the micro-robot, a plate with a 40 mm diameter hole was used to support and define the testing region.

A rich set of three-dimensional deformations was imposed, with a cycle frequency of 0.1 Hz, at three locations of the arm: the anterior right forearm; the anterior left forearm; and posterior right upper arm. Liquid cyanoacrylate adhesive was used to attach the micro-robot probe to the skin. The skin was preconditioned by performing three triangular-wave cycles. Firstly, in-plane deformations were applied, using 0° increments, in directions 0°–180° relative to the longitudinal axis of the arm. A series of out-of-plane deformations was then applied within planes oriented 0°, 45° and 90° relative to the longitudinal axis and normal to the surface of the arm. Normal indentations were also investigated. Results demonstrated that at all testing locations for all subjects, the skin exhibited nonlinear and anisotropic mechanical characteristics with significant hysteresis. Figure 8.9

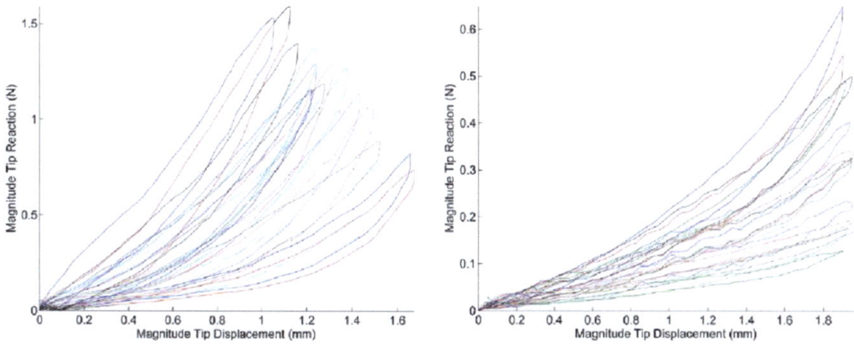

Fig. 8.9 Force-displacement curves of the anterior right forearm for all subjects tested subjected to in-plane (*left*) and out-of-plane (*right*) loading at 0° to the longitudinal axis of the arm. Reproduced with permission from Flynn et al. (2011a)

shows the force-displacement curves of the anterior right forearm for all subjects subjected to in-plane and out-of-plane loading at 0° to the longitudinal axis of the arm and normal deformations. Anisotropic effects cannot be quantified by experimental devices that apply torsion, suction or normal indentation deformations. This highlights a major advantage of the force-sensitive micro-robot, which is capable of characterising anisotropic effects by imposing both in-plane and out-of-plane deformations. There was also a wide range of variations reported across different subjects, suggesting the importance of acquiring experimental data from different body locations of a large population of individuals.

In order to interpret the experimental data, FE models coupled with appropriate constitutive relations are employed to identify constitutive parameters. In the following section, we will discuss the techniques and challenges associated with material parameter identification of soft tissues.

8.5 Constitutive Parameter Identification

Parameters of the constitutive relations chosen to describe skin are often determined by using model-based approaches where model predicted responses are fitted to experimental measurements obtained from devices such as those described in Sect. 8.4. This is usually performed using nonlinear optimisation techniques. Identifying these constitutive parameters for individual specific tissues can be valuable for determining normal vs pathological changes in a number of clinical applications. These include diagnosis of dermatological diseases, such as systemic scleroderma or other collagen related diseases, while also providing insight into improved clinical treatment and management of wounds and burns, where creation of skin scaffolds mimicking the behaviour of biological skin is required. However, determining the optimal constitutive parameters is generally a non-trivial exercise since there are many factors which can potentially confound this process. These are issues related to the large number of parameters used to describe the stress–strain responses, many of which appear to tradeoff (or be correlated) when insufficient experimental data is available for identification. It is usually difficult to ascertain this information requirement prior to performing the experiments, which can lead to difficulties in identifying these parameters.

In previous studies, both phenomenological and structurally based constitutive relations have been extensively used for fitting to specific in vitro samples of human or animal tissue. In doing so, the descriptive power of the models has also been characterised by analysing the accuracy of the fit. One benefit of in vitro testing is the ability to finely control the experimental conditions, allowing the tissues to be subjected to specific modes of deformation such as uniaxial, biaxial and shear strains. This can help focus the identification to specific attributes of the tissue, such as the biaxial response without considering shear. Such simplifications have helped the identification process by reducing the number of parameters required to be estimated. However, testing tissues in this manner becomes difficult in vivo and via non-invasive means, as often required in a clinical setting. This is due to the fact that it is difficult, if not impossible, to independently apply specific modes of deformation in vivo rendering the simplifications used with in vitro experiments unusable. While still a challenging task, phenomenologically based relations typically contain relatively few parameters, which reduces the complexity of the problem. However, the determination of constitutive parameters for structurally based relations remains a difficult task due to the complex experimental protocols required for determining their parameters (e.g. determination of fibre orientation, spread and undulation). Further modelling considerations also need to be taken into account when modelling the skin in vivo, such as the state of pre-tension and the coupling of skin to the underlying tissues. As a consequence, very few in vivo applications currently make use of structural-based constitutive relations despite their parameters directly relating to tissue microstructure and being biologically meaningful.

In this section, we highlight challenges in identifying constitutive parameters and strategies our group has employed to approach these issues. We first discuss results that highlight the difficulties associated with identifying constitutive parameters of in vitro skin using the structurally-based model described in Sect. 8.3. Next we discuss a case study performed on synthetic materials which aimed to investigate the identification issues in closer detail and look at methods for improving identifiability. We then describe more recent studies which aim to characterise in vivo skin using non-invasive techniques. Issues regarding the pre-tension and time-dependent mechanical response of skin are also discussed.

8.5.1 In Vitro Biaxial Testing of Skin

This section summarises a study that was performed to identify the mechanical properties of porcine skin using the structurally based constitutive relation described in Sect. 8.2.1. A quasi-static modelling framework was developed to simulate the experiments described in Sect. 8.4.1, whereby measured forces were applied to FE models that were created to represent the geometry and structure of the tissue samples. Parameters of the structurally based constitutive relation, which were assumed to be homogeneous across the entire specimen, were identified using nonlinear least squares optimisation. The objective function to be minimised was the sum of squared errors between experimentally tracked material points on the skin surface (described in Sect. 4.1.3), and their corresponding points embedded in the model. The optimisation was subjected to box constraints describing physiologically appropriate upper and lower bounds for each constitutive parameter which are described in the next section.

Identification Strategy

Due to the large number of parameters in the constitutive relation described in Sect. 8.3, there appeared to be 'trade-off' effects for the optimisations, thus the number of estimated parameters was limited to subsets. In the identification procedure, two different estimation strategies were investigated, as listed in Table 8.1. The parameters to be identified for *Strategy A* were mean orientation (μ_θ), collagen stiffness (K_c) and matrix stiffness (K_m); for *Strategy B*, parameters were the mean orientation (μ_θ), mean undulation (μ_x) and matrix stiffness (K_m). The fixed parameters for both strategies were fibre orientation spread (κ), fibre undulation SD (σ_x) and volume fraction (V) and were set to values previously reported in the literature. Fibre spread κ was fixed to 10 (equivalent to a circular

Table 8.1 List of parameters for a 2D a structural constitutive relation with the corresponding lower/upper bounds and values from the literature ([a]Jor et al. 2011b; [b]Manschot 1985; [c]Viidik 1980)

Type	Parameter	Description	L.B.	U.B.	Strategy A	Strategy B
Structural	Fibre orientation θ	Mean μ_θ	0	π	Free	Free
		Spread κ_θ	–	–	Fixed at 10[a]	Fixed at 10[a]
	Fibre undulation x	Mean μ_x	1.0	1.8	Fixed at 1.2[b]	Free
		SD σ_x	–	–	Fixed at 0.2[b]	Fixed at 0.2[b]
	Volume fraction V		–	–	Fixed at 0.3[b]	Fixed at 0.3[b]
Non-structural	Fibre stiffness	K_c	0.3 MPa	500 MPa	Free	Fixed at 100 MPa[c]
	Ground matrix stiffness	K_m	1 kPa	100 kPa	Free	Free

SD of 21°) based upon quantitative analysis of collagen orientation in porcine skin (Jor et al. 2011b). In order to investigate the identifiability of the parameters, both estimation strategies were performed starting from ten different sets of initial estimates (random numbers generated between the lower and upper bounds for each parameter). Determinability criteria were also used to quantitatively assess parameter identifiability by analysing the Hessian matrix (i.e. the matrix of second partial derivatives of the objective function with respect to the model parameters) within the neighbourhood of the optimal set of parameters and are described further in Babarenda Gamage et al. (2011), Lanir et al. (1996), and Nathanson and Saidel (1985). These criteria were also used in subsequent studies in order to compare proposed identification strategies and improvements.

Results and Identifiability Issues

The optimal solutions for the samples, using *Strategy A* resulted in a matrix stiffness which ranged between 5 and 32 kPa. The mean orientations were similar across the three tissue samples, ranging between 167° and 178°, indicating that the mean fibre orientations deviated only 2°–13° from the torso midline. Collagen stiffness varied greatly across the three samples, ranging from 48 to 366 MPa. The reported range of collagen stiffness for tendons is also large, ranging from 100 MPa to 1.2 GPa (van Brocklin and Ellis 1965; Wright and Rennels 1964). The mean displacement errors of the optimal models using *Strategy A* ranged between 0.64 and 0.90 mm. For *Strategy B*, the optimal values for K_m and μ_θ were similar to those identified using *Strategy A*. The mean undulation at the optimum ranged from 1.04 to 1.34 across all samples. The mean displacement errors of the models optimised using *Strategy B* ranged between 0.63 and 1.03 mm.

Of the ten different sets of initial estimates used for *Strategy A*, only initial estimates with a mean orientation μ_θ relatively close to the optimal mean orientation converged to the same optimal solution for all tissue samples, regardless of large variations in the initial estimates of the other variable parameters (K_m and K_c). The initial estimates needed to be close to the optimal minimum to be found; thus, the identification problem was highly sensitive to μ_θ. Analysis of the identifiability of the parameters from the Hessian matrix obtained at the optimal set of parameters also reflected a high sensitivity to mean fibre orientation, confirming these observations. Similar observations were made for *Strategy B*, whereby the initial estimate for the mean orientation needed to be relatively close to the optimal mean orientation in order to achieve successful convergence. Although quantitative data on fibre orientation have been extracted from confocal images in a previous study as described in Sect. 8.2, this analysis was only performed on sagittal, not parallel, sections. Furthermore, although the samples used in imaging and mechanical experiments were excised from similar body locations of the pig, it would be better to use the same specimen for both imaging and experimental purposes in any future work in order to account for inter-specimen variations.

Given that the inter-specimen variability in structural arrangement may be large, fixing structural parameters such as fibre orientation spread, undulation mean, undulation spread, and fibre volume fraction of all the tissue samples to the same values (e.g. as reported in the literature) may potentially cause problems. In light of such difficulties in identifying parameters of the structural models, a case study was undertaken to help understand the general mechanisms leading to these identifiability issues and how they can be potentially mitigated.

8.5.2 *Case Study: Highlighting Parameter Identification Issues*

In order to investigate the identifiability issues in more detail, experiments were performed on silicone gel phantoms in Babarenda Gamage et al. (2011) under controlled conditions where boundary conditions and geometry of the body could be accurately prescribed. This allowed us to focus our attention solely on the identification problem. The study involved creating a two-layered cantilever structure from silicone gel (70 × 30 × 30 mm). Eight different gravity loaded orientations of the beam digitised using a FaroArm Laser ScanArm V2[1] and a tilt table. These datasets were used to assess both the descriptive and predictive power of the constitutive parameters of the beam that were subsequently identified using a similar model based identification approach as previously described.

Modelling

The two layers of silicone gel were modelled using an ideally incompressible hyperelastic neo-Hookean constitutive relation ($\Psi = c(I_1 - 3)$, where I_1 is the first invariant of the right Cauchy–Green deformation tensor). The different layers of the cantilever beam were described by two different stiffness parameters, c_a and c_b in units of kPa. These mechanical parameters were identified using an optimization method similar to that described in Sect. 8.5.1, however, in this case, the objective function was constructed from the laser scanned data using a closest point to model surface projection approach. The objective function to be minimised was defined as $\Phi = \sum_{j=1}^{M} \sum_{i=1}^{N_j} \|Z_{ij}\|^2$, where $\|Z_{ij}\|$ was the Euclidean distance between the ith datapoint and its closest point on the surface of a model in experiment j, M is the number of experiments performed and N_j is the number of laser scanned data points in experiment j. In order to aid interpretation of the results, the root mean square error (RMSE) was used as defined in Eq. (8.26).

---

[1]Manufactured by FARO Swiss Holding GmbH, Switzerland: http://www.faro.com.

$$\text{RMSE} = \sqrt{\frac{\sum_{j=1}^{M}\sum_{i=1}^{N_j}\|Z_{ij}\|^2}{\sum_{j=1}^{M}N_j}} \tag{8.26}$$

The optimisation tolerance on the parameters was set to the nearest 1 Pa while the objective function tolerance was lowered to an RMSE value equivalent to 1e−9 mm.

Results

Figure 8.10 illustrates the optimal model fits to the experimental surface data and their associated parameter estimates when each orientation was analysed independently. The results show that even for this simple two-layered model, the stiffness parameters derived from individual experiments varied over a wide range of values even though the surface fitting errors were similar. Figure 8.11 illustrates how the use of one of these parameter sets to predict the deformations for the other orientations resulted in large surface fit errors.

A cross-validation study was performed to see whether the use of multiple orientations would improve the predictive power of the identified parameters.

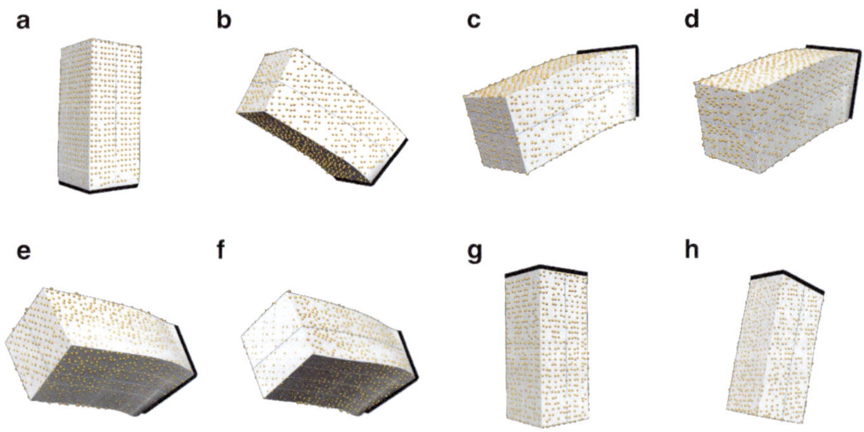

Fig. 8.10 Laser scanned datapoints from the eight cantilever beam orientations overlaid with best fit mechanics models using optimal stiffness parameters (indicated, in kPa) derived from individual gravity loading experiments. The fixed end of each model is indicated by a *thick outline*. It should be noted that model 1 (**a**) is orientated upright while model 7 (**g**) is orientated downward in the direction of gravity. Reproduced with permission from Babarenda Gamage et al. (2011). (**a**) Model 1: 1.82, 1.46 (**b**) Model 2: 4.72, 1.92 (**c**) Model 3: 4.37, 1.93 (**d**) Model 4: 4.57, 1.93 (**e**) Model 5: 4.79, 1.84 (**f**) Model 6: 4.13, 2.04 (**g**) Model 7: 4.91, 2.82 (**h**) Model 8: 3.70, 3.45

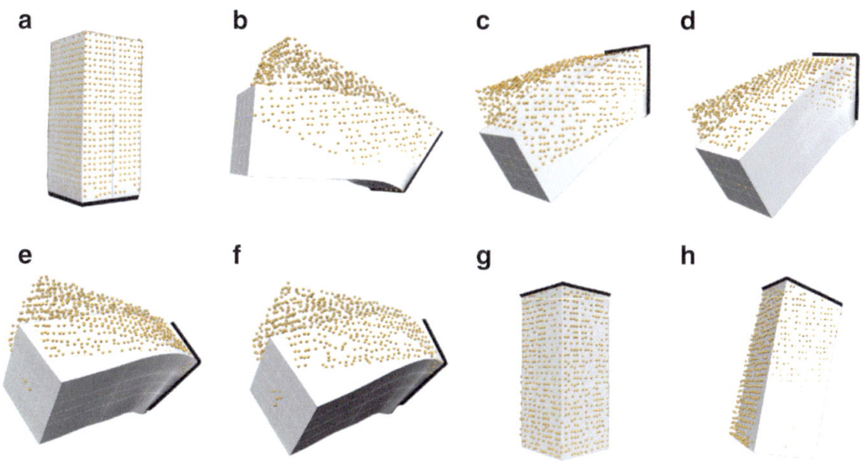

Fig. 8.11 Laser scanned data from the eight gravity loading experiments overlaid with predictions from models that used stiffness parameters derived from the upright orientation shown in Fig. 8.10. The maximum RMSE at the four corner points at the free ends of each beam was between 7.91 and 20.13 mm, respectively. Reproduced with permission from Babarenda Gamage et al. (2011). (**a**) Model 1 (**b**) Model 2 (**c**) Model 3 (**d**) Model 4 (**e**) Model 5 (**f**) Model 6 (**g**) Model 7 (**h**) Model 8

The cross-validation involved optimising a training subset of models against their respective experimental data. The resulting parameters were then used to estimate how accurately they could predict the deformation of the remaining validation subsets. The results were split into four groups, with groups 1 to 4 using 1, 2, 3 and 7 training models for identification, respectively, with eight sets of training models sampled randomly for groups 2 and 3 (for groups 1 and 4, all eight training sets were considered).

Visual interpretations of the results from groups 1 to 4 are presented in Fig. 8.12a–d, respectively, which show contour plots on the hyper-surfaces of the RMSE objective function. These hyper-surfaces were created by evaluating the objective function on the material parameter space. The markers on these plots indicate the optimal solution for each training set of models, which is the point at which the determinability of the parameters were also evaluated. The contours were plotted such that any set of parameters θ, which results in RMSE values less than γ, were inside the contour. For each contour, the value of γ was set to the optimal RMSE plus 0.1 mm to account for experimental error. Any pair of stiffness parameters inside this region can be considered as an equally valid solution within the limit of experimental error. The cross validation results showed that by increasing the number of training models used for parameter identification, the maximum RMSE error of the validation sets was reduced. The results also showed that we can satisfactorily estimate the stiffness properties of the two layers of the gel using three orientations with a maximum predictive RMSE of approximately 0.6 mm and maximum free corner error of approximately 1.9 mm. Some points

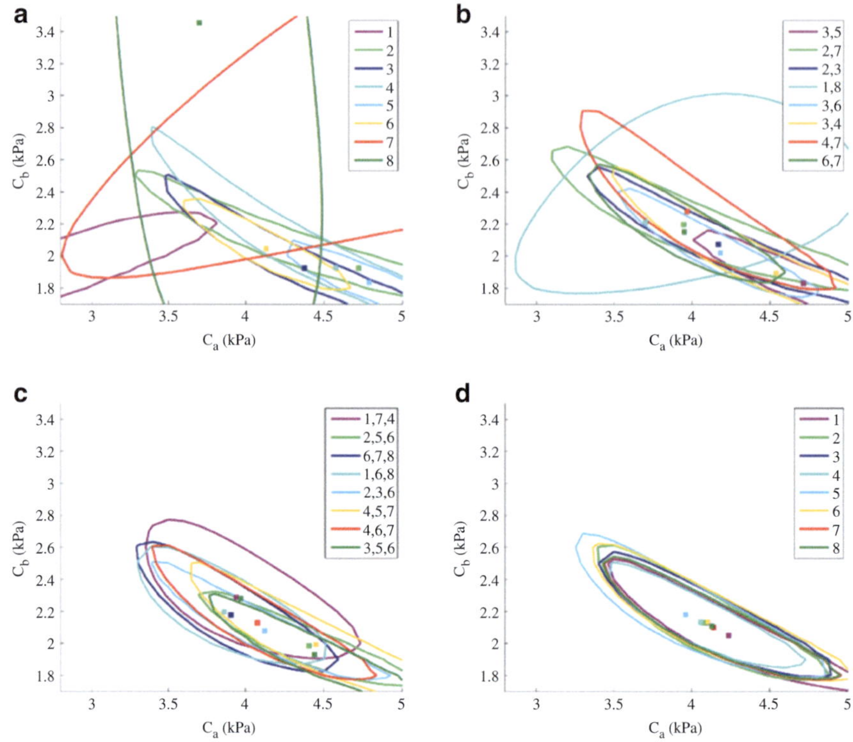

Fig. 8.12 Contour plots of the objective function (RMSE) over the stiffness parameter spaces. Panels (**a**) to (**d**) show contours as the number of training cases was increased from 1 to 2, 3 and 7, respectively, with legends indicating experiments included in the training sets (Groups 1–3), or the excluded experiment (Group 4). Areas where contours overlap represent parameter combinations that provide equally accurate predictions to within the experimental error. Reproduced with permission from Babarenda Gamage et al. (2011). (**a**) Group 1 (identification based on 1 model) (**b**) Group 2 (identification based on 2 models) (**c**) Group 3 (identification based on 3 models) (**d**) Group 4 (identification based on 7 models)

of interest related to the identification process, such as the uniqueness of the parameter estimates, are described in the following section, along with suggestions on techniques that could be used to further improve parameter identification.

Uniqueness of Parameter Estimates and Improving Identifiability

This study showed that choice of the optimality tolerance of the objective function affects the apparent uniqueness of the constitutive parameter estimates. Ideally, the optimisation should be terminated once the change in the objective function is less than the maximum error observed in the experimental measurements since this is the smallest physical quantity that can be accurately measured. Any optimisation solution which terminates under this condition should be valid (inside the contours in Fig. 8.12a–d). This gives rise to multitude of parameters sets, each of which is an equally valid solution to the identification problem. Some of these solutions can give rise to a variety of prediction errors that depend upon where the true minimum lies. This can occur if the deformation is relatively insensitive to the constitutive parameters in certain orientations, as was the case for some of the models considered here, and noise in the data could easily shift the global optimum. Thus, some model's parameter estimates appeared as outliers, but in fact were valid solutions within the experimental accuracy. Another factor that can lead to non-unique solutions is the mathematical formulation of the constitutive relation and whether the parameters of the model are correlated. However, in this study, a single parameter was used to describe the stiffness of each layer of the beam and therefore any apparent correlation between the parameters was directly related to the information contained within the data.

Setting the objective function tolerance to a very low value during the identification procedure allowed the optimisation to terminate once the constitutive parameters converged to the nearest Pa. This resulted in unique solutions being obtained independent of the initial estimate. For this reason the optimal solutions provided in Fig. 8.10 (shown as markers) are only indicative of possible sets of valid parameters. However, the prediction results when the number of orientations used for identification was increased produced contours that overlapped to a greater extent. This reduced the possibility of obtaining parameters that would cause the large prediction errors seen in Fig. 8.11. It should be noted that with constitutive equations with a large number of unknowns, and a given dataset from which these parameters are required to be identified, it may not be possible to obtain unique solutions by lowering the tolerance on the optimised parameters (Ogden et al. 2004), especially if there is high correlation between parameters. This is related to the information content present within the data used to construct each model's objective function and whether this provides sufficient information for unique identification. For example, in this study, multiple solutions could potentially be produced with the two-layered cantilever beam if data from only one side of the beam was used for identification. This underscores the importance of the quality and quantity of the data that is used for fitting. High quality data would result in a smaller region in which parameters could equally well describe the model; while a rich set of data used for fitting the model would result in parameters which increase the ability of the models to predict different deformations.

The identification methodology for assessing and improving identifiability described in this section was used in our more recent studies looking at determining mechanical properties of skin in vivo. The results of these studies will be summarised in the next section.

8.5.3 *In Vivo Identification of Skin Properties*

Previous studies from our group have focussed on identifying constitutive parameters for a simplified form of the Tong and Fung strain energy function using the biaxial rig on human forearm skin in vivo (Kvistedal and Nielsen 2009). This was achieved using a similar modelling and identification methodology as that described earlier in the previous sections. A series of multiaxial loading experiments were performed on the forearms of four age and gender matched subjects. The tissue geometry, together with recorded displacements and boundary forces, was combined with a nonlinear FE anisotropic membrane model to identify the mechanical parameters of the simplified Tong and Fung relation. Ten sets of constitutive parameters were estimated from the experiments and showed considerable differences in mechanical behaviour, both between individual subjects and symmetric locations on the body for a single subject. Variations between individuals have previously been reported as being dependent on body locations and age (Reihsner et al. 1995). The study showed that differences in the stress–strain relationship are also expected in identical body locations from age and gender matched subjects.

More recent studies from our group have utilised a micro-robot indenter (described in Sect. 8.4.2) for identifying mechanical parameters of human forearm skin. One such study will be described here, where the constitutive parameters of the Ogden and Tong and Fung constitutive models together with in vivo tensions were identified using a model-based approach (Flynn et al. 2011b). Based on the experiments on the silicon gel phantoms described in the previous section, it was shown that multiple loading conditions were important for improving identifiability of the constitutive parameters. We therefore aimed to obtain a rich set deformations, using the micro-robot indenter on the forearm for a number of volunteers, to aid subsequent identification and these are described in more detail in Sect. 8.4.2. The deformations included various in-plane and out-of-plane measurements of the force vs displacement response of the skin measured on the anterior forearm and posterior upper arm. Constitutive parameters were then identified from volunteers within this database using a model-based approach. To date, few finite element models in the literature have considered the effects of the in vivo pretension of the skin. In our studies, pretension was represented by an initial stress field imposed on the reference configuration in the X and Y directions, whereby the two parameters of the stress field were included in the optimisation procedure. When using only in-plane average forearm data to form the objective function, both the Tong and Fung and Ogden models provided good fits to the experimental data, with fitting errors of 9.9 and 11.6 %, respectively. However, they were unable to accurately predict the out-of-plane responses with overall fitting errors of 19.8 and 23.2 %, respectively. Using only the out-of-plane data in parameter optimisation, the Ogden model predicted the in-plane response with a fitting error of 17 %. The results from this study support the hypothesis that in-plane deformations alone do not provide sufficient data to identify the constitutive parameters and in vivo tensions. This hypothesis is further supported by the fact that increasing the number of deformation orientations leads to consistent improvements of the determinability criteria. It was also shown that there is a wide variability in parameter and predicted pre-stress values across the subjects, indicating the need to collect experimental data from a diverse population. A common problem in parameter estimation, especially when phenomenological models are used, is the ability to address the issue of uniqueness of the optimal solution. A limitation of this study was the use of only the force and displacement measures on the tip of the microrobot for identifying the mechanical properties.

8.6 Conclusions and Outlook

Building upon the pioneering work of Professor Yoram Lanir, increasing research efforts over the years have now resulted in a well-established structurally based constitutive modelling framework for soft connective tissues. The present article has focused on some of the work by the Auckland Bioengineering Institute contributing towards the goal of understanding the structure–function relationship of soft membranous tissue. Key aspects of our work are to (1) develop constitutive relations based on quantitative information of tissue structure; and (2) use rich sets of experimental data to aid in accurate and reliable constitutive parameter identification.

To this end, we have developed imaging techniques to analyse collagen fibre networks and represent fibre orientations with mathematical distributions. Currently, there remains a shortage of quantitative structural data that can be directly incorporated into constitutive modelling frameworks, which will help reduce the number of constitutive parameters to be identified. It has also been demonstrated in one of our studies that a priori knowledge of mean fibre orientation is important to the success of parameter identification. Due to the high inter-specimen variability in soft tissues, it is highly desirable to extract both the structural and experimental data from the *same* tissue sample. However, this becomes difficult when in vivo tissues are of interest as invasive methods, such as conventional microscopy, can no longer be considered. To address this issue, work is currently underway in our group to develop alternative methods for gathering structural data in a non-invasive manner. Furthermore, although previous research effort has been primarily devoted to the studies of collagen fibre orientations, it is important to also understand the properties and interactions of other tissue constituents, such as elastin.

The use of complex structural constitutive models—which consider the effects of anisotropy, nonlinearity, viscoelasticity, preconditioning, and multiple families of fibres—relies heavily upon the availability of rich sets of experimental data. There is clearly a need to further improve current instrumentation techniques to achieve a wide range of deformation modes. In this article, we summarised the development of a multi-axial tensile device coupled with digital image correlation techniques to track the deformation field, used to characterise the planar stress–strain response of soft tissues. Since biaxial testing does not provide a complete 3D mechanical description, a force-sensitive micro-robot has been developed to impose out-of-plane deformations. It is expected that the measurement of the strain field in the vicinity surrounding the indenter probe, using digital image correlation techniques, would improve the identification of the constitutive parameters. In light of this, we are currently investigating techniques to obtain strain measurements by tracking surface deformations using a 3D stereoscopic system (Alvares 2009; Evans and Holt 2009). In order to improve the identifiability of constitutive parameters, the design of optimal experiments (Humphrey 2003; Lanir et al. 1996) requires further development.

The merits of structurally based constitutive modelling frameworks becomes particularly apparent when the mechanics of ageing or diseased tissues are of interest. Structural changes in the microscopic level associated with ageing include the thickening of collagen fibres (which take on a much denser and sheet-like arrangement), and the fragmentation and aggregation of elastin fibres. Such geometric and mechanical changes in the individual tissue constituents are thought to contribute to the reduced extensibility and elasticity often observed in ageing skin. In order to gain improved understanding of the underlying mechanisms, structural models allow one to perturb structural constitutive parameters that are associated with time or pathological changes, and subsequently, predict the associated mechanical response. To be useful in a clinical setting, patient-specific computational models are likely to require the combination of fast and non-invasive image techniques, along with in vivo experimental data under various loading conditions.

Although recent research advancements have contributed to improved understanding of soft tissue biomechanics, many aspects of the complex interactions between microscopic structure and macroscopic mechanical response are still to be elucidated.

References

Abrahams M. Mechanical behaviour of tendon in vitro. Med Biol Eng Comput. 1967;5:433–43.

Alvares D. A 3D strain measurement system for soft materials: application to material parameter estimation. Master's Thesis, The University of Auckland; 2009.

Arbogast KB, Thibault KL, Scott Pinheiro B, Winey KI, Margulies SS. A high-frequency shear device for testing soft biological tissues. J Biomech. 1997;30(7):757–9.

Babarenda Gamage TP, Rajagopal V, Ehrgott M, Nash MP, Nielsen PMF. Identification of mechanical properties of heterogeneous soft bodies using gravity loading. Int J Numer Methods Biomed Eng. 2011;27(3):391–407.

Billiar K, Sacks M. A method to quantify the fiber kinematics of planar tissues under biaxial stretch. J Biomech. 1997;30(7):753–6.

Billiar K, Sacks M. Biaxial mechanical properties of the native and glutarldehyde-treated aortic valve cusp: part II - a structural constitutive model. J Biomech Eng. 2000;122:327–35.

Brown IA. A scanning electron microscope study of the effects of uniaxial tension on human skin. Br J Dermatol. 1973;89(4):383–93.

Carton RW, Dainauskas J, Clark JW. Elastic properties of single elastic fibers. J Appl Physiol. 1962;17(3):547–51.

Choi HS, Vito RP. Two-dimensional stress-strain relationship for canine pericardium. J Biomech Eng. 1990;112(2):153–9.

Cox H. The cleavage lines of the skin. Br J Surg. 1941;29(114):234–40.

Craik JE, McNeil IRR. Histological studies of stressed skin. In: Biomechanics and related bioengineering topics. Oxford: Pergamon Press; 1964. p. 159–64.

Daly C. The biomechanical characteristics of human skin. Ph.D. Thesis. Scotland: University of Strathclyde; 1966.

Daly CH. The role of elastin in the mechanical behavior of human skin. In: 8th International Conference on Medical and Biological Engineering; 1969.
Dokos S, LeGrice IJ, Smaill BH, Kar J, Young AA. A triaxial-measurement shear-test device for soft biological tissues. J Biomech Eng. 2000;122(5):471–8.
Dolber P, Spach M. Conventional and confocal fluorescence microscopy of collagen fibers in the heart. J Histochem Cytochem. 1993;41(3):465–9.
Eckert C, Mikulis B, Gottlieb D, Gerneke D, LeGrice I, Padera R, Mayer J, Schoen F, Sacks M. Three-dimensional quantitative micromorphology of pre- and post-implanted engineered heart valve tissues. Ann Biomed Eng. 2011;39:205–22.
Evans S, Holt C. Measuring the mechanical properties of human skin in vivo using digital image correlation and finite element modelling. J Strain Anal Eng Des. 2009;44:337–45.
Finlay B. Scanning electron microscopy of the human dermis under uni-axial strain. Biomed Eng. 1969;4(7):322–7.
Flynn C, Taberner A, Nielsen P. Measurement of the force-displacement response of in vivo human skin under a rich set of deformations. Med Eng Phys. 2011a;33(5):610–9.
Flynn C, Taberner A, Nielsen P. Modeling the mechanical response of human skin under a rich set of deformations. Ann Biomed Eng. 2011b;39:1935–46.
Freed AD, Einstein DR, Vesely I. Invariant formulation for dispersed transverse isotropy in aortic heart valves: an efficient means for modeling fiber splay. Biomech Model Mechanobiol. 2005;4(2–3):100–17.
Fung Y. Foundations of solid mechanics. Englewood Cliffs: Prentice-Hall; 1965.
Fung YC. Elasticity of soft tissues in simple elongation. Am J Physiol. 1967;213(6):1532–44.
Gasser TC, Ogden RW, Holzapfel GA. Hyperelastic modelling of arterial layers with distributed collagen fibre orientations. J R Soc Interface. 2006;3(6):15–35.
Harkness ML, Harkness RD. Effect of enzymes on mechanical properties of tissues. Nature. 1959;183:1821–2.
Holzapfel G, Ogden R. On planar biaxial tests for anisotropic nonlinearly elastic solids. a continuum mechanical framework. Math Mech Solids. 2009;14:474–89.
Holzapfel G, Gasser T, Stadler M. A structural model for the viscoelastic behavior of arterial walls: continuum formulation and finite element analysis. Eur J Mech A Solids. 2002;21:441–63.
Horowitz A, Lanir Y, Yin FCP, Perl M, Sheinman I, Strumpf RK. Structural three dimensional constitutive law for the passive myocardium. J Biomech Eng. 1988;110:200–07.
Humphrey J. Review paper: continuum biomechanics of soft biological tissues. Proc R Soc Lond A Math. 2003;459(2029):3–46.
Humphrey JD, Strumpf RK, Yin FCP. Determination of a constitutive relation for passive myocardium: I. A new functional form. J Biomech Eng. 1990;112(3):333–9.
Humphrey J, Kang T, Sakarda P, Anjanappa M. Computer-aided vascular experimentation: a new electromechanical test system. Ann Biomed Eng. 1993;21:33–43.
Hurschler C, Loitz-Ramage B, Vanderby R. A structurally based stress-stretch relationship for tendon and ligament. J Biomech Eng. 1997;119(4):392–9.
Jahne B. Practical handbook on image processing for scientific and technical applications, chap 13. 2nd ed. Boca Raton: CRC Press; 2004. p. 419–42.
Jor J, Nash M, Nielsen P, Hunter P. Estimating material parameters of a structurally based constitutive relation for skin mechanics. Biomech Model Mechanbiol. 2011a;10:767–78.
Jor JWY, Nielsen PMF, Nash MP, Hunter PJ. Modelling collagen fibre orientation in porcine skin based upon confocal laser scanning microscopy. Skin Res Tech. 2011b;17(2):149–59.
Junquiera LC, Montes GS, Sanchez EM. The influence of tissue section thickness on the study of collagen by the picrosirius-polarization method. Histochemistry. 1982;74:153–6.
Kenedi R, Gibson T, Daly C. Bio-engineering studies of the human skin: the effects of unidirectional tension. In: Structure and function of connective and skeletal tissue. London: Butterworths; 1965. p. 388–95.
Kvistedal YA, Nielsen PMF. Estimating material parameters of human skin *in vivo*. Biomech Model Mechanobiol. 2009;8(1):1–8.

Lanir Y. A structural theory for the homogeneous biaxial stress-strain relationships in flat collagenous tissues. J Biomech. 1979;12(6):423–36.
Lanir Y. Constitutive equations for fibrous connective tissues. J Biomech. 1983;16(1):1–12.
Lanir Y, Fung YC. Two-dimensional mechanical properties of rabbit skin. I. Experimental system. J Biomech. 1974a;7(1):29–34.
Lanir Y, Fung YC. Two-dimensional mechanical properties of rabbit skin. II. Experimental results. J Biomech. 1974b;7(2):171–82.
Lanir Y, Lichtenstein O, Imanuel O. Optimal design of biaxial tests for structural material characterization of flat tissues. J Biomech Eng. 1996;118:41–7.
Lokshin O, Lanir Y. Micro and macro rheology of planar tissues. Biomaterials. 2009a;30:3118–27.
Lokshin O, Lanir Y. Viscoelasticity and preconditioning of rat skin under uniaxial stretch: microstructural constitutive characterization. J Biomech Eng. 2009b;131(3):031009.
MacKenna DA, Omens JH, Covell JW. Left ventricular perimysial collagen fibers uncoil rather than stretch during diastolic filling. Basic Res Cardiol. 1996;91(2):111–22.
Malcolm D. Estimating the material properties of inhomogeneous elastic membranes. Ph.D. Thesis, University of Auckland; 2000.
Malcolm DTK, Nielsen PMF, Hunter PJ, Charette PG. Strain measurement in biaxially loaded inhomogeneous, anisotropic elastic membranes. Biomech Model Mechanobiol. 2002;1: 197–210.
Manschot JFM. The mechanical properties of human skin *In Vivo*. Ph.D. Thesis. The Netherlands: Nijmegen University; 1985.
Manschot JF, Brakkee AJ. The measurement and modelling of the mechanical properties of human skin *in vivo*–I. The measurement. J Biomech. 1986;19(7):511–5.
Marcarian HQ, Calhoun ML. Microscopic anatomy of the integument of adult swine. Am J Vet Res. 1966;27(118):765–62.
Meijer R. Characterisation of anisotropic and non-linear behaviour of human skin. Technical Report, Personal Care Institute of Philips Research Laboratories; 1997.
Melis P, Noorlander ML, van der Horst CMA, van Noorden CJF. Rapid alignment of collagen fibers in the dermis of undermined and not undermined skin stretched with a skin-stretching device. Plast Reconstr Surg. 2002;109(2):674–82.
Meyer W, Neurand K, Radke B. Collagen fibre arrangement in the skin of the pig. J Anat. 1982;134:139–48.
Mowafy M, Cassens RG. Microscopic structure of pig skin. J Anim Sci. 1975;41(5):1281–90.
Nathanson MH, Saidel GM. Multiple-objective criteria for optimal experimental design: application to ferrokinetics. Am J Physiol. 1985;248:378–86.
Nielsen PMF, Malcolm DTK, Hunter PJ, Charette PG. Instrumentation and procedures for estimating the constitutive parameters of inhomogeneous elastic membranes. Biomech Model Mechanobiol. 2002;1(3):211–8.
Noorlander ML, Melis P, Jonker A, van Noorden CJF. A quantitative method to determine the orientation of collagen fibers in the dermis. J Histochem Cytochem. 2002;50(11):1469–74.
Ogden RW, Saccomandi G, Sgura I. Fitting hyperelastic models to experimental data. Comput Mech. 2004;34(6):484–502.
Pope AJ, Sands GB, Smaill BH, LeGrice IJ. Three-dimensional transmural organization of perimysial collagen in the heart. Am J Physiol-Heart C. 2008;295(3):1243–52.
Reihsner R, Balogh B, Menzel EJ. Two-dimensional elastic properties of human skin in terms of an incremental model at the *in vivo* configuration. Med Eng Phys. 1995;17(4):304–13.
Sacks M. Biaxial mechanical evaluation of planar biological materials. J Elast. 2000;61:199–246.
Sacks MS. Incorporation of experimentally-derived fiber orientation into a structural constitutive model for planar collagenous tissues. J Biomech Eng. 2003;125(2):280–7.
Sacks M, Smith DB, Hiester E. A small angle light scattering device for planar connective tissue microstructural analysis. Ann Biomed Eng. 1997;25:678–89.
Sands GB, Gerneke DA, Hooks DA, Green CR, Smaill BH, LeGrice IJ. Automated imaging of extended tissue volumes using confocal microscopy. Microsc Res Tech. 2005;67(5):227–39.

Sands G, Goo S, Gerneke D, LeGrice I, Loiselle D. The collagenous microstructure of cardiac ventricular trabeculae carneae. J Struct Biol. 2011;173(1):110–6.
Silver F, Kato Y, Ohno M, Wasserman A. Analysis of mammalian connective tissue: relationship between hierarchical structures and mechanical properties. J Long Term Eff Med Implants. 1992;2(2–3):165–98.
Smaill B, Hunter P. Structure and function of the diastolic heart. In: Theory of heart: biomechanics, biophysics, and nonlinear dynamics of cardiac function. New York: Springer; 1991. p. 1–29.
Spurr BD, Koutbeiy MA. A comparison of various methods for estimating the parameters in mixtures of von mises distributions. Commun Stat Simul Comput. 1991;20:725–42.
Sweat F, Puchtler H, Rosenthal SI. Sirius red f3ba as a stain for connective tissue. Arch Pathol. 1964;78:69–72.
Tong P, Fung YC. The stress-strain relationship for the skin. J Biomech. 1976;9(10):649–57.
van Brocklin D, Ellis DG. A study of the mechanical behavior of toe extensor tendons under applied stress. Arch Phys Med Rehabil. 1965;46:369–73.
van Zuijlen P, Ruurda J, van Veen H, van Marle J, van Trier A, Groenevelt F, Kreis R, Middelkoop E. Collagen morphology in human skin and scar tissue: no adaptations in response to mechanical loading at joints. Burns. 2003;29(5):423–31.
Vawter DL, Fung YC, West JB. Elasticity of excised dog lung parenchyma. J Appl Physiol. 1978;45(2):261–9.
Viidik A. Biology of Collagen. Viidik A, Vuust J, editors. London: Academic Press; 1980;239–355
Wilkes GL, Brown IA, Wildnauer RH. The biomechanical properties of skin. CRC Crit Rev Bioeng. 1973;1(4):453–95.
Wright DG, Rennels DC. A study of the elastic properties of planar fascia. J Bone Joint Surg. 1964;46:482–92.
Yin FC, Strumpf RK, Chew PH, Zeger SL. Quantification of the mechanical properties of non-contracting canine myocardium under simultaneous biaxial loading. J Biomech. 1987;20(6): 577–89.
Young AA, LeGrice IJ, Young MA, Smaill BH. Extended confocal microscopy of myocardial laminae and collagen network. J Microsc. 1998;192:139–50.

Chapter 9
Structure–Function Relations in the Coronary Vasculature

Benjamin Kaimovitz, Yunlong Huo, Yoram Lanir, and Ghassan S. Kassab

The function of the coronary system is to deliver blood to the capillary network, to nourish the myocardium and to autoregulate coronary blood flow. The coronary vessels run through cyclically contracting, cardiac muscle. Consequently, intramyocardial pressures have a major influence on vascular transmural pressure and, therefore, on the flow. It is widely acknowledged that the distribution of coronary blood is spatially heterogeneous. Aside from the interaction between myocardial contraction and flow, causing impediment at the local level, the stochastic nature of the coronary tree geometry contributes to the inhomogeneity of the blood flow distribution. Additionally, variations of flow exist from epicardium to endocardium. Flow heterogeneity is further enhanced in pathologies, such as ischemia, which induces vulnerability of the subendocardium to ischemia. The coronary perfusion distribution as well as local coronary flow is difficult to measure, especially in the endocardium. Hence it is not yet fully understood as to what the specific effects cardiac contraction, local neurogenic controls, cardiac and vascular pathologies, and specific therapeutic modalities (e.g., drugs) have on the extent and distribution of coronary perfusion. For this reason, simulation is an attractive methodological alternative. Accordingly, realistic analysis of the flow distribution must be carried out within a framework of a realistic three-dimensional (3D) representation of the coronary geometry and its biological variability. Recent morphological studies facilitate realistic reconstructions of the coronary vasculature to serve as a foundation for simulation of the flow in the entire coronary system.

B. Kaimovitz • Y. Lanir • G.S. Kassab (✉)
IUPUI, Indianapolis, IN 46202, USA
e-mail: gkassab@calmi2.org

Y. Huo
Peking University, China

G.S. Kassab, M.S. Sacks (eds.), *Structure-Based Mechanics of Tissues and Organs*,
DOI 10.1007/978-1-4899-7630-7_9

9.1 Architecture of the Coronary Vasculature

The architecture of the coronary network is one of the major determinants of the coronary perfusion and oxygen supply (Ansari 2001; Loukas et al. 2009; von Lüdinghausen 2003). In term of functionality, the coronary network can be partitioned into three sub-networks (Fig. 9.1): epicardial, transmural, and perfusing. The rational for this classification is based on the differences in the functional design of these sub-networks (Berne and Rubio 1979; Fulton 1982; Grayson 1982; James 1961; Tanaka et al. 1999; Zamir 1999). Segments of the coronary vasculature act as conduits, metabolic units, and semipermeable membranes. Certain vascular segments contribute significantly to the resistance and capacitance of the coronary circulation.

The epicardial network comprises relatively large vessels which run over the surface of the heart (epicardium), giving rise to transmural branches, which in turn penetrate the heart muscle, and branch further into smaller vessels down to the capillary network.

A large variability in the anatomy of the coronary network exists at the level of the epicardial vessels, which is expressed by the diversity in the number of branches their location and dimension (diameter and length) (Baroldi et al. 1967; Berne and Rubio 1979; von Ludinghausen 1984). The epicardial coronary arteries (EPCAs), and more prominently the venous system contribute the major portion of the coronary capacitance; whereas, the small arterial transmural vessels (Chilian et al. 1989; Jones et al. 1993; Kaul and Ito 2004; Toyota et al. 2001), as well as to a lesser extent the coronary venules (Chilian et al. 1989; Jones et al. 1993; Kaul and Ito 2004; Toyota et al. 2001), play a dominant role in regulating coronary resistance. A vessel's contribution to the overall resistance depends on the network structure and the position of the vessel in the network.

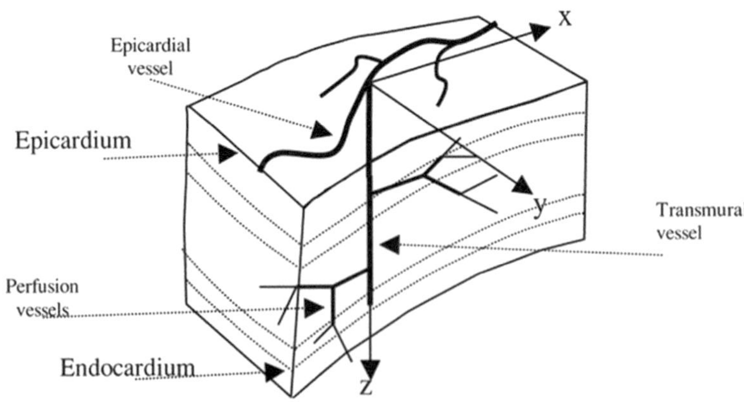

Fig. 9.1 A schematic of the classification of coronary functional sub-network hierarchy. The epicardial sub-network is confined to orders 11–9, the transmural sub-network encompasses orders 8–5 and the perfusion sub-network spans the range of order 4 to 0_a

The EPCAs give rise to intramural vessels. These vessels, in the case of the left ventricle, originate at right angles to their parent epicardial vessel and penetrate perpendicularly to the epicardial surface (Berne and Rubio 1979). In the right ventricular wall, the branches penetrate obliquely. Arterioles branch off almost at right angles from the small penetrating arteries (Kassab et al. 1993) and either take an oblique course to the nearest capillary bed or wind around a muscle fiber to proceed to a more distant muscle fiber.

Oxygen and nutrients are exchanged between blood and muscle cells at the capillary level. The capillary network is characterized by high density, which compensates for the small volumes flow in each capillary. The capillary density is species dependent, ranging between 2000 and 5000 per mm^2 cross section perpendicular to capillary direction (Bassingthwaighte et al. 1974; Olsson and Bugni 1986; Spaan 1991) with an average intercapillary distance of 20 μm (Bassingthwaighte et al. 1974; Berne and Rubio 1979). Unlike the arterial and venous networks, the coronary capillary bed is not tree-like.

The capillary network expands in the direction of the venous system, in terms of number of vessels and corresponding cross-sectional area (CSA), which serves to increase the transit time of red blood cells. The expansion rate can be calculated based on the measured data, showing that the average $0_v/0_a$ ratio is 1.61 for the left coronary artery (LCA) and 1.86 for the right coronary artery (RCA) (Kaimovitz et al. 2010).

After passing through the capillary bed, the blood drains into venules which then confluences into the larger epicardial veins and finally into the right atria. Some arterial branches (80–200 μm in diameter), accounting for approximately 5 % of LCA inflow (von Lüdinghausen 2003), drain directly into the lumens of the four cardiac chambers via the Thebesian veins (Fig. 9.2). The coronary venous system is the largest component of the coronary network in terms of volume (Grayson 1982; Hoffman and Spaan 1990; Kajiya and Goto 1999), compliance (Spaan 1985), and number of vessels (Hiramatsu et al. 1998; Hutchins et al. 1986; Kassab et al. 1994).

The major venous epicardial vessels follow the course of arteries (LAD, LCx, and RCA) (Baroldi et al. 1967; Grayson 1982; Hutchins et al. 1986; Kajiya and Goto 1999) and provide coverage of the entire epicardial surface to drain the respective arterial networks (von Lüdinghausen 2003; Mohl et al. 1984). The major veins give rise to smaller veins which run on the LV and RV surfaces and emerge from the myocardium at almost right angles. The intramural venous branches are independent of the arterial branches (Baroldi et al. 1967).

The anatomy of the coronary venous system has a complex 3D geometry. The branching pattern consists of higher asymmetry compared to the coronary arterial tree with abrupt changes in diameters. While the arterial branching is mostly dichotomous (Baroldi et al. 1967; Kalsho and Kassab 2004; Kassab et al. 1993), the venous system is characterized by abundant trifurcations, quadrifications, and quintifications with frequencies of 86 %, 12.8 %, and 0.2 %, respectively (Kassab et al. 1994). The venous branches are more numerous than the arterial ones (Bales 2004; Kassab et al. 1994; von Lüdinghausen 2003) and the branching pattern of venules is very different from that of the arterioles. The smallest venules run first in

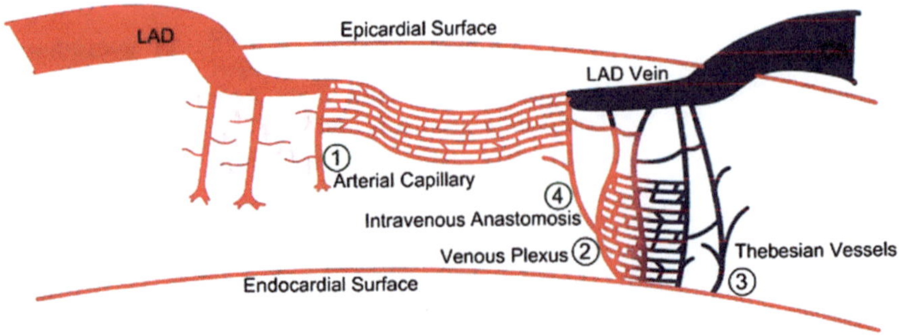

Fig. 9.2 A schematic of coronary vasculature. *Red* denotes oxygenated arterial blood flow while *blue* represents venous blood. The venous system has intervenous connections, Thebesian–Sinus connections and a venous plexus. Taken from (Kassab et al. 2008)

the direction of the capillaries they drain and then diverge obliquely toward larger veins (Kassab et al. 1994). The venous system largely determines the capacitance of the coronary circulation. Venous tone does not contribute significantly to the coronary vasculature resistance. Similar to the arterial bed, venous flow is affected by myocardial tissue stresses and strains while demonstrating out-of-phase flow patterns relative to the arterial circulation (Chilian and Marcus 1984; Hiramatsu et al. 1998; Hoffman and Spaan 1990). Coronary venules have a myogenic response (Kuo et al. 1993) and account for about 15 % of the total myocardial vascular resistance at rest (Kaul and Ito 2004; Toyota et al. 2001). Due to arteriolar–venular pairing, vasoconstrictive substances from the venous blood can diffuse and affect arteriolar resistance (Hester and Hammer 2002; Hutchins et al. 1986). The characteristics of the coronary venous system, in terms of both the structure and pressure-flow relationship, and its contribution to the capacitance of the coronary vasculature, underline the importance of the coronary venous network flow in the entire function of the coronary system (Kresh et al. 1990). Nevertheless, and despite the growing clinically motivated interest in the coronary venous system (Jain et al. 2006; Kassab et al. 2008), the venous system has received little attention relative to the arteries. There is clearly paucity of data on coronary venous pharmacology and physiology.

9.2 Diameter Asymmetry of Coronary Network

The coronary network is characterized by highly irregular and asymmetric branching patterns (Kalsho and Kassab 2004; Kassab and Fung 1994; Kassab et al. 1993, 1994, 1997; VanBavel and Spaan 1992). Due to the fractal nature of the coronary vasculature (Zamir 1999, 2001), the number of vessels increases in a geometric fashion toward the capillary vessels. This leads to an overwhelmingly high number of vessels (order of 10^9). This high vessel number, coupled with the inherent biological variability in coronary structure precludes the possibility of a deterministic description of the coronary anatomy. Consequently, a stochastic representation of the anatomy is necessary to facilitate a realistic three-dimensional (3D) reconstruction of the entire coronary vasculature.

Statistical morphometric data was obtained for the pig heart based on the diameter-defined Strahler's ordering scheme (Kassab et al. 1993). However, the process of reconstructing a tree from a statistical database requires anatomical constraints or morphometric data. In addition to coronary morphometry (diameters, lengths, number of vessels, connectivity, etc.), which has been described in great detail (Kassab and Fung 1994; Kassab et al. 1993, 1994, 1997; VanBavel and Spaan 1992), a correlation between successive vessels is required in order to impose additional bounds. Of all the morphometric parameters, the diameter is the most important hemodynamically. Based on the above statistical data, we have recently introduced local measures of diameter asymmetry at bifurcations (Kaimovitz et al. 2008, 2010). The bifurcation asymmetry data were represented by the diameter ratio of the daughters relative to mother vessel and by the area expansion ratios (AER) for entire arterial and venous trees. The diameter asymmetry at a bifurcation was quantified as a function of the order number of mother vessel for D_l/D_m and D_s/D_m; i.e., the ratios of the diameters of larger (D_l) and smaller (D_s) daughters, respectively, relative to their mother (D_m) vessel. In addition, a distinction was made between intra-element and inter-element segments, where an element consists of one or more vessel segments of the same order in series. For the intra-element case, the branching occurs along the same element, and hence the larger daughter belongs to the element. Accordingly, the D_l/D_m statistics were represented as a function of the element order number only (Fig. 9.3 for the LAD, LCx, and RCA arteries).

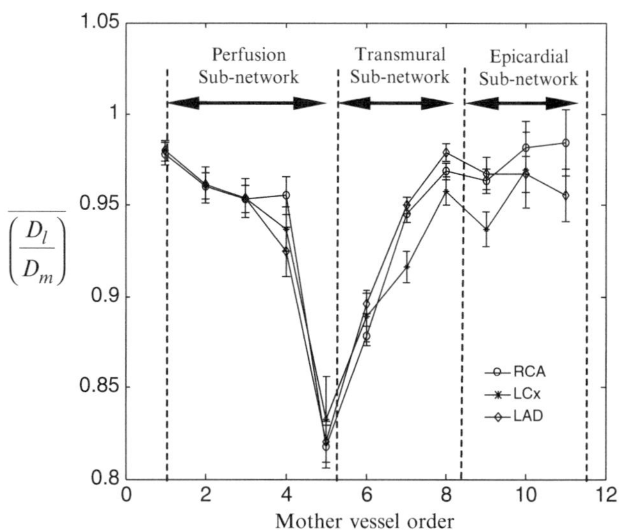

Fig. 9.3 Relationship between D_l/D_m (ratio of the diameter of larger daughters relative to mother vessel) and mother vessel order number data of left anterior descending (LAD) artery, left circumflex (LCx) artery, and right coronary artery (RCA)

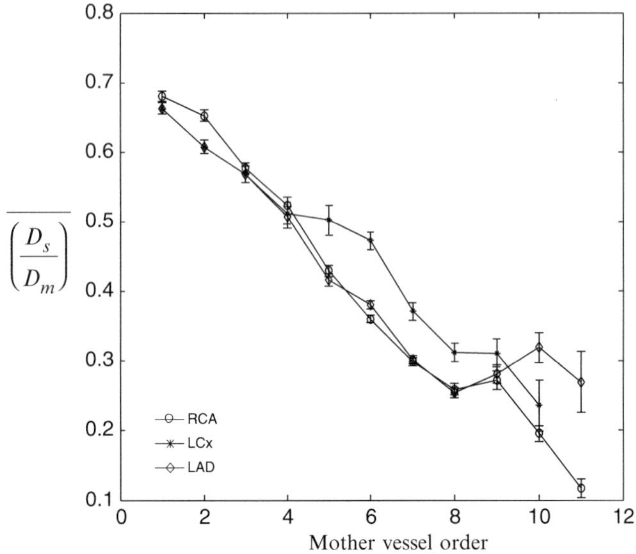

Fig. 9.4 (**a**) Polynomial fit for the D_l/D_m ratio as a function of D_m (the diameter of the mother vessel) in the manual (raw) tree data of LAD, LCx, and RCA arteries. The entire data are fitted by polynomials of *orders 4–8*. (b) D_l/D_m ratio for the Sinusal venous tree as function of the logarithm of the diameter of the mother vessel. *Blue*—raw data based on the database of Kassab et al. (Kassab et al. 1994). *Green*—fitted data using smoothing cubic-spline technique. *Red circle* indicated the location of the minimum at 70 μm, which corresponds to order number −4

For the inter-element case, each daughter belongs to a different element, and hence, the D_s/D_m ratio was formulated as an asymmetry ratio matrix (ARM), since the ratio is a function of order of mother and daughter segments. The intra-element case can be viewed as an index of vessel taper, whereas inter-element one represents branching asymmetry.

Statistical analysis of the data (Fig. 9.4) revealed that the ratio of large daughter to mother vessels has a statistically significant minimum value at order 5 and −4 (mean diameter of ~70 μm), for the arterial and venous trees, respectively, with a functional hierarchy for epicardial, transmural, and perfusion sub-networks. This result is consistent with the abrupt change in cross-sectional area and blood flow that demarcates the transition from epicardial (orders 8–11) to intramyocardial coronary arteries (transmural and perfusion vessels) (Kassab 2005).

The diameter ratio of large daughter to mother vessel (Fig. 9.3) reveals a functional hierarchy for the coronary arterial tree that discriminates epicardial, transmural, and perfusion sub-networks. The diameters ratio of small daughter to mother vessels (Fig. 9.5) decreases monotonically with order number, with the larger vessels being more asymmetric than smaller vessels. Figure 9.6 shows the relationship between AER and order number.

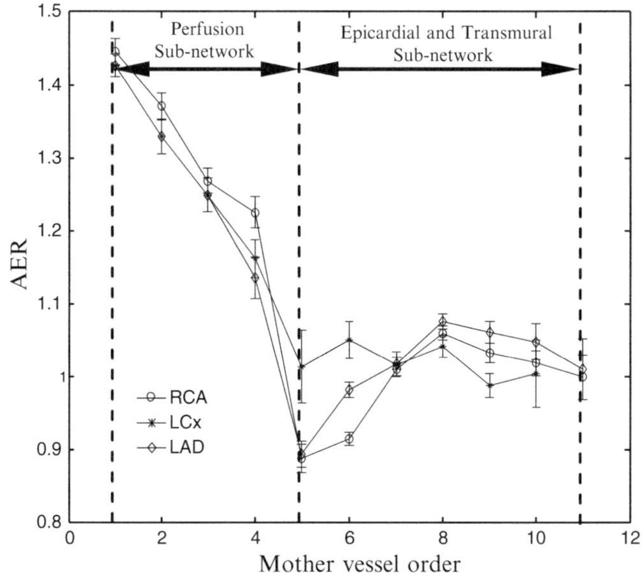

Fig. 9.5 Relationship between D_s/D_m (ratio of the diameter of smaller daughters relative to mother vessel) and mother vessel order number tree data of LAD, LCx, and RCA arteries

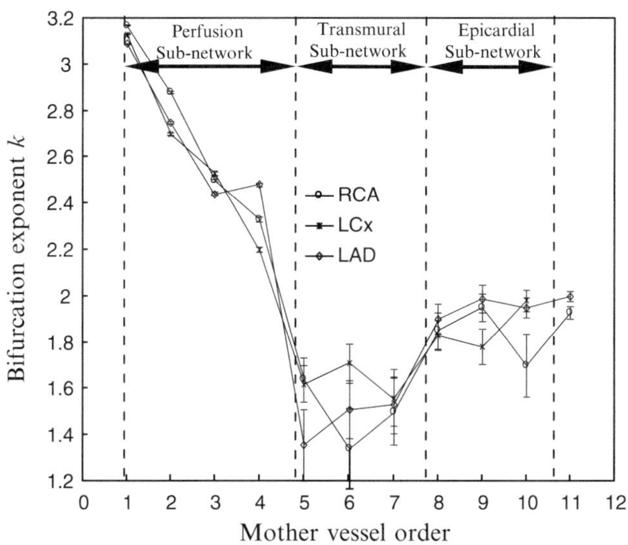

Fig. 9.6 Relationship between area expansion ratio $\left[\text{AER} = \left[D_l^2 + D_s^2\right]/D_m^2\right]$ and mother vessel order number for tree data of LAD, LCx, and RCA arteries

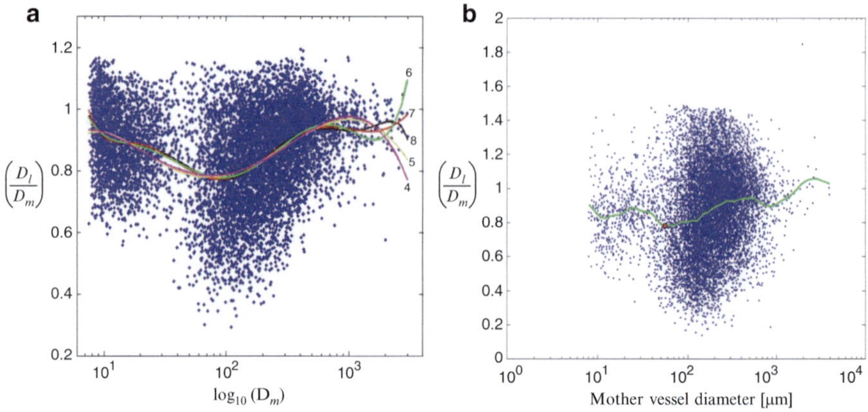

Fig. 9.7 Relationship between parameter k ($D_{\mathrm{m}}^{k} = D_{\mathrm{l}}^{k} + D_{\mathrm{s}}^{k}$) and mother vessel order number for arterial trees

$$\left[\left(\frac{D_{\mathrm{l}}}{D_{\mathrm{m}}}\right)^{2} + \left(\frac{D_{\mathrm{s}}}{D_{\mathrm{m}}}\right)^{2}\right]$$

Murray's law states that D_{m} cubed is equal to the sum of the cubes of the daughter vessels (i.e., $D_{\mathrm{m}}^{3} = D_{\mathrm{l}}^{3} + D_{\mathrm{s}}^{3}$) (Murray 1926). Figure 9.7 presents a general form of Murray's law at a bifurcation, i.e., $D_{\mathrm{m}}^{k} = D_{\mathrm{l}}^{k} + D_{\mathrm{s}}^{k}$. The bifurcation exponent k was determined for each bifurcation in the arterial trees. AER has values near unity in *orders 6–11* and monotonically increases toward *order 1*, which again reflects a functional distinction between perfusion and transmural and epicardial sub-networks.

The relationships between AER and flow velocity, and between ARM and flow distribution were considered as well (Kaimovitz et al. 2008). The hemodynamic connection to the AER can be made by using (9.1); namely:

$$\frac{U_{\mathrm{m}}}{U_{\mathrm{l}}} = \frac{U_{\mathrm{m}}}{U_{\mathrm{s}}} = \frac{A_{\mathrm{l}} + A_{\mathrm{s}}}{A_{\mathrm{m}}} = \left(\frac{D_{\mathrm{l}}}{D_{\mathrm{m}}}\right)^{2} + \left(\frac{D_{\mathrm{s}}}{D_{\mathrm{m}}}\right)^{2} = \mathrm{AER} \tag{9.1}$$

where U_{l}, U_{s}, and U_{m} are the mean velocities, and A_{l}, A_{s}, and A_{m} are the corresponding cross-section areas.

Using the flow analysis in Huo et al. (2007), the distribution of flow was calculated in the entire coronary arterial tree (Figs. 9.8 and 9.9). It is seen in Fig. 9.8 that the flow ratio (the ratio between the flows in large and small vessel, respectively) remains fairly uniform in the perfusion sub-networks and increases significantly in the epicardial and transmural vessels), which suggests that the microvasculature distal to order 5 may play a regulatory role in the arterial tree.

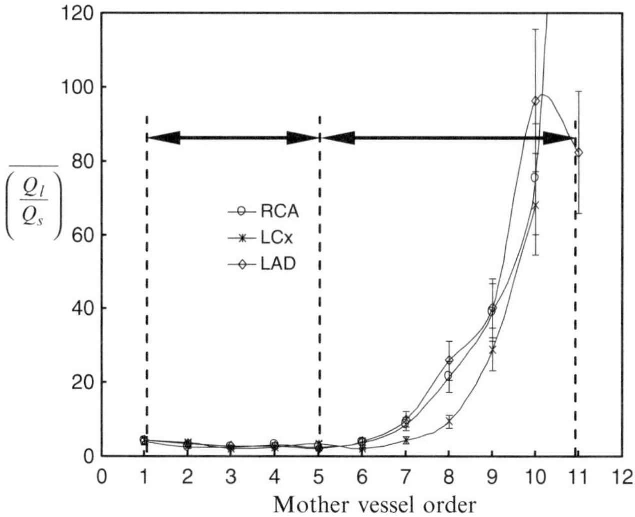

Fig. 9.8 Relation between the flow ratio ($\dot{Q}_\mathrm{l}$ and $\dot{Q}_\mathrm{s}$ are flows in large and small vessel, respectively) and mother vessel order number in the entire arterial tree of LAD, LCx, and RCA for the automatic scheme

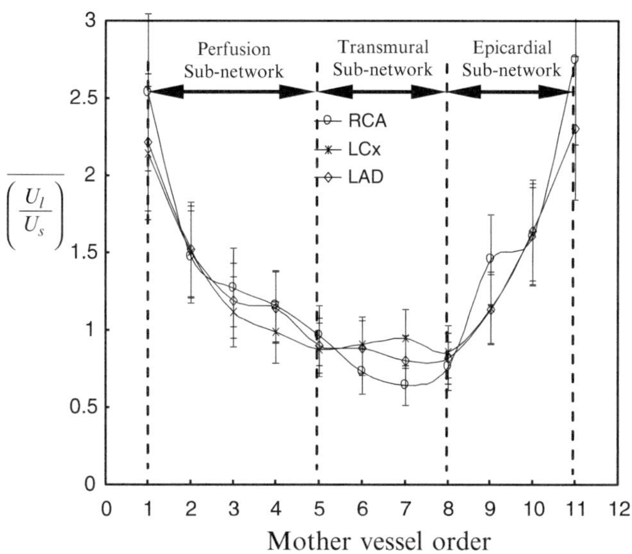

Fig. 9.9 Relation between velocity ratio (U_l and U_s are velocities in large and small vessel, respectively) and mother vessel order number in the entire arterial tree of LAD, LCx, and RCA for the automatic scheme

In comparison, the velocity ratio is about two orders of magnitude smaller than the flow ratio in all three networks. Furthermore, the transmural sub-network has a velocity ratio approximately equal to unity and increases at higher and lower orders in epicardial and perfusion sub-networks, respectively. The AER has a value close to unity in vessel orders >5 but increases toward vessel order 1 (Fig. 9.6). Figure 9.9 confirms that the velocity ratio is small such that $U_l \cong U_s$, respectively, in the transmural trees, where U_l and U_s are the large and small vessel velocities, respectively. Hence, as suggested by (9.1) (Kaimovitz et al. 2008) and Fig. 9.6, the velocity rapidly decreases from order 5 toward order 1, which is consistent with experimental measurements (Kassab 2005). This suggests that order 5 may mark a functional transition from epicardial–transmural to perfusion sub-networks.

The significance of the ratio D_l/D_m may be related to the "distributing" vessels that remain on the surface of distinct zones of the heart, as compared with the "delivering" vessels that penetrate the respective zones to implement the delivery of blood (Kassab 2005). The former constitute the epicardial vessels, whereas the latter constitute the transmural and perfusion sub-networks. The distributing arteries tend to maintain their diameter and, consequently, their flow fairly uniformly, so that the various regions of the heart can receive a similar source of blood supply (Kassab 2005). This is achieved when the vessels maintain their diameters (i.e., small tapering slope) and, accordingly, large D_l/D_m (Kassab 2005). The function of the transmural vessels is to transport blood to all layers within the myocardium, which is again accomplished with relatively large D_l/D_m values. As the transmural vessels approach their respective layers, however, their function gradually changes. The main function of order 5 (presumably the highest order of perfusion vessels) is to distribute blood equally in all directions in each layer or myocardial sheet (Nielsen et al. 1991). This requires the daughter diameters to be more similar (more symmetric), which is reflected in the decrease of D_l/D_m value toward D_s/D_m.

The wall shear stress, τ, is related to the exponent k, by $\tau \sim D_k^{-3}$ (Kaimovitz et al. 2008). If $k = 3$ (Murray's law), the wall shear stress is uniform over different size vessel. If $k = 2$ (AER = 1), the velocity is uniform, but the shear stress increases inversely with a decrease in diameter; i.e., amplifies in the microcirculation. The variation in k for different vessel sizes is shown in Fig. 9.7 and reflects the transitions in shear stress (Huo and Kassab 2007; Huo et al. 2007). Previous studies (Murray 1926; Rossitti and Lofgren 1993) proposed the minimum work and uniform shear stress as the optimization criteria for the design of arterial bifurcations. Murray postulated that an exponent k of 3 represents an optimized design. Other studies (Sherman 1981; Uylings 1977), however, showed significant scatter in the exponent. Our studies (Kassab 2006, 2007) found an exponent of ~2.3 for the entire coronary arterial tree, which deviates from Murray's prediction of 3. The results in Fig. 9.7 show that the exponent varies with different orders. This may be caused by reduced loss coefficients at bifurcations when the value of the exponent is approximately equal to 2 (Huo and Kassab 2007; Miller 1990). When AER increases in the lower orders, large loss coefficients coupled with other factors (e.g., the increase of total cross-sectional area, vessel number, and bifurcations due to the fractal nature of the coronary vasculature) may contribute to large energy losses. This loss of energy is necessary to reduce flow velocity to ensure sufficient transit time for transport of oxygen and nutrients in the capillaries. The results in Figs. 9.5 and 9.6 indicate that vessels in orders 5 and 4, for the arterial and venous trees, respectively, demarcate transition in structure and function. These findings may further support the role of microvasculature as a regulatory module or unit. In addition to the functional implications, the data are essential for faithful reconstructions of the coronary vasculature. Since the diameter is the most significant hemodynamic parameter, these data on the local diameter asymmetry can be used to physically constrain the reconstructions of full 3D coronary vascular networks.

9.3 Structure–Function Relations

The data show that there is a functional hierarchy for the design of the coronary vasculature. The relationships between the CSA plotted against the normalized cumulative length from root for the main trunk and primary branches exhibit a knee or abrupt change in trend (Fig. 9.10a). Interestingly, the transition of branching pattern for the CSA occurs at order 8 vessels ($\sim$400–500 μm in diameter).

Typically, orders 11, 10, and 9 are epicardial whereas orders $\leq$8 are intramural (Kassab 2005). Hence, the transition demarcates EPCA from intramural coronary artery (IMCA). The EPCA tend to maintain their CSA fairly uniform and serve to spread the larger branches over the surface of the heart to reach the entire surface area of the ventricles. Furthermore, the EPCA seem to maintain relatively uniform flow (Fig. 9.10b) so that the various regions of the heart can receive a similar source of blood supply (Kassab 2005).

An additional pattern can be seen for the flow velocity of the entire coronary arterial tree as shown in Fig. 9.10c. The velocity is relatively uniform throughout the larger vessels and abruptly decreases in the microcirculation. The transition corresponds to the diameter of order 2 vessels ($\sim$13 μm in diameter). This observation highlights the functional significances of various intramyocardial or delivering vessels: transition from conductive to transportive flow. The conduction vessels (orders 3–8) which maintain nearly constant CSA area (i.e., the exponent of flow-diameter relation is 2) function to conduct blood without reduction in velocity. In the smaller arterioles and capillaries (orders $\leq$2), the velocity must be reduced to ensure sufficient transit time and oxygen and nutrient transport across the capillaries. Hence, the exponent must increase to >2.

In terms of wave transition, a hybrid 1D/Womersley analysis (Huo and Kassab 2007) shows (Fig. 9.11) that the patterns of normalized flow waves are mutually similar at the major branches except for a phase angle shift in a vasodilated, potassium-arrested heart (Fig. 9.12). The observation that the normalized flow waves maintain similar patterns suggests scaling laws of coronary circulation (Huo et al. 2009).

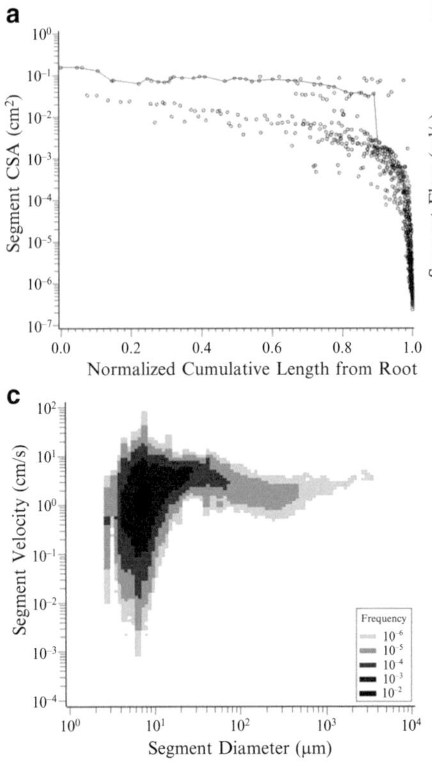

Fig. 9.10 (**a**, **b**) Relationship between segment cross-sectional area (**a**) and segment flow (**b**) and the normalized cumulative length of the segment from the root of the trunk and the primary branches of the right coronary artery (RCA) tree. The *solid line* denotes the main trunk. (**c**) Isodensity plot showing five layers of frequency between the velocity in a vessel segment and the corresponding diameter of the vessel for the RCA tree (Kassab 2005)

9.4 Morphometric Models of the Coronary Network

Several approaches have been used to model the architecture of coronary vasculature. The lumped approach relies on a single (Klocke et al. 1985; Spaan 1985) or limited number (Bruinsma et al. 1988; Chadwick et al. 1990; Kresh et al. 1990; Reneman and Arts 1985) of analog mechanical elements representing the resistance and compliance of the entire or portions of the coronary systems.

The distributive approach incorporates the network structure (Beard and Bassingthwaighte 2000; Mittal et al. 2005a, b; Smith et al. 2000, 2002). While the first approach ignores the network structure and topology but considers the dynamic aspect of coronary flow, the second allows for local perfusion analysis in a more realistic manner. Finally, the porous media approach (Vankan et al. 1997) ignores the vascular network structure and assumes that the high microvessels density can be

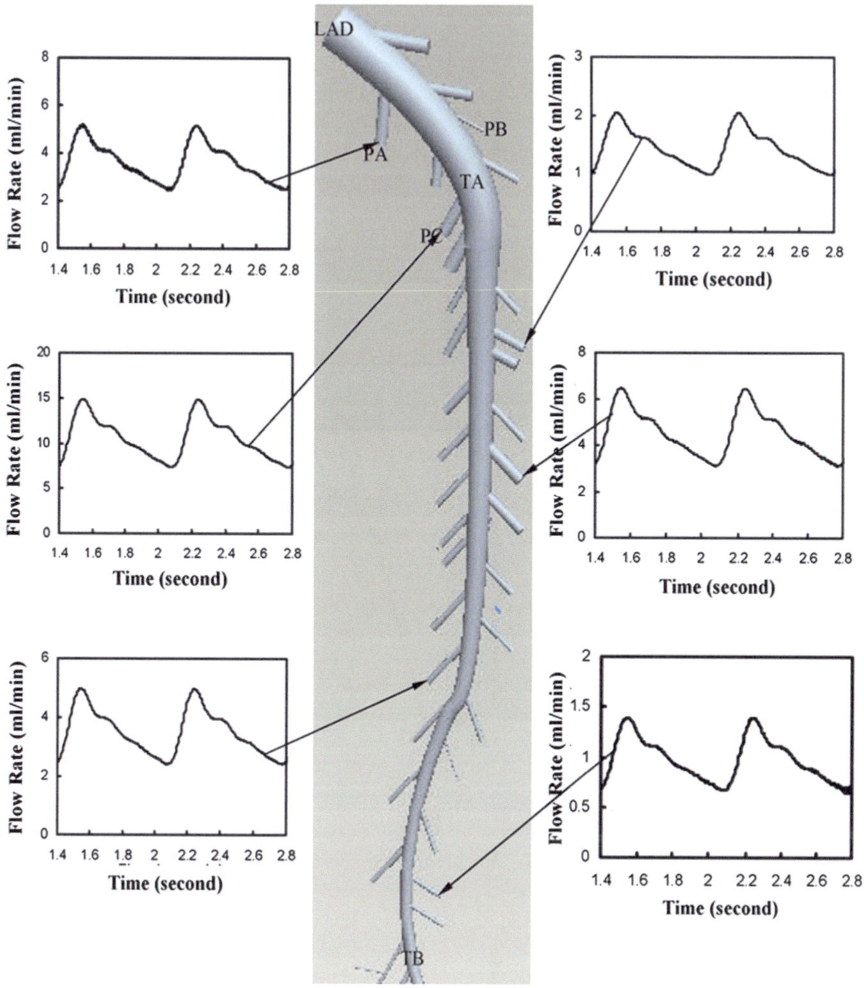

Fig. 9.11 Flow waves of the primary branches of LAD arterial trunk (Huo and Kassab 2007)

regarded as voids in porous media. Blood flow is represented by average quantities, using a hierarchical mixture model. An approach that combines the advantages of both a lumped hemodynamic model for each vessel, with the distributive network structure may significantly contribute to more realistic representation of the coronary flow.

9.5 3D Reconstruction of the Coronary Arterial Tree

For 3D reconstruction of the entire coronary arterial tree (Kaimovitz et al. 2005), the bifurcation asymmetry data was utilized to reproduce a realistic reconstruction of the coronary vascular tree. The model was based on a hierarchical scheme

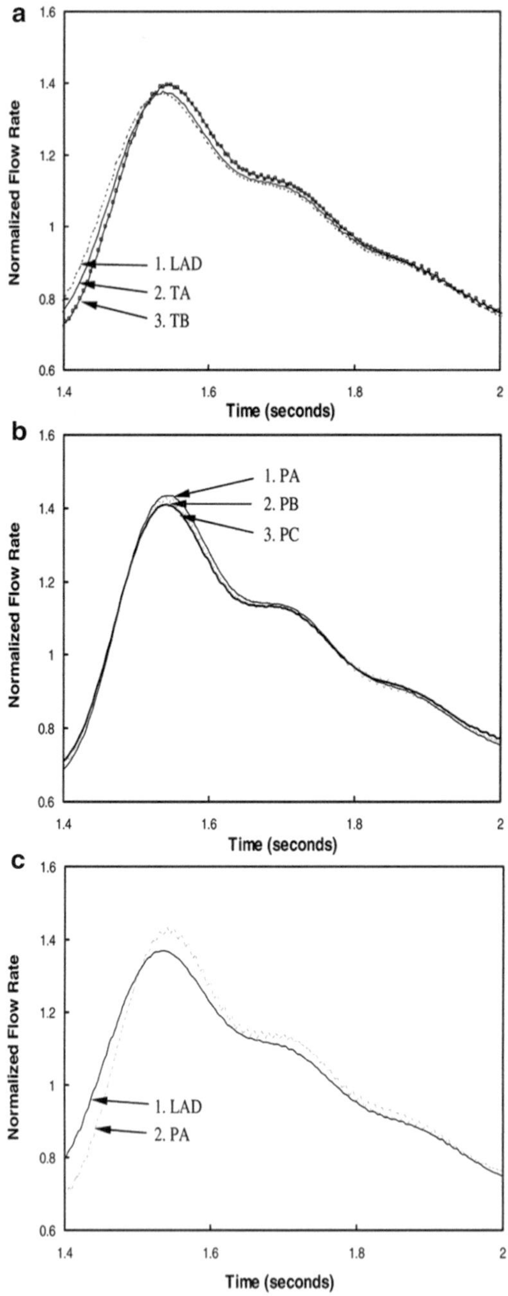

Fig. 9.12 (**a**) Flow waves normalized by time-averaged value in main trunk of LAD. (**b**) Flow waves normalized by mean value at inlet of primary branches of LAD. (**c**) Normalized flow waves at inlet of main trunk and one primary branch of LAD

consisting of epicardial, transmural, and perfusion sub-networks (Fig. 9.1), which facilitates efficient reconstruction algorithm which reconstructs each of the entities independently and hence, is amenable to parallel computing.

The epicardial sub-network was confined to orders 9–11. The selection of order 9 as the lower bound is based on the observed diameter (600–800 μm) of transmural vessels that perforate the myocardium at approximately right angles (Berne and Rubio 1979; Fulton 1982; Grayson 1982; James 1961; Tanaka et al. 1999; Zamir 1999). The transmural sub-networks were defined as vessels in the range of orders 8–5 to ensure their span over the heart wall thickness. Finally, the perfusion sub-networks spanned the remaining range starting from order 4 to order 0_a (arterial capillaries). The 3D coronary network geometry was reconstructed by implementing large-scale global optimization of each set of branching angles subject to the following: (1) published constraints on the planar branching geometry (Zamir 1978; Zamir and Brown 1982; Zamir et al. 1983, 1984), (2) avoidance rules intended to prevent intersections between segments, and (3) constraints which avoid protrusion of the myocardial wall surfaces. Based on Zamir et al. (1983), it was assumed that the arterial bifurcations are planar and satisfy the condition of minimum lateral drag force. Additionally, it was assumed that in an asymmetrical bifurcation, the branching angle of the larger daughter branch relative to its parent vessel decreases with increase in asymmetry ratio (Zamir 1978). Based on the latter observation, a set of bifurcation angular rules were formulated and implemented into the 3D network model. In order to model coronary flow dynamics, the geometric relationship between the contracting myocytes and microcirculatory vessels was ensured. Hence, the microcirculation, which encompasses the perfusion vessels down to the 0_a capillary vessels was reconstructed with the capillaries arranged in sheets along the direction of the myocytes bundles (Fulton 1982; Izumi et al. 1984) based on their location within the myocardial wall (Nielsen et al. 1991). The reconstruction procedure was common to all three coronary trees (RCA, LAD, and LCx) and was based on a hierarchical scheme by which the three types of sub-networks (epicardial, transmural, and perfusion) were generated independent of each other and meshed later.

Following the reconstruction of the network geometry, the diameter assignment procedure was generated based on the intra-element D_l/D_m and D_s/D_m inter-element mean and SD statistics (Kaimovitz et al. 2005).

The average total numbers of vessel elements obtained from the reconstructions were 2,559,721 ± 290,177; 3,928,390 ± 263,513; and 1,068,109 ± 30,358 for the RCA, LAD, and LCx trees, respectively. The corresponding epicardial branches (distributing vessels) provide coverage of the entire epicardial surface as shown in Fig. 9.13. Figure 9.14 demonstrates a view of the entire coronary arterial tree including the perfusion networks down to the 0_a capillary vessels. The transmural vessels branch from the epicardial vessels at almost right angle and course from the epicardial surface toward the endocardium to convey blood to the microcirculatory perfusion vessels at the inner layers (Baroldi et al. 1967; James 1961). Typical transmural and perfusion sub-networks are shown in Fig. 9.15 demonstrating the capillaries arrangement along the direction of the muscle fibers.

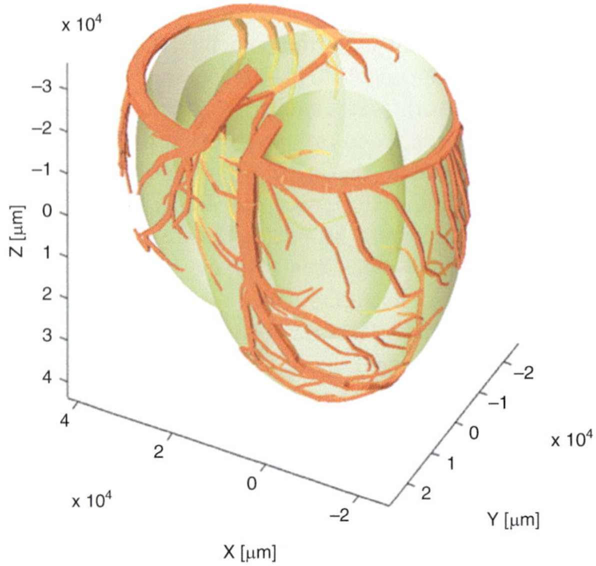

Fig. 9.13 The reconstructed epicardial branches (RCA, LAD, and LCx) following transformation to prolate spheroid surface as viewed from anterolateral oblique left aspect

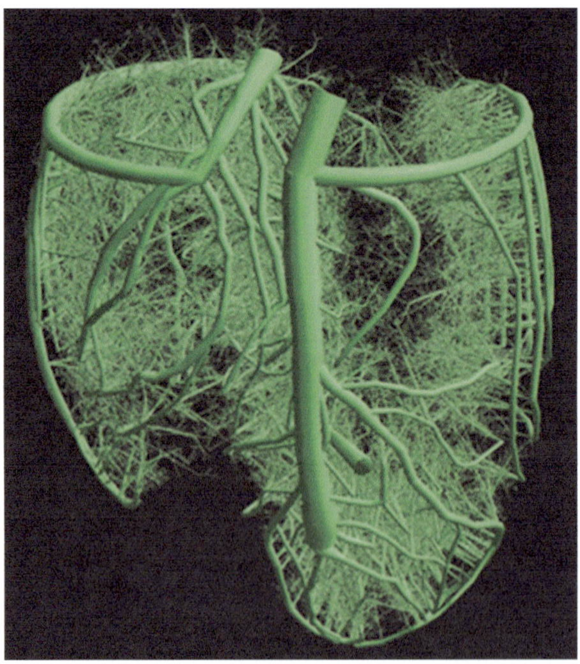

Fig. 9.14 A rendering of the entire coronary arterial tree consisting of the epicardial, transmural, and perfusion sub-networks

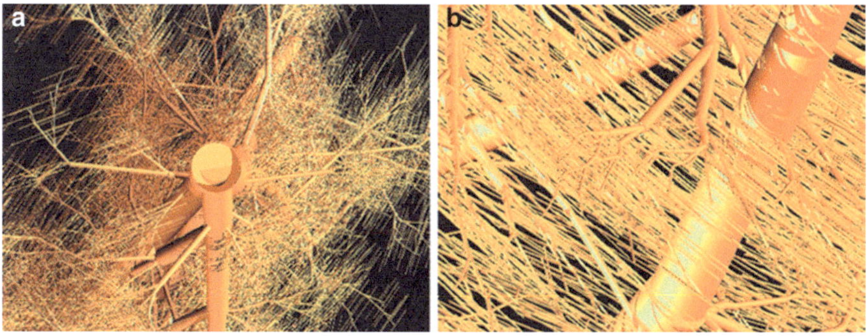

Fig. 9.15 (**a**) Inferior view of the transmural sub-tree demonstrating the change in capillary orientation along the myocardial wall. (**b**) Close-up view on the transmural sub-tree demonstrating the dense mesh of the capillaries

The arterial model is consistent with the anatomical characteristics observed in the real coronary networks. The reconstructed epicardial and perfusion sub-networks as well as the septal branches are confined to shell planes which are arranged continuously in parallel to the epicardial surface (Figs. 9.13 and 9.14). While the reconstructed epicardial vessels are aligned on the LV and RV surfaces, the septal branches which course circumferentially along the septum (Fig. 9.13) are located proximal to the right ventricular endocardium (James 1961).

In contrast to the epicardial sub-network, the reconstructed transmural sub-networks have a 3D structure. As dictated by the imposed branching geometry constraints above, the transmural reconstructed branches penetrate at nearly right angles (Berne and Rubio 1979; Fulton 1982; James 1987) (Fig. 9.14), while in the RV, they course obliquely toward the endocardium (Berne and Rubio 1979; Kassab et al. 1993). The reconstructed LV transmural branches are relatively short (total lengths range from 7 to 14 mm depending on the wall thickness of the LV and RV, respectively) since they typically course directly toward the endocardium. In few cases, the reconstructed sub-branches might be as long as 6 cm in agreement with native vessel anatomy (Baroldi et al. 1967; James 1961, 1987; McAlpine 1975). These are typically the perforating branches of the LAD artery and those that originate from the posterior descending artery (Fig. 9.13) and perfuse the anterior posterior interventricular septum (Berne and Rubio 1979; Gray et al. 1995; James 1961).

To further validate reconstructions, the blood volume and scaling exponents for the volume–length and CSA length relation were determined. It was found that the RCA and LCx arterial tree volumes to be similar to those reported by Kassab et al. (Kassab and Fung 1994; Kassab et al. 1993) as 1.4 ± 0.15 mL (RCA) and 0.57 ± 0.09 mL (LCx). The LAD volume, however, was approximately 50 % larger in the present model as compared to the previously reported value of 0.98 ± 0.21 mL (Kassab and Fung 1994; Kassab et al. 1993). It is likely that the greater blood volume for the LAD system stems from nearly twice the number of vessels obtained in the previous study (Kassab and Fung 1994; Kassab et al. 1993). The exponents of volume–length and CSA–length scaling power-laws were similar to the theoretical predictions reported by Huo and Kassab (2009, 2012).

The AER determined in the present study was similar to the relation $AER = 1.279 - 0.086 \cdot D_m$ ($r^2 = 0.024$) reported by VanBavel and Spaan (1992). The averaged value of the D_l/D_m ratio over all the orders was 0.877 ± 0.097, 0.859 ± 0.092, and 0.859 ± 0.092 for the LAD, RCA, and LCx, respectively, as compared to 0.905 ± 0.004 as reported by Van Bavel (1989).

Although the stochastic reconstruction was subject to a set of constraints, including both global (boundary avoidance) and local (bounds on branching angles), it was observed that the mean and SD of morphometric data (diameters, lengths, segment-to-element ratio, CM, and LPM) of the reconstructions were in good agreement with those of Kassab's database (Kassab and Fung 1994; Kassab et al. 1993, 1997) at each order number.

9.6 3D Reconstruction of the Entire Porcine Coronary Vasculature

The coronary arterial model was extended to include the capillaries and the entire coronary venous system. The reconstruction was based on a set of rules stemming from qualitative anatomical features (Bales 2004; Baroldi et al. 1967; Izumi et al. 1984; von Lüdinghausen 2003) and quantitative statistical morphometric data on the coronary Sinual tree (Choy and Kassab 2009; Kassab et al. 1994). The coronary venous network was partitioned into three major branches in parallel to the respective arterial LAD, LCx, and RCA sub-trees. An average $0_v/0_a$ ratio (the ratio between number of venous and arterial functional capillaries) of 1.61:1 ($2 \times 2.56/3.18$) for the LAD and LCx and 1.86:1 ($2 \times 2.56/2.75$) for the RCA was imposed. Similar to the coronary arterial tree (Kaimovitz et al. 2005), each venous branch was partitioned longitudinally into functional epicardial sub-networks (orders −9 to −12 vessels), transmural (orders −5 to −8 vessels), and perfusion networks (orders −4 to 0_v). The epicardial portion was generated by a Simulated Annealing search for the optimal coverage of the area perfused by the arterial epicardial vessels. The epicardial sub-network and the coronary arterial capillary network served as boundary conditions for the reconstruction of the in-between transmural and perfusion networks which were generated to optimize vascular homogeneity (uniform vascular density). The venous network reconstruction was formulated as a large-scale optimization process subject to local and global constraints. The optimization criteria enforced optimal coverage of the surface covered by the epicardial arterial sub-networks and drainage compatibility with the arterial capillary network subject to the constraints on the $0_v/0_a$ ratio listed above as well as capillary length and orientation. It was assumed that vessels in each of the LAD, RCA, and LCx sub-networks independently follow the respective arterial major vessels; i.e., there is no connection between the three venous sub-networks.

The diameter assignment algorithm was based on the branching asymmetry at branching points along the venous tree which was extracted from database of native trees. For this purpose, the branching asymmetry statistics (mean ± SD) was calculated from the raw database (Kassab et al. 1994) for the coronary Sinusal tree, in terms of the AER and the ratios of the diameters of larger (l) and smaller (s) daughter vessels to their mother (m) vessel (D_l/D_m and D_s/D_m), respectively, as a function of order number (Kassab et al. 1994). The assignment was done in a top-down direction starting with the network stem and moved toward the 0_v capillaries.

Five sets of full coronary venous trees down to the capillary level were reconstructed coupled to the coronary arterial trees (Kaimovitz et al. 2005). The total number of reconstructed venous segments was 17,148,946 ± 1,049,498 ($n = 5$) which spans orders −12 (coronary sinus) to 0_v capillaries (first segment of venous capillary). Combined with the reconstructed arterial network, the number of vessel segments for the entire coronary network adds up to 27,307,376 ± 1,155,359 ($n = 5$). The reconstructed network agrees with the gross anatomy of coronary vasculature in terms of structure, location of major vessels (Fig. 9.16), and measured morphometric statistics of native coronary networks.

The epicardial and septal vessels are aligned along the prolate spheroid surfaces (Streeter 1979) of the LV and RV (Fig. 9.16). The major venous epicardial vessels (Fig. 9.16) follow the correct course and provide coverage of the entire epicardial surface perfused by corresponding reconstructed arterial epicardial vessels (LAD, LCx, and RCA) (Baroldi et al. 1967; Grayson 1982; Hutchins et al. 1986; Kajiya and Goto 1999) and drain their respective networks (von Lüdinghausen 2003; Mohl et al. 1984). This includes the small cardiac vein (SCV) which follows the route of the RCA artery and branches off to the posterior interventricular vein (PIV) which accompanies the posterior interventricular artery. Similarly, the great cardiac vein (GCV) follows the LCx and gives rise to the lateral LV veins, the posterior vein of the left ventricle (PVLV), and the anterior interventricular vein (AIV), which runs in parallel to the anterior interventricular artery. Finally, the coronary sinus, which drains most of the epicardial ventricular veins (von Lüdinghausen 2003) is situated at the posterior region of the coronary sulcus and drains the GCV, SCV, PVLV, and PIV. The septal veins originate anteriorly and posteriorly from the AIV and PIV, respectively (Baroldi et al. 1967).

All the major veins collect the smaller veins which run on the LV and RV surfaces (order −12 to −9) and run through the myocardium at almost right angles to reach orders −8 to −5 (Fig. 9.16c). The reconstructed transmural venous branches are independent of their arterial counterparts in terms of both their location and disposition while only some of the principle epicardial vessels coincide with the arrangement of the principle arteries (Fig. 9.16a, b), similar to anatomical observations (Baroldi et al. 1967). The venous branches are more numerous than the arterial ones (Bales 2004; Kassab et al. 1994; von Lüdinghausen 2003) as evident from both the opaque appearance of the combined network (Fig. 9.16), in line

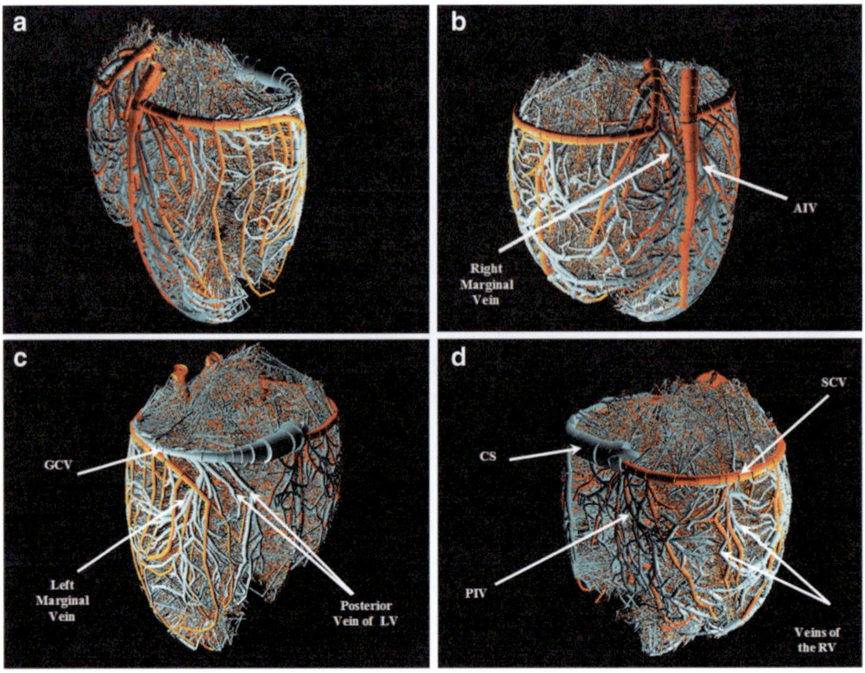

Fig. 9.16 Rendering of the reconstructed arterial and venous trees (orders 1–11 and −1 to −12) as viewed from four different aspects: (**a**) anterolateral left, (**b**) anterolateral right, (**c**) posterolateral left, and (**d**) posterolateral right. The rendering was done using POVWIN™ Raytracer. *Orange*—arterial, *cyan*—venous

with anatomical observations (Baroldi et al. 1967; James 1961; Mohl et al. 1984), and from quantitative evidence. For example, the reconstructed venous network encompasses 1.69 ± 0.10 times the number of vessels than in the reconstructed coronary arterial tree. The average and SD of the $0_v/0_a$ capillaries ratio for the entire reconstructed venous network was found to be 1.63 ± 0.02:1 and 1.87 ± 0.005:1 for the venous LAD/LCx and RCA, respectively, compared to the estimated values of 1.61:1 and 1.86:1. The percentage of branching types was found to be 85.4 ± 0.34 %, 13.3 ± 0.29 %, 1.10 ± 0.04 %, and 0.18 ± 0.01 % for bifurcations, trifurcations, quadrifications, and quintifications, respectively, compared to 86 %, 12.8 %, 1 %, and 0.2 % in the native trees (Kassab et al. 1994).

The coronary system was found to have fractal characteristics (Zamir 1999). A system which demonstrates self-similarity is termed "fractal" and obeys Horton's law, which suggests that a geometric relationship exists between system attributes. In the reconstructed venous tree, the values of the reconstructed venous networks of R_L (mean element length ratio) were 2.1 ± 0.91 and 1.16 ± 0.009, and the values obtained for R_N (mean element branching ratio) and R_D (mean diameters of successive generations) were 3.30 ± 0.03 and 1.67 ± 0.17, respectively, which were in good agreement with measure values for the Sinusal network (Kassab et al. 1994).

The total volume for the venous and arterial trees was obtained by adding the capillary volume to the volumes of the venous and arterial trees. Accordingly, the total mass normalized CBV obtained for the venous and arterial trees was 14.6 ± 0.56 mL/100 g and 7.32 ± 0.31 mL/100 g, respectively, and accordingly, the total myocardial mass normalized CBV was 21.9 ± 0.64 mL/100 g. This estimate, which is within the bounds of reported data (Feigl 1983; Hoffman and Spaan 1990; Olsson and Bugni 1986; Spaan 1985), is nearly twice the CBV estimated from the morphometric database (Kassab et al. 1994). The difference, in absolute volume, between the reconstruction and the database is consistent with the 1.8 difference in number of vessels between the reconstruction and the database. Reports on measured volumes of the coronary system have large variation (Feigl 1983; Hoffman and Spaan 1990; Olsson and Bugni 1986; Spaan 1985) which may stem from biological variability and experimental error. Similarly, large variability exists in estimates of the number of coronary vessels (Kassab et al. 1993, 1994; Spaan 1991). In addition, some variability is expected in stochastic reconstructions, depending on range of distribution parameters in the statistical database.

Finally, the LV was divided into slabs with an average volume of 0.017 mL. The number of capillaries in each slab was computed and averaged over the subepicardium, midwall, and subendocardium to yield mean transmural values. The calculated ratio of the mean number of capillaries between the subendocardium and the subepicardium was 1.33 ± 0.11 ($n = 5$). This value agrees with the reports of greater coronary blood volume in the subendocardial than the subepicardial layer in the LV, and is similar to reported values of 1.10–1.55 for endo-to-epi ratios of red blood cell content (Crystal et al. 1981; Howe and Winbury 1973; Myers and Honig 1964; Weiss and Winbury 1974).

In summary, the reconstructed venous trees demonstrate similarity in their native 3D structure. Furthermore, they show good fit to the measured data both quantitatively and qualitatively; i.e., the overall fit follows the general behavior of the database, and usually falls within ± 1 SD of the measured data. The effect of these small deviations on local hemodynamics and flow distribution is likely to be small since these deviations are primarily affected by the asymmetry of the network (D_l/D_m and D_s/D_m), a feature of the reconstructed tree which demonstrates excellent fit to the database. All these results serve as validation of the anatomical model.

9.7 Physiological Significance of the 3D Model

Based on the reconstructed arterial 3D model, the regional myocardial flow heterogeneity was calculated using flow simulation in the RCA, LAD, and LCx arterial branches (Huo et al. 2009). The flow model predictions were validated experimentally with the use of nonradioactive fluorescent microspheres of 15 μm

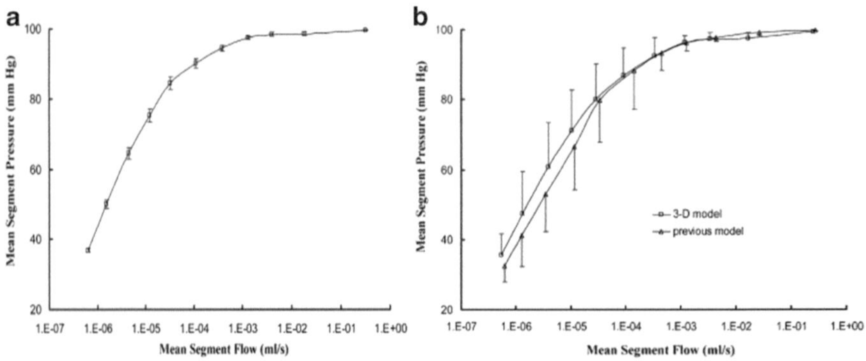

Fig. 9.17 (**a**) Mean ± 1 SD for pressure-flow relationship in five reconstructions of 3D models (both mean pressure and mean flows over each order are averaged over five reconstructions). (**b**) Comparison of pressure-flow relationships of blood using the LCX in one representative 3D model and a previous LCX arterial tree model (Mittal et al. 2005b)

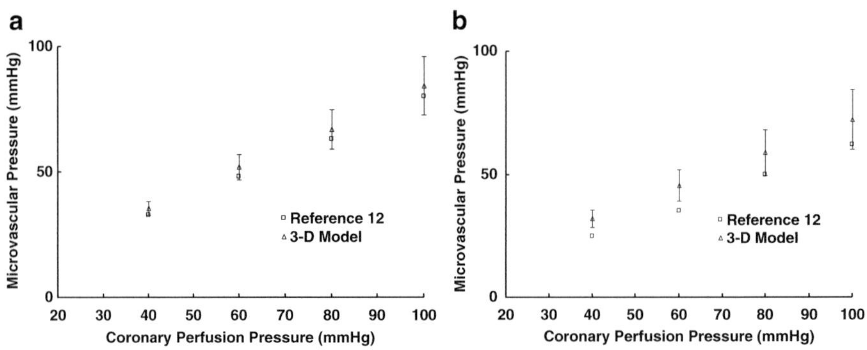

Fig. 9.18 Arteriolar pressures in the (**a**) subepicardial and (**b**) subendocardial microcirculation at orders 5 and 6 (with vessel diameter range of 64.4∼150 μm) of LAD arterial tree at the different coronary perfusion pressures, which agrees reasonably with the data (Chilian 1991)

diameter (Huo et al. 2009). Figure 9.17 shows on panel (a) the mean ± 1 SD (both mean pressure and mean flow over each order are averaged in five reconstructions) for the pressure-flow relationship of blood in five reconstructions of the 3D tree models. The 3D model longitudinal pressure-flow predictions were validated by comparison to the results of the topologic model of Mittal et al. (2005b) (panel (b)) (as indicated above, this analysis can be done realistically based on the topological 2D primitive tree), and were found to be similar (small SD) in longitudinal distribution of pressure-flow.

Further validation included comparison of the arteriolar pressures predictions of the 3D model in the subepicardial and subendocardial to the measured data of Chilian (1991). The comparison results are depicted in Fig. 9.18a, b, which reveal an excellent match.

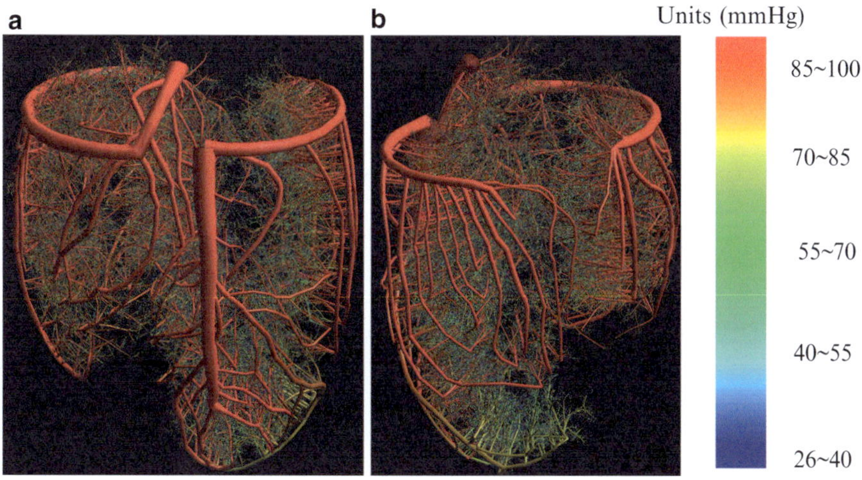

Fig. 9.19 Pressure distribution in two views (lateral left and posterolateral oblique left) in the 3D entire coronary arterial trees consisting of the epicardial, transmural, and perfusion sub-networks: (**a**) lateral-left-view pressure and (**b**) posterolateral-oblique-left-view pressure

These comparisons demonstrate the capability of the arterial 3D model to predict longitudinal pressure-flow distributions. Figure 9.19a, b illustrates the pressure distribution, in two views (lateral left and posterolateral oblique left), of the entire coronary arterial tree model down to the first capillary segments. It is clear that the longitudinal pressure distribution is fairly uniform in larger vessels and changes significantly in smaller vessels (<100 μm).

Since the 2D primitive model prediction is limited to longitudinal distribution of pressure-flow, validation of the 3D model capability to predict spatial flow distribution was done by comparison of the model analysis results with the experimental data (Huo et al. 2009). The current 3D model predicted spatial heterogeneity of flow and showed good agreement with the experimental data. The average myocardial flow prediction of the model was 2.39 mL/min/g compared to 2.17 mL/min/g in the experiment. The relative flow dispersion, defined as RD = SD/Mean, was found to be 44 % and 48 % for the numerical model and the experiment, respectively. The numerical flow dispersion was found to have fractal characteristics (Fig. 9.20) with fractal dimension D of 1.25 vs. 1.27 in the experiment and compared to 1.23 as reported by Bassingthwaighte et al. (1989) where D is defined by (9.2).

$$\mathrm{RD}(m) = \mathrm{RD}\left(m_{\mathrm{ref}}\right) \cdot \left[\frac{m}{m_{\mathrm{ref}}}\right]^{1-D} \tag{9.2}$$

An important attribute of the 3D reconstruction is to predict the coronary flow heterogeneity, based on a mathematical model of the 3D coronary arterial tree in the myocardium. Results of the numerical flow analysis (Huo et al. 2009) clearly

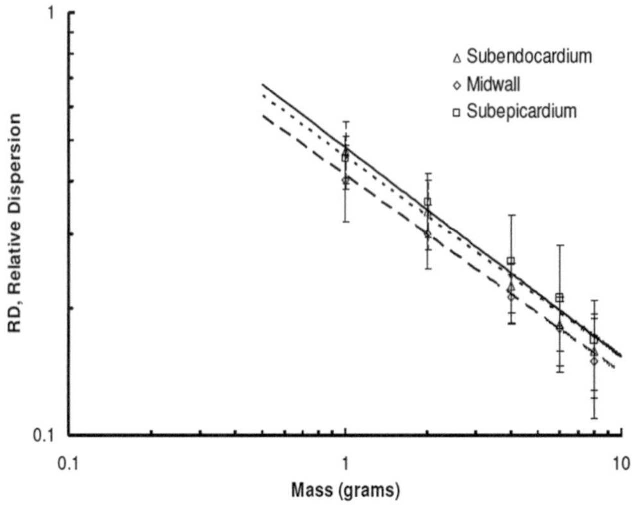

Fig. 9.20 The fractal regression for spatial flow in three layers of the LV and septum (subepicardium, midwall, and subendocardium) obtained from experiments (mean ± 1 SD) and the 3D model. The *solid*, *dashed*, and *dotted lines* represent the results of the 3D model for the subendocardium, midwall, and subepicardium, respectively. A power fit for fractal regression in the 3D model shows exponents of 1.47 ($R^2 = 0.99$), 1.47 ($R^2 = 0.98$), and 1.49 ($R^2 = 0.98$) for the subepicardium, midwall, and subendocardium, respectively. These agree well with the experimental values (six hearts) of 1.47, 1.45, and 1.51 for the subepicardium, midwall, and subendocardium, respectively

demonstrate that the fractal nature of regional myocardial blood flow heterogeneity can be simulated by the 3D model. As such, the reconstruction can serve as a basic platform for detailed, realistic large-scale distributive analysis of the coronary flow. Since the 3D venous network is based on the structure of the 3D arterial network, the integrative arterio-venous reconstructed network is expected to preserve the spatial heterogeneity characteristics of flow and pressure.

The reconstruction, which includes a detailed, distributive description of the 3D geometry and morphometry of the coronary network (Fig. 9.21), constitutes an excellent platform to include flow control in future studies, with local, global and autoregulation mechanisms. Furthermore, the reconstruction not only provides a predictive model of coronary circulation, which can be used to test quantitative hypotheses and address basic issues in coronary physiology and pathophysiology, but it can also serve as a valuable educational tool. The reconstruction, which has a strong visual component, can be set up in an interactive, user-friendly mode which will become a valuable tool for cardiovascular scientists, physiologists, bioengineers, cardiologists, cardiac surgeons, and students.

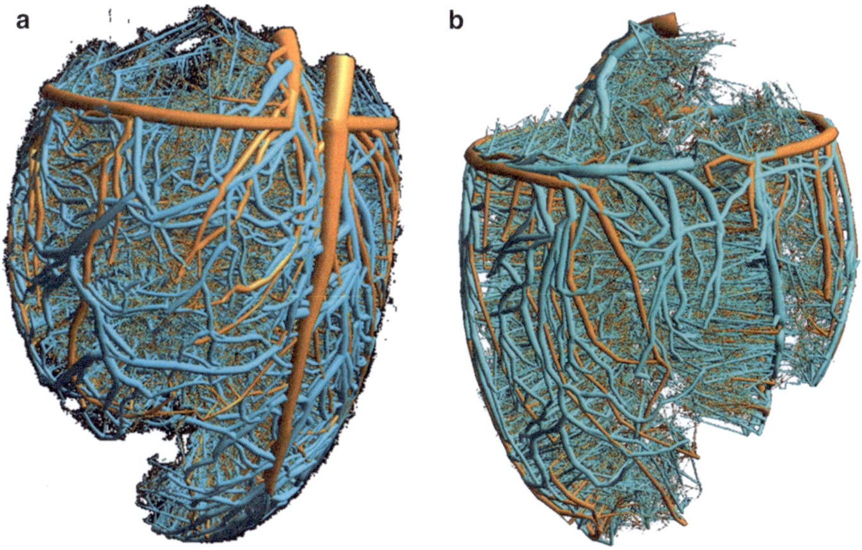

Fig. 9.21 Rendering of the reconstructed arterial and venous trees, covering orders 0_a to 11 and 0_v to -12 as viewed from: (**a**) right anterolateral aspect, (**b**) posterior aspect. Overall, about 27 million vessels were rendered

References

Ansari A. Anatomy and clinical significance of ventricular Thebesian veins. Clin Anat. 2001;14:102–10.

Bales GS. Great cardiac vein variations. Clin Anat. 2004;17:436–43.

Baroldi G, Scomazzoni, G. Coronary circulation in the normal and the pathologic heart. Washington, DC: Office of the surgeon general. Department of the army; 1967.

Bassingthwaighte JB, Yipintsoi T, Harvey RB. Microvasculature of the dog left ventricular myocardium. Microvasc Res. 1974;7:229–49.

Bassingthwaighte JB, King RB, Roger SA. Fractal nature of regional myocardial blood flow heterogeneity. Circ Res. 1989;65:578–90.

Beard DA, Bassingthwaighte JB. The fractal nature of myocardial blood flow emerges from a whole-organ model of arterial network. J Vasc Res. 2000;37:282–96.

Berne R, Rubio R. Coronary circulation. In: Berne RM, editor. Handbook of physiology. Section 2. The cardiovascular system, The heart, vol. 1. Baltimore: American Physiological Society; 1979.

Bruinsma P, Arts T, Dankelman J, Spaan JA. Model of the coronary circulation based on pressure dependence of coronary resistance and compliance. Basic Res Cardiol. 1988;83:510–24.

Chadwick RS, Tedgui A, Michel JB, Ohayon J, Levy BI. Phasic regional myocardial inflow and outflow: comparison of theory and experiments. Am J Physiol. 1990;258:H1687–98.

Chilian WM. Microvascular pressures and resistances in the left ventricular subepicardium and subendocardium. Circ Res. 1991;69:561–70.

Chilian WM, Marcus ML. Coronary venous outflow persists after cessation of coronary arterial inflow. Am J Physiol. 1984;247:H984–90.

Chilian WM, Layne SM, Klausner EC, Eastham CL, Marcus ML. Redistribution of coronary microvascular resistance produced by dipyridamole. Am J Physiol. 1989;256:H383–90.

Choy JS, Kassab GS. Wall thickness of coronary vessels varies transmurally in the LV but not the RV: implications for local stress distribution. Am J Physiol Heart Circ Physiol. 2009;297:H750–8.

Crystal GJ, Downey HF, Bashour FA. Small vessel and total coronary blood volume during intracoronary adenosine. Am J Physiol. 1981;241:H194–201.

Feigl EO. Coronary physiology. Physiol Rev. 1983;63:1–205.

Fulton W. Morphometry of the myocardial circulation in microcirculation of the heart: theoretical and clinical problems. New York: Springer; 1982.

Gray H, Williams PL, Bannister LH. Gray's anatomy: the anatomical basis of medicine and surgery. New York: Churchill Livingstone; 1995.

Grayson J. Functional morphology of the coronary circulation in the coronary artery. New York: Oxford University Press; 1982.

Hester RL, Hammer LW. Venular-arteriolar communication in the regulation of blood flow. Am J Physiol Regul Integr Comp Physiol. 2002;282:R1280–5.

Hiramatsu O, Goto M, Yada T, Kimura A, Chiba Y, Tachibana H, Ogasawara Y, Tsujioka K, Kajiya F. In vivo observations of the intramural arterioles and venules in beating canine hearts. J Physiol. 1998;509(Pt 2):619–28.

Hoffman JI, Spaan JA. Pressure-flow relations in coronary circulation. Physiol Rev. 1990;70:331–90.

Howe BB, Winbury MM. Effect of pentrinitrol, nitroglycerin and propranolol on small vessel blood content of the canine myocardium. J Pharmacol Exp Ther. 1973;187:465–74.

Huo Y, Kassab GS. A hybrid one-dimensional/Womersley model of pulsatile blood flow in the entire coronary arterial tree. Am J Physiol Heart Circ Physiol. 2007;292:H2623–33.

Huo Y, Kassab GS. The scaling of blood flow resistance: from a single vessel to the entire distal tree. Biophys J. 2009;96:339–46.

Huo Y, Kassab GS. Intraspecific scaling laws of vascular trees. J R Soc Interface. 2012;9:190–200.

Huo Y, Linares CO, Kassab GS. Capillary perfusion and wall shear stress are restored in the coronary circulation of hypertrophic right ventricle. Circ Res. 2007;100:273–83.

Huo Y, Kaimovitz B, Lanir Y, Wischgoll T, Hoffman JI, Kassab GS. Biophysical model of the spatial heterogeneity of myocardial flow. Biophys J. 2009;96:4035–43.

Hutchins GM, Moore GW, Hatton EV. Arterial-venous relationships in the human left ventricular myocardium: anatomic basis for countercurrent regulation of blood flow. Circulation. 1986;74:1195–202.

Izumi T, Yamazoe M, Shibata A. Three-dimensional characteristics of the intramyocardial microvasculature of hypertrophied human hearts. J Mol Cell Cardiol. 1984;16:449–57.

Jain AK, Smith EJ, Rothman MT. the coronary venous system: an alternative route of access to the myocardium. J Invasive Cardiol. 2006;18:563–8.

James TN. Anatomy of the coronary arteries. New York: P.B. Hoeber; 1961.

James T. Anatomy and pathology of small coronary arteries in coronary circulation: from basic mechanisms to clinical implications. Leiden: Nijhoff; 1987.

Jones CJ, Kuo L, Davis MJ, Chilian WM. Distribution and control of coronary microvascular resistance. Adv Exp Med Biol. 1993;346:181–8.

Kaimovitz B, Lanir Y, Kassab GS. Large-scale 3-D geometric reconstruction of the porcine coronary arterial vasculature based on detailed anatomical data. Ann Biomed Eng. 2005;33:1517–35.

Kaimovitz B, Huo Y, Lanir Y, Kassab GS. Diameter asymmetry of porcine coronary arterial trees: structural and functional implications. Am J Physiol Heart Circ Physiol. 2008;294:H714–23.

Kaimovitz B, Lanir Y, Kassab GS. A full 3-D reconstruction of the entire porcine coronary vasculature. Am J Physiol Heart Circ Physiol. 2010;299:H1064–76.

Kajiya F, Goto M. Integrative physiology of coronary microcirculation. Jpn J Physiol. 1999;49:229–41.

Kalsho G, Kassab GS. Bifurcation asymmetry of the porcine coronary vasculature and its implications on coronary flow heterogeneity. Am J Physiol Heart Circ Physiol. 2004;287:H2493–500.

Kassab GS. Functional hierarchy of coronary circulation: direct evidence of a structure-function relation. Am J Physiol Heart Circ Physiol. 2005;289:H2559–65.
Kassab GS. Scaling laws of vascular trees: of form and function. Am J Physiol Heart Circ Physiol. 2006;290:H894–903.
Kassab G. Design of coronary circulation: a minimum energy hypothesis. Comput Methods Appl Mech Eng. 2007;196:3033–42.
Kassab GS, Fung YC. Topology and dimensions of pig coronary capillary network. Am J Physiol. 1994;267:H319–25.
Kassab GS, Rider CA, Tang NJ, Fung YC. Morphometry of pig coronary arterial trees. Am J Physiol. 1993;265:H350–65.
Kassab GS, Lin DH, Fung YC. Morphometry of pig coronary venous system. Am J Physiol. 1994;267:H2100–13.
Kassab GS, Pallencaoe E, Schatz A, Fung YC. Longitudinal position matrix of the pig coronary vasculature and its hemodynamic implications. Am J Physiol. 1997;273:H2832–42.
Kassab GS, Navia JA, March K, Choy JS. Coronary venous retroperfusion: an old concept, a new approach. J Appl Physiol. 2008;104:1266–72.
Kaul S, Ito H. Microvasculature in acute myocardial ischemia: part I: evolving concepts in pathophysiology, diagnosis, and treatment. Circulation. 2004;109:146–9.
Klocke FJ, Mates RE, Canty Jr JM, Ellis AK. Coronary pressure-flow relationships. Controversial issues and probable implications. Circ Res. 1985;56:310–23.
Kresh JY, Fox M, Brockman SK, Noordergraaf A. Model-based analysis of transmural vessel impedance and myocardial circulation dynamics. Am J Physiol. 1990;258:H262–76.
Kuo L, Arko F, Chilian WM, Davis MJ. Coronary venular responses to flow and pressure. Circ Res. 1993;72:607–15.
Loukas M, Bilinsky S, Bilinsky E, el-Sedfy A, Anderson RH. Cardiac veins: a review of the literature. Clin Anat. 2009;22:129–45.
McAlpine W. Heart and coronary arteries: an anatomical atlas for clinical diagnosis, radiological investigation, and surgical treatment. New York: Springer; 1975.
Miller DS. Internal flow systems. Cranfield: BHRA; 1990.
Mittal N, Zhou Y, Ung S, Linares C, Molloi S, Kassab GS. A computer reconstruction of the entire coronary arterial tree based on detailed morphometric data. Ann Biomed Eng. 2005a;33(8):1015–26.
Mittal N, Zhou Y, Linares C, Ung S, Kaimovitz B, Molloi S, Kassab GS. Analysis of blood flow in the entire coronary arterial tree. Am J Physiol Heart Circ Physiol. 2005b;289:H439–46.
Mohl W, Wolner E, Glogar D. The coronary sinus: proceedings of the 1st international symposium on myocardial protection via the coronary sinus. Darmstadt: Steinkopff; 1984.
Murray CD. The physiological principle of minimum work applied to the angle of branching of arteries. J Gen Physiol. 1926;9:835–41.
Myers WW, Honig CR. Number and distribution of capillaries as determinants of myocardial oxygen tension. Am J Physiol. 1964;207:653–60.
Nielsen PM, Le Grice IJ, Smaill BH, Hunter PJ. Mathematical model of geometry and fibrous structure of the heart. Am J Physiol. 1991;260:H1365–78.
Olsson RAB, Bugni WJ. Coronary circulation. In: Fozzard HA, editor. The heart and cardiovascular system: scientific foundations. New York: Raven; 1986. p. 2 v. (1694p.).
Reneman RS, Arts T. Dynamic capacitance of epicardial coronary arteries in vivo. J Biomech Eng. 1985;107:29–33.
Rossitti S, Lofgren J. Vascular dimensions of the cerebral arteries follow the principle of minimum work. Stroke. 1993;24:371–7.
Sherman TF. On connecting large vessels to small. The meaning of Murray's law. J Gen Physiol. 1981;78:431–53.
Smith NP, Pullan AJ, Hunter PJ. Generation of an anatomically based geometric coronary model. Ann Biomed Eng. 2000;28:14–25.
Smith N, Pillan AJ, Hunter PJ. Ana anatomically based model of transient coronary blood flow in the heart. SIAM J Appl Math. 2002;62:990–1018.

Spaan JA. Coronary diastolic pressure-flow relation and zero flow pressure explained on the basis of intramyocardial compliance. Circ Res. 1985;56:293–309.

Spaan JAE. Coronary blood flow: mechanics, distribution, and control. Dordrecht: Kluwer Academic Publishers; 1991.

Streeter DJ. Gross morphometry and fiber geometry of the heart. In: Berne RM, Sperelakis N, Geiger SR, editors. Handbook of physiology. Section 2. The cardiovascular system, The heart, vol. 1. Baltimore: American Physiological Society; 1979.

Tanaka A, Mori H, Tanaka E, Mohammed MU, Tanaka Y, Sekka T, Ito K, Shinozaki Y, Hyodo K, Ando M, Umetani K, Tanioka K, Kubota M, Abe S, Handa S, Nakazawa H. Branching patterns of intramural coronary vessels determined by microangiography using synchrotron radiation. Am J Physiol. 1999;276:H2262–7.

Toyota E, Koshida R, Hattan N, Chilian WM. Regulation of the coronary vasomotor tone: what we know and where we need to go. J Nucl Cardiol. 2001;8:599–605.

Uylings H. Optimization of diameter and bifurcation angles in lung and vascular tree structures. Bull Math Biol. 1977;39:509–19.

Van Bavel E. Metabolic and myogenic control of blood flow studied on isolated small arteries. Ph.D., University of Amsterdam; 1989.

VanBavel E, Spaan JA. Branching patterns in the porcine coronary arterial tree. Estimation of flow heterogeneity. Circ Res. 1992;71:1200–12.

Vankan WJ, Huyghe JM, Slaaf DW, van Donkelaar CC, Drost MR, Janssen JD, Huson A. Finite-element simulation of blood perfusion in muscle tissue during compression and sustained contraction. Am J Physiol. 1997;273:H1587–94.

von Lüdinghausen M. The venous drainage of the human myocardium. New York: Springer; 2003.

von Ludinghausen M. Anatomy of the coronary arteries and veins. In: Mohl W, Wolner E, Glogar D, editors. The coronary sinus: proceedings of the 1st international symposium on myocardial protection via the coronary sinus. Darmstadt: Steinkopff; 1984. p. 5–7.

Weiss HR, Winbury MM. Nitroglycerin and chromonar on small-vessel blood content of the ventricular walls. Am J Physiol. 1974;226:838–43.

Zamir M. Nonsymmetrical bifurcations in arterial branching. J Gen Physiol. 1978;72:837–45.

Zamir M. On fractal properties of arterial trees. J Theor Biol. 1999;197:517–26.

Zamir M. Arterial branching within the confines of fractal L-system formalism. J Gen Physiol. 2001;118:267–76.

Zamir M, Brown N. Arterial branching in various parts of the cardiovascular system. Am J Anat. 1982;163:295–307.

Zamir M, Wrigley SM, Langille BL. Arterial bifurcations in the cardiovascular system of a rat. J Gen Physiol. 1983;81:325–35.

Zamir M, Phipps S, Langille BL, Wonnacott TH. Branching characteristics of coronary arteries in rats. Can J Physiol Pharmacol. 1984;62:1453–9.

Chapter 10
Biomechanical Basis of Myocardium/Vessel Interaction: Implications for Pathophysiology and Therapy

Dotan Algranati, Ghassan S. Kassab, and Yoram Lanir

10.1 Introduction

Ischemic heart disease is a major cause of morbidity and mortality worldwide. Interestingly, the onset of ischemia is transmurally heterogeneous, where the deeper (subendocardial) layers are more vulnerable to ischemia than the more superficial (subepicardial) ones (Hoffman 1987). This observation is especially puzzling in light of the opposite manifestation of coronary artery disease, which exclusively affects the epicardial coronary arteries, whereas intramural arteries are athero-protected (Geiringer 1951). Initiation of both atherosclerosis and ischemia depend highly on flow conditions; therefore, investigation of the hemodynamic determinants of both pathologies requires comprehension of the local coronary flow conditions, which are measured in the beating heart. Computer simulation is an attractive approach to study local coronary flow conditions. For hemodynamic simulation to be realistic, however, it must incorporate both a realistic description of the coronary network and the manner by which the contracting myocardium affects coronary flow—the myocardium/vessel interaction (MVI). Such an approach has several inherent challenges: First, the vast number of coronary blood vessels (Kaimovitz et al. 2005) is associated with an extensive computational cost to solve the network dynamic flow. To circumvent this difficulty, previous flow models (Bruinsma et al. 1988; Cornelissen et al. 2000; Flynn et al. 1992; Klocke et al. 1985; Manor et al. 1994) used lumped representations for the coronary vasculature. Although this approach is useful to reveal basic flow characteristics, it cannot

---

D. Algranati • Y. Lanir
Faculty of Biomedical Engineering, Technion, Haifa 3200003, Israel

G.S. Kassab (✉)
California Medical Innovations Institute, Roselle St, Rm 211 11107, San Diego, CA 92121, USA
e-mail: gkassab@calmi2.org

G.S. Kassab, M.S. Sacks (eds.), *Structure-Based Mechanics of Tissues and Organs*,
DOI 10.1007/978-1-4899-7630-7_10

address the physical relation between structure, vessel mechanics, and blood flow. Moreover, validation of a lumped model with experimental data is limited due to the inability of the model to describe flow conditions in specific vessels. A second challenge stems from paucity of experimental data required for both the flow model representation and for validation. Finally, the physical origins of the MVI, a key determinant in coronary flow analysis, are under a long-standing dispute and hitherto unknown. In fact, none of the mechanisms previously proposed (Downey and Kirk 1975; Krams et al. 1989a; Rabbany et al. 1989; Spaan et al. 1981; Zinemanas et al. 1994) to describe this mechanical interaction predict all of the characteristics of coronary flow (Westerhof et al. 2006), i.e., the blood flow velocities, pressures, and vascular diameters that correspond with the measured data.

To address this issue, a recently developed, structure-based computational framework for coronary distributed flow analysis was developed. This analysis allows for elucidation of (a) the MVI and its biomechanical origins, (b) the origins of subendocardial vulnerability to ischemia under conditions of epicardial artery stenosis, and (c) evaluation of the consistency of indices that are presently used in the clinic to quantify the severity of coronary stenosis.

10.2 Flow Analysis Computational Platform

To investigate the transmural, distributive, and dynamic features of coronary flow, a computational platform (Fig. 10.1) was developed (Algranati et al. 2010).

The platform consists of a number of sub-models and auxiliary computational processes developed to reconstruct the network anatomy (Fig. 10.1a) to determine the vessel mechanics (in reference to the pressure–diameter relationship, PDR, Fig. 10.1b) and to analyze the flow in the entire coronary network (Fig. 10.1c) based on MVI mechanisms (Fig. 10.1d). The network flow boundary conditions (Fig. 10.1e) were adopted from measured data. The computational scheme consists of the following elements:

10.2.1 Network Anatomy

Reconstruction of the network anatomy was based on detailed, statistical morphometric data of Kassab and coworkers (Kassab and Fung 1994; Kassab et al. 1993b, 1994, 1999) and was carried out in three stages: (1) reconstruction of microvascular networks, (2) integration into an intramyocardial coronary network, and (3) linking of the network to a large epicardial arterial tree. The latter stage was performed for the analysis of the effect of epicardial stenosis on the flow.

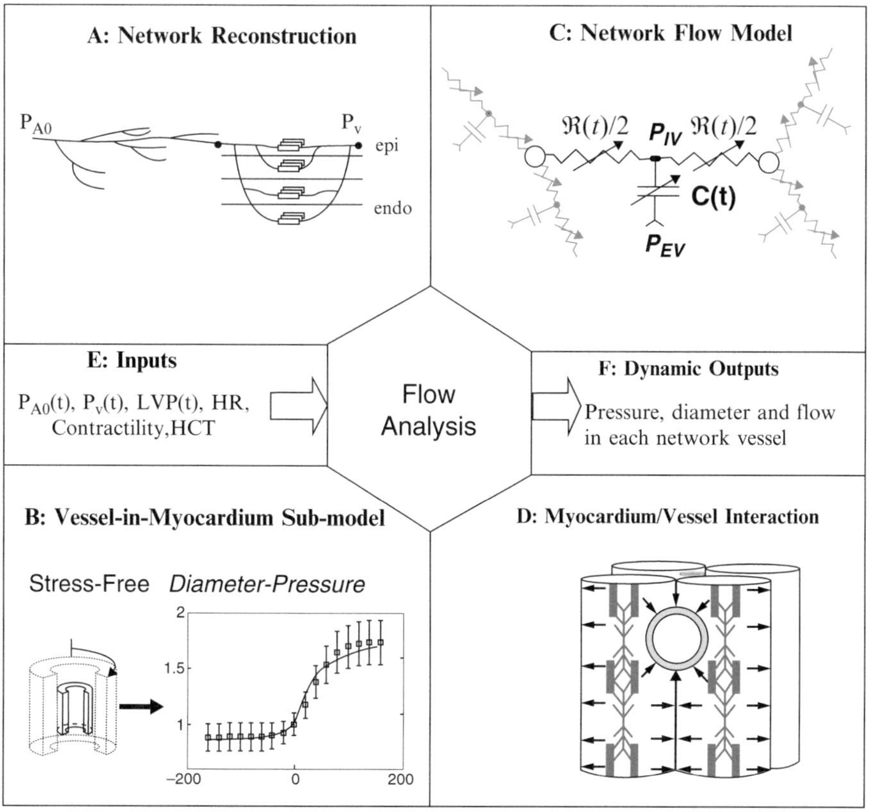

Fig. 10.1 Flow analysis computational platform. The platform consists of a number of sub-models aimed at reconstructing the network anatomy (**a**) to determine each vessel reference and loaded diameters (**b**) and to analyze network flow (**c**) based on myocardium/vessel interaction mechanisms (**d**). Platform hemodynamic inputs (**e**) include the aortic, venous, and left ventricle dynamic pressure waveforms (P_{Ao}, P_v, and LVP, respectively) the heart rate (HR), contractility, and the blood hematocrit (HCT). Based on these inputs, the dynamic flow conditions in terms of pressure, diameter, and flow are calculated in each vessel (**f**)

10.2.2 In Situ Vessel Mechanics

Vessel diameters are of primary significance since they determine the resistance to the flow (Spaan 1995). In situ data (Hamza et al. 2003) show that vascular diameter–pressure relations are sigmoid in shape (Fig. 10.1b), namely:

$$D_{(\Delta P)} = (D_+ - D_-) \Big/ \left[1 + \left(\frac{D_+ - D_0}{D_0 - D_-} \right)^{1 - \frac{\Delta P}{\Delta P_{1/2}}} \right] + D_- \tag{10.1}$$

where $\triangle P$ is the vessel trans-vascular pressure, i.e., the difference between intravascular and extravascular pressures (P_{IV} and P_{EV}, respectively). D, D_+, D_0, and D_- denote the loaded, maximum inflated, 0-pressure, and maximum unloaded diameters, respectively, and $\triangle P_{1/2}$ is the trans-vascular pressure corresponding to a diameter equal to the average of D_+ and D_-.

The values of these four parameters were obtained for each large epicardial vessel (>0.7 mm in diameter) from measured data (Hamza et al. 2003). For smaller intramyocardial vessels, where such data are unavailable, the parameters were evaluated from a mechanical stress analysis of a vessel-in-myocardium (Fig. 10.1b, details in Ref. (Algranati et al. 2010)).

10.2.3 Dynamic Flow Analysis

Conservation of mass requires that the difference between inlet and outlet flows of each vessel, Q_{in} and Q_{out}, respectively, should equal the time derivative of the vessel volume (V), namely:

$$Q_{\mathrm{in}}(t) - Q_{\mathrm{out}}(t) = \frac{P_{\mathrm{in}}(t) - P_{\mathrm{IV}}(t)}{\Re(t)/2} + \frac{P_{\mathrm{out}}(t) - P_{\mathrm{IV}}(t)}{\Re(t)/2} = \frac{dV}{dt} = \frac{d}{dt}\left(\frac{\pi D^2(t)L(t)}{4}\right) \tag{10.2}$$

where P_{in} and P_{out} denote vessel inlet and outlet pressures, respectively (Fig. 10.1c), P_{IV} is the vessel midlength pressure, and $\Re$ is the vessel flow resistance. For intramural vessels (diameters <0.7 mm), the instantaneous hydraulic resistance $\Re$ was calculated from Poiseuille's law as:

$$\Re(t) \equiv \frac{Q(t)}{P_{\mathrm{in}}(t) - P_{\mathrm{out}}(t)} = \frac{128\mu(t)L(t)}{\pi D(t)^4} \tag{10.3}$$

where L, D, and μ are the vessel length, diameter, and the blood apparent viscosity, respectively, with the latter as a function of diameter and hematocrit (Pries et al. 1994). The predictions of such a lumped, single vessel flow model were shown (Jacobs et al. 2008) to match those of a more comprehensive distributive model (Fibich et al. 1993) under dynamic flow conditions.

For stenosis in epicardial vessels (diameters >0.7 mm), the longitudinal pressure gradient stems from blood viscosity, flow pulsatility, and the inertial pressure losses at the exit of a stenosis (Young 1979), namely:

$$P_{\mathrm{in}} - P_{\mathrm{out}} = \frac{k_v \mu Q}{D^3} + k_i \rho L \frac{d\left(Q/D^2\right)}{dt} + \frac{k_e \rho \left[D^2/D_s^2 - 1\right]^2 |Q|\, Q}{D^4} \tag{10.4}$$

where Q is the instantaneous flow (assumed to equal the mean value of Q_{in} and Q_{out}), and blood density ρ was taken as 1060 kg/m^3. The magnitude of each mechanism was determined by the respective coefficients k_v, $k_{i,}$ and k_e, evaluated in detail elsewhere (Young 1979). Though thoroughly validated (Gould 1985; Siebes et al. 2004; Young 1979), the major drawback of (10.4) is that it relies on empirical coefficients, whereas recent studies (Huo et al. 2012) propose analytical formulation for pressure drop across coronary stenoses.

For network flow analysis, mass conservation was imposed at each network bifurcation. This implies that the sum of discharges Q^{jk} should vanish, i.e.,

$$\sum_{j=1}^{3} Q^{jk} = \sum_{j=1}^{3} \frac{P_{\text{IV}}^{j} - P_{\text{bif}}{}^{k}}{\Re^{j}/2} = 0 \quad k = 1,\ 2, \ldots, \ \text{Bifurcation No.} \qquad (10.5)$$

where P_{IV}^{j} denotes the intravascular pressure in each of the three vessels that compose the kth bifurcation, and $P_{\text{bif}}{}^{k}$ is the bifurcation pressure.

10.2.4 Myocardium/Vessel Interaction

The computational platform presented above (details in (Algranati et al. 2010, 2012)) was applied first to examine several different MVI mechanisms. This was done to determine which of the previously proposed MVI mechanisms (or any of their combinations) can predict the measured data and thus is suitable for further analyses of coronary flow. The analysis of MVI mechanisms (Algranati et al. 2010) is briefly presented in the following section.

10.3 Physical Mechanisms of Myocardium/Vessel Interaction

Cardiac contraction has significant impact on the blood flow in the distensible coronary vessels. Myocardial contraction applies pressure on and reduces the diameters of intramyocardial vessels, thus increasing their resistance to flow. This "systolic impediment" is the subject of extensive research as it relates to: (1) the nature and distribution of the extravascular forces (often termed intramyocardial pressure, IMP) exerted by the myocardium on the coronary vessels and (2) the manner by which these forces affect the flow in the embedded coronary vessels. The elucidation of these issues will allow for a deeper insight into the MVI which is seminal to understanding coronary phasic flow (Rogers et al. 2006) and transmural flow distribution.

The interpretation of the spatial and temporal distribution of IMP from direct in vivo tissue pressure measurements (Heineman and Grayson 1985; Mihailescu and Abel 1994) is complex, due to the cardiac motion, the distortion of the tissue microstructure by the pipette tip (Westerhof 1990), and the heterogeneity of this pressure in various micro-compartments (Spaan 1995). Downey and Kirk (1975) and Spaan et al. (1981) assumed IMP to be equal to left ventricular pressure (LVP) at the endocardium and to decrease linearly to zero at the epicardium. Krams et al. (1989b) found that LVP plays only a partial role and proposed (Krams et al. 1989a) the "time-varying elastance concept" to account for the full extent of systolic flow impediment. The elastance concept by itself, however, does not explain why epicardial flows are uninhibited to the same degree as endocardial ones (Spaan 1991). More recent studies suggest that both LVP and myocardial contraction have significant effect on the time-varying coronary flow (Kouwenhoven et al. 1992). Studies of epicardial lymph pressure (VanTeeffelen et al. 1998) and of coronary arterial pressure and flow (Kouwenhoven et al. 1992) suggest that rather than inducing flow impediment, systolic stiffening shields intramural vessels from the effects of LVP during a portion of systole (Spaan 1995). Another hypothesis (Rabbany et al. 1989) attributes IMP to a combination of LVP-derived interstitial fluid pressure, and a stress induced by fiber contraction.

Some models have been developed (Vis et al. 1995, 1997) to study the separate and combined effects of MVI mechanisms on the local vessel diameter at peak systole and diastole. These models were not integrated into a dynamic network flow analysis, however, and thus, cannot provide a complete account of the temporal/spatial coronary flow features. In a recent review (Westerhof et al. 2006), several specific MVI mechanisms were conceptually proposed but were neither modeled nor tested quantitatively. Hence, the mechanisms underlying MVI, their interplay, and the effect on flow conditions remain unclear.

10.3.1 Methodological Approach

Since there is still no agreement on the nature of the MVI mechanism, a coronary flow analysis was applied to establish the most likely hypothesis for the MVI mechanisms (Algranati et al. 2010). The flow analysis under each MVI mechanism was carried out several times to account for the various experimental conditions reported for each data base (e.g., heart rate, LVP; Fig. 10.1e). The MVI mechanisms studied were as follows:

Varying Elasticity (*VE*, Fig. 10.2a): Under this mechanism, contractility was taken to affect coronary flow through activation-dependent changes of myocardial stiffness (Vis et al. 1995).

Shortening-Induced Intracellular Pressure (*SIP*, Fig. 10.2b): Myocytes were modeled as membrane-contained fluid compartments that surround the vessels. During shortening, myocyte thickening (lateral expansion of their membranes) is due to internal volume preservation and is accompanied by pressure elevation. This intra-myocyte pressure is transmitted to the vessels due to their impingement on the vessels. This mechanism is based on data and model (Rabbany et al. 1994) showing linear increase in pressure with contractile shortening.

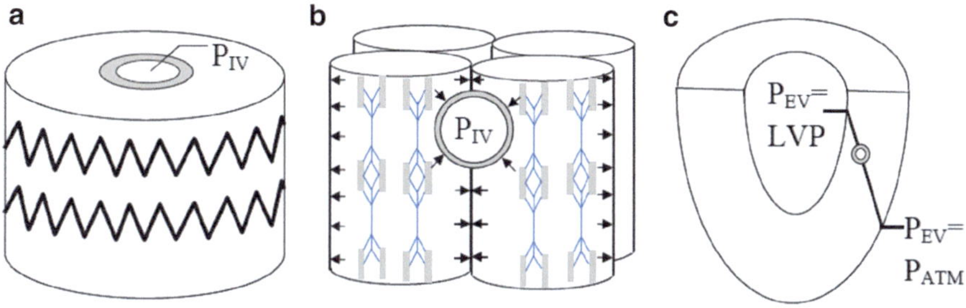

Fig. 10.2 The studied myocardium/vessel interaction mechanisms. (**a**) Varying elasticity (VE): flow is affected by the activation-mediated changes in myocardial stiffness (represented by a spring). (**b**) Shortening-induced intracellular pressure (SIP): flow is regulated by the difference between intravascular and contraction-induced myocyte intracellular pressures (*arrows*), (**c**) Cavity-induced extracellular pressure (CEP): extravascular pressure (P_{EV}) is the interstitial pressure which varies linearly (Heineman and Grayson 1985) from cavity pressure (LVP) at the endocardium to atmospheric pressure (P_{atm}) at the epicardium

Cavity-Induced Extracellular (Interstitial) Pressure (CEP, Fig. 10.2c): This mechanism relies on the underlying assumption of both the intramyocardial pump (Spaan et al. 1981) and the vascular waterfall (Downey and Kirk 1975) models. P_{EV} was assumed to stem from the LVP alone and is taken to vary linearly (Heineman and Grayson 1985) with transmural position (expressed by MRD, myocardial relative depth, that ranges between 0 at epicardium and 1 at endocardium).

CEP + VE: The extracellular pressure and varying elasticity were combined in this model. The material law of surrounding myocardium was activation dependent (as in VE) and the extracellular pressure was applied to the myocardial external surface.

CEP + SIP: The myocytes were assumed to contract within an LVP-derived pressurized interstitium. Hence, the assigned P_{EV} equals the algebraic sum of extracellular pressure and of the shortening-dependent intracellular pressure, namely:

$$P_{EV}(t) = \mathrm{LVP}(t) \times \mathrm{MRD} + \alpha \times \mathrm{MRS}(t) \quad (10.6)$$

where MRS is the myocytes relative shortening, and α is a coefficient that relaxes this shortening to intracellular pressure (Algranati et al. 2010).

10.3.2 Summary of Results

The major observed coronary flow features were compared with the model predictions under each of the tested MVI mechanisms. A summary of the results is listed in Table 10.1. These results clearly point to the deficiency of hypotheses that exclude CEP. Although CEP predictions generally fit the data well, this mechanism alone cannot account for the observed reduced coronary flow under unchanged LVP

Table 10.1 Comparison of tested (Algranati et al. 2010) MVI mechanisms with experimental observations

	Mechanism				
Observation	VE	SIP	CEP	CEP + VE	CEP + SIP
Transmural distribution of perfusion under contraction is close to homogeneous	−	−	+	+	+
Elevated contractility attenuate total perfusion under similar LVP values	+	+	−	−	+
Elevated heart rate attenuates both total perfusion and endo/epi perfusion ratio	−	−	+	+	+
Reduced inlet pressure attenuates both total perfusion and endo/epi perfusion ratio	−	−	+	+	+
Epicardial arteriolar waveforms follow aortic pressure	−	+	+	−	+
Predicted systolic/diastolic diameters change follows measured data	−	−	+	+	+
Predicted velocity waveforms follow measured data	−	−	+	−	+

levels but increased contractility (Marzilli et al. 1979). The CEP + VE mechanism can account for this observation only under higher than normal inlet pressures (Algranati et al. 2010). Yet under CEP + VE, the predicted subepicardial pressure and subendocardial velocity waveforms were found to differ significantly from experimental results. Most notably, only the combination of CEP + SIP mechanisms is in good agreement with all observations.

10.3.3 Biophysical Implications

Through cardiac contraction, myocytes intrinsically generate contraction during which their intracellular pressure is increased (Rabbany et al. 1994). In light of the condensed packing of myocytes in myocardium (Spaan 1991), it is reasonable to assume that myocytes contact directly with both the interstitial fluid and with the adjacent vessels during cardiac contraction (Caulfield and Borg 1979). Thus, the overall extravascular pressure (P_{EV}) consists of both the interstitial pressure (CEP mechanism) and intra-myocyte pressure (SIP mechanism) (Fig. 10.3a).

Unfortunately, experimental determination of IMP is unreliable and subject to large variance (Heineman and Grayson 1985; Mihailescu and Abel 1994; Rabbany et al. 1989). Some of the confounding factors are that when inserting a small-bore pressure-gauge tip into the myocardial tissue, the measured pressure depends on the

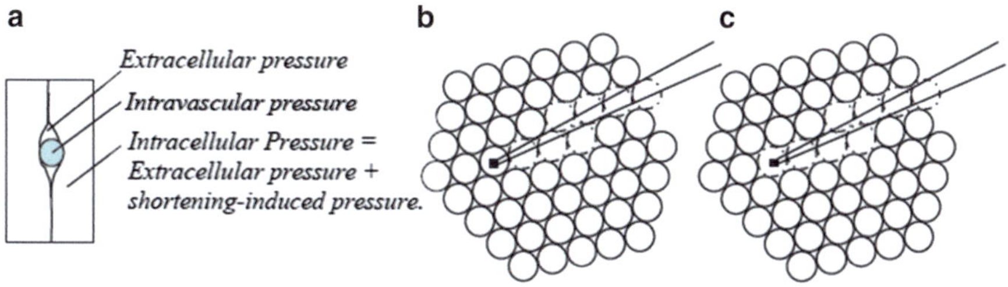

Fig. 10.3 (**a**) Structural illustration of PEV origins. The coronary vessel is exposed both to extracellular (interstitial) pressure (where vessel and myocyte membrane have no surface-contact) and to intracellular pressure, the latter being a combination of shortening-induced intracellular pressure at atmospheric pressure and the myocyte environmental pressure, which equals to the extracellular pressure. (**b**), (**c**) Micropipette inserted into myocardium. The transducer (*black square*) measures the myocyte intracellular pressure, which although perforated still generates SIP to a certain extent. Perforation of the target myocyte (illustrated in **c**) inhibits SIP generation, thus the measured pressure is the extracellular one. Broken lines represent myocytes that were perforated during micropipette insertion

tissue compartment in which the tip resides (whether within a myocyte, small artery or vein, or extracellular fluid). Even if within a myocyte, the measured pressure depends on whether its membrane perforation was mild enough as to permit SIP generation or not, as illustrated in Fig. 10.3b, c. In addition, perforation invariably induces microstructural distortion which has a significant effect on the measured pressure.

The varying elastance concept (Vis et al. 1995) assumes that the myocardium, being a continuum solid, affects the intravascular pressure through change in myocardial mechanical properties. Such a model applies to the left ventricle cavity, where the myocytes annularly circumvent the cavity thus applying tangential "hoop" forces in the cavity wall which, by Laplace's law, resist the cavity pressure. In the coronary vasculature, however, the myocytes configuration is oblique or tangential to the vessel (Fig. 10.3a), and they do not circumvent the vessels. During contraction, myocytes do not apply tangential hoop forces and are thus incapable of resisting vascular pressure. Hence, myocytes stiffening per se is not expected to exert significant extravascular forces on intramyocardial vessels. It is the myocytes internal pressure that exerts radial forces on the vessels which impinge them.

10.4 Subendocardial Vulnerability to Ischemia

Myocardial ischemia is a major cause of morbidity and mortality. It is transmurally heterogeneous where the subendocardium is at a higher risk than the midwall or epicardium (Hoffman 1987). Despite its significant clinical relevance, the physical determinants of subendocardial vulnerability remain controversial. Although subendocardial metabolic demand is somewhat higher than subepicardial (Hoffman 1987), data suggest that it is the lower blood supply to the subendocardium rather than the higher demand that induces subendocardial vulnerability (Hoffman et al. 1985). For example, under partial occlusions (Bache and Schwartz 1982) or otherwise decreased perfusion pressure in a non-auto-regulated coronary circulation (Chilian and Layne 1990), the subendocardial blood supply has been observed to be compromised to a higher extent than subepicardial. Conversely, an increase of perfusion pressure was found to improve primarily the subendocardial blood supply (Boatwright et al. 1980). Hence, changes in perfusion pressure induce transmural flow redistribution, i.e., changes in subendocardial-to-subepicardial perfusion ratio (endo/epi).

Subendocardial vulnerability to ischemia has been previously attributed to several mechanisms. The greater subendocardial systolic compression was proposed to induce one or more of the following: (1) higher subendocardial vessel resistance (Buckberg et al. 1972; Moir 1972), (2) systolic backflow from endocardial to epicardial vessels (Flynn et al. 1992), induced by systolic–diastolic interaction (Hoffman et al. 1985; Spaan et al. 1981), or (3) transient vessel collapse (Downey and Kirk 1975), which would result in effectively higher subendocardial back pressures (Bache and Schwartz 1982). The first two mechanisms account for a preferentially epicardial blood flow (i.e., endo/epi reduction) under *increased systolic compression*. The predicted effect of these mechanisms on endo/epi was not shown to increase, however, under *reduced perfusion pressure*. It is, therefore, unclear whether these mechanisms alone can account for the measured endo/epi reduction under such conditions (Bache and Schwartz 1982). The proposed transmural differences in back pressures (Bache and Schwartz 1982) may, alternatively, account for the measured change in endo/epi under reduced perfusion pressure, but vessel collapse has not been demonstrated in the coronary vasculature (Hamza et al. 2003; Hiramatsu et al. 1998; Hoffman and Spaan 1990). It is still unknown whether back pressure differences exist or are large enough to account for substantial endo/epi changes (Hoffman 1987; Hoffman et al. 1985). Another proposed mechanism relates to anatomically induced higher resistance of subendocardial conduit vessels or lower resistance of subendocardial microcirculation, which was suggested (Chilian 1991) to reduce the driving pressure of the subendocardial microcirculation. This mechanism cannot account for flow redistribution under reduced perfusion pressure, as discussed below. Moreover, the pressure drop over conduit vessels was shown to be small in magnitude (Mittal et al. 2005).

We recently proposed (Algranati et al. 2011) that the measured flow redistribution, and the consequent subendocardial vulnerability under reduced perfusion pressure, results from the sigmoidal shape of the pressure–diameter relation (PDR, Fig. 10.1b), where there is a higher compliance of subendocardial as compared to the respective subepicardial vasculature. This hypothesis was based on the premise that since both subendocardial and subepicardial flows are driven by the same pressure source, a difference in transmural flow distribution may only stem from a difference in the respective vascular resistances. Flow resistance depends on vascular diameters, and diameters depend on the pressure through vascular compliance. Hence, only a higher compliance of subendocardial vasculature can induce a greater increase in resistance to flow there, as compared to the subepicardium. The underlying determinants of endo-to-epi compliance differences are presented below.

10.4.1 Methodological Approach

The flow analysis scheme described above was used (Algranati et al. 2011) to predict the time-averaged ΔP and the endo/epi with and without the effect of myocardial contraction (i.e., not accounting for myocardium/vessel interaction), and under a wide range of physiologic and pathologic conditions.

To study the effect of contraction on subendocardial vulnerability, flow was compared between the cases of the contracting and non-contracting heart. To that end, contraction effects were abolished by zeroing P_{EV} (namely (10.6) becomes $P_{\mathrm{EV}} = 0$).

10.4.2 Summary of the Results

The analysis predicts that myocardial contractions induce significant transmural heterogeneity in the mean trans-vascular pressure, namely to induce lower (more compressive) pressures in subendocardial vessels. Systolic LVP was found to be the most important determinant of this contraction-induced heterogeneity; whereas, changes in diastolic LVP, heart rate, cardiac contractility, and perfusion pressure and its pulsatility were shown to be of lower significance.

Moreover, the model predicts that in beating hearts, a reduction in perfusion pressure inhibits subendocardial blood flow to a higher extent than epicardial flow, a prediction that is in-line with observed data (Bache and Schwartz 1982; Boatwright et al. 1980; Chilian and Layne 1990). The differences in trans-vascular pressures and the nonlinear (sigmoid) vessel PDR result in more compliant arteries in the subendocardium than in the subepicardium (Fig. 10.4). Hence, the response of subendocardial diameters (and resistance) to changes in perfusion pressure is more prominent, thus explaining the predicted higher subendocardial flow drop under reduced perfusion pressure.

In contrast, the mean trans-vascular pressures are similar between subendocardial and epicardial vessels in diastole, and so is the PDR slope (Fig. 10.4). Consequently, flow is predicted to be evenly inhibited in subendocardium and epicardium under reduced perfusion pressure.

10.4.3 Clinical Implications

The predicted higher subendocardial vulnerability depends on a number of clinical factors such as heart rate, LVP, and contractility. The analysis results imply that pharmacological reduction of heart rate, in addition to reducing myocardial metabolic demand, enhances subendocardial blood supply by (1) increase of total coronary flow under a given perfusion pressure, (2) increasing endo/epi under a given perfusion pressure; and (3) decreasing the redistribution of blood away from the subendocardium under perfusion pressure reduction.

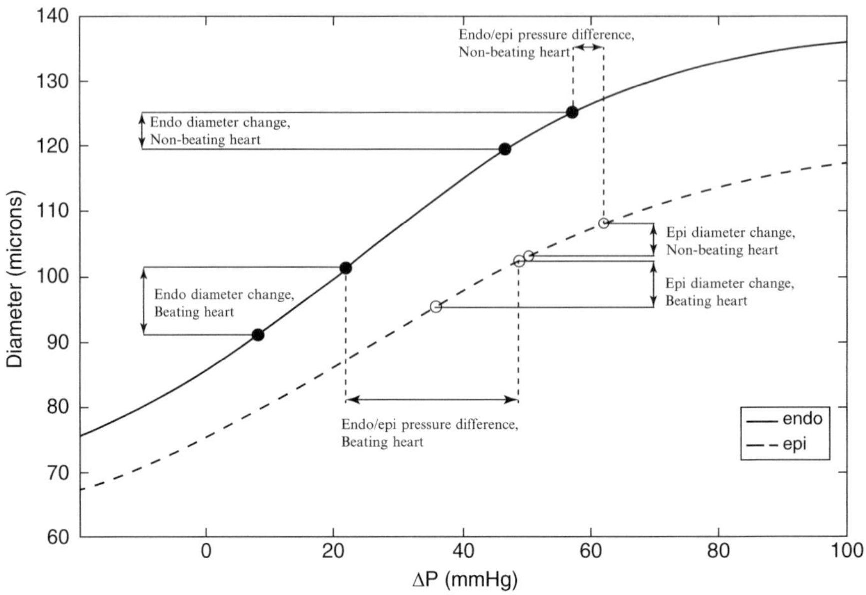

Fig. 10.4 Origins of subendocardial vulnerability: The time-averaged trans-vascular pressure predicted in 100 μm diameter subendocardial (*full*) and subepicardial (*blank*) arterioles in the beating (*circles*) and non-beating (*diamonds*) heart are indicated on the vessels' pressure–diameter curves (subendocardial—*solid*, subepicardial—*dash*). Results are under baseline (95 mmHg) and reduced (35 mmHg) perfusion pressure, and under otherwise baseline conditions (Algranati et al. 2011). *X*-axis—trans-vascular pressure (mmHg). *Y*-axis—vessel diameter (μm). In the beating heart, a reduced perfusion pressure induces a more substantial vessel narrowing (*double arrows*) in the subendocardium than in the subepicardium, due to the higher subendocardial slope (i.e., its lower stiffness). In the non-beating heart, diameter change is similar in subendocardial and subepicardial vessels

Under severe stenosis of the aortic valve, systolic LVP is considerably higher than aortic (i.e., coronary perfusion) pressure. Treatment of valve stenosis should thus result in both lower systolic LVP and possibly also in higher coronary perfusion pressure. Both effects were predicted (Algranati et al. 2011) to enhance subendocardial coronary flow, consistent with experimental results (Iwanaga et al. 1995).

In contrast, pharmacological reduction of afterload reduces simultaneously both the aortic (coronary perfusion) pressure, and the systolic LVP. The effect on subendocardial perfusion is more complex as the former diminishes subendocardial supply while the latter increases it. The effect of aortic pressure is more significant than that of the systolic LVP, i.e., when both pressures were simultaneously reduced from their baseline level, the total coronary flow was predicted to drop substantially (Algranati et al. 2011).

Elevated diastolic LVP, often observed in patients with congestive cardiomyopathy, was predicted to compromise both total and especially the subendocardial flow, consistent with previous observations (Hoffman 1987).

10.5 Consistency of Stenosis Severity Indices

Percutaneous coronary interventions and coronary artery bypass grafts focus on improving myocardial blood supply, i.e., to restore blood flow downstream from the stenosed vessel (Qs) to non-stenotic value (Qn). Stenosis functional severity depends on the ratio Qs/Qn. Clinically, the stenosis severity determines the treatment strategy. Since, however, Qn is unknown prior to treatment, a number of indices are used to estimate the patient stenosis severity. Since both Qn and Qs depend on coronary autoregulation, the study presented below focused on indices measured under full vasodilation. But even under this condition, there are a number of structural, hemodynamic, and mechanical factors which may significantly affect both Qn and Qs. Below, we address the question of how consistent are the indices used in the clinic in predicting the flow ratio Qs/Qn under variability of the hemodynamic determinants.

The three indices considered were: the relative area of stenotic occlusion (%AS), which relates to the stenosis geometry; the hyperemic stenosis resistance, HSR (Hoffman 1987); and the pressure-based fractional flow reserve (FFR), which attempts to represent the flow ratio Qs/Qn.

Previous clinical studies of stenosis indices (Kini et al. 2008; Tonino et al. 2009) did not include prior and post-interventional flow data (in mL/min) which can shed light on the reliability of the indices. In silico studies of coronary circulation, on the other hand, are impeded by the vast number of network vessels (Kassab and Fung 1994; Kassab et al. 1993a, 1994) and by the complex myocardial/vessel interaction (Algranati et al. 2010). Hence, previous in silico studies of clinical indices used lumped network structures, lumped linear coronary pressure–flow relations (Siebes et al. 2002), and empirically derived concepts of the effect of myocardial contraction on coronary flow (van den Wijngaard et al. 2008). Although such approaches facilitate understanding of basic behavior under specific empirical conditions, a more general and realistic distributive analysis is required to gain both mechanistic and quantitative insight into the physical factors that affect the consistency of stenosis indices. In the following section, it is shown how the above computational platform for coronary flow facilitates an understanding of various indices of stenosis severity (Algranati et al. 2012).

Table 10.2 Reference hemodynamic conditions and associated perturbations used for analysis of indices consistency. Pao, Pv, and LVP denote (time-varying) aortic, venous, and left ventricle pressures, respectively, and are listed as systolic/diastolic levels. High and low values represent the borders of physiological levels

Determinant	Reference	High	Low
Pao (mmHg)	120/80	140/90	90/60
Heart rate (beats/min)	80	100	60
LVP (mmHg)	120/5	140/5	90/5
Hematocrit (%)	45	53	39
Pv (mmHg)	15/5	30/10	0/0
Vascular stiffness	Average PDR based on measured data (Hamza et al. 2003)	3 standard deviations stiffer than measured (Hamza et al. 2003) PDR	3 standard deviations more compliant than measured (Hamza et al. 2003) PDR

10.5.1 Methodological Approach

In the clinic, each index has its own experience-based cutoff value, used to distinguish between severe and non-severe stenosis. The values are 0.75 for FFR (Tonino et al. 2009), 75 % for %AS (Baptista et al. 1994), and 0.8 mmHg s/cm for HSR (Hoffman and Spaan 1990). We define the index consistency as a measure of its robustness in predicting Qs/Qn against variations in hemodynamic conditions. In other words, for a given cutoff level of the index, how sensitive is the "real" Qs/Qn (as predicted by the flow model) to possible variations in the patient hemodynamic conditions (Table 10.2). For clinical purposes, index consistency is most important at its cutoff level. Hence, an index is reliable if its cutoff level is associated with a consistent value of Qs/Qn in pressure of hemodynamic variability. In practice, Qs/Qn was found to have dispersion and can have a range of several percent in a single subject under the same index level (Pijls et al. 1993). Hence, we (Algranati et al. 2012) considered a threshold level of ±5 % change from reference in Qs/Qn under physiological hemodynamic variations to be an acceptable range of consistency.

10.5.2 Summary of the Results

The model predictions (Fig. 10.5) indicate that FFR is more consistent than both %AS and HSR under the physiological variations tabulated in Table 10.2; and that FFR is only inconsistent under variability in the vascular stiffness (Algranati et al. 2012). As predicted by the theory (Spaan et al. 2006), FFR and Qs/Qn values tend to equalize only in rigid vasculature (Fig. 10.6) but may differ substantially in nonrigid tubes.

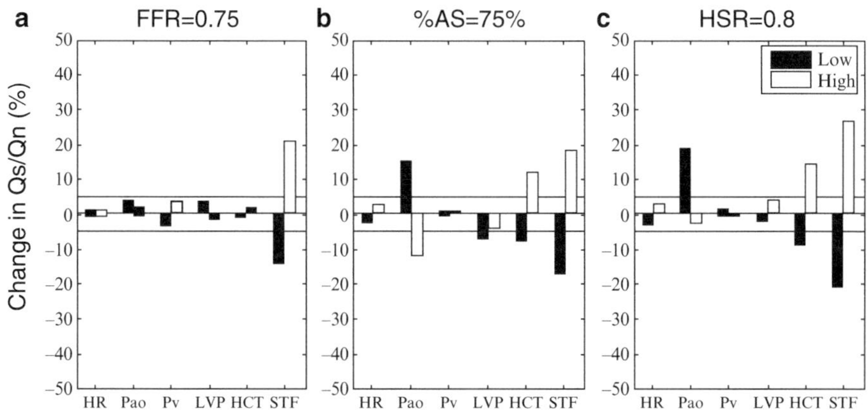

Fig. 10.5 Hemodynamic effects on indices-predicted Qs/Qn under the indices cutoff values of FFR = 0.75 (**a**), %AS = 75 % (**b**), and HSR = 0.8 mmHg s/cm (**c**). The low and high values of aortic blood pressure (Pao), heart rate (HR), left ventricle pressure (LVP), vascular stiffness (STF), and blood hematocrit (HCT) are within physiological range (Table 10.2). *Y*-axis: Qs/Qn levels normalized by the value of Qs/Qn under baseline conditions (Table 10.2). The FFR predicted Qs/Qn changes (**a**) are lower than 10 % (*horizontal lines*) under all but vascular stiffness variation and are lower than the predicted Qs/Qn changes under %AS (**b**) and HSR (**c**) cutoff levels

Fig. 10.6 Effect of vascular stiffness on Qs/Qn under FFR = 0.75. The predicted Qs/Qn (*Y*-axis) increases with vascular stiffness (*X*-axis), until reaching FFR value of 0.75 in perfectly rigid vasculature. Under stiffness variation of ±3 standard deviation of the measured (Hamza et al. 2003) vascular stiffness (*vertical dash*), Qs/Qn changes by more than 10 % (*horizontal dash*)

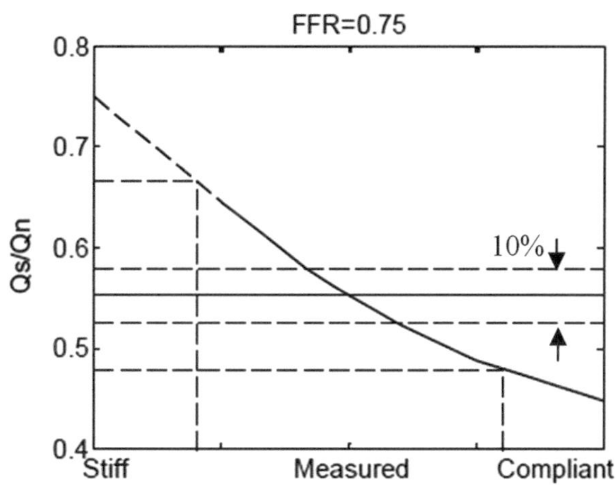

10.5.3 Clinical Implications

The predictions (Fig. 10.6) imply that under a given value of FFR, Qs/Qn is higher in patients with higher vessel stiffness (e.g., diabetic, hypertensive, smokers). Hence, the maximal flow improvement (Qn/Qs-1) due to a coronary intervention is expected to be lower under higher vascular stiffness. This is of clinical significance since it provides mechanistic insights into the suboptimal outcome of percutaneous coronary intervention in patients with diabetes mellitus (Heineman and Grayson 1985). This result can be readily understood since stenosis treatment reduces resistance to flow, which leads to an elevation of the pressure downstream from the stenosis (Pd) and an FFR increase. In healthy compliant vessels, this Pd elevation increases the vessel diameters downstream from the stenosis and thereby increases Qn. In a stiffer vasculature, a similar Pd (FFR) elevation is not expected to increase the vessel diameters as much as in compliant vessels. Hence, the flow improvement due to stenosis treatment is expected to be lower when compared to the normal population. As Spaan et al. (2006) pointed out, FFR describes Qs/Qn exactly only if the coronary resistance downstream to the occlusion is pressure independent. Physically, vessel resistance depends highly on diameter (10.2–10.4), and hence, a pressure-independent resistance can only occur in rigid vessels. Since passive coronary vessels are compliant (albeit less so in diabetics) (Cornelissen et al. 2000; Hamza et al. 2003), FFR underestimates Qs/Qn (Fig. 10.6) in agreement with measured data (Pijls et al. 1993; Spaan et al. 2006). A lower flow improvement in a stiffer coronary vasculature supports the view (Yanagisawa et al. 2002) that stratification of patients which includes consideration of pathologies known to affect vascular stiffness may enhance the consistency of a priori prediction of an interventional success.

10.6 Future Directions

The presented structure-based computational platform for coronary flow analysis provides a wide spectrum of novel information regarding the dynamic local flow conditions in vessels from different transmural myocardial layers and of various diameters, thus providing opportunity to investigate additional open questions in the coronary field. Some examples are:

10.6.1 Coronary Vascular Stiffness

An obvious aim for future research arising from the present results and conclusions would be to quantify coronary vascular stiffness, as this stiffness is a key for deeper insight into the coronary flow (Fig. 10.4), and for clinical interpretation of measured indices of stenosis severity (Fig. 10.6). Unfortunately, with the exception of the swine large epicardial (Hamza et al. 2003) coronary arteries, previous studies focused on isolated vessels, thus disregarding their in vivo stiffness. As discussed in Chap. 5, a method for evaluating patient-specific vascular stiffness in the clinic is required. At present, it is unclear whether reliable evaluation of the patient's vascular stiffness may be achieved by combining current clinical indices of stenosis (e.g., combining FFR and %AS). An alternative approach may be to measure functional stenosis indices (i.e., FFR) in a single individual under several loading conditions. The functional characteristics (e.g., resistance, stiffness) of the stenosis may then be separated from those of its downstream microvasculature to allow for patient-specific understanding of the effect of the stenosis on the coronary flow.

10.6.2 Coronary Autoregulation

Autoregulation is achieved by vasoactivity of the smooth muscle cell (SMC) in the walls of coronary arterioles. The mechanisms of coronary autoregulation are complex and only partially understood. It has been shown (Jones et al. 1995; Kuo et al. 1988, 1995; Liao and Kuo 1997) that coronary arterioles (<200 μm diameter) actively respond to changes in intravascular pressure (myogenic response) in shear stresses (flow-mediated dilation) and to concentrations of metabolites (metabolic response). The magnitude of response to each stimulus depends, however, on the vessel diameter (Kuo et al. 1995; Liao and Kuo 1997) and on the transmural location (Kuo et al. 1988). Unfortunately, data regarding the magnitude of responses are highly sparse. Furthermore, the available data were obtained only under steady flow conditions, whereas coronary flow is highly oscillatory. Hence, a major research question relates to the driving stimuli of each autoregulation response in the in vivo dynamic flow, e.g., is it the time-averaged pressure/shear stress or other characteristics of the phasic pattern such as the stimulus frequency, or its extremum levels? As an example, the myogenic response of arteries in the rat kidneys was observed (Loutzenhiser et al. 2002) to be controlled by the highest (systolic) blood pressure.

10.6.3 Coronary Flow and Ventricular Energetics

An interesting outcome of the summarized studies relates to the combined effect of elevation in both aortic and left ventricle pressure (P_{Ao} and LVP, respectively) on coronary flow. Such combined elevation is a characteristic response to sympathetic stimulus (afterload elevation) and is predicted by the model to have a complex effect on coronary flow perfusion: P_{Ao} elevation increases perfusion, whereas LVP elevation reduces it. Our preliminary results indicate that under such conditions, which are associated with an elevation of myocardial work, the total flow to all layers is significantly increased. These results suggest a mechanical control mechanism that compensates over the increased myocardial metabolic demand by increasing coronary flow. Suga (1979) showed a linear relation between the myocardial pressure-volume area and oxygen consumption, where the slope of this relation is the myocardium work efficiency. Hence our computational framework, if combined with a model of the entire heart system, is likely to facilitate quantitative evaluation of the above suggested control mechanism or an alternative hypothesis.

10.6.4 Determinants of Intramural Artery Athero-Protection

Atherosclerosis was shown (Caro 2009) to be highly affected by the dynamic, local flow conditions. However, only epicardial arteries are prone to atherogenesis, whereas intramural coronaries were observed to be athero-protected (Geiringer 1951; Robicsek and Thubrikar 1994). The underlying mechanistic reason for this epicardium/intramyocardium difference is yet unknown, though it is commonly accepted (Zhang et al. 2007) that the dynamic nature of blood flow induces morphological changes in large arteries. Hence, consideration of the dynamic local trans-vascular pressure and shear stress in vessels from different transmural myocardial layers and of the associated vessel responses to them are expected to provide insights into the determinants of athero-protection of intramural arteries.

10.7 Discussion

Comprehension of the underlying mechanisms of coronary flow dates back to the seventeenth century, when Scaramucci (1696), commonly regarded as “the father of coronary physiology” (Kajiya et al. 2008), hypothesized that deep myocardial vessels are squeezed during systole and filled during diastole. However, only the later introduction of mathematical models allowed for quantitative flow analysis. This resulted in a significant advancement in the understanding of coronary flow, especially when the advance in models was accompanied by advances in flow measurement techniques. Although coronary flow models became increasingly sophisticated, the coronary system is still simplified using lumped, linear mathematical models (Spaan 1991). This simplification is justified mainly by the ease of conception and simulation. It is not justified, however, in line with the highly nonlinear vessel mechanical and biological behavior, nor with the substantial effect of flow conditions in remote vessels along the coronary tree.

Here, we advanced the current understanding of coronary flow under normal and pathological conditions by removing ad hoc assumptions. This was achieved by a physics-based analysis of the dynamic, transmural flow in a distributive coronary network. The analysis is unique in several aspects: First, it is based on fundamental mechanical properties and on measured morphometry. Secondly, it provides unique information about the dynamic flow and its mechanistic origins in coronary vessels of all diameters and in all myocardial transmural locations. The predictions are validated with both qualitative and quantitative available data (Algranati et al. 2010), resulting in good agreement. Importantly, the agreement with the data is obtained with no parameter adjustment. An important attribute of the present analysis is the extensive model testing and validation against both qualitative and quantitative data (Algranati et al. 2010). The results show (Table 10.1) that although several previous linear models may be acceptable when validating against a narrow scope or limited ranges of data, they cannot account for a wide scope of different data. This points to the importance of using as many data sources as possible for model validation.

In summary, the model-based predictions are as follows:

(1) The extravascular loading applied on coronary vessels by the contracting myocardium (myocardium/vessel interaction) results from a combined effect of LVP-derived interstitial pressure and contraction-induced intramyocyte pressure; (2) subendocardial vulnerability to stenosis-induced reduction in perfusion pressure stems from the combined effects of cardiac contraction and a sigmoidal PDR that gives rise to transmural compliance; and (3) the reliability of the FFR as an index of stenosis severity is higher than that of other commonly used indices such as the degree of occlusion (%-stenosis) and the hyperemic stenosis resistance (HSR), but its reliability is compromised by dependency on coronary vascular stiffness.

References

Algranati D, Kassab GS, Lanir Y. Flow restoration post revascularization predicted by stenosis indexes: sensitivity to hemodynamic variability. Am J Physiol Heart Circ Physiol. 2013;305:H145–154.

Algranati D, Kassab GS, Lanir Y. Why is the subendocardium more vulnerable to ischemia? A new paragidm. Am J Physiol Heart Circ Physiol. 2011;300:H1090–100.

Algranati D, Kassab KG, Lanir Y. Consistency of stenosis severity indices. Am J Physiol Heart Circ Physiol. 2012.

Bache RJ, Schwartz JS. Effect of perfusion pressure distal to a coronary stenosis on transmural myocardial blood flow. Circulation. 1982;65:928–35.

Baptista J, Arnese M, Roelandt JR, Fioretti P, Keane D, Escaned J, Boersma E, di Mario C, Serruys PW. Quantitative coronary angiography in the estimation of the functional significance of coronary stenosis: correlations with dobutamine-atropine stress test. J Am Coll Cardiol. 1994;23:1434–9.

Boatwright RB, Downey HF, Bashour FA, Crystal GJ. Transmural variation in autoregulation of coronary blood flow in hyperperfused canine myocardium. Circ Res. 1980;47:599–609.

Bruinsma P, Arts T, Dankelman J, Spaan JA. Model of the coronary circulation based on pressure dependence of coronary resistance and compliance. Basic Res Cardiol. 1988;83:510–24.

Buckberg GD, Fixler DE, Archie JP, Hoffman JI. Experimental subendocardial ischemia in dogs with normal coronary arteries. Circ Res. 1972;30:67–81.

Caro CG. Discovery of the role of wall shear in atherosclerosis. Arterioscler Thromb Vasc Biol. 2009;29:158–61.

Caulfield JB, Borg TK. The collagen network of the heart. Lab Invest. 1979;40:364–72.

Chilian WM. Microvascular pressures and resistances in the left ventricular subepicardium and subendocardium. Circ Res. 1991;69:561–70.

Chilian WM, Layne SM. Coronary microvascular responses to reductions in perfusion pressure. Evidence for persistent arteriolar vasomotor tone during coronary hypoperfusion. Circ Res. 1990;66:1227–38.

Cornelissen AJ, Dankelman J, VanBavel E, Stassen HG, Spaan JA. Myogenic reactivity and resistance distribution in the coronary arterial tree: a model study. Am J Physiol Heart Circ Physiol. 2000;278:H1490–9.

Downey JM, Kirk ES. Inhibition of coronary blood flow by a vascular waterfall mechanism. Circ Res. 1975;36:753–60.

Fibich G, Lanir Y, Liron N. Mathematical model of blood flow in a coronary capillary. Am J Physiol. 1993;265:H1829–40.
Flynn AE, Coggins DL, Goto M, Aldea GS, Austin RE, Doucette JW, Husseini W, Hoffman JI. Does systolic subepicardial perfusion come from retrograde subendocardial flow? Am J Physiol. 1992;262:H1759–69.
Geiringer E. The mural coronary. Am Heart J. 1951;41:359–68.
Gould KL. Quantification of coronary artery stenosis in vivo. Circ Res. 1985;57:341–53.
Hamza LH, Dang Q, Lu X, Mian A, Molloi S, Kassab GS. Effect of passive myocardium on the compliance of porcine coronary arteries. Am J Physiol Heart Circ Physiol. 2003;285:H653–60.
Heineman FW, Grayson J. Transmural distribution of intramyocardial pressure measured by micropipette technique. Am J Physiol. 1985;249:H1216–23.
Hiramatsu O, Goto M, Yada T, Kimura A, Chiba Y, Tachibana H, Ogasawara Y, Tsujioka K, Kajiya F. In vivo observations of the intramural arterioles and venules in beating canine hearts. J Physiol. 1998;509(Pt 2):619–28.
Hoffman JI. Transmural myocardial perfusion. Prog Cardiovasc Dis. 1987;29:429–64.
Hoffman JI, Spaan JA. Pressure-flow relations in coronary circulation. Physiol Rev. 1990;70:331–90.
Hoffman JI, Baer RW, Hanley FL, Messina LM. Regulation of transmural myocardial blood flow. J Biomech Eng. 1985;107:2–9.
Huo Y, Svendsen M, Choy JS, Zhang ZD, Kassab GS. A validated predictive model of coronary fractional flow reserve. J R Soc Interface. 2012;9(71):1325–38.
Iwanaga S, Ewing SG, Husseini WK, Hoffman JI. Changes in contractility and afterload have only slight effects on subendocardial systolic flow impediment. Am J Physiol. 1995;269:H1202–12.
Jacobs J, Algranati D, Lanir Y. Lumped flow modeling in dynamically loaded coronary vessels. J Biomech Eng. 2008;130:054504.
Jones CJ, Kuo L, Davis MJ, DeFily DV, Chilian WM. Role of nitric oxide in the coronary microvascular responses to adenosine and increased metabolic demand. Circulation. 1995;91:1807–13.
Kaimovitz B, Lanir Y, Kassab GS. Large-scale 3-D geometric reconstruction of the porcine coronary arterial vasculature based on detailed anatomical data. Ann Biomed Eng. 2005;33:1517–35.
Kajiya F, Yada T, Hiramatsu O, Ogasawara Y, Inai Y, Kajiya M. Coronary microcirculation in the beating heart. Med Biol Eng Comput. 2008;46:411–9.
Kassab GS, Fung YC. Topology and dimensions of pig coronary capillary network. Am J Physiol. 1994;267:H319–25.
Kassab GS, Imoto K, White FC, Rider CA, Fung YC, Bloor CM. Coronary arterial tree remodeling in right ventricular hypertrophy. Am J Physiol. 1993a;265:H366–75.
Kassab GS, Rider CA, Tang NJ, Fung YC. Morphometry of pig coronary arterial trees. Am J Physiol. 1993b;265:H350–65.
Kassab GS, Lin DH, Fung YC. Morphometry of pig coronary venous system. Am J Physiol. 1994;267:H2100–13.
Kassab GS, Le KN, Fung YC. A hemodynamic analysis of coronary capillary blood flow based on anatomic and distensibility data. Am J Physiol. 1999;277:H2158–66.
Kini AS, Kim MC, Moreno PR, Krishnan P, Ivan OC, Sharma SK. Comparison of coronary flow reserve and fractional flow reserve in patients with versus without diabetes mellitus and having elective percutaneous coronary intervention and abciximab therapy (from the PREDICT Trial). Am J Cardiol. 2008;101:796–800.
Klocke FJ, Mates RE, Canty Jr JM, Ellis AK. Coronary pressure-flow relationships. Controversial issues and probable implications. Circ Res. 1985;56:310–23.
Kouwenhoven E, Vergroesen I, Han Y, Spaan JA. Retrograde coronary flow is limited by time-varying elastance. Am J Physiol. 1992;263:H484–90.
Krams R, Sipkema P, Westerhof N. Varying elastance concept may explain coronary systolic flow impediment. Am J Physiol. 1989a;257:H1471–9.
Krams R, Sipkema P, Zegers J, Westerhof N. Contractility is the main determinant of coronary systolic flow impediment. Am J Physiol. 1989b;257:H1936–44.

Kuo L, Davis MJ, Chilian WM. Myogenic activity in isolated subepicardial and subendocardial coronary arterioles. Am J Physiol. 1988;255:H1558–62.
Kuo L, Davis MJ, Chilian WM. Longitudinal gradients for endothelium-dependent and -independent vascular responses in the coronary microcirculation. Circulation. 1995;92:518–25.
Liao JC, Kuo L. Interaction between adenosine and flow-induced dilation in coronary microvascular network. Am J Physiol. 1997;272:H1571–81.
Loutzenhiser R, Bidani A, Chilton L. Renal myogenic response: kinetic attributes and physiological role. Circ Res. 2002;90:1316–24.
Manor D, Sideman S, Dinnar U, Beyar R. Analysis of flow in coronary epicardial arterial tree and intramyocardial circulation. Med Biol Eng Comput. 1994;32:S133–43.
Marzilli M, Goldstein S, Sabbah HN, Lee T, Stein PD. Modulating effect of regional myocardial performance on local myocardial perfusion in the dog. Circ Res. 1979;45:634–41.
Mihailescu LS, Abel FL. Intramyocardial pressure gradients in working and nonworking isolated cat hearts. Am J Physiol. 1994;266:H1233–41.
Mittal N, Zhou Y, Linares C, Ung S, Kaimovitz B, Molloi S, Kassab GS. Analysis of blood flow in the entire coronary arterial tree. Am J Physiol Heart Circ Physiol. 2005;289:H439–46.
Moir TW. Subendocardial distribution of coronary blood flow and the effect of antianginal drugs. Circ Res. 1972;30:621–7.
Pijls NH, van Son JA, Kirkeeide RL, De Bruyne B, Gould KL. Experimental basis of determining maximum coronary, myocardial, and collateral blood flow by pressure measurements for assessing functional stenosis severity before and after percutaneous transluminal coronary angioplasty. Circulation. 1993;87:1354–67.
Pries AR, Secomb TW, Gessner T, Sperandio MB, Gross JF, Gaehtgens P. Resistance to blood flow in microvessels in vivo. Circ Res. 1994;75:904–15.
Rabbany SY, Kresh JY, Noordergraaf A. Intramyocardial pressure: interaction of myocardial fluid pressure and fiber stress. Am J Physiol. 1989;257:H357–64.
Rabbany SY, Funai JT, Noordergraaf A. Pressure generation in a contracting myocyte. Heart Vessels. 1994;9:169–74.
Robicsek F, Thubrikar MJ. The freedom from atherosclerosis of intramyocardial coronary arteries: reduction of mural stress–a key factor. Eur J Cardiothorac Surg. 1994;8:228–35.
Rogers PA, Kiyooka T, Chilian WM. Is there a need for another model on the pulsatile nature of coronary blood flow? Am J Physiol Heart Circ Physiol. 2006;291:H1034–5.
Scaramucci J. Theoremata familiaria viros eruditos consulentia de variis physico-medicis lucubrationibus juxta leges mecanicas. *Apud Joannem Baptistam Bustum.* 1696;70–81.
Siebes M, Chamuleau SA, Meuwissen M, Piek JJ, Spaan JA. Influence of hemodynamic conditions on fractional flow reserve: parametric analysis of underlying model. Am J Physiol Heart Circ Physiol. 2002;283:H1462–70.
Siebes M, Verhoeff BJ, Meuwissen M, de Winter RJ, Spaan JA, Piek JJ. Single-wire pressure and flow velocity measurement to quantify coronary stenosis hemodynamics and effects of percutaneous interventions. Circulation. 2004;109:756–62.
Spaan JA. Coronary blood flow: mechanics, distribution, and control. Dordrecht: Kluwer; 1991.
Spaan JA. Mechanical determinants of myocardial perfusion. Basic Res Cardiol. 1995;90:89–102.
Spaan JA, Breuls NP, Laird JD. Diastolic-systolic coronary flow differences are caused by intramyocardial pump action in the anesthetized dog. Circ Res. 1981;49:584–93.
Spaan JA, Piek JJ, Hoffman JI, Siebes M. Physiological basis of clinically used coronary hemodynamic indices. Circulation. 2006;113:446–55.
Suga H. Total mechanical energy of a ventricle model and cardiac oxygen consumption. Am J Physiol. 1979;236:H498–505.
Tonino PA, De Bruyne B, Pijls NH, Siebert U, Ikeno F, van't Veer M, et al. Fractional flow reserve versus angiography for guiding percutaneous coronary intervention. N Engl J Med. 2009;360:213–24.

van den Wijngaard JP, Kolyva C, Siebes M, Dankelman J, van Gemert MJ, Piek JJ, Spaan JA. Model prediction of subendocardial perfusion of the coronary circulation in the presence of an epicardial coronary artery stenosis. Med Biol Eng Comput. 2008;46:421–32.
VanTeeffelen JW, Merkus D, Bos LJ, Vergroesen I, Spaan JA. Impairment of contraction increases sensitivity of epicardial lymph pressure for left ventricular pressure. Am J Physiol. 1998;274:H187–92.
Vis MA, Sipkema P, Westerhof N. Modeling pressure-area relations of coronary blood vessels embedded in cardiac muscle in diastole and systole. Am J Physiol. 1995;268:H2531–43.
Vis MA, Bovendeerd PH, Sipkema P, Westerhof N. Effect of ventricular contraction, pressure, and wall stretch on vessels at different locations in the wall. Am J Physiol. 1997;272:H2963–75.
Westerhof N. Physiological hypotheses–intramyocardial pressure. A new concept, suggestions for measurement. Basic Res Cardiol. 1990;85:105–19.
Westerhof N, Boer C, Lamberts RR, Sipkema P. Cross-talk between cardiac muscle and coronary vasculature. Physiol Rev. 2006;86:1263–308.
Yanagisawa H, Chikamori T, Tanaka N, Hatano T, Morishima T, Hida S, Iino H, Amaya K, Takazawa K, Yamashina A. Correlation between thallium-201 myocardial perfusion defects and the functional severity of coronary artery stenosis as assessed by pressure-derived myocardial fractional flow reserve. Circ J. 2002;66:1105–9.
Young DF. Fluid mechanics of arterial stenoses. J Biomech Eng. 1979;101:157–75.
Zhang W, Liu Y, Kassab GS. Viscoelasticity reduces the dynamic stresses and strains in the vessel wall: implications for vessel fatigue. Am J Physiol Heart Circ Physiol. 2007;293:H2355–60.
Zinemanas D, Beyar R, Sideman S. Relating mechanics, blood flow and mass transport in the cardiac muscle. Int J Heat Mass Transf. 1994;37:191–205.

Chapter 11
Microstructure-Based Constitutive Models for Coronary Artery Adventitia

Huan Chen, Xuefeng Zhao, Xiao Lu, and Ghassan S. Kassab

Abstract A structure-based constitutive model can accurately predict mechanical behaviors of blood vessels and enables a better understanding vascular pathophysiology. Most microstructural models assume affine deformation, i.e., fiber constituent deforms as the same as the tissue, and employ idealized microstructure due to the limited morphological data on vessel constituents. The goal of this chapter is to (1) introduce a new microstructural mechanical model that removes the affine deformation assumption and (2) to obtain quantitative microstructural data of coronary arteries. We develop a micromechanics-based constitutive model of fibrous tissue to take into consideration non-affine deformation that intrinsically induced by heterogeneous interactions between the constituents. Elastin fibers, cells, and ground substance are collectively considered as a solid-like matrix while collagen fibers is a reinforced phase. The model accounts for the waviness, orientation and spatial distributions of collagen fibers and provides a good prediction of macroscopic responses of the tissue which agree well with the finite element simulation results as a golden standard. We then use multiphoton microscopy to quantify the geometrical features of elastin and collagen fibers under mechanical loads. Simultaneous loading-imaging of the coronary adventitia allows measurements of the morphometry and in situ deformation of individual fibers. The population of fibers geometrical parameters including orientation angle, waviness, width and area fraction were measured at no-load state and the mechanical loading–deformation relation of fiber geometrical parameter were obtained as well. The present model and experimental studies are seminal for structural models and will lead to a better understanding of vascular biomechanics.

H. Chen • X. Zhao • X. Lu • G.S. Kassab (✉)
California Medical Innovations Institute, Roselle St, Rm 211 11107,
San Diego, CA 92121, USA
e-mail: gkassab@calmi2.org

G.S. Kassab, M.S. Sacks (eds.), *Structure-Based Mechanics of Tissues and Organs*,
DOI 10.1007/978-1-4899-7630-7_11

11.1 Introduction

Hemodynamics, wave propagation and distensibility of arteries, plaque stability and rupture, and vascular growth and remodeling are strongly affected by the stress or strain of the cells in vessel walls. An understanding of mechanical properties of blood vessels is thus fundamental to clarify the initiation and progression of diseases such as atherosclerosis (Vito and Dixon 2003). The vascular mechanical properties largely stem from microstructural components such as elastin and collagen fibers, cells, and ground substance (Oka and Azuma 1970; Azuma and Hasegawa 1971; Azuma and Oka 1971; Oka 1972). Thus, the relation between the microstructure and macroscopic mechanical properties of the vessel is essential in both biomedical research and clinical practice. Accurate prediction of microstructural deformation and loading, and in turn function, will result in a new level of understanding of the tissue.

The arterial wall is composed of three distinct layers: intima, media, and adventitia. The intima is normally very thin and contributes negligibly to the mechanical properties of the artery, while the media serves as the most important mechanical layer as it bears the majority of the load under physiological state (Lu et al. 2004). The media consists of three mechanically significant constituents: smooth muscle cells, elastin, and collagen fibers. The smooth muscles, when activated, achieve active response to physiological loads by altering the circumferential mechanical properties (Dobrin 1984; Tanaka and Yamada 1990). The passive state of smooth muscle cell may also contribute to the arterial behavior but the extent of this contribution is not currently well known. Elastin fibrils have relatively lower stiffness and larger deformability, which helps to maintain blood flow through a windkessel effect in elastin vessels. The outer layer, the adventitia, consisting of dense collagen fibers, some elastin fibers and some fibroblasts, contributes to the mechanical properties mainly by facilitating tethering to the surrounding connective tissue (Clark and Glagov 1985). At physiological pressures, the adventitia is less stiff than the media and its mechanical function is mainly to support the vessel. At higher pressures, such as in hypertension, however, the collagen fibers reach their straightened lengths and the adventitia becomes a stiff tube which prevents the artery from overstretch and ruptures (Holzapfel et al. 2000; Humphrey and Na 2002).

11.1.1 Phenomenological Constitutive Model

The vast majority of constitutive models of blood vessel assumes a single layer homogeneous wall and is phenomenological in nature. A 2-D exponential strain energy function (SEF) was introduced by Fung et al. (1979) to describe highly nonlinear mechanical behavior of arteries, and later was generalized into a 3-D form (Chuong and Fung 1983), in which the formulation applied to axisymmetric deformation of the vessel where the principal directions of the stress and strain tensors coincide with the radial, circumferential, and axial directions, i.e., assuming zero shear deformation. An extension of this model was proposed by Deng et al. (1994)

by adding a radial–circumferential strain to analyze axial torsion experiment and to determine the respective shear parameter. Humphrey (1995) subsequently developed a more general form of Fung-type SEF for arbitrary 3-D deformations. Other forms of hyperelastic constitutive models include the four-parameter logarithmic SEF of Takamizawa and Hayashi (1987), and the polynomial SEFs developed by Vaishnav et al. (1973) with 3, 7, or even 12 parameters. Since the 1990s, a number of studies have considered the artery to be a two-layered structure theoretically and experimentally (Von Maltzahn et al. 1981, 1984; Demiray and Vito 1991; Xie et al. 1995; Matsumoto and Sato 2002; Lu et al. 2003, 2004). The vessel wall is assumed to be made of two layers, of which each has its own elasticity constants and its own state of zero-stress resultants and zero-stress moments. These single- or two-layer phenomenological models, however, cannot predict the microenvironment of vessel wall and hence cannot elucidate the underlying mechanisms of vessel behavior. The parameters in these models have no physical meaning and are obtained by fitting experimental data, which show large variability in material constants (Zulliger et al. 2004).

11.1.2 Microstructure-Based Constitutive Model

The idea of relating the macroscopic mechanical properties of arteries to the arterial microstructures, including elastin and collagen fibers and cells, was first demonstrated by Burton and Yamada (1951). Roach and Burton (1957) made a quantitative study by differential digestion of elastin or collagen and measured the mechanical properties of the digested artery. Oka (1972) proposed a theoretical analysis of arterial wall, culminating in several well-known papers (Oka and Azuma 1970; Azuma and Hasegawa 1971; Azuma and Oka 1971). Azuma and Hasegawa (1971) discussed the rheological properties of arteries and veins in terms of the networks of collagen, elastin, and smooth muscle cells. Their works showed that the mechanical properties of the vessels are intimately associated with microstructural components such as elastin and collagen fibers, cells, and ground substance. Based on this idea, many efforts have been made to derive the constitutive model of the soft tissue from the geometry, distribution, and the mechanical properties of the individual microstructures.

Two major classes of micromechanical models, in relation to the matrix material (cells and ground substance) of the tissue, have been reported in literature. The first class of models proposed by Lanir (1979, 1983) considers the tissue as a composite of elastin and collagen fibers embedded in a fluid-like matrix. Thus, the fibers are the only constituent phases that sustain non-hydrostatic loading such as tension and shear, while the contribution of the fluid-like matrix is a hydrostatic pressure. This assumption leads to a simplification that all the microstructures deform identically to the macroscopic deformation of the tissue since no fiber interactions are considered, such that the macroscopic SEF is the volumetric sum of the individual fibers' SEF. On the basis of this assumption and thermodynamic consideration, Lanir developed a general multi-axial theory for the constitutive relations in fibrous connective tissues (Lanir 1979, 1983). Similarly, Decraemer et al. (1980) proposed a parallel wavy fibers model for soft biological tissues in uniaxial tension, followed by Wuyts et al. (1995). More recent developments of fluid-like matrix based models can be referred to in Humphrey and Yin (1987), Dahl et al. (2008), and Lokshin and Lanir (2009).

A second class of micromechanical models assume the tissue as a collagen fiber reinforced composite, whose matrix is a solid-like material that can take up loading. This assumption is motivated by the fact that the elastin, which is part of the matrix, becomes straightened and starts to take the load in the early deformation of the tissue. For example, the experimental study of Gundiah et al. (2007) suggested that the elastin is described with a neo-Hookean constitutive model. Based on this solid-like matrix assumption, Holzapfel and Weizsäcker (1998) and Holzapfel et al. (2000) modeled the arterial wall as a two-layer fiber-reinforced composite, where the macroscopic SEF of soft tissue stems from two sources: (1) an isotropic part associated with the mechanical response of the non-collagenous matrix material (elastin fibers, cells, and ground substance) and (2) an anisotropic part due to the deformation of two classes of collagen fibers symmetrically disposed with respect to the axis of the vessel. Successive developments of this model can be found in Zulliger et al. (2004), Kroon and Holzapfel (2008), and Li and Robertson (2009). Specifically, Zulliger et al. (2004) made further refinement to account for the distribution of the waviness of collagen fibers and different SEF of the matrix and collagen fibers. Mechanical predictions of these models are more accurate than that of phenomenological models as they account for heterogeneity of material properties and geometrical features of vessel components. These models, however, cannot accurately predict microenvironments of vessel, i.e., strain and stress of individual fiber or cell, since they all assume *affine* deformation in tissue, i.e., the deformation of the collagen fibers and the matrix are identical to the macroscopic deformation of the tissue (Chen et al. 2011b), and the microstructure is not based on histological measurements in arteries (Chen et al. 2011a).

Recently, Chen et al. (2011b) developed a finite-strain homogenization approach based on the second-order estimate (SOE) theory to predict the macroscopic stress–strain relation and microstructural deformation of vascular tissue, and showed significant improvements over previous microstructure models when compared to finite-element (FE) simulations. This micromechanical model considers measured histological geometrical features and material properties of vessel constituents, and allows more flexible deformation in each component. Hence, the model provides a better prediction of macro- and microscopic mechanical behavior of vascular tissue.

11.1.3 Quantitative Data of Coronary Artery Microstructure

The majority of microstructure-based constitutive models simplify the microstructure by assuming fibers to be symmetrically disposed with respect to the axis of the vessel (with a preferred orientation) to yield a macroscopic orthotropic constitutive law for each layer as aforementioned (Holzapfel and Weizsäcker 1998; Holzapfel et al. 2000; Zulliger et al. 2004). Other models assume the geometrical features of fibers follow a typical continuous distribution such as beta distribution with primarily planar array (Lanir 1983; Lokshin and Lanir 2009; Chen et al. 2011b; Hollander et al. 2011). The lack of accurate quantitative data of microstructure and deformation of elastin and collagen fibers in structural models is a barrier to accurately predict a vessel microenvironment. Therefore, comprehensive geometrical data of tissue microstructure are needed for understanding the morphology of arteries and for the development of mechanical modeling.

The development of nonlinear imaging modality (multi-photon microscopy, MPM) has enabled noninvasive measurements of biological tissues and cells. MPM imaging is driven by two primary types of nonlinear interaction between ultra-fast laser light and biological tissues: two-photon excited fluorescence (TPEF) for elastin and second-harmonic generation (SHG) for collagen. MPM is well developed and widely implemented for imaging cells (Campagnola et al. 2001; Mansfield et al. 2009), thin tissue sections (Campagnola et al. 2002; Garcia and Kassab 2009), thick unstained biological specimens (Zoumi et al. 2004; Arkill et al. 2010), and engineered tissues (Zoumi et al. 2002; Raub et al. 2008). Although the MPM studies of microstructure of arteries have advanced during the past decade, quantitative morphological data for elastin and collagen fibers have not been well established as the fibers are well knit, close to each other and dense.

The mechanical loading–deformation relation of elastin and collagen fibril bundles is fundamental to understand the microstructural properties of arteries. In situ deformation studies of fibers under physiological state have been limited, however, due to difficulties in implementation of simultaneous mechanical loading–imaging of fresh unfixed and unstained tissue at a microscopic level. Some studies, based on the use of the custom planar mechanical devices (Hu et al. 2009; Timmins et al. 2010; Keyes et al. 2011a), required splaying a vessel open that introduces undesired stress to the specimens. Recently, simultaneous mechanical loading-imaging on a fresh unfixed, unstained lymphatic vessel was conducted to investigate the microstructure and their responses to external mechanical loading (Arkill et al. 2010; Keyes et al. 2011b). They investigated the reorientation of collagen and elastin fibers under pressurized conditions and provided the corresponding statistical data. These studies, however, lacked the capability of tracking the scan area between loading states and were not able to measure in situ deformation of individual fibers. A recent study overcame this limitation by using fluorescent microspheres as markers to track the scan area as well as deformation of individual fibers (Chen et al. 2011a). This study quantified the geometrical data of collagen and elastin fibers as well as their loading–deformation behavior on a unstained fresh coronary artery adventitia.

In this chapter, we will provide some details on micromechanical models and measurements of artery microstructure. The motivation and basis of the recently proposed SOE homogenization approach are introduced in Sect. 11.2, and discussion of microstructural models based on affine deformation assumption is made as well. In Sect. 11.3, quantitative geometrical data of the collagen and elastin fibers of arterial wall under in situ deformation using MPM are described. The limitations and future directions are elaborated in the last section.

11.2 A Finite-Strain Micromechanics Model of Fibrous Tissue

All microstructural constituents contribute to the mechanical response of vascular tissue to loads. The macroscopic effective strain–stress relation of vessel wall is associated with the geometrical feature and mechanical properties of elastin, collagen fibers, cells, and ground substance. Statistical deformation or stress of each component (such as a single fiber) depends on its own stiffness, geometry, and interaction with each other. Thus, accurate description of geometrical and mechanical properties of microstructure is essential to a microstructure-based constitutive model.

11.2.1 The Geometrical and Mechanical Features of Tissue Constitutes

In the adventitia, dense and wavy collagen fibers form an interwoven network that tangles with elastin fibers (Wolinsky and Glagov 1967; Rhodin 1980). Histologically, a collagen or elastin fiber is well described as a bundle of loosely bounded fibrils (Fratzl et al. 1998; Ottani et al. 2001; Chen et al. 2011a). In an undulating state, such a fiber can deform with very little stress. When straightened, it can sustain a significant amount of stress. Experimental observations showed that elastin fibers are much less undulated than collagen at zero-stress state, so they gradually extend (Lu et al. 2004; Chen et al. 2011a) to take up load together with the fibroblasts and ground substance at very low strain level. At this level of deformation, the stress–strain behavior of the tissue exhibits only weak nonlinearity since the stiffness of elastin fibers is not significantly higher than the fibroblasts and ground substance. When the artery is distended beyond a certain stretch ratio, the collagen fibers are gradually straightened and begin to take up increasing loads in a manner, which depends on the deformation of the network. Since the collagen fibers have much higher tensile stiffness than elastin fibers and ground substance, full engagement of collagen leads to the highly nonlinear overall mechanical properties of the adventitia.

Material Property of Matrix

Noncollagenous matrix material (elastin and cells) is associated with an isotropic mechanical response of vessel wall, while anisotropic deformation is almost entirely due to collagen fibers. So, elastin, fibroblasts, and ground substance make up the effective isotropic matrix and the collagen fibers are considered as the reinforcing phases. This approximation is consistent with the framework of Holzapfel and Weizsäcker (1998), Holzapfel et al. (2000), and Gundiah et al. (2007), and the matrix material is described by an approximate incompressible ($\kappa \gg \mu$) neo-Hookean SEF as:

$$W^{(0)}(\mathbf{F}) = W_{\text{iso}}(\mathbf{F}) = \mu\,(I_1 - 3)/2 - \mu \ln I_3 + (3\kappa - 2\mu)(I_3 - 1)^2/6, \quad (11.1)$$

where $\mathbf{F}$ is the deformation gradient, $I_1 = \operatorname{tr}\mathbf{C}$ ($\mathbf{C} = \mathbf{F}^{\mathrm{T}} \cdot \mathbf{F}$), and $I_3 = \sqrt{\det(\mathbf{C})}$ are the first and the third invariants of the right Cauchy-Green deformation tensor. κ and μ is bulk and shear modulus, respectively.

Geometry and Material Property of Collagen Fibers

The SEF of a single collagen fiber is associated with its material property as well as structure. The waviness of a single undeformed collagen fiber is defined as the ratio of the end-to-end distance S_0 and the straight length l_0 as $\lambda_0 = S_0/l_0$. It should be noted that the fiber becomes straightened to take up loads when it is stretched along the fiber direction with a ratio $> \lambda_0$. The orientation of a fiber is described by the overall direction $\mathbf{N}$. The shape of the fiber can be described by a geometric tensor $\mathbf{Z}_s = 2l_0^{-1}\mathbf{N} \otimes \mathbf{N} + \overline{r}_L^{-1}\overline{\mathbf{n}}_L \otimes \overline{\mathbf{n}}_{LL} + \overline{r}_s^{-1}\overline{\mathbf{n}}_S \otimes \overline{\mathbf{n}}_S$, where $(\overline{r}_L,\ \overline{\mathbf{n}}_L)$ is the cross-sectional long axis and $(\overline{r}_S,\ \overline{\mathbf{n}}_S)$ is the short axis. So each fiber has its own geometric properties $(\lambda_0, \mathbf{Z}_s)$.

The orientation and waviness of fibers were found to follow certain continuous distribution functions (Sacks 2003; Chen et al. 2011a). In the present homogenization modeling, the collagen fibers are categorized into N phases for convenience, such that the r-th phase has the same geometric properties $(\lambda_0, \mathbf{Z}_s)$ and occupies a volume fraction $c^{(r)}$ in the composite. Correspondingly, the volume fraction of the matrix, denoted as phase 0, is $c^{(0)} = 1 - \sum_{r=1}^{N} c^{(r)}$. Furthermore, we use another geometric tensor $\mathbf{Z}_d$ to characterize the spatial distribution of fibers centers (Willis 1977; Ponte Castañeda and Willis 1995).

Given that a collagen fiber deforms the same as the soft matrix before straightening and becomes stiffer when straightened, the constitutive SEF for a single fiber is assumed to have the form:

$$W^{(r)}(\mathbf{F},\ \mathbf{X}) = \begin{cases} W^{(0)} & \lambda < \lambda_0^{(r)}, \quad (2a) \\ W^{(0)} + W_{\text{fiber}}^{(r)}(\lambda) & \lambda \geq \lambda_0^{(r)}, \quad (2b) \end{cases}$$

where $\lambda = \sqrt{\mathbf{N}^{(r)} \cdot (\mathbf{F}^T \cdot \mathbf{F}) \cdot \mathbf{N}^{(r)}}$ is the stretch of the r-th fiber along its direction $\mathbf{N}^{(r)}$ and $\lambda_0^{(r)}$ is the waviness of the r-th fiber. This approximation implies that when the tissue is at the reference state (macroscopic deformation gradient $\overline{\mathbf{F}} = \mathbf{I}$), all the r-th phase fibers are wavy so that they deform with SEF $W^{(0)}$. When $\overline{\mathbf{F}}$ increases, some fibers are straightened to take up loads, and the tissue is considered as a composite with reinforced fibers with additional SEF $W_{\text{fiber}}^{(r)}$, of which macroscopic SEF is $\overline{W}\left(\overline{\mathbf{F}}\right)$. The anisotropic term $W_{\text{fiber}}^{(r)}(\lambda)$ in the SEF of fibers is selected as:

$$W_{\text{fiber}}^{(r)}(\lambda) = E_1\left(\lambda - \lambda_0^{(r)}\right)^2/2 + E_2\left(\lambda - \lambda_0^{(r)}\right)^3/3 \tag{11.3}$$

This is a generalization of the linear model employed in other works (Lanir 1979, 1983; Decraemer et al. 1980; Wuyts et al. 1995) where only the first term $\left(\lambda - \lambda_0^{(r)}\right)^2$ was included. In Principle, the homogenization model employed in this work can use any well-defined constitutive model of the matrix and fibers.

11.2.2 Homogenization Approach Based on Uniform-Field Assumption

Based on thermodynamics, the heterogeneous hyperelastic SEF of a composite is determined as $W(\mathbf{X}, \mathbf{F}) = \sum_{r=0}^{N} \chi^{(r)}(\mathbf{X}) W^{(r)}(\mathbf{F})$, where $\chi^{(r)} = 1$ when $\mathbf{X} \in \Omega^{(r)}$ (the volume occupied by phase r), and 0 otherwise. The classical works of Hill (1972), Hill and Rice (1973) and Ogden (1978) proved that the macroscopic constitutive property for a microscopically inhomogeneous hyperelastic material can be described with macroscopic or effective SEF:

$$\overline{W}\left(\overline{\mathbf{F}}\right) = \min_{\mathbf{F}\in\kappa\left(\overline{\mathbf{F}}\right)} \langle W(\mathbf{X}, \mathbf{F})\rangle = \min_{\mathbf{F}\in\kappa\left(\overline{\mathbf{F}}\right)} \sum_{r=0}^{N} c^{(r)} \left\langle W^{(r)}(\mathbf{F})\right\rangle^{(r)} \tag{11.4}$$

where $\kappa\left(\overline{\mathbf{F}}\right)$ denotes all admissible deformation gradient field in the representative volume element (RVE) of the composite such that the displacement $\mathbf{u}$ on the boundary $\partial\Omega$ of RVE is $\mathbf{u}(\mathbf{X}) = \left(\overline{\mathbf{F}} - \mathbf{I}\right) \cdot \mathbf{X}$ (where rigid body motion is excluded). The notation $\langle\cdot\rangle$ denotes volumetric average in the RVE and $\langle\cdot\rangle^{(r)}$ denotes volumetric average in the r-th phase. This equation represents the principle of minimum strain energy in micromechanics, and the efforts to seek an approximate minimizing field $\mathbf{F}(\mathbf{X})$ have resulted in several classes of finite-strain micromechanics models.

Some homogenization models of soft tissue assume that deformation of reinforced collagen fibers is identical to that of solid-like matrix as well as bulk tissue, i.e., assuming a uniform deformation field in the RVE $\mathbf{F}(\mathbf{X}) = \overline{\mathbf{F}}$. Thus, the macroscopic SEF $\overline{W}\left(\overline{\mathbf{F}}\right)$ is simply the volumetric sum of the constituents:

$$\overline{W}\left(\overline{\mathbf{F}}\right) \approx \overline{W}_U\left(\overline{\mathbf{F}}\right) = \sum\nolimits_{r=0}^{N} c^{(r)} W^{(r)}\left(\overline{\mathbf{F}}\right) \tag{11.5}$$

Since $\mathbf{F}(\mathbf{X}) = \overline{\mathbf{F}}$ is an admissible deformation field that does not minimize the total strain energy, $\overline{W}_U\left(\overline{\mathbf{F}}\right)$ is an upper bound of the exact $\overline{W}\left(\overline{\mathbf{F}}\right)$. Consequently, the macroscopic stress field $\overline{\mathbf{S}}_U = \partial\overline{W}_U/\partial\overline{\mathbf{F}}$ ($\overline{\mathbf{S}}$ is the transpose of the first Piola–Kirchhoff stress and is related to the Cauchy stress $\overline{\boldsymbol{\sigma}}$ by $\overline{\mathbf{S}} = \overline{\boldsymbol{\sigma}} \cdot \overline{\mathbf{F}}^{-T}$) is the upper bound of the exact macroscopic stress for a given $\overline{\mathbf{F}}$. A number of works (Holzapfel and Weizsäcker 1998; Holzapfel et al. 2000; Zulliger et al. 2004; Kroon and Holzapfel 2008; Li and Robertson 2009) employ this uniform-field approximation $\mathbf{F}(\mathbf{X}) = \overline{\mathbf{F}}$ in various forms, and thus are all upper bounds.

11.2.3 Second-Order Microstructural Model

An SOE homogenization method for fibrous tissue in finite-strain deformation was implemented (Chen et al. 2011b) to consider the statistical microstructural geometry of the tissue and provide a better constitutive model than the uniform-field upper bound recently. This SOE homogenization approach has been well developed and applied for composites in the past decade (Ponte Castañeda 2002; Lopez-Pamies and Ponte Castañeda 2004a).

The key premise of SOE method is to introduce a linear thermoelastic comparison composite (LTCC) that has the same microstructures as the nonlinear composites, but each of the phases is linearly thermoelastic with SEF (Ponte Castañeda 2002; Lopez-Pamies and Ponte Castañeda 2004a):

$$\begin{aligned} W_{\mathrm{T}}^{(r)}(\mathbf{F}) &= W^{(r)}\left(\mathbf{F}^{*(r)}\right) + \boldsymbol{\rho}^{*(r)}\left(\mathbf{F} - \mathbf{F}^{*(r)}\right) \\ &\quad + \frac{1}{2}\left(\mathbf{F} - \mathbf{F}^{*(r)}\right)\mathbf{L}^{(r)}\left(\mathbf{F} - \mathbf{F}^{*(r)}\right) \quad (r = 0,\ 1,\ \ldots,\ N), \end{aligned} \tag{11.6}$$

in which the reference modulus $\mathbf{L}^{(r)}$ and virtual residual deformation gradient $\mathbf{F}^{*(r)}$ are to be determined later for conditional minimization, i.e., $\overline{W}\left(\overline{\mathbf{F}}\right) = \min\limits_{\mathbf{F}\in\kappa(\overline{\mathbf{F}})} \langle W(\mathbf{X}, \mathbf{F})\rangle$, and $\boldsymbol{\rho}^{*(r)} = \left(\partial W^{(r)}/\partial\mathbf{F}\right)_{\mathbf{F}=\mathbf{F}^{*(r)}}$. According to generalized Legendre transform of the strain energy as in Ponte Castañeda (2002) and Chen et al. (2011b), it is shown that the exact macroscopic SEF can be approximated by:

$$\overline{W}\left(\overline{\mathbf{F}}\right) = \min_{\mathbf{F}\in\kappa\left(\overline{\mathbf{F}}\right)} \langle W\left(\mathbf{X},\ \mathbf{F}\right)\rangle \leq \min_{\left\{\mathbf{F}^{*(s)},\mathbf{L}^{(s)}\right\}} \left\{\overline{W}_{\mathrm{T}}\left(\overline{\mathbf{F}};\left\{\mathbf{F}^{*(s)},\ \mathbf{L}^{(s)}\right\}\right)\right.$$
$$\left.+\sum\nolimits_{r=0}^{N} c^{(r)} V^{(r)}\left(\mathbf{F}^{*(r)},\ \mathbf{L}^{(r)}\right)\right\} \tag{11.7}$$

where $\overline{W}_T$ is the effective strain energy for the LTCC, and $V^{(r)}\left(\mathbf{F}^{*(r)},\ \mathbf{L}^{(r)}\right) = \sup_{\widehat{\mathbf{F}}}\left\{W^{(r)}\left(\widehat{\mathbf{F}}\right) - W_{\mathrm{T}}^{(r)}\left(\widehat{\mathbf{F}}\right)\right\}$. This derivation translates the optimization over continuous deformation field $\mathbf{F} \in \kappa\left(\overline{\mathbf{F}}\right)$ to that of respective tensors $\mathbf{L}^{(r)}$ and $\mathbf{F}^{*(r)}$. In theory, the field $\mathbf{F} \in \kappa\left(\overline{\mathbf{F}}\right)$ must satisfy continuity and compatibility conditions, while $\mathbf{L}^{(r)}$ and $\mathbf{F}^{*(r)}$ can be arbitrary within the physically meaningful ranges. Equation (11.7) is identical to the principle of minimum energy, provided that $\mathbf{L}^{(r)}$ and $\mathbf{F}^{*(r)}$ are allowed to change from point to point in the material. In standard nonlinear micromechanics (Willis 1977; Ponte Castañeda 2002; Lopez-Pamies and Ponte Castañeda 2004a), $\mathbf{L}^{(r)}$ and $\mathbf{F}^{*(r)}$ are assumed to be uniform for the r-th phase and equation (11.7) is an estimate of the exact SEF.

As has been derived in (Kailasam et al. 1997), the effective strain energy for the LTCC $\overline{W}_{\mathrm{T}}$ is estimated by taking into account the shape and distribution tensors $\mathbf{Z}_s^{(r)}$ and $\mathbf{Z}_d$, as

$$\overline{W}_{\mathrm{T}}\left(\overline{\mathbf{F}}\right) = \frac{1}{2}\left(\overline{\mathbf{F}}-\mathbf{I}\right)\tilde{\mathbf{L}}\left(\overline{\mathbf{F}}-\mathbf{I}\right) + \tilde{\mathbf{s}}\left(\overline{\mathbf{F}}-\mathbf{I}\right) + \frac{1}{2}\tilde{g} + \sum\nolimits_{r=0}^{N} c^{(r)} w_0^{(r)} \tag{11.8}$$

where $\tilde{\mathbf{L}}$ is the effective stiffness of LTCC, $\tilde{\mathbf{s}}$, $w_0^{(r)}$, and $\tilde{g}$ are the macroscopic residual stress and energies. These quantities all depend on the phase modulus $\mathbf{L}^{(r)}$, residual deformation gradient $\mathbf{F}^{*(r)}$ as well as geometric property $\mathbf{Z}_s^{(r)}$ and microstructural spatial distribution function $\mathbf{Z}_d$ (Chen et al. 2011b).

The stationary procedures involved in the above derivation, maximization of $V^{(r)}$ and the minimization in (11.7), yield a set of nonlinear tensorial equations for determination of the unknown reference modulus $\mathbf{L}^{(r)}$ and reference deformation gradient $\mathbf{F}^{*(r)}$ (Ponte Castañeda 2002; Lopez-Pamies and Ponte Castañeda 2004a, b, 2006). These equations have multiple solutions that lead to various estimations of the macroscopic SEF $\overline{W}\left(\overline{\mathbf{F}}\right)$. A *tangent* solution was employed here, in which the reference deformation gradient $\mathbf{F}^{*(r)}$ is taken to be the average deformation gradient $\overline{\mathbf{F}}^{(r)}$ in the r-th phase of the LTCC and $\mathbf{L}^{(r)}$ is the tangent stiffness tensor evaluated at $\overline{\mathbf{F}}^{(r)}$. This solution leads to $V^{(r)}\left(\mathbf{F}^{*(r)},\ \mathbf{L}^{(r)}\right) = 0$ and an estimation of $\overline{W}\left(\overline{\mathbf{F}}\right)$, as:

$$\overline{W}\left(\overline{\mathbf{F}}\right) \approx \overline{W}_S\left(\overline{\mathbf{F}}\right) = \sum\nolimits_{r=0}^{N} c^{(r)}\left\{W^{(r)}\left(\overline{\mathbf{F}}^{(r)}\right) + \frac{1}{2}\boldsymbol{\rho}^{(r)}\left(\overline{\mathbf{F}}^{(r)}\right)\left(\overline{\mathbf{F}}-\overline{\mathbf{F}}^{(r)}\right)\right\}. \tag{11.9}$$

11.2.4 Comparison of Modeling Predictions

Both uniform-field (UF) approach and tangent second-order micromechanics model were applied to 2-D material with distributed collagen fibers, respectively. The planar tissue stretches along the principal direction of the fibers with stretch ratio $\overline{\lambda}$, but shrinks in the transverse direction to maintain material incompressibility. The material property of matrix is directly taken from Holzapfel et al. (2000) and the bulk modulus is set as $k \gg \mu$ in simulation to account for material incompressibility. The material parameters E_1 and E_2 in the present $W_{\text{fiber}}^{(r)}(\lambda)$ were fitted to the exponential form in Holzapfel et al. (2000) with $\lambda_0 = 1.6$ (the mean value of distribution of the waviness).

To simplify the simulations, all the fibers were assumed to be isotropically distributed in space ($\mathbf{Z}_d = \mathbf{I}$) and have the same undeformed cross-sectional geometry with $\overline{r}_L = \overline{r}_S = 0.05l_0$. The fiber orientation was assumed to follow a beta distribution (Sacks 2003), and the waviness was also assumed to be beta distributed but direction dependent in line with tissue ultrastructure (Brown 1973; Sverdlik and Lanir 2002). The orientation distribution was discredited into 13 regions and the waviness was discredited into five regions, so that the collagen fibers are categorized into 65 phases according to their orientation and waviness. The total volume fraction of collagen fibers is taken as 20 % and that of the matrix was 80 %. The volume fraction of every fiber phase was determined by its cumulative probabilities of the orientation and waviness. For example, the r-th fiber phase with mean orientation angle $\theta = 10°$ and waviness $\lambda_0 = 1.6$ is 0.15 % (Chen et al. 2011b). FE simulations were used for the purpose of model validation. The FE model contained ~200 randomly distributed fibers whose orientation angle and waviness were randomly assigned according to the beta distribution. The average Cauchy stress was outputted and compared with the model prediction.

The model predictions of the macroscopic SEF $\overline{W}$ and the tensile Cauchy stress $\overline{\sigma}_{11}$ of fibrous tissue are plotted in Fig. 11.1, in comparison with FE computational results. At low stretch level ($\overline{\lambda} < 1.6$), most fibers are still undulated or have just been straightened and deform as the same as matrix so that effective macroscopic SEF and Cauchy stress of tissue predicted by UF and SOE models are very similar and approximate to that of pure matrix material. When the macroscopic deformation $\overline{\lambda}$ is >1.6, predictions of the two models become different. The UF predictions of SEF and Cauchy stress increase rapidly with higher macroscopic stretch ratio while SOE results increase slowly and predicted Cauchy stress is consistent with FE simulation as shown in Fig. 11.1b. At this loading level, collagen fibers become straighten completely to take up loads and deform less than matrix. The UF model, however, overestimates the deformation field of stiffer fibers (assuming it is identical to that of matrix) and hence approaches upper bounds of the exact SEF and Cauchy stress. The SOE model statistically accounts for the heterogeneous deformation of the matrix and fibers, which is due to the heterogeneities of microstructural geometries and material properties as well as matrix–fiber and fiber–fiber interactions. Thus, the deformation field employed by SOE is more realistic and leads to lower estimates of the macroscopic SEF and stress than the UF upper bounds.

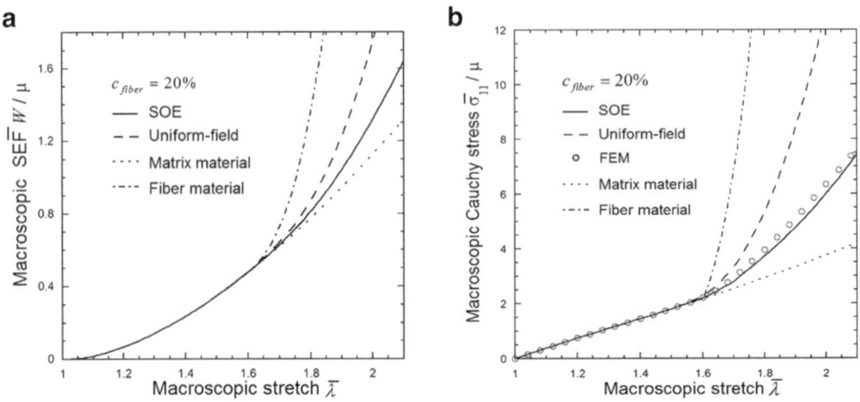

Fig. 11.1 The macroscopic strain energy function (**a**) and Cauchy stress–stretch relation (**b**) of fibrous tissue with 20 % randomly distributed collagen fibers. Values are normalized with material parameter μ. The reference fiber material has waviness $\lambda_0 = 1.6$

In the SOE prediction, the statistical deformation is flexible and non-affine, i.e., fibers with different orientation and waviness deform differently once they are straightened and become stiffer to take up the load. Figure 11.2a describes the influence of waviness λ_0 on the microscopic stretch of the fibers. The axial stretch of collagen fibers that are parallel to the loading axis ($\theta = 0°$) and with waviness $\lambda_0 = 1.5,\ 1.6,\ 1.7$ and 1.8 are plotted. In UF method, the deformation gradient of the matrix and all fibers are identical to the applied macroscopic deformation, i.e., $\overline{\lambda}$. In the SOE model, the microscopic deformation is non-affine once some fibers are straightened and become stiffer to take up the load. For fibers with a low waviness, such as $\lambda_0 = 1.5$, their stretch shows a tendency of becoming straight at large macroscopic deformation. As a result, the matrix phase must deform slightly more to accommodate the applied macroscopic deformation, as indicated by the solid line in Fig. 11.2a, which is noticeably higher than the UF prediction. Figure 11.2b plots the stretch of fibers with orientation angle $\theta = 0°,\ 10°$ and 20°, and waviness $\lambda_0 = 1.6$, showing the influence of the orientation angle θ on the statistical average microscopic stretch of the fibers. The UF predictions depend on the orientation but not the waviness. The SOE predictions also show similar trend that larger orientation angle θ leads to later engagement of the fiber.

The macroscopic stretch $\overline{\lambda}$ that straightens the fibers with $\theta = 0°,\ 15°$ and 25°, and λ_0 from 1.4 to 1.8, denoted as the macroscopic straightening stretch, is plotted in Fig. 11.2c. The macroscopic straightening stretches predicted by SOE model are consistently lower than the UF predictions, indicating earlier engagement of the fibers. The difference is more significant for fibers with larger angle and/or higher waviness. As discussed above, the earlier fiber engagement is due to the heterogeneous microscopic deformation predicted by SOE method. Once a fiber is straightened, it deforms less than those undulated fibers and the macroscopic deformation, due to the engagement of higher SEF. This mechanism prevents over

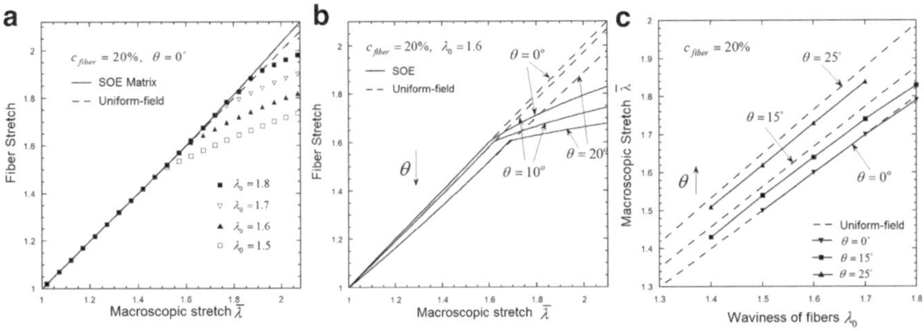

Fig. 11.2 The influence of waviness λ_0 and orientation angle θ on the statistical average microscopic stretch of the fibers. (**a**) The microscopic stretch of matrix phase (*solid line*) and collagen fibers (*symbols*) with orientation angle $\theta = 0°$ and different waviness; (**b**) the microscopic stretch of collagen fibers with waviness $\lambda_0 = 1.6$ and different orientation angle. (**c**) The macroscopic stretch that straightens the undulated collagen fibers with different orientation angle

stretch of the fibers with low waviness λ_0 in the tissue under large macroscopic deformation (Fig. 11.2). Since there is limit for the collagen fibers to stretch beyond the straightening value (Mohanaradhakrishnan et al. 1970), this prediction is qualitatively consistent with the protective role of fibers in the tissue.

It is important to investigate the predictive capability of UF and SOE models under different conditions. Figure 11.3 shows model predictions of macroscopic Cauchy stress of two tissues with the same waviness $\lambda_0 = 1.6$ but with different span of the fiber orientation distribution (span $\Delta = 0°$ and $60°$, respectively). These tissues are stretched along the principal direction of the fibers ($\theta = 0°$), but contracted in the transverse direction. While $\Delta = 60°$, the UF estimates are significantly higher than both SOE and FE. For $\Delta = 0°$, the UF estimates, however, are similar to both FE and SOE results. The span Δ of fiber orientation distribution plays a very important role in FE and SOE predictions, but is much less significant in UF model. It seems that the UF prediction of the macroscopic stress is considered accurate for tissue with very narrow distribution of fiber orientation.

11.3 Quantitative Morphometric Data of Coronary Adventitia

In order to quantify the geometrical features of fibers under deformation, unstained fresh adventitia specimens of arteries were imaged simultaneously under various mechanical loading, and the in situ deformation of elastin and collagen fibers were measured for a given fiber. Statistical analysis was also carried out to describe the population of fibers in terms of geometrical features, such as orientation angle, waviness, width, and area fraction. These data establish a microstructural foundation

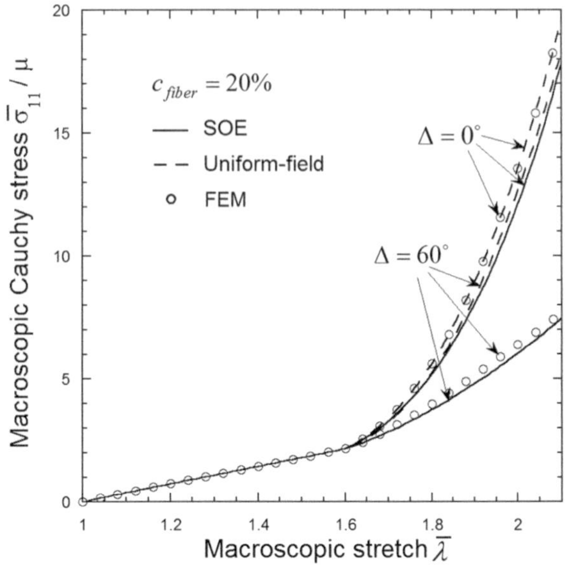

Fig. 11.3 The macroscopic tensile stress–strain curve of fibrous tissues with 20 % randomly distributed collagen fibers. The beta distribution of the fiber orientation angle θ has mean $M = 0°$ and span $\Delta = 0°$ and $60°$, respectively

for vascular biomechanics and may lead to a better understanding of vascular function and dysfunction.

11.3.1 Material and Method

Sample preparation: The swine left anterior descending (LAD) arteries ($n = 7$) were dissected carefully from hearts in saline and were inverted (turned inside out). The exposed intima-media layers were then peeled off carefully (Chen et al. 2011a). Since the penetration depth of the MPM is limited and the loose tissue on external adventitia layer cannot be removed thoroughly, inverted adventitia segment was scanned. The adventitia at no-load state is rather loose and the opening angle is relatively small (Lu et al. 2004), so the mechanical deformation is not significantly altered by the inversion.

Simultaneous mechanical loading–imaging: MPM images (Z-stack) from the adventitia specimen were obtained using a combined SHG/TPEF setup (Chen et al. 2011a). The specimen was first mounted on a custom-made organ bath chamber, and an elastic balloon was inserted to conduct distension and extension testing for vessel segments. The specimen was preconditioned several times before testing. Diameter of a segment was recorded by a commercial camera (Canon T1i), and fluorescent microspheres (excitation/emission wavelength: 540/584 *nm*) were dispersed on the outer surface of specimen to track the scan area as well as the deformation of individual fiber during loading. When increasing distension load, the elastic balloon was inflated slowly to follow the scan area at all times during the scan. Five mechanical loading conditions were considered: $\lambda_\theta = 1.0,\ 1.2,\ 1.4,\ 1.6$ and 1.8.

Image processing and statistical analysis: The elastin and collagen were found to form multiple concentric densely packed fiber sheets, with thickness of 3–7 μm and alternating principal directions. MPM images of the first five layers of specimens were collected (total average thickness of 30 μm) due to penetration limitation of MPM (Fig. 11.4). Image processing was then implemented on selected MPM images (Chen et al. 2011a). Four geometrical parameters were measured: the orientation angle θ (circumferential direction $\theta = 0°$), the waviness λ_0, the width D of a single fiber and the area fractions of fibers in images. The significance of the differences between the variables under various mechanical loading was evaluated by one-way ANOVA test, and the results were considered statistically different when $P < 0.05$.

11.3.2 Morphometric Data of Elastin and Collagen Fibers at No-Distension State

Both elastin and collagen fibers in inner adventitia formed layered structures (Haas et al. 1991; Finlay et al. 1995), where orientation, width, and area fraction of fibers revealed significant transmural variation as shown in Fig. 11.4. Such sublayer structure was about 3–7 μm thick and occupied 30–40 % of the total adventitial thickness. This structure did not present towards the exterior adventitia where collagen fibers were highly random and elastin fibers were largely absent.

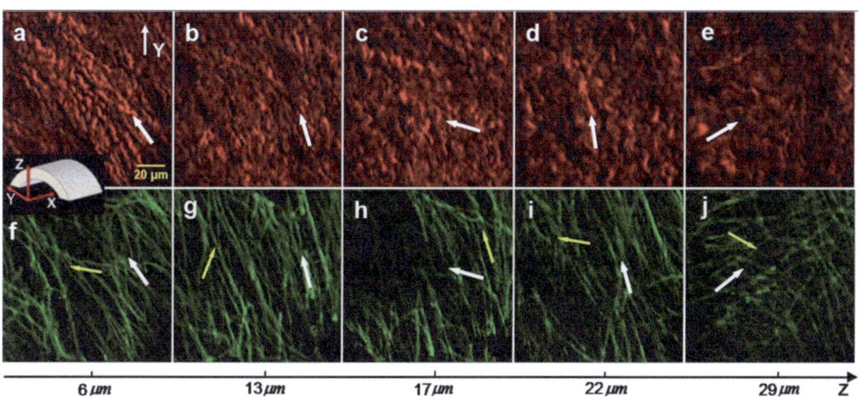

Fig. 11.4 Collagen and elastin fibers in inner adventitia form layered structures. Images (**a–e**) are five collagen layers at different depth *Z*, and images (**f–j**) are five elastin layers at the exactly same depth. In each layer, *white arrow* indicates the main orientation of fibers and *yellow arrow* indicates the second principal direction for elastin (*X*, *Y*, *Z*: axial, circumferential, and radial directions)

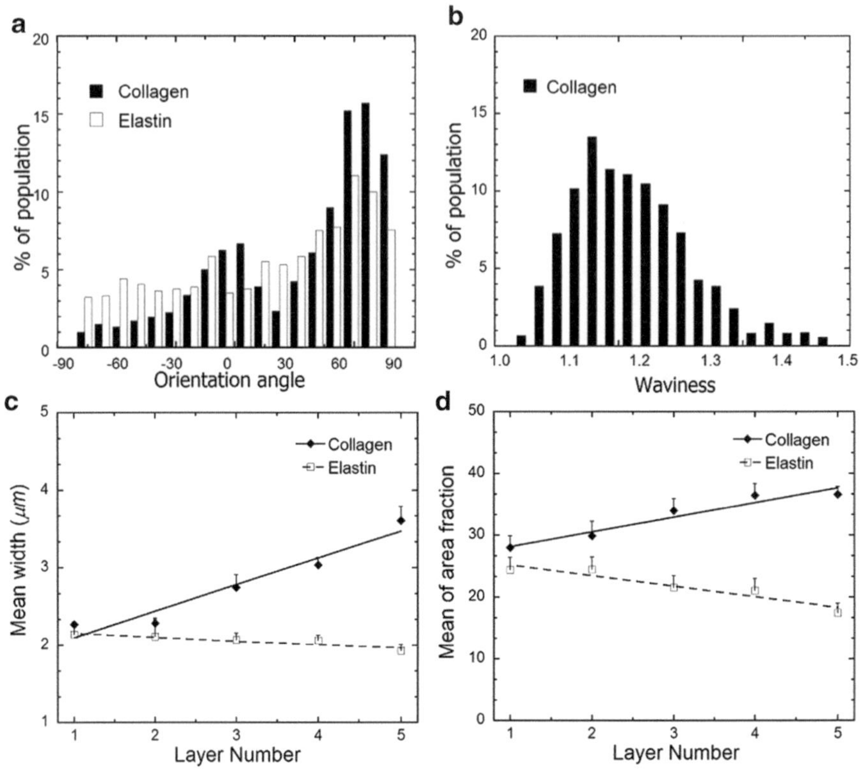

Fig. 11.5 Statistical data of microstructure of collagen and elastin fibers. (**a**) The orientation distribution of fibers of all specimen layers; (**b**) the *bell-shaped* distribution of the collagen fiber waviness of all layers; (**c**) layer-to-layer heterogeneity of the fiber width; (**d**) layer-to-layer heterogeneity of area fraction of fiber

In each adventitia layer, there was a main orientation of collagen and elastin fibers, and a secondary principal direction for elastin fibers as shown in Fig. 11.4. The orientation angles of collagen fibers followed a bell-shaped distribution (truncated normal distribution) in each layer. Furthermore, the main orientation angles varied in subsequent layers randomly as shown in Fig. 11.4. The orientation distribution of collagen fibers was shown in Fig. 11.5a, which demonstrated that most of collagen fibers aligned at about 60° in the no-distension state. The main orientation angles of elastin fibers were parallel with collagen fibers but with secondary principal directions. As a result, elastin fibers form a net-like structure and collagen fibers tend to uniformly align in a layer, which confirms that isotropic mechanical response of vessel wall predominates by non-collagenous matrix material (mainly elastin fibers) while anisotropic deformation is almost entirely due to collagen fibers in most models (Holzapfel et al. 2000; Zulliger et al. 2004; Kroon and Holzapfel 2008; Li and Robertson 2009; Chen et al. 2011b).

Table 11.1 The layer-to-layer heterogeneity of mean width and area fraction of fibers are assumed as linear function: $P = \alpha N + \beta$, where N is the layer order

Geometric parameter	Collagen			Elastin		
	α	β	R	α	β	R
Width	0.34	1.76	0.97	−0.05	2.20	0.92
Area fraction	2.38	25.8	0.97	−1.73	27.0	0.95

This function is determined by least-squares method, and R is the correlation coefficient between predicted data value and experimental measurement

The waviness of the collagen fibril bundle was also found to follow a bell-shaped distribution as shown in Fig. 11.5b (mean waviness is about 1.24). The morphometric features of elastin fiber differed significantly from collagen where elastin was rod-like rather than a spiral, and more than 80 % of elastin fiber was straight (waviness $\lambda_0 \approx 1.0$) at no-distension state (Fig. 11.4f–j). Collagen fibers became thicker towards the exterior of adventitia, while elastin fibers became thinner as shown in Fig. 11.5c. The width-layer relationships of fibers were curve fitted using linear regression (solid line in Fig. 11.5c, Table 11.1). The width of collagen changed more than that of elastin in inner adventitia. The average width of collagen fibers of all layers was about 2.8 μm and that of elastin fibers was 2.0 μm. Similarly, the area fraction of collagen fibers was found to increase significantly in deeper layers while that of elastin fibers decreased moderately as shown in Fig. 11.5d and Table 11.1. The average area fraction of collagen and elastin over all layers were 33 % and 22 %, respectively.

The images showed that the elastin fibers were rarely found in exterior adventitia, where collagen fibers become much thicker and orient randomly. The outer adventitia occupies about 60–70 % of adventitia thickness and the main function of thicker collagen fibers is to support the vessel and connect with surrounding tissue rather than to resist the transmural pressure. On the other hand, the inner adventitia which consists of alternating elastin and collagen fiber layers plays an important role in opposing the transmural pressure and prevents overdistension of vessel at high loading. The aforementioned microstructure-based model (Chen et al. 2011b) can consider the protective function of adventitia of coronary arteries with the present measured geometry and deformation of microstructures to accurately predict both macroscopic responses and microenvironment of blood vessel.

11.3.3 In Situ Deformation of Elastin and Collagen Fibers

In situ deformation of individual fibers under various mechanical loading was tracked in reference to the fluorescent microspheres and the extent of fiber deformation depended on the initial orientation angle at no-distension state. Fibers oriented towards the circumferential direction (distension direction) were stretched much more than those aligned more axially, such that the former became straightened or thinner earlier than the latter. For collagen fibers, the orientation angle and the

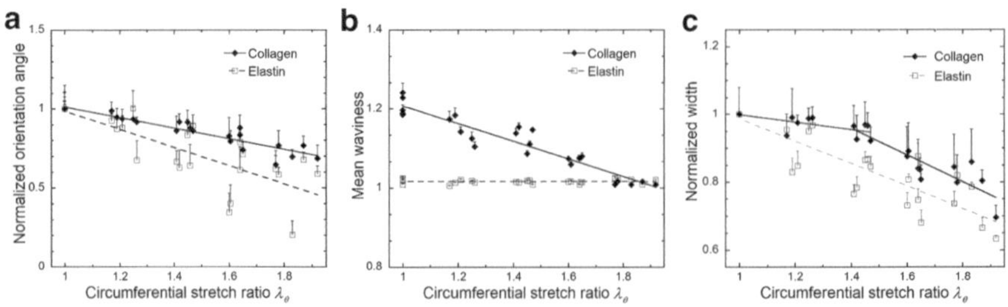

Fig. 11.6 In situ deformation of collagen and elastin fibers. (**a**) Change of mean orientation angle normalized by initial orientation angle; (**b**) change of mean waviness and waviness of elastin fibers is about 1.0 constantly; (**c**) change of mean width normalized by initial width

Table 11.2 The mechanical loading–deformation relation of fiber geometrical parameter is assumed as linear function: $P = \alpha\,\lambda_\theta + \beta$, where λ_θ is circumferential stretch ratio of vessels

Geometric parameter		Collagen			Elastin		
		α	β	R	α	β	R
Waviness		−0.22	1.43	0.95			
Width	$1.0 \le \lambda_\theta < 1.4$	−0.10	1.11	0.70	−0.33	1.32	0.87
	$1.4 \le \lambda_\theta < 1.8$	−0.40	1.51	0.88			

waviness were approximately uniform in a layer so that fiber deformation tended to be homogeneous while elastin fiber deformation was much more heterogeneous due to the various fiber orientations.

Both collagen and elastin fibers gradually oriented towards the circumferential direction (Fig. 11.6a). The orientation–distension relationships of collagen and elastin were curve fitted by linear least-squares method in Table 11.2 (solid line in Fig. 11.6a). Since collagen fibers aligned uniformly in a layer, the change of their mean orientation angles was more homogeneous. The orientation angles of elastin were heterogeneous with multiple directions as shown in Fig. 11.6a, i.e., the orientation of collagen shifted towards the circumferential direction more gradually than the elastin fibers. Furthermore, collagen fibers in a layer tended to align towards the principal direction so that the bell-shaped distribution of orientation angles became narrower (Chen et al. 2011a).

In Fig. 11.6b, a linear least-squares fit was used to describe the waviness–distension relation of collagen (Table 11.2), which showed that the collagen fibers were stretched gradually at elevated distension and became completely straightened to take up loads at $\lambda_\theta \approx 1.8$. Although the initial waviness of collagen fiber slightly varied in different layers, the waviness became uniform under higher loads. As shown in Fig. 11.6b (error bars are smaller under higher loads), the relative dispersion (RD = SD/mean) of waviness of collagen fiber decreased gradually with increase of loading conditions (i.e., RD = 6.0–2.9 λ_θ; $R = 0.78$ and $P < 0.05$). This suggests that the layered structure adventitia has heterogeneous mechanical response to loads given the observed significant transmural heterogeneity. It should be noted that the RDs of orientation angle and width of fibers were independent of circumferential loads.

The change of collagen width was more homogeneous than that of elastin fiber (Fig. 11.6c) consistent with the change of fiber orientation angle. The width change of collagen fibers was insignificant at lower loading ($\lambda_\theta = 1.0$, 1.2 and 1.4, and $P > 0.05$), but decreased at $\lambda_\theta > 1.4$ ($P < 0.05$). Hence, a piecewise linear function was used to describe the width–distension relationship of collagen fibers (Table 11.2). On the other hand, elastin fibers became thinner continuously and gradually at higher loading as shown in Fig. 11.6c.

The nearly uniform alignment of collagen fibers makes their deformation more homogeneous than elastin fibers which have multiple directions in each layer. As a result, collagen fibers were uniformly stretched and shifted towards the circumferential direction of adventitia and elastin fibers gradually resembled a network. Additionally, change of the fiber orientation and waviness depends on both the longitudinal stretch and luminal pressure. The statistical data (Fig. 11.5a) showed that most of elastin and collagen fibers oriented towards axial direction rather than the circumferential direction at no-distension state with physiological axial stretch ratio $\lambda_{\text{axial}} = 1.3$. When the specimens were returned to their relaxed axial length with luminal pressure remaining zero, elastin and collagen fibers reoriented and aligned in the circumferential direction (Arkill et al. 2010), and collagen fibers became slightly more undulated.

The collagen fibers thinned down significantly (Fig. 11.6c) at $\lambda_\theta \approx 1.4$ albeit they were not completely straightened at this state. This is consistent with the diameter-pressure curve of coronary adventitia that shows the adventitia becomes stiffer after $\lambda_\theta \approx 1.4$ (Chen et al. 2011a). In other words, collagen fibers in adventitia were gradually recruited after $\lambda_\theta \approx 1.4$ to take up loads, and then gradually predominate in the mechanical function of adventitia and became completely straightened when loaded at $\lambda_\theta \approx 1.8$. This finding suggests that collagen contributes mainly to the flat region of the nonlinear stress–strain curve whereas elastin mainly contributes to the toe portion of the stress–strain curve (Roach and Burton 1957; Zoumi et al. 2004). This underscores the function of the adventitia in a normal vessel (with physiological blood pressure $\lambda_\theta \approx 1.5$) is to support the vessel rather than to take up loads.

All of the deformation patterns were described by linear least-squares fit as shown in Table 11.2. However, these linear relationships were only determined between distension loading $\lambda_\theta = 1.0$ to $\lambda_\theta = 1.8$. When considering distension loading beyond $\lambda_\theta = 1.8$, these relationships may become nonlinear. For instance, since collagen fibers became completely straightened ($\lambda_0 \approx 1.0$) at $\lambda_\theta = 1.8$, the waviness of collagen fibers did not change further, and remained 1.0 at higher loading $\lambda_\theta \geq 1.8$. Similarly, the change of orientation angle of collagen and elastin fibers decreased when they approached the circumferential direction. In future studies, a larger range of loading should be considered to determine the extent of linearity.

11.4 Limitations

There are some limitations that deserve discussion. Firstly, the present model takes into consideration two functional microstructures in the vascular tissue: an effective matrix and distributed collagen fibers. The elastin fibers, which are also undulated, are considered as part of the effective matrix. This simplification is motivated by the fact that the elastin, while being stiffer than the ground substance and cells, are much softer than the collagen fibers and become straightened in the early deformation of the tissue. Thus, $W^{(0)}(\mathbf{F})$ in (11.1) should be considered as a macroscopic SEF of the effective matrix, and thus the presently employed neo-Hookean form of $W^{(0)}$ is phenomenological. For more microstructural modeling of $W^{(0)}$ and accurate description of the deformation of the cells and elastin, a multiscale homogenization method is needed, in which the effective matrix is considered as an elastin fiber-reinforced composite, with the ground substance and cells as the ground material. Lanir (1983) and Lokshin and Lanir (2009) considered both collagen and elastin as fibrous entities but embedded in a fluid-like matrix. When considering solid-like matrix, the effective SEF of matrix is first estimated with use of homogenization methods which takes into account the constitutive relation, geometries, and distributions of the elastin. Additionally, the homogenization approaches should be adjusted to reflect the microstructures of soft tissue. For example, in the media layer of the vessel, the elastin forms a complex network connected to vascular smooth muscle cells, in which the cytoskeleton (actin, myosin, tubulin, etc.) contributes to mechanical loading, that can sustain non-hydrostatic loading such as tension and shear. Therefore, the assumption of solid-like matrix may be an accurate description for artery media. In the adventitial layer, however, the non-fiber constituents are fibroblasts, macrophage, and amorphous gel-like ground substances that do not take much non-hydrostatic loading. Thus, the fluid-like matrix as employed in the models of Lanir (1979, 1983) may be appropriate.

Secondly, the most significant advantage of the UF method is the lower computational cost, i.e., it only requires a simple summation as in (11.5). In comparison, the SOE method requires higher computational expense, albeit it is much more efficient than direct FE simulation. Therefore, it is useful to investigate the predictive abilities of UF and SOE methods. In the illustrative simulation, it is found that the UF prediction of the macroscopic stress is considered accurate for tissue with very narrow distribution of fiber orientation. As shown in Sect. 11.3, arterial adventitia forms layered structure, and alignment of collagen fiber in each sublayer follows a narrow bell-shaped distribution (almost uniformly align), for which UF model may work well. Elastin fibers, however, are found to distribute randomly compared with collagen fibers. Hence, it is important to test predictive power of UF and SOE models with inclusion of realistic microstructure of arteries.

Finally, it was observed that some collagen fibers were out of the imaging plane in the z-direction and formed coils, i.e., 3-D helix structures (Arkill et al. 2010), which resulted in false impression of discontinuities of collagen networks in 2-D images. Therefore, the present 2-D image processing may underestimate some of the geometrical features of collagen fibers (e.g., for the waviness and area fraction) and any apparent discontinuities were excluded from measurements. This limitation suggests the need for 3-D reconstruction of fibers which can render more accurate morphology of fibers in vascular tissue in the future.

11.5 Summary

The developed microstructural model takes into consideration a full set of statistical microstructural geometries, the interactions between fibers and matrix, and the nonlinearity of constituent phases. As a result, this approach shows comparative advantage over the classical uniform-field theory, and enables more realistic predictions of the macroscopic stress and the statistical deformation of the wavy collagen fibers, when compared with direct FE numerical simulations. Meanwhile, the distribution and transmural heterogeneity of fiber geometrical parameters reflect realistic microstructures of adventitial wall, and statistical loading–deformation relation of fibers reveals partially mechanical properties of individual fibers. The present modeling and experimental studies will lead to a better understanding of vascular biomechanics and will contribute to future models of blood vessels.

Additionally, extension of these studies is necessary. For instance, material properties of single fibers are significant in microstructure models but remain unknown. The microstructural model has potential to identify the stress–strain relation of individual elastin and collagen fibers. Given MPM measured microstructure data of arteries, the predicted constitutive relation can be integrated with the FE model to predict both macro- and micro-scopic mechanical response of blood vessel wall under various mechanical loadings. Simultaneous mechanical loading-imaging, implemented on fresh unstained arteries, provides the loading–deformation relation of individual fibers. Hence, material properties of fibers can be identified by comparing model predictions and experimentally measured deformation of the fibers. While the microstructural morphometric data may be different from vessel to vessel, the microstructural material parameters are expected to be unique since in theory they are independent of morphometric data. It is also expected that identified material parameters will be physically meaningful and reflect the realistic mechanical behavior of individual fibers.

References

Arkill KP, Moger J, Winlove CP. The structure and mechanical properties of collecting lymphatic vessels: an investigation using multimodal nonlinear microscopy. J Anat. 2010;216:547–55.

Azuma T, Hasegawa M. A rheological approach to the architecture of arterial walls. Jpn J Physiol. 1971;21:27–47.

Azuma T, Oka S. Mechanical equilibrium of blood vessel walls. Am J Physiol. 1971;221:1310–8.

Brown IA. A scanning electron microscope study of the effects of uniaxial tension on human skin. Br J Dermatol. 1973;89:383–93.
Burton AC, Yamada S. Relation between blood pressure and flow in the human forearm. J Appl Physiol. 1951;4:329–39.
Campagnola PJ, Clark HA, Mohler WA, Lewis A, Loew LM. Second-harmonic imaging microscopy of living cells. J Biomed Opt. 2001;6:277–86.
Campagnola PJ, Millard AC, Terasaki M, Hoppe PE, Malone CJ, Mohler WA. Three-dimensional high-resolution second-harmonic generation imaging of endogenous structural proteins in biological tissues. Biophys J. 2002;82:493–508.
Chen H, Liu Y, Slipchenko MN, Zhao XF, Cheng JX, Kassab GS. The layered structure of coronary adventitia under mechanical load. Biophys J. 2011a;101:2555–62.
Chen H, Liu Y, Zhao XF, Lanir Y, Kassab GS. A micromechanics finite-strain constitutive model of fibrous tissue. J Mech Phys Solids. 2011b;59:1823–37.
Chuong CJ, Fung YC. Three-dimensional stress distribution in arteries. J Biomech Eng. 1983;105:268–74.
Clark JM, Glagov S. Transmural organization of the arterial media. The lamellar unit revisited. Arterioscler Thromb Vasc Biol. 1985;5:19–34.
Dahl SLM, Vaughn ME, Hu J-J, Driessen NJB, Baaijens FPT, Humphrey JD, Niklason LE. A microstructurally motivated model of the mechanical behavior of tissue engineered blood vessels. Ann Biomed Eng. 2008;36:1782–92.
Decraemer WF, Maes MA, Vanhuyse VJ. An elastic stress-strain relation for soft biological tissues based on a structural model. J Biomech. 1980;13:463–8.
Demiray H, Vito RP. A layered cylindrical shell model for an aorta. Int J Eng Sci. 1991;29:47–54.
Deng SX, Tomioka J, Debes JC, Fung YC. New experiments on shear modulus of elasticity of arteries. Am J Physiol Heart Circ Physiol. 1994;266:H1.
Dobrin PB. Mechanical behavior of vascular smooth muscle in cylindrical segments of arteries in vitro. Ann Biomed Eng. 1984;12:497–510.
Finlay HM, McCullough L, Canham PB. Three-dimensional collagen organization of human brain arteries at different transmural pressures. J Vasc Res. 1995;32:301–12.
Fratzl P, Misof K, Zizak I, Rapp G, Amenitsch H, Bernstorff S. Fibrillar structure and mechanical properties of collagen. J Struct Biol. 1998;122:119–22.
Fung YC, Fronek K, Patitucci P. Pseudoelasticity of arteries and the choice of its mathematical expression. Am J Physiol. 1979;237:H620–31.
Garcia M, Kassab GS. Right coronary artery becomes stiffer with in-crease in elastin and collagen in right ventricular hypertrophy. J Appl Physiol. 2009;106:1338–46.
Gauderon R, Lukins PB, Sheppard CJ. Optimization of second-harmonic generation microscopy. Micron. 2001;32:691–700.
Gundiah N, Ratcliffe MB, Pruitt LA. Determination of strain energy function for arterial elastin: experiments using histology and mechanical tests. J Biomech. 2007;40:586–94.
Haas KS, Phillips SJ, Comerota AJ, White JV. The architecture of adventitial elastin in the canine infrarenal aorta. Anat Rec. 1991;230:86–96.
Hill R. On constitutive macro-variables for heterogeneous solids at finite strain. Proc R Soc Lond Ser A. 1972;326:131–47.
Hill R, Rice JR. Elastic potentials and the structure of inelastic constitutive laws. SIAM J Appl Math. 1973;25:448–61.
Hollander Y, Durban D, Lu X, Kassab GS, Lanir Y. Experimentally validated microstructural 3D constitutive model of coronary arterial media. J Biomech Eng. 2011;133:031007.
Holzapfel GA, Weizsäcker HW. Biomechanical behavior of the arterial wall and its numerical characterization. Comput Biol Med. 1998;28:377–92.
Holzapfel GA, Gasser TC, Ogden RAYW. A new constitutive framework for arterial wall mechanics and a comparative study of material models. J Elast. 2000;61:1–48.
Hu J-J, Humphrey JD, Yeh AT. Characterization of engineered tissue development under biaxial stretch using nonlinear optical microscopy. Tissue Eng Part A. 2009;15:1553–64.

Humphrey JD. Mechanics of the arterial wall: review and directions. Crit Rev Biomed Eng. 1995;23:1–162.
Humphrey JD, Na S. Elastodynamics and arterial wall stress. Ann Biomed Eng. 2002;30:509–23.
Humphrey JD, Yin FC. A new constitutive formulation for characterizing the mechanical behavior of soft tissues. Biophys J. 1987;52:563–70.
Kailasam M, Ponte Castañeda P, Willis JR. The effect of particle size, shape, distribution and their evolution on the constitutive response of nonlinearly viscous composites. I. Theory. Philos Trans A Math Phys Eng Sci. 1997;355:1835–52.
Keyes JT, Borowicz SM, Rader JH, Utzinger U, Azhar M, Vande Geest JP. Design and demonstration of a microbiaxial optomechanical device for multiscale characterization of soft biological tissues with two-photon microscopy. Microsc Microanal. 2011a;17:167–75.
Keyes JT, Haskett DG, Utzinger U, Azhar M, Geest JPV. Adaptation of a planar microbiaxial optomechanical device for the tubular biaxial microstructural and macroscopic characterization of small vascular tissues. J Biomech Eng. 2011b;133:075001–8.
Kroon M, Holzapfel GA. A new constitutive model for multi-layered collagenous tissues. J Biomech. 2008;41:2766–71.
Lanir Y. A structural theory for the homogeneous biaxial stress-strain relationships in flat collagenous tissues. J Biomech. 1979;12:423–36.
Lanir Y. Constitutive equations for fibrous connective tissues. J Biomech. 1983;16:1–12.
Li D, Robertson AM. A structural multi-mechanism constitutive equation for cerebral arterial tissue. Int J Solids Struct. 2009;46:2920–8.
Lokshin O, Lanir Y. Micro and macro rheology of planar tissues. Biomaterials. 2009;30:3118–27.
Lopez-Pamies O, Ponte Castañeda P. Second-order estimates for the macroscopic response and loss of ellipticity in porous rubbers at large deformations. J Elast. 2004a;76:247–87.
Lopez-Pamies O, Ponte Castañeda P. Second-order homogenization estimates incorporating field fluctuations in finite elasticity. Math Mech Solids. 2004b;9:243–70.
Lopez-Pamies O, Ponte Castañeda P. On the overall behavior, micro-structure evolution, and macroscopic stability in reinforced rubbers at large deformations: I—theory. J Mech Phys Solids. 2006;54:807–30.
Lu X, Yang J, Zhao JB, Gregersen H, Kassab GS. Shear modulus of porcine coronary artery: contributions of media and adventitia. Am J Physiol Heart Circ Physiol. 2003;285:H1966–75.
Lu X, Pandit A, Kassab GS. Biaxial incremental homeostatic elastic moduli of coronary artery: two-layer model. Am J Physiol Heart Circ Physiol. 2004;287:H1663–9.
Mansfield J, Yu J, Attenburrow D, Moger J, Tirlapur U, Urban J, Cui Z, Winlove P. The elastin network: its relationship with collagen and cells in articular cartilage as visualized by multiphoton microscopy. J Anat. 2009;215:682–91.
Matsumoto T, Sato M. Analysis of stress and strain distribution in the artery wall consisted of layers with different elastic modulus and opening angle. JSME Int J Ser C. 2002;45:906–12.
Mohanaradhakrishnan V, Ramanathan N, Nayudamma Y. Strength of collagen fibres obtained from different sources. Biorheology. 1970;7:119–24.
Ogden RW. Extremum principles in non-linear elasticity and their application to composites—I: theory. Int J Solids Struct. 1978;14:265–82.
Oka S. Some theoretical studies on hemorheology. Adv Biophys. 1972;3:97–160.
Oka S, Azuma T. Physical theory of tension in thick-walled blood vessels in equilibrium. Biorheology. 1970;7:109–17.
Ottani V, Raspanti M, Ruggeri A. Collagen structure and functional implications. Micron. 2001;32:251–60.
Ponte Castañeda P. Second-order homogenization estimates for nonlinear composites incorporating field fluctuations: I—theory. J Mech Phys Solids. 2002;50:737–57.
Ponte Castañeda P, Willis JR. The effect of spatial distribution on the effective behavior of composite materials and cracked media. J Mech Phys Solids. 1995;43:1919–51.
Raub CB, Unruh J, Suresh V, Krasieva T, Lindmo T, Gratton E, Trom-berg BJ, George SC. Image correlation spectroscopy of multiphoton images correlates with collagen mechanical properties. Biophys J. 2008;94:2361–73.

Rhodin JAG. Architecture of the vessel wall. In: Bohr DF, Somlyo AM, Sparks HV, editors. Handbook of physiology. Section 2: the cardiovascular system. Bethesda: American Physiology Society; 1980. p. 1–31.
Roach MR, Burton AC. The reason for the shape of the distensibility curves of arteries. Can J Biochem Physiol. 1957;35:681–90.
Sacks MS. Incorporation of experimentally-derived fiber orientation into a structural constitutive model for planar collagenous tissues. J Biomech Eng. 2003;125:280.
Sverdlik A, Lanir Y. Time-dependent mechanical behavior of sheep digital tendons, including the effects of preconditioning. J Biomech Eng. 2002;124:78–84.
Takamizawa K, Hayashi K. Strain energy density function and uniform strain hypothesis for arterial mechanics. J Biomech. 1987;20:7–17.
Tanaka E, Yamada H. An inelastic constitutive model of blood vessels. Acta Mech. 1990;82:21–30.
Timmins LH, Wu Q, Yeh AT, Moore JE, Greenwald SE. Structural inhomogeneity and fiber orientation in the inner arterial media. Am J Physiol Heart Circ Physiol. 2010;298:H1537–45.
Vaishnav RN, Young JT, Patel DJ. Distribution of stresses and of strain-energy density through the wall thickness in a canine aortic segment. Circ Res. 1973;32:577–83.
Vito RP, Dixon SA. Blood vessel constitutive—1995–2002. Annu Rev Biomed Eng. 2003;5:413–39.
Von Maltzahn WW, Besdo D, Wiemer W. Elastic properties of arteries: a nonlinear two-layer cylindrical model. J Biomech. 1981;14:389–97.
Von Maltzahn WW, Warriyar RG, Keitzer WF. Experimental measurements of elastic properties of media and adventitia of bovine carotid arteries. J Biomech. 1984;17:839–47.
Willis JR. Bounds and self-consistent estimates for the overall properties of anisotropic composites. J Mech Phys Solids. 1977;25:185–202.
Wolinsky H, Glagov S. A lamellar unit of aortic medial structure and function in mammals. Circ Res. 1967;20:99–111.
Wuyts FL, Vanhuyse VJ, Langewouters GJ, Decraemer WF, Raman ER, Buyle S. Elastic properties of human aortas in relation to age and atherosclerosis: a structural model. Phys Med Biol. 1995;40:1577.
Xie J, Zhou J, Fung YC. Bending of blood vessel wall: stress-strain laws of the intima-media and adventitial layers. J Biomech Eng. 1995;117:136–45.
Zoumi A, Yeh A, Tromberg BJ. Imaging cells and extracellular matrix in vivo by using second-harmonic generation and two-photon excited fluorescence. Proc Natl Acad Sci U S A. 2002;99:11014–9.
Zoumi A, Lu X, Kassab GS, Tromberg BJ. Imaging coronary artery microstructure using second-harmonic and two-photon fluorescence microscopy. Biophys J. 2004;87:2778–86.
Zulliger MA, Fridez P, Hayashi K, Stergiopulos N. A strain energy function for arteries accounting for wall composition and structure. J Biomech. 2004;37:989–1000.

Chapter 12
Structural-Based Models of Ventricular Myocardium

Lik Chuan Lee, Jonathan Wenk, Doron Klepach, Ghassan S. Kassab, and Julius M. Guccione

12.1 Introduction

Ventricular wall stress is an important determinant of myocardial oxygen consumption (Sarnoff et al. 1958; Strauer et al. 1977), ventricular remodeling, and hypertrophy (Grossman 1980), and is necessary for an understanding of both physiological and pathological ventricular mechanics (Yin 1981). Despite recent advancement in measurement techniques and computational modeling, a detailed knowledge of ventricular wall stress (particularly in patients diagnosed with heart disease) remains elusive.

As pointed out by Huisman et al. (1980), forces or stresses cannot be measured directly in the intact left ventricle (LV) because of the uncertainty arising from the

L.C. Lee
Department of Mechanical Engineering, Michigan State University, 428 S. Shaw Lane, Room 2445, East Lansing, MI 48824, USA

J. Wenk
Department of Mechanical Engineering and Surgery, University of Kentucky, 269 Ralph G. Anderson Building, Lexington, KY 40506, USA

D. Klepach
Department of Mechanical or Industrial Engineering, Altair, Tel Aviv University, Multiscale Design Systems, LLC, UCSF, UC Berkeley, University of California, Berkeley, USA

G.S. Kassab (✉)
California Medical Innovations Institute, Roselle St, Rm 211 11107, San Diego, CA 92121, USA
e-mail: gkassab@calmi2.org

J.M. Guccione
Department of Surgery, University of California at San Francisco, 1657 Scott St., Brunn Bldg., Room 219, San Francisco, CA 94143, USA

G.S. Kassab, M.S. Sacks (eds.), *Structure-Based Mechanics of Tissues and Organs*,
DOI 10.1007/978-1-4899-7630-7_12

degree of coupling between the transducer and the muscle wall. In-vivo ventricular wall stress, therefore, has to be determined indirectly through measurements of the LV cavity pressure and the laws of mechanics (Yin 1981).

Closed-form analytical equations have been developed to estimate LV wall stress (e.g., Janz 1982), but they are accompanied by assumptions which include axis-symmetry of the LV, as well as material isotropy and homogeneity within the LV. These assumptions can be removed with more versatile approaches such as mathematical modeling using the finite element (FE) method (e.g., Zhang et al. 2011). However, the use of mathematical modeling to predict ventricular wall stress requires accurate constitutive models of the ventricular material, at least, at the tissue-level.

Two approaches have been used to construct constitutive models of the ventricular material at the tissue-level based on: (1) phenomenological experimental measurements of the material behavior at the tissue-level and (2) structural composition of the ventricular constituents and their micromechanical behavior. The first approach was pioneered by Fung (1967), who showed that under uniaxial tension, the load-deflection curve of a rabbit mesentery is highly nonlinear and can be described using an exponential function of the extension ratio (stretch). This result has formed the basis of many phenomenological constitutive models in soft tissues, including the ventricular myocardium (Humphrey and Yin 1987; Guccione et al. 1991; Costa et al. 2001; Schmid et al. 2006; Holzapfel and Odgen 2009[1]). Other functions have also been used to describe the monotonic increase in stiffness of the myocardial tissue as it stretches, including the "pole-zero" model proposed by Hunter et al. (1997).

12.2 Structural-Based Constitutive Models

In contrast to phenomenological constitutive models, structural-based constitutive models take into account the inherent microstructural compositions of the tissue, as well as their arrangement, interactions, and individual mechanical behavior of the constituents. Consequently, parameters of structural-based constitutive models can be directly related to physical features of the tissue. Structural-based constitutive model of the tissue was first introduced by Lanir for flat collagenous tissue (Lanir 1979) and was later expanded and tailored to the myocardium (Horowitz et al., 1988a, b, c; Nevo and Lanir 1989, 1994). To understand such model, it is appropriate to first describe the structure of the ventricular wall and the microstructure of the heart muscle.

---

[1]Although the constitutive model by Holzapfel and Odgen (2009) was constructed by taking into account structural features of the myocardium in the sense of an orthotropic configuration, it is still (strictly speaking) phenomenological as the constitutive model did not distinguish individual contribution from the tissue constituents and the model's parameters did not relate directly to the microstructure of the tissue.

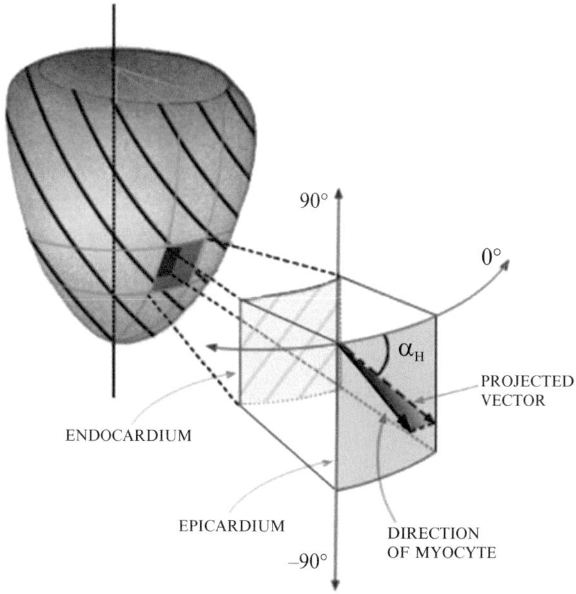

Fig. 12.1 Helical structure of the muscle fiber orientation in the ventricular wall (Reproduced from Walker et al. 2005)

12.2.1 Structure of the Ventricular Wall

At a macroscopic level, the ventricular wall exhibits a regular arrangement of cardiomyocytes (or muscle cells) that has been observed across species. The arrangement of the cardiomyocytes orientation was first quantified by Streeter et al. (1969) that showed a smooth transition in the orientation of cardiomyocytes from a left-handed helix at the epicardium to a right-handed helix at the endocardium (Fig. 12.1).

Other studies have confirmed these results in man (Greenbaum et al. 1981), rabbit (Vetter and McCulloch 1998), and many other species. Such arrangement of the helical myocytes orientation is widely believed to augment the ventricular pumping efficiency (Guccione et al. 1991; Vendelin et al. 2002; Ingels. 1997).

12.2.2 Structure of the Cardiac Muscle

At a microscopic level, the structure of cardiac muscle bears some resemblance to other two muscle types; i.e., skeletal muscle and smooth muscle. Cardiomyocytes are embedded by an extracellular matrix (ECM) comprising of endomysium that separates individual cardiomyocytes, and perimysium that separates between tracts

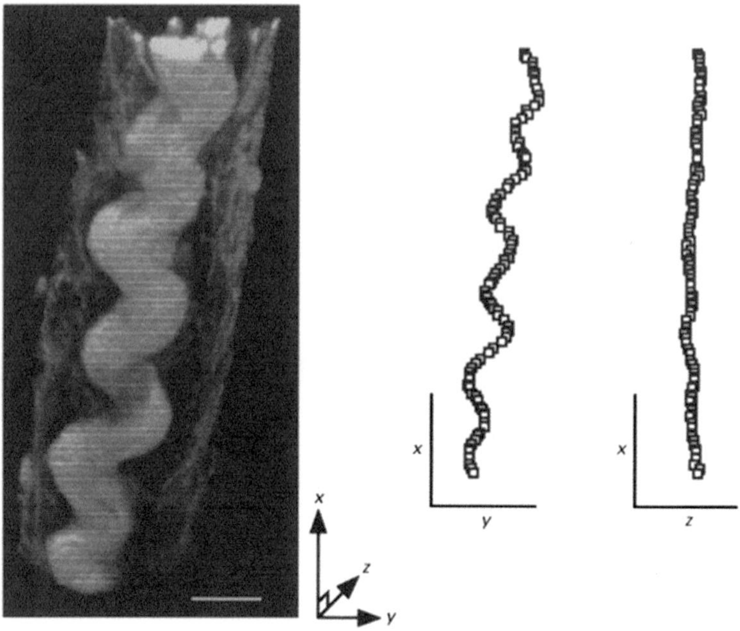

Fig. 12.2 Planar wavy perimysial collagen fiber from confocal laser scanning microscopy of the right ventricular trabecula of rat (Reproduced from Hanley et al. 1999)

of cardiomyocytes. The ECM consists of blood vessels, collagen fibers, connective tissue cells and interstitial fluid (Frank and Langer 1974). Collagen fibers have been observed to be periodic planar wavy-like structures in the endomysium and perimysium (Hanley et al. 1999) (Fig. 12.2). On the other hand, Robinson et al. (1988) described that the perimysial fibers of rat heart papillary muscle as helical coil structures as opposed to 2-D wavy-like structures.

While the mechanical role of the cardiomyocytes is to generate a contractive force that propels blood and drives the systemic and pulmonary circulation, the role of ECM is to provide mechanical support for the tissue. As such, the ECM is a major contributor to the tissue's mechanical behavior that affects the ventricular pumping function. To describe the mechanical behavior of the ventricular myocardium, it is therefore important to account for the mechanical effects of major cardiac tissue constituents (i.e., ECM and the cardiomyocytes). Structural-based constitutive models enable one to do that elegantly by individually accounting for the mechanical effects of the cardiac tissue constituents, as well as their interactions.

Constitutive Models

A structural-based constitutive model of fibrous tissue was developed by Lanir (1983) where he took into account two important features found in collagen and muscle fibers, namely, the fiber waviness and the distribution of fiber orientation in the tissue.

Fiber Waviness

The nonlinearity of stress-strain relationship found typically in cardiac tissue is widely believed to be caused by the unraveling of initially wavy fibers as the tissue stretches. To model this effect, a strain energy function w_k of type k fiber was assumed to exist for a straightened fiber. The uniaxial stress for the straightened fiber is given as

$$f_k = \frac{\partial w_k}{\partial E_{ff}}, \tag{12.1}$$

where E_{ff} is defined as the fiber strain. For a given waviness function $D_{k,n}(x)$, where $D_{k,\mathbf{n}}(x)\Delta x$ can be interpreted as the distribution of fibers aligned in the direction n and having a straightening strain between x and $x + \Delta x$, the uniaxial stress of this collection of wavy fibers has the form:

$$f_k{}^*\left(E_{ff}\right) = \int_0^{E_{ff}} D_{k,\mathbf{n}}(x) f_k\left(E_{ff}^t\right) dx, \tag{12.2}$$

where E_{ff} is the total fiber strain and E_{ff}^t is the true fiber strain with the straightened fiber as the reference configuration. The true strain E_{ff}^t can be expressed as a function of E_{ff} and x (see Eq. (5) in Horowitz et al. 1988c). Figure 12.3a shows the resultant stress-strain relationship from (12.2) when f_k is a linear function of the true strain with stiffness (or slope) c and $D_{k,\mathbf{n}}(x)$ is a normal distribution function with mean $\overline{x}$ and standard deviation σ_x. This relationship is nonlinear and captures the stiffening effects of the originally wavy fibers as it unravels and becomes stretched. We also note that as the total strain of the fiber $E_{ff} \to \infty$, all the fibers becomes straightened and the stiffness approaches an asymptotic value.

It should be noted that in addition to modeling the nonlinear stress-strain relationship through the use of a given waviness function, the nonlinear stress-strain relationship of a (perimysial) collagen fiber has been modeled by a helical coil consisting of a curved, inextensible rod that is capable of bending and twisting (MacKenna et al. 1997). The origin of the nonlinear stress-strain relationship of this model is different from that proposed by Lanir (1983), where the nonlinear behavior is caused by the straightening of the collagen fibers with different straightening

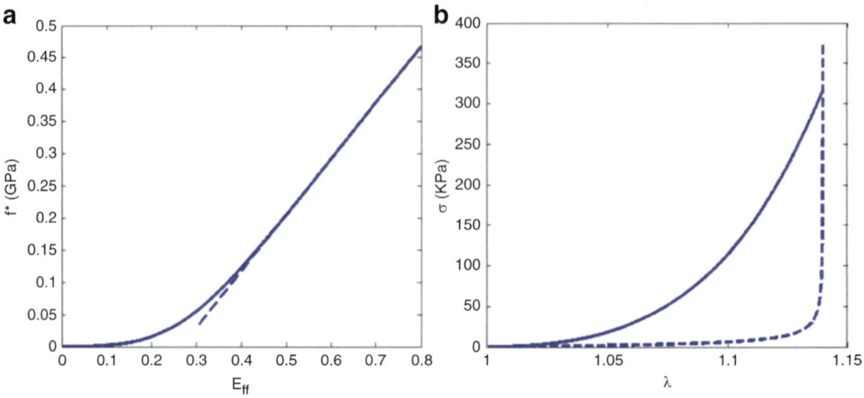

Fig. 12.3 (**a**) Stress-strain relationship (*solid line*) calculated from (12.2). *Dotted line*: asymptotic value of the stiffness calculated using Leibniz's rule of differentiation and is given here as $\lim_{E_{ff}\to\infty}\int_0^{E_{ff}} \boldsymbol{D}_{\bar{k},\mathbf{n}}(x)\partial f_k\left(\boldsymbol{E}_{ff}^t\right)/\partial \boldsymbol{E}_{ff}dx$. Parameters of c, $\bar{x}$ and σ_x are based on biaxial testing of canine myocardial strips and have values 35 GPa, 0.28 GPa and 0.11 GPa, respectively. (**b**) Comparison of collagen material models. *Solid line*: Lanir (1983) with same material constants as in (**a**). *Dotted line*: MacKenna et al. (1997) with material constants given in the text. *Note*: for comparison purposes, the uniaxial stress-strain relationship in (12.2) was multiplied by the collagen volumetric ratio $\boldsymbol{S}_k = 0.05$ in the myocardial tissue

strain. Instead, the nonlinearity found in the model by MacKenna et al. (1997) is inherited directly from the stress-strain relationship of a helical coil. This relationship was described by the following equation:

$$\sigma = -\frac{\pi D^4 N_d E}{64 P_0^2 V_0 \lambda}\left(1 - \frac{\sqrt{{V_0}^2 - 1}}{\sqrt{{V_0}^2 - \lambda}}\right), \tag{12.3}$$

where D is the fiber diameter, N_d is the number density of fiber per unit area, E is the Young's modulus, P_0 is the coil period, V_0 is the convolution index and λ is the stretch of the fiber. These parameters were determined based on histological measurements of the collagen geometry and 2-D measurements of the strain at the LV midwall (Omens et al. 1993). For a canine LV, these values were found to be $D = 3.44 \pm 0.18\,\mu\text{m}$, $P_0 = 15.7 \pm 0.1\,\mu\text{m}$, $N_d = 2030 \pm 290$ fibers/mm^2, $V_0 = 1.14 \pm 0.02$ and $E = 160$ MPa. The resultant stress-strain relationship is shown in Fig. 12.3b by the dotted line together with that from the model by Lanir (1983) (solid line) using material parameters obtained from the biaxial testing of canine myocardial strip. Because the rod is inextensible, the stress $\sigma \to \infty$ as the stretch $\lambda \to V_0$ (or when the helical coil is straightened). Compared to the model by Lanir (1983), the helical coil model by MacKenna et al. (1997) is significantly softer. It should be noted that even though the same species was modeled, the experimental data used to determine the material parameters is different between the two models.

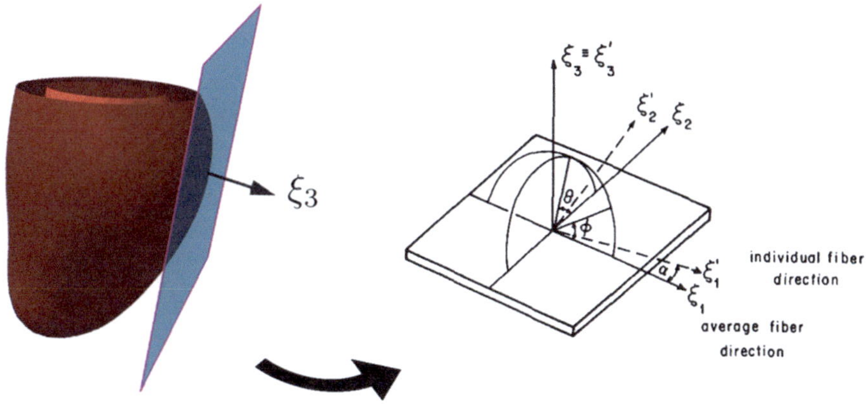

Fig. 12.4 Orientation of the fiber direction. *Left*: Tangential plane of the LV subepicardial slice defined by normal ξ_3. *Right*: Local coordinate systems in the slice (reproduced from Horowitz et al. 1988a, b, c)

Distribution of the Fiber Orientation

To account for the fiber distribution found in the myocardium, a density distribution function $R_k(\mathrm{n})$ was employed. Based on histological studies where the muscle fiber ($k = 1$) was found to vary by about 10–15° along the predominant fiber direction ξ_1 in subepicardial layer of the LV wall (Yin et al. 1987), the muscle fiber was assumed to vary uniformly in the subepicardial myocardial sheet (Fig. 12.4) by the distribution function

$$R_1(\alpha) = 12/\pi \tag{12.4}$$

On the other hand, the collagen fiber ($k = 2$) direction was assumed to vary based on the following distribution function defined in terms of a spherical coordinate system (r', θ', $\varnothing'$), where the tangent to the muscle fiber serves as a polar axis (Fig. 12.4):

$$R_2\left(\theta', \varnothing'\right) = R_{2a}\left(\theta'\right) R_{2b}\left(\varnothing'\right) \tag{12.5}$$

Because the muscle fiber was assumed to serve as an axis of symmetry for the collagen fiber, the distribution function in the circumferential direction is $R_{2a}(\theta') = 1/2\pi$. The distribution of collagen fiber in the meridional $R_{2b}(\varnothing')$ direction was assumed to be a double-peaked normal distribution with standard deviation $\sigma_\varnothing$ and mean angles $\overline{\varnothing}'$ and $-\overline{\varnothing}'$ the choice of a double-peak normal distribution is to ensure that there is no preference between the fiber direction and its opposite direction. Consequently, the model is effectively transversely isotropic.

By summing the contributions of the muscle and collagen fibers to the tissue strain energy, and accounting for the arrangement of these fibers, the total strain energy of the tissue can be expressed as

$$W = \sum_{k=1}^{2} s_k \int_{\Omega} R_k (\mathbf{n}) \; w_k^* \left(E_{ff}\right) d\Omega, \tag{12.6}$$

where Ω is the tissue domain, s_k is the volumetric fraction of fiber type k in the tissue and w_k^* is the resultant strain energy function from the stress-strain relationship in (12.2). The effects of interstitial fluid in the ECM are considered by enforcement of the incompressibility condition in the strain energy function.

Validation of Constitutive Model

This model has been validated and the material parameters have been found by fitting the data from the biaxial testing of canine myocardial strip taken from the subepicardial layer of the left ventricular wall (Horowitz et al. 1988a, b, c). Because the contribution of muscle fiber to the tissue overall stress-strain relationship is small compared to that from the collagen fiber (which has a stiffness ~1 GPa that is three order of magnitude greater than that from the muscle fiber), the mechanical effects of muscle fiber were neglected. This leaves only five essential material parameters; namely, the collagen fiber stiffness c_2, the waviness function's mean and variance $\tilde{x}$, σ_x and the mean and variance of its distribution in the meridional direction $\varnothing'$, $\sigma_{\varnothing}$ to be determined. The average values of these parameters over three specimens were $c_2 = 1.99$ GPa, $\tilde{x} = 0.27$, $\sigma_x = 0.1$, $\varnothing' = 0.69$ and $\sigma_{\varnothing} = 0.18$. These values suggests that the average straightening strain of the collagen is about 0.27 and the average angle between collagen fibers and the muscle fiber is roughly 40° (Horowitz et al. 1988c).

In Fig. 12.5, we compare the results from equi-biaxial testing of a canine myocardial strip described in Horowitz et al. (1988a, b, c) using the structural-based constitutive model and a phenomenological transversely isotropic model (Guccione et al. 1991). The parameters from the model described in Guccione et al. (1991) was determined to match the epicardial strains measured in an intact canine heart preparation (McCulloch et al. 1989). Although both models show that the tissue is stiffer in the muscle fiber direction than in the transverse directions, the phenomenological model is significantly softer than that from the structural-based model. One possible reason for the apparently stiffer structural-based model (when compared to the phenomenological model) is because the experiment specimens used in the biaxial testing (Yin et al. 1987) from which the model was fitted was under partial contracture (Humphrey et al. 1990).

More recently, the structural-based constitutive model (Lanir 1983) was validated using direct measurements on the microstructure of planar collagenous tissue. Specifically, Sacks (2003) used the collagen fiber angular distribution (i.e., $R_2(\mathbf{n})$) from native bovine pericardium measured by a small angle light scattering device as direct input into the structural-based constitutive model (Lanir 1983). The rest of the

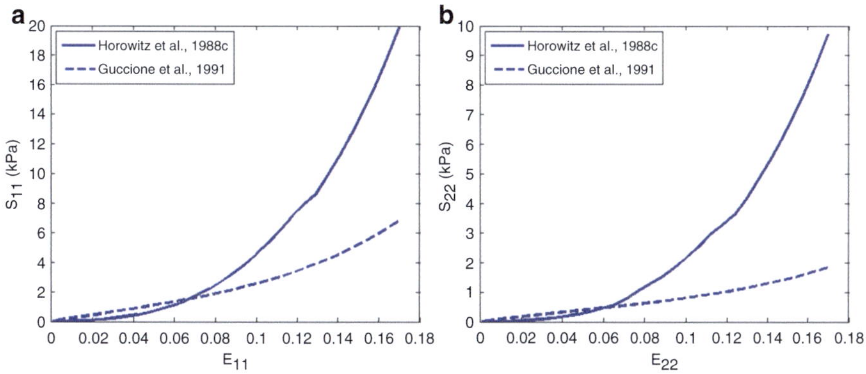

Fig. 12.5 Comparison of structural-based constitutive model (Horowitz et al. 1988c) with the transversely isotropic phenomenological constitutive model (Guccione et al. 1991) based on equi-biaxial testing protocol as described in Horowitz et al. (1988c). Material parameters of the structural-based constitutive model were taken from specimen one in Horowitz et al. (1988c)

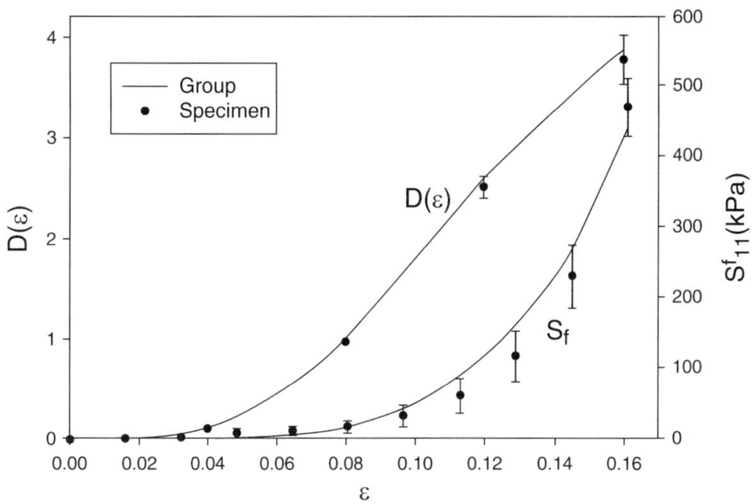

Fig. 12.6 Comparison of the effective fiber stress-strain curve S_{11}^{f} obtained from biaxial stretching experimental results with that obtained from the fitted structural-based constitutive model using directly measured angular distribution of collagen fibers $\boldsymbol{R}_2(\mathbf{n})$ in bovine pericardium. Note that the waviness function $\boldsymbol{D}(\varepsilon)$ used in that study is a Gamma distribution as opposed to a Normal distribution described in Sect. 12.2.3 (Reproduced from Sacks 2003)

parameters (i.e., elastic modulus and waviness properties) were then determined by fitting the constitutive model to the results of biaxial testing on the same specimens. This analysis showed that the resultant stress-strain relationship derived from the constitutive model closely matches that obtained from biaxial testing ($R^2 > 0.9$) (Fig. 12.6).

12.3 Future Directions

Structural-based constitutive models describing the myocardium are clearly limited in the literature when compared to phenomenological models, which are typically simpler and easier to implement numerically. Despite the complexity, the ability of structural-based constitutive model to directly relate the material parameters to the microstructure of myocardial tissue (as opposed to phenomenological models) is an extremely useful feature that can be used to deepen our understanding of the functional behavior. The structurally based model parameters may also serve as useful noninvasive biomarkers for myocardial infarction, particularly if the material parameters are quantified in-vivo and noninvasively, which has been done for phenomenological constitutive model (see Sect. 12.3.1).

Two potentially useful parameters that can help serve as useful biomarkers for myocardial infarction are the volume ratio of collagen and the "waviness" of the collagen fibers. Ventricular remodeling as a result of myocardial infarction typically leads to an increase in interstitial collagen content (to replace the apoptotic myocytes) in the infarcted regions and in the borderzone (a region separating the infarct and the non-infarcted tissue) (Marijianowski et al. 1997). As a consequence of the increase in collagen contents, the ventricle becomes less compliant and diastolic dysfunction may follow. The increase in collagen content as a result of myocardial infarction can be directly modeled in the structural-based constitutive model through the volumetric ratio of the collagen s_2. Horowitz et al. (1988a, b, c) showed that, by doubling the collagen volumetric ratio s_2, the stresses are considerably higher in a thick ventricular wall segment under equi-biaxial tension. In addition to determining the stresses found in the stiffer infarct region (or in the entire ventricles), structural-based constitutive model can also be used to determine volumetric ratio of the collagen in individual patients by fitting the model using in-vivo strain data from medical imaging and LV pressure measurements. Patient-specific collagen content can be helpful in determining the extent of myocardial infarction. Given that the apoptotic cardiomyocytes are replaced through the accumulation of collagen, a measure of the increase in collagen content can help determine the decrease in cardiomyocytes at the infarct/borderzone and the resultant reduction in active contraction in these regions to predict the depressed function of the ventricles.

The waviness of collagen fibers has been found to be slightly smaller in the infarct than in the remote regions rat hearts. This difference becomes increasingly significant as the ventricular cavity pressure increases. With an increase in pressure, the collagen fibers become significantly less wavy than those found in the remote region (Omens et al. 1997). The result that a difference in waviness was found only to be significant at a higher pressure suggests the presence of a different kind of uncoiling or straightening process in the infarct and in the normal tissue (Omens et al. 1997). Little is known about the in-vivo waviness (or crimp period) of collagen fibers found in normal or infarcted cardiac tissue, however, especially in humans. Since a difference in collagen waviness was observed in infarct and in remote regions of the rat's left ventricle, it is possible that such a difference also exists in human ventricles. If such difference does exist in human, the waviness of collagen fiber can potentially serve as a biomarker for the extent of myocardial infarction. Structural-based constitutive model can readily be used to determine the in-vivo waviness of collagen fibers as the extent of waviness and the distribution of waviness are directly quantified by the mean straightening strain and the variance of straightening strain found in the collagen fibers. For instance, a large mean straightening strain of the collagen fibers implies that the collagen are more wavy as opposed to a small mean straightening strain. The study by Sacks (2003) has shown that it is feasible to determine the collagen waviness in the myocardial tissue through the use of structural-based constitutive model. In that study, the mean straightening Green-Lagrange strain of collagen tissue was determined by fitting the constitutive model to the results from biaxial testing of bovine pericardial tissue and was found to be about 0.24. This result was close to the ex-vivo measurements by (Liao et al. 2005), where they estimated the waviness or crimp amplitude and period in similar bovine pericardial tissue to be ~15 μm and ~30 μm, respectively. Subsequent calculation using these results show that the resultant straightening strain is 0.22 (Meyers et al. 2008).

12.3.1 In-Vivo Characterization of Myocardial Microstructure

Material parameters of structural-based constitutive model have been determined largely using ex-vivo biaxial testing results, which can be potentially different from that in-vivo. The significant progress made in medical imaging (e.g., magnetic resonance imaging and 3D echocardiography) has enabled accurate and noninvasive quantification of in-vivo myocardial strain. A 3D optical flow method (Xu et al. 2010) applied to tagged magnetic resonance imaging (MRI), in combination with mathematical modeling using finite element method (FE), has been recently used to determine regional myocardial diastolic material parameters based on a transversely isotropic phenomenological constitutive model (Guccione et al. 1991) of a sheep with LV posterolateral infarct (Fig. 12.7).

The strain energy function describing the transversely isotropic constitutive model is given as follows:

$$W=\frac{C}{2}\left(\exp\left[b_f E_{ff}^{\,2}+b_t\left(E_{nn}^{\,2}+E_{ss}^{\,2}+E_{ns}^{\,2}+E_{sn}^{\,2}\right)+b_{fs}\left(E_{fn}^{\,2}+E_{nf}^{\,2}+E_{fs}^{\,2}+E_{sf}^{\,2}\right)\right]-1\right) \tag{12.7}$$

where C, b_f, b_t and b_{fs} are material parameters and E_{ij} is the Green-Lagrangian strains with subscripts f denoting the fiber direction and s, n denoting two other directions perpendicular to f. Specifically, a systematic search methodology (successive response surface method) was used to determine the material parameters by minimizing the mean square error between the measurements of 7 LV cavity

Table 12.1 Resultant material parameters determined from FE and MRI

	C (kPa)	b_f	b_t	b_{fs}
Remote	2.19	8.65	7.5	5.5
Infarct	6.71	25	22.43	1.11

volumes during filling and 1080 end-diastolic strain components (from MRI) to those predicted by FE

The resultant material parameters are shown in Table 12.1 and there is good agreement between the FE-predicted end-diastolic strains and the MRI-measured strains (Fig. 12.8). Overall, the larger values of b_f and b_t at the infarct implies that the infarct was stiffer in the fiber and cross-fiber directions compared to the remote region, as observed frequently across species.

The fact that the phenomenological model is able to quantify the mechanical effects of LV remodeling after myocardial infarction as described in this section is useful, as such information can help us better understand the effects of myocardial infarction and to create more realistic mathematical models of infarcted LV. It is difficult to infer from these parameters, however, the mechanisms for the observed mechanical behavior at the infarct; i.e., whether and how much the increase in stiffness is caused by an increase in collagen content or by a decrease in the collagen fiber waviness. Given that these physical attributes are directly quantified by the material parameters in structural-based models, it is possible to use the methodology described in this section (in combination with histological studies) to quantify the physical properties at the infarct and at the remote regions of the myocardium.

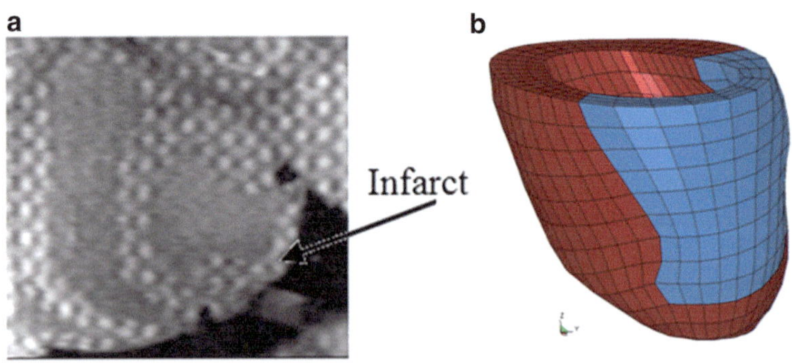

Fig. 12.7 (**a**) Short-axis view from 3D tagged magnetic resonance image (infarct is between the black markers on the epicardium). (**b**) Finite element model of the infarcted LV. *Red* and *blue* denotes the remote and infarcted regions, respectively

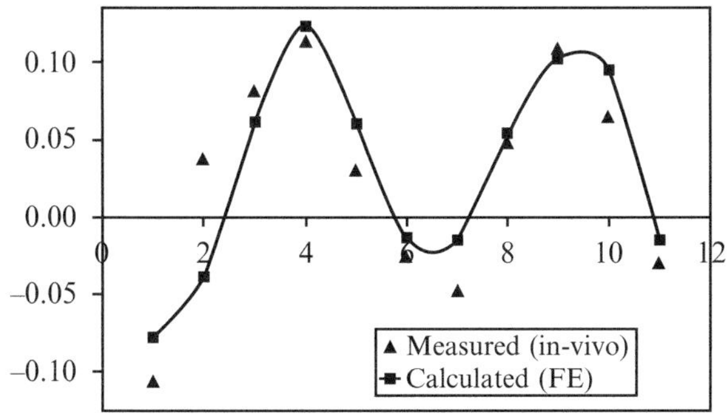

Fig. 12.8 Comparison of strain at 11 points around the circumference of a short-axis slice at mid-ventricle

References

Costa KD, Holmes JW, McCulloch AD. Modeling cardiac mechanical properties in three dimensions. Philos Trans R Soc Lond A. 2001;359:1233–50.

Frank JS, Langer GA. The myocardial interstitium: its structure and its role in ionic exchange. J Cell Biol. 1974;60:596–601.

Fung YC. Elasticity of soft tissues in simple elongation. Am J Physiology–Legacy Content 1967;213(6):1532–1544.

Greenbaum RA, Ho SY, Gibson DG, Becker AE, Anderson RH. Left ventricular fibre architecture in man. Br Heart J. 1981;45(3):248–63.

Grossman W. Cardiac hypertrophy: useful adaption or pathologic process? Am J Med. 1980;69(4):576–84.

Guccione JM, McCulloch AD, Waldman LK. Passive material properties of intact ventricular myocardium determined from a cylindrical model. J Biomech Eng. 1991;113:42–55.

Hanley PJ, Young AA, LeGrice IJ, Edgar SG, Loiselle DS. 3-Dimensional configurations of permysial collagen fibres in rat cardiac muscle at resting and extended sarcomere lengths. J Physiol. 1999;517:831–7.

Holzapfel GA, Odgen RW. Constitutive modelling of passive myocardium: a structurally based framework for material characterization. Philos Trans R Soc Lond A. 2009;367:3445–75.

Horowitz A, Sheinman I, Lanir Y, Perl M, Sideman S. Nonlinear incompressible finite element for simulating loading of cardiac tissue–part I: two dimensional formulation for thin myocardial strips. J Biomech Eng. 1988a;110(1):57–61.

Horowitz A, Sheinman I, Lanir Y. Nonlinear incompressible finite element for simulating loading of cardiac tissue–part II: three dimensional formulation for thick ventricular wall segments. J Biomech Eng. 1988b;110(1):62–8.

Horowitz A, Lanir Y, Yin FCP, Perl M, Sheinman I, Strumpf RK. Structural three-dimensional constitutive law for the passive myocardium. J Biomech Eng. 1988c;110(3):200–7.

Huisman RM, Elzinga G, Westerhof N, Sipkema P. Measurement of left ventricular wall stress. Cardiovasc Res. 1980;14:142–53.

Humphrey JD, Yin FC. On constitutive relations and finite deformations of passive cardiac tissue: I. A pseudostrain-energy function. J Biomech Eng. 1987;109(4):298–304.

Humphrey JD, Strumpf RK, Yin FCP. Determination of a constitutive relation for passive myocardium: I a new functional form. J Biomech Eng. 1990;112:333–9.

Hunter PJ, Nash MP, Sands GB. Computational electromechanics of the heart. In: Panfilov AV, Holden AV, editors. Computational biology of the heart. Chichester: Wiley; 1997. p. 345–407.

Ingels Jr NB. Myocardial fiber architecture and left ventricular function. Technol Health Care. 1997;5(1–2):45–52.

Janz RF. Estimation of local myocardial stress. Am J Physiol. 1982;242(5):H875–81.

Lanir Y. A structural theory for the homogeneous biaxial stress-strain relationship in flat collagenous tissues. J Biomech. 1979;12:423–36.

Lanir Y. Constitutive equations for fibrous connective tissues. J Biomech. 1983;16:1–12.

Liao J, Yang L, Grashow J, Sacks MS. Molecular orientation of collagen in intact planar connective tissues under biaxial stretch. Acta Biomater. 2005;1(1):45–54.

MacKenna DA, Vaplon SM, McCulloch AD. Microstructural model of perimysial collagen fibers for resting myocardial mechanics during ventricular filling. Am J Physiol. 1997;273:H1576–86.

Marijianowski MMH, Teeling P, Becker AE. Remodeling after myocardial infarction in humans is not associated with interstitial fibrosis of noninfarcted myocardium. J Am Coll Cardiol. 1997;30:76–82.

McCulloch AD, Smaill BH, Hunter PJ. Regional left ventricular epicardial deformation in the passive dog heart. Circ Res. 1989;64:721–33.

Meyers MA, Chen PY, Lin AYM, Seki Y. Biological materials: structure and mechanical properties. Prog Mater Sci. 2008;53:1–206.

Nevo E, Lanir Y. Structural finite deformation model of the left ventricle during diastole and systole. J Biomech Eng. 1989;111(4):342–9.

Nevo E, Lanir Y. The effect of residual strain on the diastolic function of the left ventricle as predicted by a structural model. J Biomech. 1994;27(12):1433–46.

Omens JH, MacKenna DA, McCulloch AD. Measurement of strain and analysis of stress in resting rat left ventricular myocardium. J Biomech. 1993;26:665–76.

Omens JH, Miller TR, Covell JW. Relationship between passive tissue strain and collagen uncoiling during healing of infarcted myocardium. Cardiovasc Res. 1997;33:351–8.

Robinson TF, Geraci MA, Sonnenblick EH, Factor SM. Coiled perimysioal fibres of papillary muscle in rat heart: morphology, distribution and changes in configuration. Circ Res. 1988;63:577–92.

Sacks MS. Incorporation of experimentally-derived fiber orientation into a structural constitutive model for planar collagenous tissues. J Biomech Eng. 2003;125:280–7.

Sarnoff SJ, Braunwald E, Welch Jr GH, Case RB, Stainsby WN, Macruz R. Hemodynamic determinants of oxygen consumption of the heart with special reference to the tension time index. Am J Physiol. 1958;192:148–56.

Schmid H, Nash MP, Young AA, Hunter PJ. Myocardial material parameter estimation—a comparative study for simple shear. J Biomech Eng. 2006;128:742–50.

Strauer BE, Beer K, Heitlinger K, Höfling B. Left ventricular systolic wall stress as a primary determinant of myocardial oxygen consumption: comparative studies in patients with normal left ventricular function, with pressure and volume overload and with coronary heart disease. Basic Res Cardiol. 1977;72:306–13.

Streeter DD, Spotnitz HM, Patel DP, Ross Jr J, Sonnenblick EH. Fiber orientation in the canine left ventricle during diastole and systole. Circ Res. 1969;24:339–47.

Vendelin M, Bovendeerd PH, Engelbrecht J, Arts T. Optimizing ventricular fibers: uniform strain or stress, but not ATP consumption, leads to high efficiency. Am J Physiol Heart Circ Physiol. 2002;283(3):H1072–81.

Vetter FJ, McCulloch AD. Three-dimensional analysis of regional cardiac function: a model of rabbit ventricular anatomy. Prog Biophys Mol Biol. 1998;69(2–3):157–83.

Walker JC, Guccione JM, Jiang Y, Zhang P, Wallace AW, Hsu EW, Ratcliffe MB. Helical myofiber orientation after myocardial infarction and left ventricular surgical restoration in sheep. J Thorac Cardiovasc Surg. 2005;129(2):382–90.

Xu C, Pilla JJ, Isaac G, Gorman 3rd JH, Biom AS, Gorman RC, et al. Deformation analysis of 3D tagged cardiac images using an optical flow method. J Cardiovasc Magn Reson. 2010;12(1):19.

Yin FC, Strumpf RK, Chew PH, Zeger SL. Quantification of the mechanical properties of noncontracting canine myocardium under simultaneous biaxial loading. J Biomech. 1987;20(6): 577–589.

Yin FC. Ventricular wall stress. Circ Res. 1981;49:829–42.

Zhang Z, Tendulkar A, Sun K, Saloner DA, Wallace AW, Ge L, Guccione JM, Ratcliffe MB. Comparison of the Young-Laplace law and finite element based calculation of ventricular wall stress: implications for postinfarct and surgical ventricular remodeling. Ann Thorac Surg. 2011;91(1):150–6.

Chapter 13
Structure-Based Constitutive Model of Coronary Media

Yaniv Hollander, David Durban, Xiao Lu, Ghassan S. Kassab, and Yoram Lanir

13.1 Introduction

Blood vessel biomechanics is an essential and interdisciplinary research topic, incorporating disciplines ranging from nonlinear solid mechanics to anatomy, physiology, and pathology. Vascular mechanics is imperative for predicting vascular physiology and the way blood vessels interact mechanically with other organs. In addition, a number of vascular pathologies (e.g., *atherosclerosis*, *hypertension*, *arterial aneurysms*) have a pivotal biomechanical basis since they are initiated and propagated as a result of non-homeostatic mechanical loadings. Knowledge of the stress field in the blood vessel wall and its relation to strains and to loading imposed on the vessel cells will aid in understanding these diseases.

Blood vessels are anisotropic materials, exhibiting a global orthotropic behavior. As in other soft tissues, blood vessels have a nonlinear stress–strain response and are viscoelastic. A typical large blood vessel is made of three layers: *intima*, *media*, and *adventitia*. Each layer has a different *fibrous* structure, and therefore, behaves differently in response to applied loads. Importantly, blood vessels are capable of actively changing their diameter and length. This is achieved by activation of smooth muscle cells in the vessel wall.

Y. Hollander • D. Durban
Faculty of Aerospace Engineering, Technion, Israel Institute of Technology, Haifa 32000, Israel

X. Lu • G.S. Kassab (✉)
California Medical Innovations Institute, 11107 Roselle St., Rm. 211, San Diego, CA 92121, USA
e-mail: gkassab@calmi2.org

Y. Lanir
Faculty of Biomedical Engineering, Technion, Israel Institute of Technology, Haifa 32000, Israel

G.S. Kassab, M.S. Sacks (eds.), *Structure-Based Mechanics of Tissues and Organs*, DOI 10.1007/978-1-4899-7630-7_13

Over the last 40 years, several models were proposed for constitutive laws of passive blood vessel elastic response. Most of these models are phenomenological, attempting mainly to fit mathematical expressions to experimental data. The main disadvantage of phenomenological models is that their material constants have no direct physical interpretation and thus do not pave the way for understanding the connection between tissue architecture and the inherent mechanical characteristics. Recently, an increasing number of models have attempted to incorporate structural information into the constitutive relations. These data are usually incorporated phenomenologically, however, implying that the associated material constants still have little physical significance. Structural models for soft tissues, in general, and for blood vessels, in particular, are essential for analysis of tissue mechanical response and understanding the individual role of each of the tissue constituents. Since several vascular pathologies are related to the degradation of tissue fibers, the prediction of the onset of these diseases can be done only by using a model which adequately incorporates the influence of each fiber type.

This chapter describes the development of a structural model for the passive blood vessel, with focus on the coronary media which is a major load-bearing layer in the wall. The model, which is based on histological information, contains independent parameters representing both the three-dimensional (3D) inner fibrous structure of the media and the fiber properties. The model also includes the effects of residual stresses and osmotic swelling. Model parameter estimation was done using the *Genetic Algorithm Method*, and was based on mechanical data of porcine left anterior descending (LAD) coronary media. The data, provided by Kassab's group, includes measurements of vessel response to combined radial inflation, axial extension, and twist. Although the full model has 12 parameters, results show that a reduced model containing only four parameters that represent key structural and mechanical features is sufficient to reliably predict the passive mechanical properties. It is also shown that the model provides good predictions of the LAD media response to data under protocols not used for the parameter estimation.

The constitutive model constructed in this work has noteworthy novelties when compared to current models in the literature. It is the first fully structural model, which considers the effect of fibers embedded in a fluid-like matrix. Secondly, the model includes the influence of osmotic swelling which affects the nonlinear mechanical interaction between the tissue solid fibers and the fluid-like matrix. Also, the predictive capabilities of the model were verified, in contrast to previous models, which were purely curve fitting of existing data and not predictive of additional experiments. Finally, the present model was validated with 3D data which includes twist in addition to inflation and extension. To date, no model found in the literature was validated with shear data.

To assess whether the improvements proposed by the present model are significant, a multi-faceted comparison was carried out between the present structural model and representatives of two major classes of arterial models (phenomenological and semi-structural). A comparison was conducted in terms of model descriptive and predictive powers, using identical 3D mechanical data. The results show that the fully structural model performed best, followed by the structure-motivated and the phenomenological models, when considering the descriptive accuracy to the given data as well as the predictive capabilities to data not used for estimation.

An important conclusion from this comparison is that incorporation of tissue structure, as well as other physical properties, such as tissue swelling, is crucial for a reliable description and prediction of the vessel mechanical behavior.

The work described in this chapter is focused on constructing a model for only the coronary media layer. Since the blood vessel wall is composed of two mechanically significant layers, future investigation is needed to formulate a model for the adventitia layer. The adventitia layer is responsible for tethering the vessel to surrounding tissues and, more importantly, protects against rupture due to hypertension. The degradation of adventitial fibers is responsible for the onset of several pathologies, including arterial aneurysms. The adventitia layer has a different structure than the media, and hence, behaves differently in response to mechanical loads.

13.2 Mathematical Model

The present media model was compared and validated with data of inflation, extension, and twist tests, under internal pressure P_i, external axial load F, and external torque M. The media was initially loaded by residual stress (RS) and by tissue osmotic swelling. For these protocols, the media was considered a hollow cylindrical tube, and the following kinematical assumptions were adopted: (1) deformations are axis-symmetric and independent of axial position (Green and Adkins 1970; Humphrey et al. 1989); (2) incompressible vessel media; (3) transverse sections remain planar (no warping); (4) both the twist angle and axial displacement are independent of radial position; (5) quasi-static and elastic response; (6) no luminal flow and associated shear stress; and (7) there is a unique stress-free reference configuration, which can be obtained through a combination of cutting the tissue radially and unswelling it by immersion in a hyper-osmotic solution (Chuong and Fung 1986; Fung 1991; Guo et al. 2007; Azeloglu et al. 2008; Lanir 2009).

13.2.1 Kinematics and Statics

The general deformation field describing vessel kinematics can be written as

$$r = r(R), \quad \theta = \theta\left(\Theta, Z\right), \quad z = z(Z), \tag{13.1}$$

where (r, θ, z) and (R, Θ, Z) are the radial, tangential, and axial cylindrical coordinates in the deformed and reference configurations, respectively.

To incorporate all physiological mechanisms affecting the wall response, such as residual stress and osmotic swelling, four distinct kinematical configurations are considered (Fig. 13.1): (1) *stress-free* (SF), with inner and outer radii, R_i and R_o, length L and an opening angle Θ_0; (2) *swollen* (SW), with $\widehat{R}_i$, $\widehat{R}_o$, $\Lambda_0 L$ as inner and

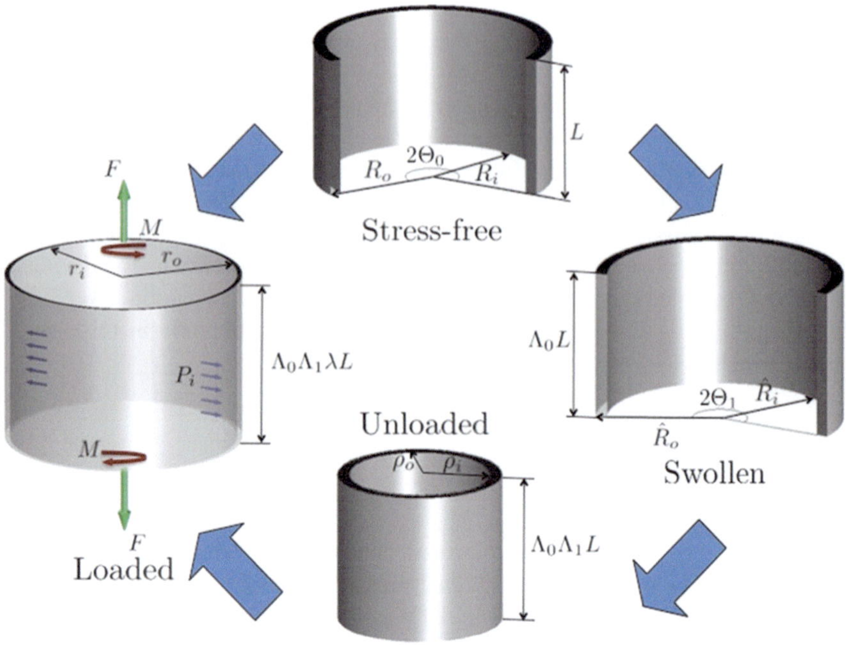

Fig. 13.1 Kinematical configurations. A schematic description of the mappings from the open stress-free (SF) sector configuration, through the open sector swollen (SW) and closed unloaded (UL) states, to the loaded (L) configuration

outer radii, and length, and Λ_0 and Θ_1 as SF–SW stretch ratio and opening angle; (3) *unloaded* (UL), with dimensions ρ_i, ρ_o, and $\Lambda_0\Lambda_1 L$, with Λ_1 as the SW–UL stretch ratio; and (4) *loaded* (L) with dimensions: r_i, r_o, and $\lambda\Lambda_0\Lambda_1 L$, where λ is the UL-L stretch ratio, and γ as the twist per unit length. The general deformation field, from the stress-free to the loaded states, is expressed as (Green and Adkins 1970; Humphrey 2002):

$$r = r(R), \quad \theta = \frac{\pi}{\Theta_0}\Theta + \gamma\Lambda_0\Lambda_1 Z, \quad z = \lambda\Lambda_0\Lambda_1 Z, \tag{13.2}$$

where $r(R)$ is given by:

$$r = \sqrt{r_i^2 + \frac{J_{\mathrm{SW}}\Theta_0}{\pi\Lambda_0\Lambda_1\lambda}\left(R^2 - R_i^2\right)}. \tag{13.3}$$

The swollen state is obtained by osmotic swelling, which induces inner stresses to the reference stress-free configuration, due to balance between the tissue fibers and the fluid-like solid matrix. The swelling affects the opening angle of the open sector-shaped tissue (Guo et al. 2007). Volume swelling ratio between the stress-free and swollen configurations is denoted in (13.3) by J_{SW}.

Static equilibrium is obtained by satisfying the radial equation

$$\frac{\partial \sigma_{\mathrm{rr}}}{\partial r} + \frac{\sigma_{\mathrm{rr}} - \sigma_{\theta\theta}}{r} = 0, \tag{13.4}$$

where σ_{rr} and $\sigma_{\theta\theta}$ are the radial and tangential components of the Cauchy stress tensor. The boundary conditions associated with (13.4) are given by:

$$\sigma_{\mathrm{rr}}\left(r = r_i\right) = -P_i, \quad \sigma_{\mathrm{rr}}\left(r = r_o\right) = 0,$$

$$F = 2\pi \int_{r_i}^{r_o} \sigma_{zz} r dr - P_i \pi r_i^2, \quad M = 2\pi \int_{r_i}^{r_o} \sigma_{\theta z} r^2 dr, \tag{13.6}$$

where σ_{zz} and $\sigma_{\theta z}$ are the two additional axial and shear stress components of the Cauchy stress tensor.

Hyperelastic Constitutive Model Formulation

The passive media is assumed to be a hyperelastic material, with strain energy function (SEF) $W(\mathbf{E})$, where $\mathbf{E} = \left(\mathbf{F}^{\mathrm{T}} \cdot \mathbf{F} - \mathbf{I}\right)/2$ is the Green-Lagrange strain, and $\mathbf{F}$ is the deformation gradient tensors. If we assume that incompressibility holds, the Cauchy stress tensor can be expressed as, $\sigma = J_{\mathrm{SW}}^{-1} \mathbf{F} \cdot \partial W / \partial \mathbf{E} \cdot \mathbf{F}^{\mathrm{T}} - p\mathbf{I}$, where p is the matrix hydrostatic pressure.

In the microstructural approach (Lanir 1983), W reflects the tissue inner microstructure and associated changes during deformation. The basic assumptions are (Lanir 1983; Nevo and Lanir 1989; Billiar and Sacks 2000; Lokshin and Lanir 2009): (1) the passive tissue total strain energy is the sum of the strain energies of the fibrous constituents, (2) each fiber can respond to stretch and buckles in response to compressive loads; (3) the fluid-like ground matrix contribution to the overall stress field is given by hydrostatic pressure; and (4) the strain of each fiber can be computed from the tensorial transformation of the global strain tensor (affine deformation). Following these assumptions, $W(\mathbf{E})$ is written as

$$W(\mathbf{E}) = \sum_i \phi_i^0 \iint \Re_i^* (\psi, \theta)\, w_i(e) \sin \psi \, d\psi \, d\theta, \tag{13.7}$$

where ϕ_i^0 is the reference (stress-free) volume fraction of type i fibers, $\Re_i^*(\psi, \theta)$ is the fibers orientation density distribution function, whose arguments are the spherical angles ψ and θ, and w_i(e) is the type i fiber energy function which depends

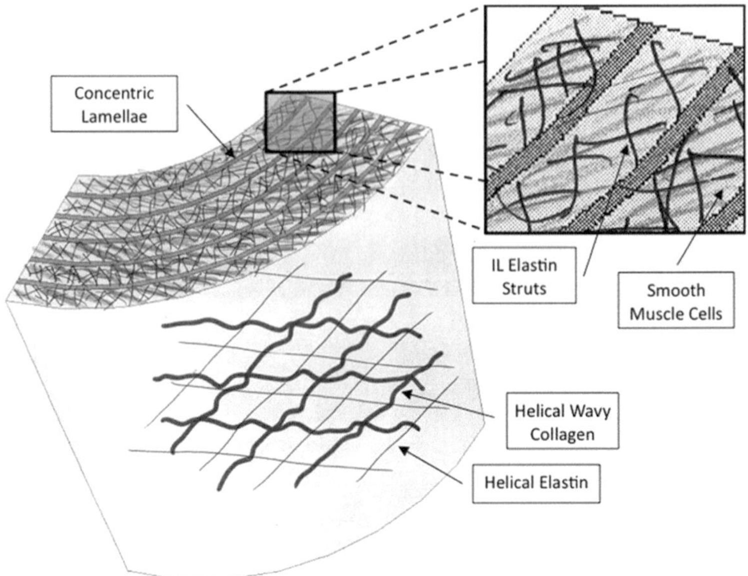

Fig. 13.2 Model illustration. A scheme of the vessel wall microstructure including the lamellae helical elastin-collagen fibers, the interlamellar strut networks, and the smooth muscle cells

on the fibers axial Lagrangian strain e. The assumption of affine deformation suggests that e can be determined from the global Green strain tensor **E** by $e = \mathbf{E} : \mathbf{NN}$, where **N** is the fiber group orientation vector in the reference configuration.

Histological surveys of the arterial media suggest that the media is made up of an elastin scaffold in the shape of thick fenestrated concentric laminae, interconnected by a network of thin and short struts (Wolinsky and Glagov 1967; Wasano and Yamamoto 1983; Clark and Glagov 1985; O'Connell et al. 2008), which are uniformly dispersed over orientation space with a mean radial direction. The concentric lamellae are composed by two families of helical fibers with a mean symmetric pitch angle (Canham et al. 1989, 1997; Dahl et al. 2007). The fibers' orientations are distributed around the mean symmetric angle. The first family consists of straight elastin fibers, and the other, of wavy collagen fibers arranged in parallel with the elastin fibers. The mathematical model developed for the media (Hollander et al. 2011b) is based on this interpretation of histological studies and can be best described schematically by the illustration in Fig. 13.2. The model includes a summation of the contribution of the following families of fibers:

1. Helical wavy collagen fibers aligned in two groups with mean symmetrical polar angles, around which the fibers are dispersed.
2. Straight helical elastin fibers that run in parallel to the helical collagen network. Both the helical collagen and elastin fibers form concentric lamellae.

Table 13.1 The 12 parameters of the general structural model and their physical significance: Those underlined are of significant importance for the LAD media model

Parameter	Physical meaning
k_h^0	Helical elastin stiffness
k_h	Helical collagen stiffness
N_h	Power law term representing gradual recruitment of collagen fibers
e_{01}, e_{02}	Collagen lowest and highest straightening strains
m_h	Exponent of symmetric beta function related with helical fiber distribution
β_h^0	Mean polar angle of helical fibers
$\Delta\beta_h$	Range of polar angle dispersion of helical fibers
k_{il}	Inter-lamellar struts stiffness
m, n, c_0	Anisotropic inter-lamellar network

3. Interlamellar short elastin struts interconnect adjacent media lamellae. Fibers in this group have a 3D anisotropic dispersion with a mean radial direction.

The general structural media model based on the above description includes 12 parameters (Hollander et al. 2011a, b). The model parameters represent the different fiber stiffnesses and include the effect of the increasing stiffness of collagen fibers recruited gradually with stretch (Table 13.1). The model also takes into account the effect of fiber dispersion. In this study, it was found that only four parameters are needed for the LAD media model in order to adequately describe the given experimental data. The formulations for this four parameter reduced model are given below:

Interlamellar Fiber Energy Function

$$W_{\mathrm{il}}\left(\mathbf{E}\right) = \phi_{\mathrm{il}}^{0} \iint \frac{w_{\mathrm{il}}\left(\mathrm{e}_{\mathrm{il}}\right)}{\pi^2} d\alpha d\beta, \quad \frac{\partial w_{\mathrm{il}}}{\partial e_{\mathrm{il}}} = \begin{cases} k_{\mathrm{il}}\mathrm{e}_{\mathrm{il}}, & \mathrm{e}_{\mathrm{il}} \geq 0 \\ 0, & \mathrm{e}_{\mathrm{il}} < 0 \end{cases}, \tag{13.8}$$

where k_{il} is the model parameter representing interlamellar fiber stiffness. The axial strain of an interlamellar fiber e_{il} is found by the tensorial projection of the Green-Lagrange strain tensor on the fiber orientation in the reference configuration, which can be described by the polar angles α and β. The fiber family volume fraction is denoted by ϕ_{il}^{0}.

Helical Fiber Energy Function

$$W_h(\mathbf{E}) = \frac{\phi_h^0}{2}\left\{w_h\left[\mathrm{e}_h\left(\beta_h^0\right)\right] + w_h\left[\mathrm{e}_h\left(-\beta_h^0\right)\right]\right\},$$

$$\frac{\partial w_h}{\partial \mathrm{e}_h} = \begin{cases} k_h \mathrm{e}_h^{N_h}, & \mathrm{e}_h \geq 0 \\ 0, & \mathrm{e}_h < 0 \end{cases}. \tag{13.9}$$

Here, the additional model parameters are β_h^0, representing the angle between the two symmetrical helical fiber groups, and k_h and N_h, parameters that describe the increasing stiffness of collagen fibers with gradual recruitment by stretch. The axial strain of a helical fiber e_h is found in a similar manner to the one described above for e_{il}.

13.3 Testing the Model

13.3.1 Experimental Database

Model validation was carried out based on triaxial mechanical data (Lu et al. 2003; Wang et al. 2006) of porcine coronary arteries. The database includes 5 LAD coronary media specimen, which were separated from the intact blood vessel before the experiment (Lu et al. 2003). The cylindrical media was cannulated to a triaxial machine and preconditioned prior to mechanical testing. Measurements were made of the outer vessel radius, axial force, and torque, in response to a series of applied stepwise luminal pressures (0–85 mmHg), axial stretches (1.2–1.4), and twist (−25° to 25°, which reflects a range of twist angle per unit length $\gamma = \pm 0.04\,\mathrm{rad/mm}$). In addition, the geometry of the open sector swollen configuration was recorded for each vessel and used as input to the model.

13.3.2 Torsional Stiffness

Mechanical testing of blood vessel twist response (Deng et al. 1994; Lu et al. 2003) shows a linear relation between torque and twist angle per unit length (denoted in this study as M and γ, respectively). The torsional stiffness μ^* relates these two measures by

$$M = \mu^*(P_i,\ \lambda)\,\gamma, \tag{13.10}$$

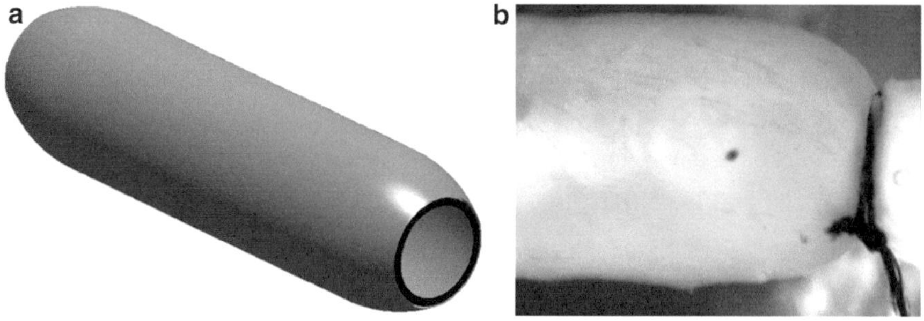

Fig. 13.3 Bulging effect of the inflated vessel. (**a**) 3D scheme of the actual loaded vessel axial profile, showing how the vessel radius remains constant throughout most of the vessel length. A short transition zone, over which the vessel radius decreases from its mid-length level to the cannula radius, is seen around the vessel ends. (**b**) An image of pig RCA media, inflated by luminal pressure of 100 mmHg

Unlike linear elasticity, where the torsional stiffness is a constant quantity, here $\mu*$ is a function of both luminal pressure and axial stretch ratio. Using M from (13.6) $\mu*$ is given by:

$$\mu^* = 2\pi \int_{r_i}^{r_o} \sigma_{\theta z,\gamma} r^2 dr. \tag{13.11}$$

During the mechanical testing, the luminal pressure is gradually increased to inflate the blood vessel. Since the test specimen have their ends tied to cannulas, the vessel bulges as demonstrated schematically as well as for an actual pig RCA media in Fig. 13.3.

This bulging violates the kinematic assumption of an axis-symmetric deformation field. For simplicity, both the twist angle per length and axial stretch ratio where taken to be functions of z, hence $\gamma(z)$ and $\lambda(z)$. Each axial segment satisfies

$$\gamma(z) = \frac{M}{\mu^*(z)}, \tag{13.12}$$

while $\lambda(z) = \partial z / \partial \zeta$ (ζ denotes the axial coordinate in the unloaded configuration) needs to conform with the global constraint of known vessel length

$$l = \int_0^{L_{\mathrm{UL}}} \lambda(\zeta)\, d\zeta, \tag{13.13}$$

with L_{UL} and l as the vessel unloaded and loaded lengths, respectively. For the varying axial vessel profile, the axial force at each axial position has the form

$$F = 2\pi \int_{r_i(z)}^{r_o(z)} \sigma_{zz} r dr - P_i \pi r_i^2(z), \tag{13.14}$$

where $r_i(z)$ and $r_o(z)$ are the inner and outer radii. The torque M remains constant between adjacent axial segments as well as between the two cannulas, namely

$$M = \mu \frac{\theta(l)}{l}, \tag{13.15}$$

where $\theta(l)$ is the global twist angle of the vessel. If we equate M from (13.10) and (13.15), we obtain an expression for the *effective torsional stiffness* μ as:

$$\mu = \frac{1}{\frac{1}{l}\int_0^l \frac{dz}{\mu^*(z)}}. \tag{13.16}$$

13.3.3 Parameter Estimation

Parameters were optimized by least squares fit to the data by minimizing an objective function consisting of the sum of squared errors (SSE) between model predictions and experimental data. The objective function incorporates the averaged normalized data of all three protocols and was defined as

$$\mathrm{SSE} = \frac{1}{n}\sum_i \left[\left(\frac{r_o^i - \widehat{r}_o^i}{\sigma_{\widehat{r}_o}} \right)^2 + \left(\frac{F^i - \widehat{F}^i}{\sigma_{\widehat{F}}} \right)^2 + \left(\frac{\mu^i - \widehat{\mu}^i}{\sigma_{\widehat{\mu}}} \right)^2 \right], \tag{13.17}$$

where n is the number of data points, $(\bullet)^i$ and $(\hat{\bullet})^i$ are the i'th point model prediction and measured data, respectively, and $\sigma_{\hat{\bullet}}$ are the standard deviations of the experimental data for the three testing protocols (inflation, extension, and shear). A similar, objective function was used by Zulliger et al. (2004), but for 2D estimations, which included only inflation and extension (without the contribution of torsion). The search for the optimal parameter set was carried out using the *Genetic Algorithm* (GA) method (Goldberg 1989), using an MPI parallel computation based version of the C-code GAUL package (Adcock 2004).

13.3.4 Aspects of Model Examination

The examination of the proposed model was done in two aspects. First, the model descriptive power was tested by estimating its parameters and comparing the simulated response with known data. Second, model predictive capabilities were evaluated by simulating the behavior of known data which was not used for parameter estimation. An example of this is prediction of the vessel twist response while using only the data of inflation–extension for parameter estimation and comparing with the known experimental twist behavior.

13.4 Comparison with Known Models

The current model has been compared with two models published in literature, which represent major classes of arterial constitutive laws: (1) phenomenological (Chuong and Fung 1983; Deng et al. 1994) and (2) semi-structural (Holzapfel et al. 2000, 2002, 2004b, 2005). The comparison was made for the same aspects as those specified for the structural model, i.e., model descriptive and predictive powers, requisite data dimensionality for reliable parameter estimation, and the requisite number of model parameters.

The first phenomenological model is based on an extension of a 1D model proposed by Fung (1972). A generalization of this model for the 2D and 3D cases was given by Fung et al. (1979), Chuong and Fung (1983), Deng et al. (1994). The model is based on an exponential strain energy function that conforms with the experimental observations, suited for many types of soft tissues. The general form of the Fung-type SEF can be written as Humphrey (1995)

$$W = \frac{1}{2}C\left[\exp(Q) - 1\right], \tag{13.18}$$

$$\begin{aligned} cQ = {} & c_1 E_{\Theta\Theta}^2 + c_2 E_{ZZ}^2 + c_3 E_{RR}^2 + 2c_4 E_{\Theta\Theta} E_{ZZ} + 2c_5 E_{ZZ} E_{RR} \\ & + 2c_6 E_{\Theta\Theta} E_{RR} + c_7 E_{R\Theta}^2 + c_8 E_{\Theta Z}^2 + c_9 E_{RZ}^2, \end{aligned} \tag{13.19}$$

denoting E_{ij} $(i,\ j = R,\ \Theta,\ Z)$ as the Green-Lagrange strain tensor components and C and c_i $(i = 1, \ldots, 9)$ as material parameters.

The second semi-structural model considers a more realistic representation of the helical collagen fibers. The model combines the contribution of these fibers and the matrix embedding them. The SEF for this model (Holzapfel et al. 2000, 2005) reads

$$W = W_{\text{iso}}\left(I_1\right) + W_{\text{aniso}}\left(I_4, I_6\right), \tag{13.20}$$

$$W_{\text{iso}}\left(I_1\right) = C_{I_1}\left(I_1 - 3\right), \tag{13.21}$$

$$W_{\text{aniso}}\left(I_4, I_6\right) = \frac{k_1}{2k_2}\sum_{\alpha=4,6}\left\{\exp\left[k_2(I_\alpha - 1)^2\right] - 1\right\}, \tag{13.22}$$

where C_{I_1}, k_1, k_2, and ϕ are model parameters. The invariants in (13.4) are $I_1 = \text{tr}\,\mathbf{C}$, $I_4 = \mathbf{C} : \mathbf{MM}$, and $I_6 = \mathbf{C} : \mathbf{M}'\mathbf{M}'$, where $\mathbf{C}$ is the right Cauchy-Green tensor, and

M and **M**′ are the directions of the two symmetric collagen helices. In order to account for possible collagen fiber buckling, the anisotropic energy function W_{aniso} is active only when either $I_4 \geq 0$ or $I_6 \geq 0$. It should be noted, however, that the model does not account for possible buckling of interlamellar fibers, which may occur as the blood vessel is inflated and its wall gets thinner.

13.5 Examples

13.5.1 Model Descriptive Power

The fit to the data of the three models for the combined 3D protocols (inflation-extension-twist) is plotted in Fig. 13.4.

The Fung-type model provides good fit to the inflation data (Fig. 13.4b) and satisfactory fit to the axial force (Fig. 13.4e) and torsional stiffness (Fig. 13.4h) data under extension ratios of 1.2 and 1.3. The fit under extension ratio of 1.4 is not

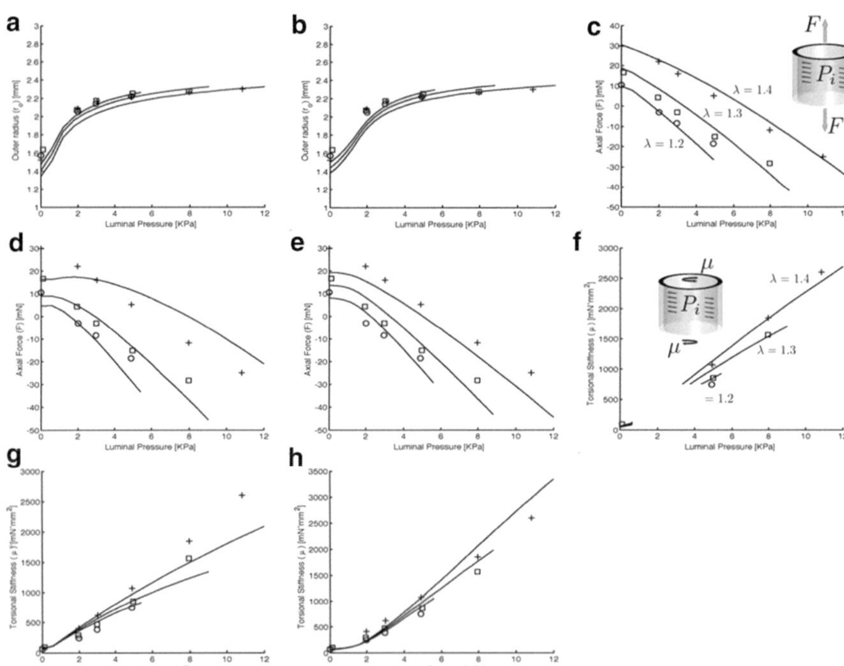

Fig. 13.4 Model descriptive power. Predictions (*lines*) compared with experimental data (*symbols*) of vessel outer radius r_o (**a–c**), axial force F (**d–f**), and torsional stiffness μ (**g–i**) vs. inner luminal pressure P_i, under three axial stretch ratios (λ), for (**a**, **d**, **g**) structural model, compared with (**b**, **e**, **h**) Fung type, and (**c**, **f**, **i**) semi-structure models

as good since the model underestimates the torsional stiffness and overestimates the axial force in the medium to high pressure range, while it underestimates it in the low pressure range. In addition, the predicted axial force–pressure relationship (Fig. 13.4e) indicates leveling of the curve and a changing trend (at curves maxima under low pressures), which is not born out by the data. The Holzapfel semi-structural model provides good fit to the inflation data (Fig. 13.4c) and to the axial force (Fig. 13.4f) data at extensions 1.2 and 1.3, while it slightly underestimates the 1.4 axial force data. For the torsional stiffness (Fig. 13.4i), the model is seen to slightly overestimate the data at the medium to high pressure ranges. In addition, the model predicts leveling of axial force data at low pressures, a feature which is not confirmed by the data. The fully structural model provides good fit to the inflation (Fig. 13.4a), axial force (Fig. 13.4d), and torsional stiffness data (Fig. 13.4g) under all three extension ratios. It should be noted that the qualitative analysis presented here is supported by a more quantitative investigation (Hollander et al. 2011b) which shows similar results (Hollander et al. 2011a).

13.5.2 Model Predictive Capabilities

One example of the prediction power of the different models is their ability to predict the behavior of the torsional stiffness based on model estimations using only 2D data of inflation–extension. Figure 13.5 shows such a prediction and compares both the fully and semi-structural models. As can be deduced from (13.4), the Fung-type model torsional response is inherently decoupled from inflation-extension. For this model, estimation based on 2D protocol cannot be used for simulation of twist. However, for the other two models, their structural basis allows the prediction of the coupled inflation–extension–twist response. It can be seen (Fig. 13.5) that the fully structural prediction of twist outperforms that of the semi-structural model.

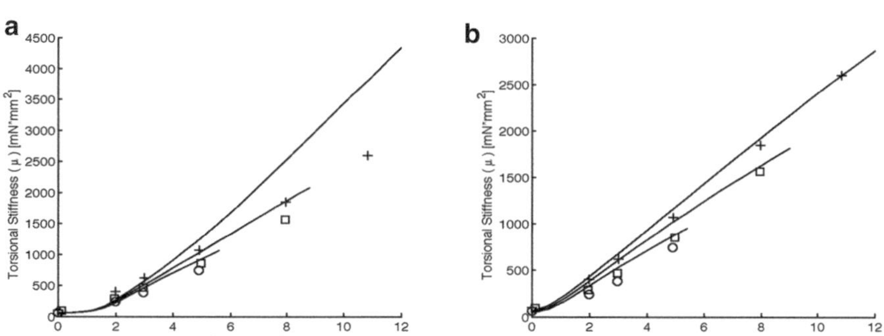

Fig. 13.5 2D-to-3D predictive power. Comparison between predicted twist apparent stiffness (*lines*) and measured data (*symbols*) at three axial stretch ratios (λ) of 1.2 (*open circle*), 1.3 (*open square*), and 1.4 (+), for (**a**) the semi-structural, and (**b**) the fully structural models

13.6 Concluding Remarks and Future Research

In the study described in this chapter, a new constitutive model has been developed for the coronary arterial media. The model is based on the structural approach (Lanir 1983) and incorporates a mixture of various fibers in different morphologies, embedded within a fluid-like matrix. These morphologies follow general structural features, studied in a number of histological observations. The proposed model is the first 3D fully structural model developed for blood vessels, while most previous models were either purely or partially phenomenological.

Blood vessels in vivo are loaded by inner luminal pressure, affecting their diameter, and axial force, which affects their length and stretch. Some blood vessels also twist. The coronaries, for example, twist following normal contraction of the heart. Experimental data of blood vessel response to twist is rare (Deng et al. 1994; Lu et al. 2003). Consequently, while many known models can be used for simulation of twist, as demonstrated in Holzapfel et al. (2000) and Humphrey (2002), no work was found which experimentally validates the shear properties of blood vessels. The model studied in this research is the first to be validated against shear data, in addition to inflation and extension data.

In general, a soft tissue is a fibrous structure composed of various fibers in different morphologies, embedded within a ground substance assumed to be a fluid-like matrix (Lanir 1983; Nevo and Lanir 1989). The fibers can sustain only tension, while the fluid carries compressive loads. In normal conditions, there is an osmotic balance between the fibers and fluid-like matrix. Osmotic pressure is therefore an important factor which affects the tissue mechanical response, an influence which was also observed in experiments of blood vessels and other soft tissues like the myocardium (Lanir et al. 1996; Guo et al. 2007; Azeloglu et al. 2008; Lanir 2009). Unlike previous models for blood vessels, the model in the present work also includes the effect of osmotic pressure and tissue swelling, and introduces two open sector configurations: open unswollen and swollen sectors. The swollen sector is not stress-free—an inner balance exists between stretched fibers and compressed fluid. A true stress-free state, characterized by an opening angle, is found in unswelling the tissue.

In past research of arterial hyperelastic constitutive modeling, two classes of material laws were proposed: purely phenomenological and structural models. The latter combines a structurally oriented characteristic along with a phenomenological contribution. Two representatives of these classes were chosen for a quantitative comparison with the fully structural model proposed in this work. The first is the 3D version of the Fung-type model (Chuong and Fung 1983; Deng et al. 1994) formulated in (13.4), and the second is the semi-structural model (Holzapfel et al. 2000, 2002, 2004b, 2005) given in (13.4). The models were compared with the reduced form of the fully structural model, formulated in (13.8). The three model performances were characterized in two ways: (1) their ability to describe given data of radial inflation, axial extension, and twist; and (2) their ability to predict twist response, which was not used for parameter estimation. It was found that the fully structural model outperformed the other two in descriptive and predictive powers. This finding was illustrated graphically in Figs. 13.4 and 13.5. A more thorough investigation, which also includes a numerical study done by Hollander et al. (2011b), showed that the SSE values for the fully structural model were the lowest among the three for all tests performed. Hollander et al. (2011b) also expanded the examination to other tests not covered in this chapter, which support similar conclusions.

Phenomenological models in general, and Fung-type in particular, involve a number of parameters which are estimated to fit experimental data but have no physical meaning or relation to the tissue inner structure. These models also cannot account for other tissue properties such as osmotic swelling and fiber buckling under compression. For the semi-structural model, the representation of the elastin scaffold as a neo-Hookean solid is unrealistic and contradicts histological observations (Wasano and Yamamoto 1983). It considers the matrix as a continuum, where in fact, it is a fibrous structure embedded within a possibly *fluid-like matrix*. The neo-Hookean solid can sustain compression, whereas within the tissue, fibers buckle under compression taken by the fluid matrix. These drawbacks of the phenomenological models limit their ability to describe certain data and may explain the comparative results described above.

Some limitations can be attributed to structural models (Chen et al. 2011a, b). Firstly, such models require large computational cost since they involve many integral computations (sums of fiber contributions in many directions). Phenomenological models, by contrast, are much less computationally costly. Secondly, to account for all histological features, structural models require a large number of model parameters. This increases the computational effort of the parameter estimation process and requires advance estimation techniques (e.g., Genetic Algorithms). It should be noted, however, that for the specific model in this work, it was found that four parameters were sufficient for reliable description of experimental data. This number is on par with the number of parameters of other arterial constitutive models, including the Fung and Holzapfel model types. We also note that any process of structural model buildup involves parsimony examinations, which aim at ensuring a minimal set of required parameters. If the tissue microstructure is very complex, however, the high number of parameters results directly from the need to take this complexity into account. The ability to consider complex structures is not possessed by phenomenological models. Evidently, the Fung-type model included nine parameters, but nevertheless failed to achieve as good descriptive capabilities as those demonstrated by the fully structural model. The third model limitation is related to the validation of the model which was based on mechanical data of global properties (outer radius, axial force, and torsional stiffness). A validation based on direct quantitative measurement of histological data would be more ideal. The quantitative measurement of histological data is, however, very difficult, and there are very few techniques available to quantify complex 3D structural features, such as that of the interlamellar network. In recent work by Chen et al. (2011a, b), the authors used multi-photon microscopy, a relatively new technique for quantitative imaging of fibers in unfixed, unstained tissue, to measure the inner structure of the adventitia. Such a technique may decrease the uncertainty of model structure and decrease the number of parameters actually estimated using global mechanical data. It should be noted that past experience with other types of tissues (e.g., skin, myocardium, cartilage, pericardium, heart valves) showed that validation of microstructural features based on appropriate global mechanical data is sufficient, in most cases, for good model descriptive and predictive capabilities.

13.6.1 Future Work: Incorporation of the Adventitia Layer

The research described in this chapter focused on establishing and validating a 3D structural constitutive model for the coronary media. The coronaries, as well as other blood vessels, are composed of two mechanically significant layers, the *media* and *adventitia*. The adventitia tethers the vessel to the surrounding tissues, and has an important role in preventing vessel burst during pressure-overload (hypertension). When the vessel is inflated, collagen and elastin fibers within the adventitia are stretched and thus stiffen the vessel. Degradation of adventitial fibers is related with the onset of pathologies like arterial aneurysms Humphrey (2003).

Both arterial layers are composed of the same passive fibrous structures, elastin and collagen. These are arranged differently in the two layers, however, and hence, induce two different mechanical responses for the media and adventitia. These differences are enhanced by the fact that the two layers have mutually different reference configurations with two unequal opening angles (Lu et al. 2003; Wang et al. 2006; Holzapfel and Gasser 2007) and swelling ratios (Fung and Liu 1992). An example of the difference between the mechanical response of media and adventitia is given in Fig. 13.6, which presents stress-stretch curves obtained from rectangular slices cut from human LADs. It is apparent that the adventitia is stiffer than the media in both the circumferential and axial directions. Note also that the media is stiffer in the circumferential direction, whereas the adventitia is stiffer in the axial direction, consistent with (Wang et al. 2006). The strength differential between the two materials, and their different reference configurations, has an important effect on the mechanical response of the two-layer intact structure.

Some studies treated blood vessels as a one-layer tube (Vaishnav et al. 1972; Fung et al. 1979), whereas others treated the layers separately (Holzapfel et al. 2004a, b; Holzapfel et al. 2005; Wang et al. 2006; Holzapfel and Gasser 2007). The studies which considered both layers used the *same* form of the constitutive law for each layer. For example, Wang et al. (2006) used the same Fung-type constitutive relation for both the media and adventitia. The strength differential between the layers was taken into account by introducing separate sets of material parameters for each layer (resulting in 14 parameters, 7 for the media, and 7 for the adventitia). Moreover, structurally motivated invariant-based models (Holzapfel et al. 2005) used the same law described in (13.4) for both layers, where the only difference between the layers was the values of their parameters. As demonstrated, a fully structural model has a better potential to predict experimental data not used to estimate its parameters. These results are mainly attributed to the fact that structural models include an accurate representation of the true tissue microstructure. It is

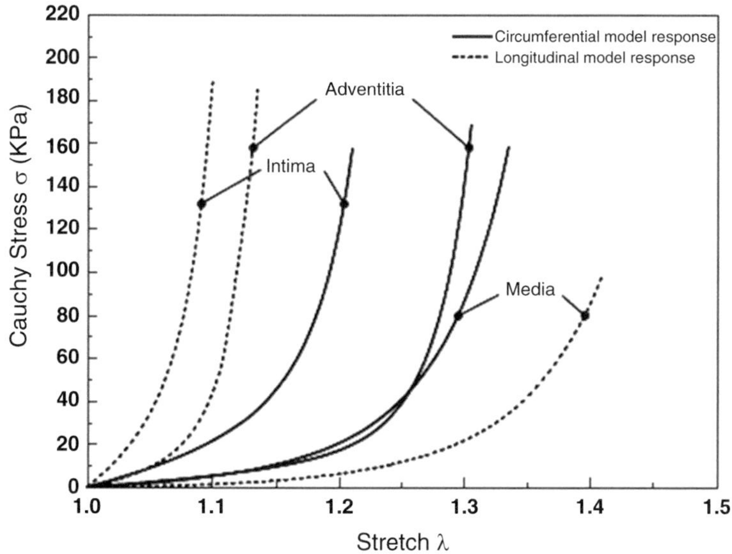

Fig. 13.6 Average stress-stretch curves in circumferential and axial directions for the media and adventitia layers—data of the human left anterior descending coronary artery (taken from Holzapfel et al. 2005)

expected, therefore, that a fully structural model for the adventitia, which will include its various structural features found in histological studies, will be fruitful. Such a model will also incorporate tissue swelling, similar to the media. A search of literature on vascular mechanics revealed that, like in the media case, no constitutive model based solely on tissue microstructure is available for the adventitia.

To conclude, a constitutive law for the adventitia is important as a complement to the media model developed in this work. This formulation should be constructed and validated using a 3D mechanical data, similar to the data used for the estimation of the media model. It would be ideal if the experimental database will include many loading combinations (as in Fig. 13.6) to increase statistical reliability of validation. The adventitia model can enable the analysis of the mechanical significance of this layer constituents and its microstructure. In addition, incorporation of the strength and reference differentials between the two layers would facilitate a better understanding of the behavior of the intact blood vessel.

References

Adcock S. Gaul, the genetic algorithm utility library. http://gaul.sourceforge.net (2004).

Azeloglu EU, Albro MB, Thimmappa VA, Ateshian GA, Costa KD. Heterogeneous transmural proteoglycan distribution provides a mechanism for regulating residual stresses in the aorta. Am J Physiol Heart Circ Physiol. 2008;294(3):H1197–205.

Billiar KL, Sacks MS. Biaxial mechanical properties of the native and glutaraldehyde-treated aortic valve cusp: Part II-a structural constitutive model. J Biomech Eng. 2000;122:327–35.
Canham PB, Finlay HM, Dixon JG, Boughner DR, Chen A. Measurements from light and polarized light microscopy of human coronary arteries fixed at distending pressure. Cardiovasc Res. 1989;23(11):973–82.
Canham PB, Finlay HM, Boughner DR. Contrasting structure of the saphenous vein and internal mammary artery used as coronary bypass vessels. Cardiovasc Res. 1997;34(3):557–67.
Chen H, Lin Y, Zhao X, Lanir Y, Kassab GS. A micromechanics finite-strain constitutive model of fibrous tissue. J Mech Phys Solids. 2011a;59:1823–37.
Chen H, Liu Y, Slipchenko MN, Zhao X, Cheng J-X, Kassab GS. The layered structure of coronary adventitia under mechanical load. Biophys J. 2011b;101(11):2555–62.
Chuong CJ, Fung YC. Three dimensional stress distribution in arteries. J Biomech Eng. 1983;105(3):268–74.
Chuong CJ, Fung YC. On residual stresses in arteries. J Biomech Eng. 1986;108(2):189–92.
Clark JM, Glagov S. Transmural organization of the arterial media. The lamellar unit revisited. Arteriosclerosis. 1985;5:19–34.
Dahl SLM, Vaughn ME, Niklason LE. An ultrastructural analysis of collagen in tissue engineered arteries. Ann Biomed Eng. 2007;35(10):1749–55.
Deng SX, Tomioka J, Debes JC, Fung YC. New experiments on shear modulus of elasticity of arteries. Am J Physiol. 1994;266(1 Pt 2):H1–10.
Fung YC. Stress-strain history relations of soft tissues in simple elongation. In: Fung YC, Perrone N, Anliker M, editors. Biomechanics—its foundations and objectives. Englewood Cliff: Prentice-Hall; 1972. p. 181–208.
Fung YC. What are the residual stresses doing in our blood vessels? Ann Biomed Eng. 1991;19(3):237–49.
Fung YC, Liu SQ. Strain distribution in small blood vessels with zero-stress state taken into consideration. Am J Physiol. 1992;262(2 Pt 2):H544–52.
Fung YC, Fronek K, Patitucci P. Pseudoelasticity of arteries and the choice of its mathematical expression. Am J Physiol Heart Circ Physiol. 1979;237:H620–31.
Goldberg DE. Genetic algorithms in search, optimization, and machine learning. Reading: Addison-Weseley; 1989.
Green AE, Adkins JE. Large elastic deformations. 2nd ed. Oxford: Oxford University Press; 1970.
Guo X, Lanir Y, Kassab GS. Effect of osmolarity on the zero-stress state and mechanical properties of aorta. Am J Physiol Heart Circ Physiol. 2007;293(4):H2328–34.
Hollander Y, Durban D, Lu X, Kassab GS, Lanir Y. Constitutive modeling of coronary arterial media: comparison of three model classes. J Biomech Eng. 2011a;133(6):0610008.
Hollander Y, Durban D, Lu X, Kassab GS, Lanir Y. Experimentally validated microstructural 3D constitutive model of coronary arterial media. J Biomech Eng. 2011b;133(3):031007.
Holzapfel GA, Gasser TC. Computational stress-deformation analysis of arterial walls including high pressure response. Int J Cardiol. 2007;116(1):78–85.
Holzapfel GA, Gasser TC, Ogden RW. A new constitutive framework for arterial wall mechanics and comparative study of material models. J Elast. 2000;61:1–48.
Holzapfel GA, Gasser TC, Stadler M. A structural model for the viscoelastic behavior of arterial walls: continuum formulation and finite element simulation. Eur J Mech A Solids. 2002;21(3):441–63.
Holzapfel GA, Gasser TC, Ogden RW. Comparison of a multi-layer structural model for arterial walls with a fung-type model, and issues of material stability. J Biomech Eng. 2004a;126:264–75.
Holzapfel GA, Sommer G, Regitnig P. Anisotropic mechanical properties of tissue components in human atherosclerosis plaques. J Biomech Eng. 2004b;126:657–65.
Holzapfel GA, Sommer G, Gasser CT, Regitnig P. Determination of layer-specific mechanical properties of human coronary arteries with nonatherosclerosis intimal thickening and related constitutive modeling. Am J Physiol Heart Circ Physiol. 2005;289:2048–58.

Humphrey JD. Mechanics of the arterial wall: review and directions. Crit Rev Biomed Eng. 1995;23(1–2):1–162.
Humphrey JD. Cardiovascular solid mechanics: cells, tissues, and organs. New York: Springer; 2002.
Humphrey JD. Review paper: continuum biomechanics of soft biological tissues. Proc R Soc Lond A. 2003;459(2029):3–46.
Humphrey JD, Strumpf RK, Yin FC. A theoretically-based experimental approach for identifying vascular constitutive relations. Biorheology. 1989;26(4):687–702.
Lanir Y. Constitutive equations for fibrous connective tissues. J Biomech. 1983;16(1):1–12.
Lanir Y. Mechanisms of residual stress in soft tissues. J Biomech Eng. 2009;131(4):044506.
Lanir Y, Hayam G, Abovsky M, Zlotnick Y, Uretzky G, Nevo E, Ben-Haim SA. Effect of myocardial swelling on residual strain in the left ventricle of the rat. Am J Physiol Heart Circ Physiol. 1996;39:H1736–43.
Lokshin O, Lanir Y. Micro and macro rheology of planar tissues. Biomaterials. 2009;30(17):3118–27.
Lu X, Yang J, Zhao JB, Gregersen H, Kassab GS. Shear modulus of porcine coronary artery: contributions of media and adventitia. Am J Physiol Heart Circ Physiol. 2003;285:1966–75.
Nevo E, Lanir Y. Structural finite deformation model of the left ventricle during diastole and systole. J Biomech Eng. 1989;111(4):342–9.
O'Connell MK, Murthy S, Phan S, Xu C, Buchanan J, Spilker R, Dalman RL, Zarins CK, Denk W, Taylor CA. The three-dimensional micro- and nanostructure of the aortic medial lamellar unit measured using 3d confocal and electron microscopy imaging. Matrix Biol. 2008;27(3):171–81.
Vaishnav RN, Young JT, Janicki JS, Patel DJ. Nonlinear anisotropic elastic properties of the canine aorta. Biophys J. 1972;12(8):1008–27.
Wang C, Garcia M, Lu X, Lanir Y, Kassab GS. Three-dimensional mechanical properties of porcine coronary arteries: a validated two-layer model. Am J Physiol Heart Circ Physiol. 2006;291(3):H1200–9.
Wasano K, Yamamoto T. Tridimensional architecture of elastic tissue in the rat aorta and femoral artery—a scanning electron microscope study. J Electron Microsc (Tokyo). 1983;32(1):33–44.
Wolinsky H, Glagov S. A lamellar unit of aortic medial structure and function in mammals. Circ Res. 1967;20:99–111.
Zulliger MA, Fridez P, Hayashi K, Stergiopulos N. A strain energy function for arteries accounting for wall composition and structure. J Biomech. 2004;37(7):989–1000.

Chapter 14
Biomechanics of the Cornea and Sclera

Thao D. Nguyen

Abstract The cornea and sclera are soft fibrous tissues that serve important structural and visual functions. The tissues protect the eye from external injury, provide mechanical support for critical internal ocular structures, and preserve an optimal shape for vision. The mechanical properties of the cornea and sclera supporting these functions are achieved through the collagen, elastin, and proteoglycans architecture of the extracellular matrix. This review discusses recent advances in experimental characterization of the structure and mechanical properties of the cornea and sclera, as well as theoretical and computational modeling studies of the relationship between the structure, mechanical properties, and physiologic function of the tissues.

14.1 Introduction

The transparent cornea and opaque sclera together form the tough outer layer of the eye wall, which provides mechanical support and protection from external injuries to the more delicate internal structures of the eye, such as the lens, retina, and optic nerve head. The tissues also function to maintain an optimal shape for vision in the presence of physiological fluctuations in the intraocular pressure. The domed shape cornea produces approximately 75 % of the unaccommodated power of the eye (Ruberti et al. 2011), while the sclera positions the retina relative to the focal plane. Myopia and hyperopia, also known as nearsightedness and farsightedness, are caused by axial lengthening and shortening of the sclera.

The mechanical properties of the cornea and sclera arise from the fibrous microstructure of the stroma, which contains densely stacked lamellae primarily of type I ($\sim$90 %) and type III ($\leq$5 %) collagen fibrils embedded in a hydrated matrix of proteoglycans (Watson and Young 2004). The collagen fibrils in the central cornea have a uniform diameter ($\sim$32 nm) and interfibrillar spacing ($\sim$55 nm) (Boote et al. 2003). This highly ordered fibrillar structure is needed to provide the cornea with its

T.D. Nguyen (✉)
Department of Mechanical Engineering, The Johns Hopkins University, Baltimore, MD 21218, USA
e-mail: vicky.nguyen@jhu.edu

G.S. Kassab, M.S. Sacks (eds.), *Structure-Based Mechanics of Tissues and Organs*,
DOI 10.1007/978-1-4899-7630-7_14

optical transparency (Maurice 1957). The sclera possesses a wider range and larger spatial variation in the collagen fibril diameter and interfibrillar spacing. These are also on average larger in the sclera, where the mean collagen fibril diameter is 70 nm in the peripapillary region and twice as much in the equatorial region (Quigley et al. 1991).

The stroma in the human cornea is comprised of approximately 250 lamellae, also referred to as fibers, which run mainly in the plane of the tissue (Kaufman and Alm 2003). Collagen fibrils lie parallel within a lamella, but are orientated at widely different angles in adjacent lamellae. The collagen fiber structure of the human cornea has been investigated extensively using wide-angle X-ray scattering (WAXS) by Meek and coworkers (Boote et al. 2004; Aghamohammadzadeh et al. 2005; Boote et al. 2006; Meek and Boote 2009). In WAXS, X-rays passing through the ordered collagen structure are scattered to produce a diffraction pattern, which provides a thickness-averaged measurement of the orientation distribution of the collagen lamellae. Both the preferred orientation and degree of fiber alignment of the human sclera display substantial regional variation. In the human cornea, the collagen lamellae are weakly aligned along the orthogonal nasal-temporal and inferior–superior directions in the central region, but are strongly aligned in the circumferential direction in the limbus (Fig. 14.1). Pijanka et al. (2012)

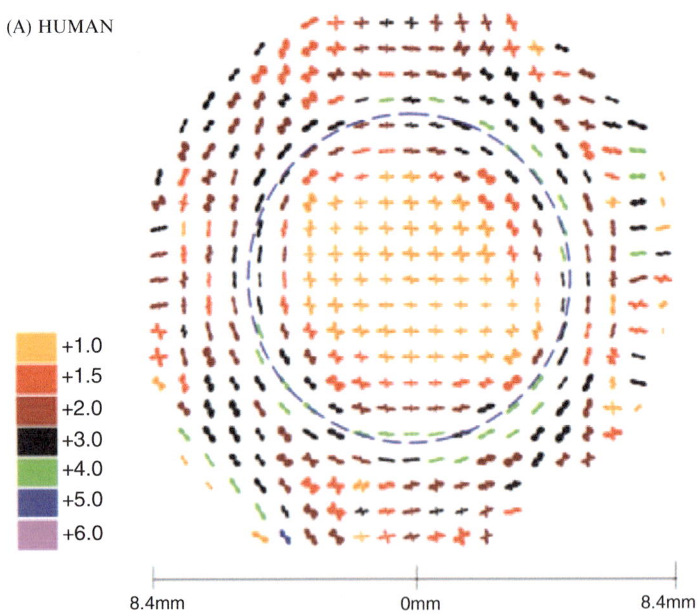

Fig. 14.1 Polar plot map of the WAXS scattering intensity of a human cornea, limbus (*solid line*), and adjacent sclera. Reproduced from Hayes et al. (2007) with permission from copyright holder. The orthogonal arrangement in the central region changes to a more highly aligned circumferential arrangement in the limbus and sclera

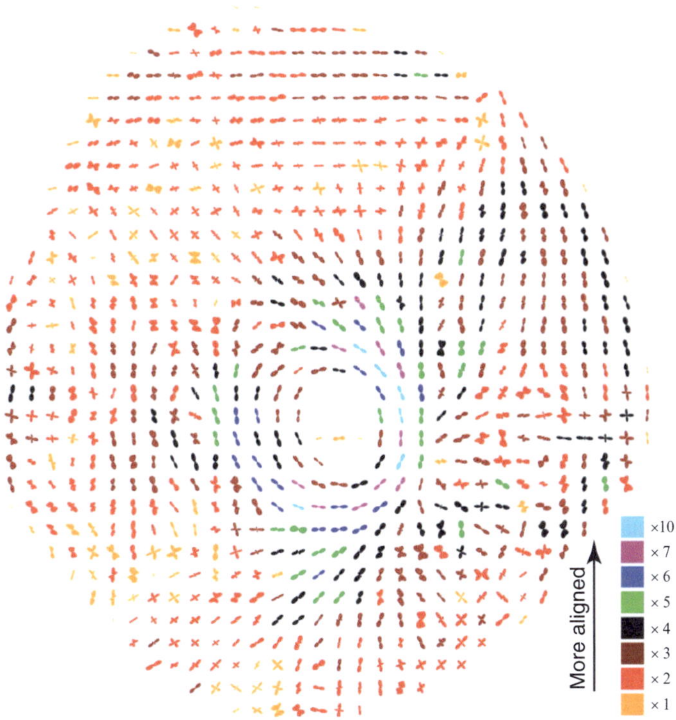

Fig. 14.2 Polar plot map of the scattering intensity measured by WAXS of a human posterior sclera. Reproduced from Pijanka et al. 2012 with permission of copyright holder. The strong circumferential orientation in the peripapillary region decays away from the optic nerve region

recently mapped the collagen structure of the scleral stroma using WAXS. In the peripapillary region, the collagen lamellae display a uniform circumferential orientation, ringing the optic nerve head, but a spatially heterogeneous degree of fiber alignment. The degree of fiber alignment decreases further away from the peripapillary region, and the fiber structure loses any discernible long range preferred orientation (Fig. 14.2). The sclera contains additional reinforcements from elastic fibers, the presence of which are concentrated in the peripapillary region, immediately adjacent to the optic nerve head (Quigley et al. 1996; Gelman et al. 2010), and towards the inner lamellae of the tissue adjacent to the choroid (Kamma-Lorger et al. 2010).

Mechanical experiments have shown that the cornea (Nyquist 1968; Woo et al. 1972; Andreassen et al. 1980; Jue and Maurice 1991; Hoeltzel et al. 1992; Hjortdal 1996; Shin et al. 1997; Kampmeier et al. 2000; Wollensak et al. 2003; Elsheikh and Anderson 2006; Boyce et al. 2007, 2008; Elsheikh et al. 2008, 2010b) and the sclera (Curtin 1969; Woo et al. 1972; Friberg and Lace 1988; Phillips and McBrien 1995; Phillips et al. 2000; Downs et al. 2001, 2003; Girard et al. 2009a,b; Elsheikh et al. 2010a; Girard et al. 2011; Coudrillier et al. 2012) exhibit a nonlinear strain-stiffening, anisotropic, and viscoelastic stress response. The mechanical anisotropy is attributed generally to the anisotropic collagen fiber structure; however, the structural origins of the viscoelastic behavior of the tissues are not well understood. Transmission electron microscopy studies of scleral sections labeled with cationic dyes, such as cuprolinic blue, show proteoglycans binding to adjacent collagen fibrils at specific sites of the D-periodic structure (Young 1985). The transient interactions of the collagen fibrils and proteoglycans may be a significant contributor to the rate-dependent stress response of the tissue.

The cornea and sclera are dynamic biological tissues that undergo physiological changes with age and pathological changes with diseases, such as keratoconus and glaucoma. Studies have shown that the human and monkey sclera stiffen (Friberg and Lace 1988; Girard et al. 2009b; Coudrillier et al. 2012) and creep more slowly with age (Siegwart and Norton 1999). The same has been observed for human corneas in uniaxial strip and inflation tests (Elsheikh et al. 2007, 2010b). In keratoconus corneas, a degenerative disease caused by progressive thinning of the cornea, the collagen structure of the central cornea loses the orthogonal fiber structure characteristic of healthy corneas (Meek et al. 2005). Chick eyes afflicted by globe-enlarged retinopathy have a meridional fiber structure rather than circumferential structure in the limbus (Boote et al. 2008). Regional differences in the scleral collagen structure of glaucoma and age-matched normal eyes have also been measured by WAXS. Glaucoma is a blinding disease characterized by progressive degeneration of the axons of retinal ganglion cells. The degree of fiber alignment is significantly lower in the superior-temporal and inferior-nasal quadrants in the peripapillary region, immediately adjacent to the optic nerve head, of glaucoma eyes (Pijanka et al. 2012). Uniaxial strip tests and inflation tests of the sclera consistently measure a stiffer mechanical response (Downs et al. 2005; Hommer et al. 2008; Girard et al. 2011; Coudrillier et al. 2012; Steinhart et al. 2012) and altered viscoelastic properties (Downs et al. 2005; Coudrillier et al. 2012) for glaucoma eyes in humans and animal models of the disease. Dramatic alterations in the mechanical behavior and structure have also been measured in myopia, where the sclera thins and elongates in response to the quality of the image focus on the retina. The sclera of induced-myopia chick and tree shrew eyes are significantly more compliant and exhibit faster creep than the sclera of contralateral control eyes (Phillips et al. 2000; McBrien et al. 2001; McBrien and Gentle 2003).

The mechanical behavior of the cornea is of central importance to a number of biomedical applications, including the development of tonometric procedures for glaucoma screening and corneal biomaterials for prosthetics and wound healing. Mechanical evaluation and computational modeling of the cornea is needed for surgical planning and design of therapeutic procedures, such as collagen crosslinking to stabilize the mechanical properties of keratoconus cornea. Research in corneal biomechanics has increasingly focused on establishing a relationship between the collagen structure and mechanical behavior to understand the effect of pathological alterations in structure and properties on the tissue function and disease progression.

In the same manner, scleral biomechanics is driven to understand the structural origins of mechanical properties and their alterations and aberrations in myopia and glaucoma.

This paper provides a review of recent advances towards understanding the relationship between the collagen structure, mechanical properties, and physiologic function of the cornea and sclera. The following sections will review first the development of inflation experiments with full-field displacement mapping and associated analytical methods to determine the anisotropic mechanical properties of the tissues. Next, constitutive models for the anisotropic stress response based on the collagen fiber structure will be presented. The final sections will focus on the author's recent work investigating the effect of collagen anisotropy on the deformation response of the cornea, sclera, and adjacent optic nerve region.

14.2 Inflation Experiments and Analytical Methods for Strains and Stresses

14.2.1 A Comparison of Uniaxial Strip Testing and Inflation Testing

Mechanical characterization of the cornea and sclera has relied primarily on either uniaxial strip testing or inflation testing. A few studies, such as those of Eilaghi et al. (2010), have applied biaxial mechanical testing to the sclera, but none to the author's knowledge has been applied to the cornea. Biaxial testing is challenging for the cornea and sclera because of the small size of the tissue, large natural curvature, and large spatial variation in the thickness and collagen structure, all of which makes it difficult to prepare a uniform gage section. These issues also pose challenges for uniaxial strip tests. Conventional uniaxial tensile strip tests can probe the anisotropic response of the central cornea and midposterior sclera, but would face difficulties in the limbus region of the cornea and peripapillary region of the sclera. The natural curvature and small dimensions of these regions prevent cutting sufficiently large and straight samples for uniaxial testing of different orientations. Uniaxial strip tests also require preconditioning to characterize viscoelastic properties such as hysteresis (Fung 1993); however, the process of preconditioning, where the tissue sample is repeatedly loaded and unloaded to obtain repeatable measurements, produces large permanent deformation and alterations in the stress response (Boyce et al. 2007). El-Sheikh et al. (2006) showed that the stress-strain response of the cornea measured by uniaxial tests is significantly stiffer than measured by inflation tests. In contrast, Lari et al. (2012) comparing whole globe inflation tests and uniaxial testing of the sclera showed a more extended toe region of the load-deformation curve and a more compliant response for uniaxial testing.

14.2.2 Basic Experimental Setup

Inflation experiments measure the mechanical response of the cornea and sclera to controlled pressure changes, and the method has been developed for both excised tissue structures and whole globes. As the limitations of biaxial and uniaxial strip testing become increasingly recognized, inflation testing has become the preferred method to obtain more detailed, physiological characterization of the material properties of the cornea and sclera. Compared to uniaxial and biaxial tests, the experimental apparatus for inflation testing is relatively simple and inexpensive to construct. A typical inflation setup includes a syringe injection system, which can be a simple column of water (Woo et al. 1972; Jue and Maurice 1991; Hjortdal 1996) or a commercial syringe pump; a feedback controller for pressure; an inflation chamber to hold the tissue with inlets for syringe injection and pressure measurement; and a displacement measurement system (see, for example, Elsheikh and Anderson 2006; Girard et al. 2007). In our experimental setup for the cornea and the posterior sclera, an MTS universal tension tester (Fig. 14.3) provides both the syringe injection system and feedback controller for pressure and pressure rate (Boyce et al. 2008; Myers et al. 2010a; Coudrillier et al. 2012). Whole globe systems directly inject fluid into the eye (e.g., via a syringe inserted through the optic nerve head Bisplinghoff

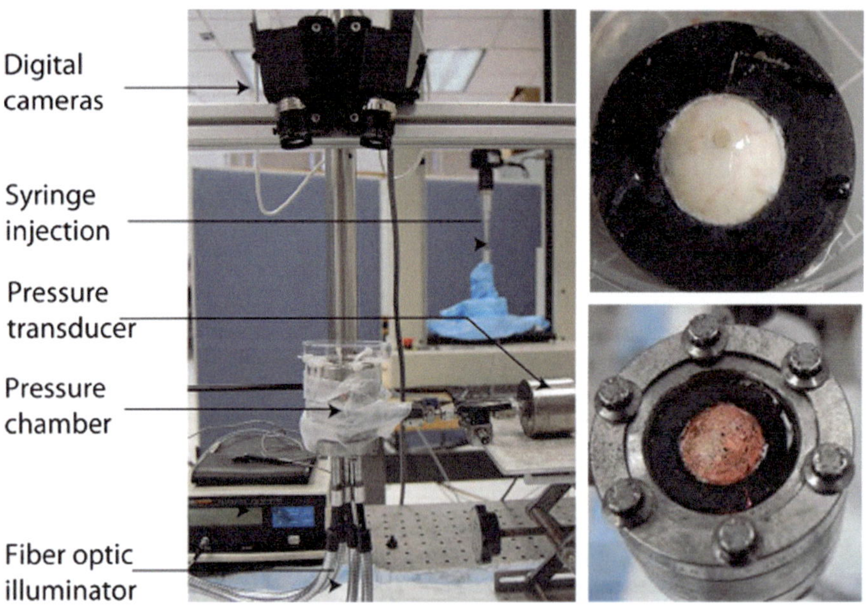

Fig. 14.3 Inflation testing of human eyes showing the experimental setup, the posterior sclera mounted on a custom plastic holder, and the posterior scleral cup speckled with graphite powder and transilluminated by fiber optic lights through the inflation chamber. Reproduced from Coudrillier et al. 2012, 2013 with permission of copyright holder

et al. 2009 or near the corneoscleral junction Thornton et al. 2009; Myers et al. 2010b; Lari et al. 2012) and require only a holder rather than an inflation chamber to secure the globe during pressurization.

Significant advances in inflation testing have been made to the displacement measurement system and to analysis methods used to determine the stress response and mechanical properties. Woo et al. (1972) used a laser system to measure the motion of markers placed near the apex. An inverse finite element analysis (IFEA) was used determine the nonlinear isotropic stress-strain response of the tissues that reproduces the deformation of the markers. El-Sheikh et al. (2006) used a laser sensor system to measure the apex displacement of the corneal specimen, then applied linear elastic shell theory to determine the instantaneous (linearized) Young's modulus of an isotropic nonlinear constitutive relation from the pressure-displacement response. A number of works digitally imaged the deformation of makers and analytically calculated local strains from the relative displacements of the makers (Jue and Maurice 1991; Hjortdal 1996; Bisplinghoff et al. 2009). The deformed markers were also used to calculate the principal curvatures of the pressurized tissues and principal stresses using membrane models for a thin spherical shell or shells of revolution.

Recent works have taken advantage of advances in full-field, non-contact measurements to spatially map the deformation response of the specimen surface. These methods bring the ability to characterize the spatial heterogeneity of the material response. This is of particular importance to the cornea and sclera, which exhibit large variations in the collagen structure and anatomic dimensions, both of which lead to large gradients in the strain response (Coudrillier et al. 2012; Fazio et al. 2012). Girard et al. (2009a) and subsequent works (Girard et al. 2011; Fazio et al. 2012) used electronic speckle pattern interferometry (ESPI), a laser holographic technique, to map the three-dimensional (3D) displacement field of the scleral surface. In ESPI, laser light is scattered off the specimen surface to produce a speckle pattern, where the phase, amplitude, and intensity are related to the specimen surface structure. This is superimposed onto a reference beam from the same laser source to produce an interferogram. Deformation of the object is visualized as changes in the fringe separations of the interferogram. The method can be extended to 3D by illuminating the object with two laser beams coming from different orientations. ESPI is a highly accurate method for displacement measurement with submicron accuracy; however, the time requirements for the laser scanning of the specimen restrict its measurements to the equilibrium (long-time) elastic response.

The authors and coworkers have used an optical method, stereoscopic digital image correlation (DIC), to map the surface deformation field of the cornea and sclera (Boyce et al. 2008; Myers et al. 2010a,b; Coudrillier et al. 2012). In stereoscopic DIC, a speckle pattern is applied to the specimen surface to determine the motion of material points during inflation. We have used both fine graphite powder for the transparent cornea and white human sclera as well as India ink to even out the surface contrast of the heavily pigmented bovine sclera. Spray-painting India ink provides more control of the particle size, uniformity of the speckle

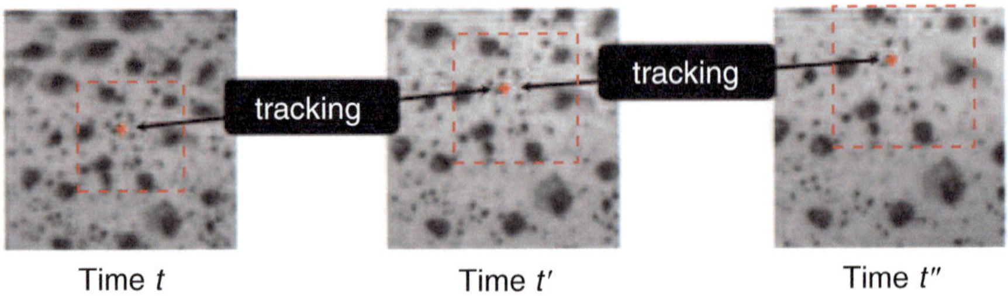

Fig. 14.4 Illustration of the motion tracking of material points by DIC for a bovine sclera specimen speckled with graphite powder during an inflation test. Reproduced from Myers et al. 2010b with permission of copyright holder. The material point is illustrated by a *red dot* and the *dashed line* highlights the pixel subset

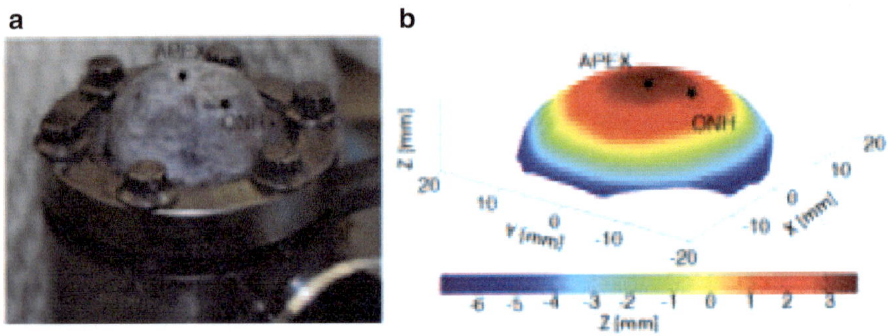

Fig. 14.5 Illustration of the 3D surface reconstruction of an undeformed specimen of the bovine posterior sclera using DIC shown (**a**) the specimen mounted on the pressurization chamber and (**b**) a contour plot of the vertical (Z) position. Reproduced from Myers et al. 2010b with permission of copyright holder

pattern, and surface contrast. However, extensive surface coverage with the ink can dry out the surface of the tissue over long testing periods. The deforming surface is imaged at regular intervals by two CCD cameras arranged in stereo (Fig. 14.3). The images are post-processed and the displacement field is determined by comparing the deformed images to a reference image taken at the nominally zero baseline pressure. The textbook of Sutton (2009) provides a comprehensive description of the various manifestations of digital image correlation methods. In brief, the algorithm compares the distribution of gray values of a subset of pixels from the reference image of one camera to those of subsequent deformed images, essentially optimizing the parameters of a cross-correlating function, to track the motion of material points over time (Fig. 14.4). The algorithm then performs a cross-camera correlation to reconstruct the 3D surface of the deforming specimen (Fig. 14.5). Stereo DIC provides both the undeformed 3D configuration of the specimen surface and a time-dependent deformation map of the displacement field. The accuracy of displacement and strain measurements depends on a variety of factors, including the quality of the speckling pattern, lighting, and stereo angle (Ke et al. 2011) but the method has the potential for high accuracy. Spatial resolution of 1 % of the field of view can be realized (Sutton et al. 2007). In our method, we were able to achieve displacement uncertainties of $<10\,\mu\text{m}$ (Coudrillier et al. 2012), which was consistent with the findings of Ke et al. (2011). The time resolution of DIC is limited by the speed of the imaging system, which allows the method to be applied to measure the time-dependent and rate-dependent viscoelastic behavior of tissues.

14.2.3 Strain Analysis

The preferred fiber orientation of the collagen structure for the most part follows the lines of curvature, e.g. circumferential and meridional, in the cornea and sclera. Thus, it is convenient to calculate the stress and strain state for a coordinate system following the lines of curvature of the undeformed tissues from the measured displacement field (Coudrillier et al. 2012; Nguyen et al. 2013). The undeformed surface of the cornea and sclera can be described well by an ellipsoidal and spherical geometry, respectively. The coordinates of the specimen surface and displacements measurements are commonly provided by the imaging system, such as DIC, on a Cartesian grid and in a Cartesian coordinate system, e.g. $u_X(X, Y, Z), u_Y(X, Y, Z), u_Z(X, Y, Z)$. To determine the strain components in the circumferential-meridional plane of the sclera, the coordinates of points on the undeformed surface can be fit to sphere to provide an analytical representation of the reference configuration. This allows a spherical grid to be created for the specimen surface on which to interpolate the Cartesian displacement components

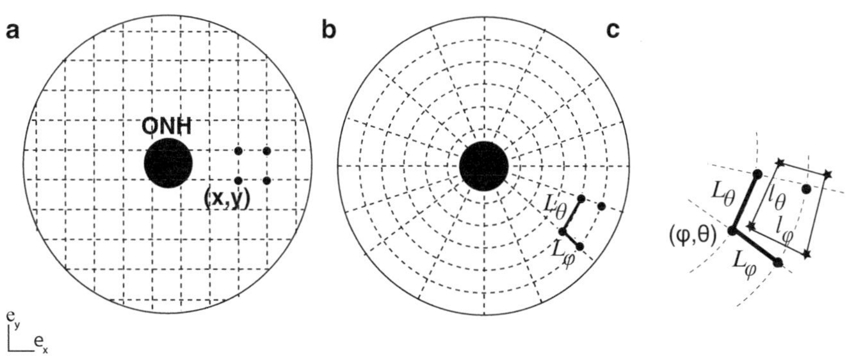

Fig. 14.6 Grids for numerical strain calculations from DIC displacement data showing (**a**) the Cartesian grid native to the DIC algorithm, (**b**) a 2D polar projection of a spherical grid, where φ is the meridional direction and θ is the circumferential direction, and (**c**) the deformed and undeformed material lines along the meridional and circumferential orientations. Reproduced from Coudrillier et al. (2013) with permission from copyright holder

(Fig. 14.6). These are transformed to the spherical coordinate system to provide the circumferential u_θ and meridional u_φ displacement components

$$\begin{aligned} u_r(\theta,\varphi) &= u_X \sin\varphi\cos\theta + u_Y \sin\varphi\sin\theta + u_Z\cos\varphi, \\ u_\varphi(\theta,\varphi) &= u_X \cos\varphi\cos\theta + u_Y\cos\varphi\sin\theta - u_Z\sin\varphi, \\ u_\theta(\theta,\varphi) &= -u_X\sin\theta + u_Y\cos\theta. \end{aligned} \quad (14.1)$$

The tangent vectors of the deformed circumferential and meridional directions can be calculated from the displacement gradients as follows:

$$\mathbf{t}_\theta = \mathbf{T}_\theta + \frac{1}{R\sin\varphi}\frac{\partial\mathbf{u}}{\partial\theta}, \quad \mathbf{t}_\varphi = \mathbf{T}_\varphi + \frac{1}{R}\frac{\partial\mathbf{u}}{\partial\varphi}, \quad (14.2)$$

where $\mathbf{T}_\theta = \mathbf{e}_\theta$ and $\mathbf{T}_\varphi = \mathbf{e}_\varphi$ are tangents of the undeformed circumference and meridian. It follows that the components of the surface Lagrange strain tensor can be determined as follows:

$$\begin{aligned} E_{\theta\theta} &= \frac{1}{2}\left(\mathbf{t}_\theta\cdot\mathbf{t}_\theta - 1\right), \\ E_{\varphi\varphi} &= \frac{1}{2}\left(\mathbf{t}_\varphi\cdot\mathbf{t}_\varphi - 1\right), \\ E_{\varphi\theta} &= E_{\theta\varphi} = \frac{1}{2}\left(\mathbf{t}_\varphi\cdot\mathbf{t}_\theta\right), \end{aligned} \quad (14.3)$$

where the shear strain $E_{\varphi\theta}$ describes the distortion of the initially orthogonal angle spanning the circumferential and meridional directions. The transverse shear strains, $E_{r\varphi}$ and $E_{r\theta}$, remain unknown because the radial gradients cannot be calculated from the surface measurements, while the radial strain E_{rr} can be estimated from in-plane strain components in Eq. (14.3) by assuming incompressibility.

The above analysis can be applied to the cornea, though the reference positions of the corneal surface would be fitted to an ellipsoid rather than a sphere. Alternatively, the reference configuration can be reconstructed directly from the coordinates of the reference position. In this method, a 2D polar grid is created then projected into a 3D surface using the measured Z-position interpolated to the polar grid (Fig. 14.6). The Cartesian displacement components are interpolated to the polar grid. This allows the stretch to be calculated from the ratio of the deformed, l, and undeformed, L, lengths between two adjacent grid points. For example, the meridional and circumferential stretch are $\lambda_\varphi = l_\varphi/L_\varphi$ and $\lambda_\theta = l_\theta/L_\theta$. This method were also employed to calculate the surface strain field for the human sclera. A comparison between the two methods found no noticeable differences for a sufficiently small grid size.

Figure 14.7 plots the strain field for a representative human sclera specimen tested to 30 mmHg, showing tremendous spatial gradients. Measurements for the

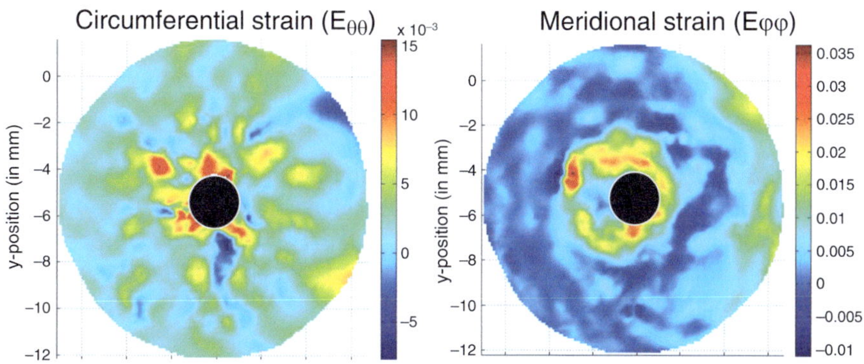

Fig. 14.7 The circumferential and meridional Lagrange strains calculated for the human posterior sclera. Note the strain concentration surrounding the optic nerve head (*black*) (Coudrillier et al. 2012, 2013)

optic nerve head has been omitted for clarity. The strain components increased dramatically in the peripapillary region, ringing the optic nerve head. This strain concentration is caused by the presence of the more compliant tissues of the nerve head. The meridional strain in the peripapillary region was also significantly larger than the circumferential strain, which reflects the strong circumferential collagen fiber reinforcement. Large gradients was also observed for the surface shear strain component $E_{\varphi\theta}$ (not shown), which is to be expected from the spatial heterogeneity of the collagen structure. However, when averaged over a circumference, the average shear strain was an order of magnitude smaller than the average meridional and circumferential strains.

14.2.4 Stress Analysis

Inflation testing does not directly measure the stress response of the tissue to pressurization. Rather, a model of the pressurized specimen is needed to describe the stress response and ultimately determine the material properties of the tissue. One strategy is to model the tissues as a thin shell that exhibits negligible bending under physiological pressures. The thin assumption neglects the radial stress component and variations of the in-plane stress components through the thickness. Under these conditions, and conditions of axisymmetry, the principal stress resultants N_φ and N_θ can be determined from the pressure loading solely using equilibrium. The equilibrium equations for a thin shell under uniform pressure loading p simplifies to (Tomoshenko 1959)

$$\frac{\mathrm{d}}{\mathrm{d}r}\left(rN_\varphi\right) - N_\theta = 0, \quad \frac{N_\theta}{r_2} + \frac{N_\varphi}{r_1} = p, \tag{14.4}$$

where r_1 and r_2 are the principal curvatures in the meridian and transverse plane of the deformed shape and

$$r = r_2 \sin \varphi \tag{14.5}$$

is the radius of the planar projection at the position φ measured from the apex. The stress components can be calculated from the resultants for a thin shell by dividing through by the deformed thickness t (i.e., $\sigma_{\varphi\varphi} = N_\varphi / t$ and $\sigma_{\theta\theta} = N_\theta / t$). Assuming an axisymmetric surface, the equilibrium equations can be solved analytically for the cornea (Hjortdal 1996) and sclera (Coudrillier et al. 2012) as follows:

$$\sigma_{\varphi\varphi} = \frac{pr_2}{2t}, \quad \sigma_{\theta\theta} = pr_2 \left(1 - \frac{r_2}{2r_1}\right). \tag{14.6}$$

For a spherical shell, $r_1 = r_2 = r$ and the solution for the stress response reduce to that of a thin-walled spherical pressure vessel:

$$\sigma_{\varphi\varphi} = \sigma_{\theta\theta} = \frac{pr}{2t}. \tag{14.7}$$

The local principal radii of curvature for an ellipsoid of revolution for the position φ can be determined from the major and minor axis, a and b of the meridian as,

$$r_1 = \frac{a^2 b^2}{\left(a^2 \sin^2 \varphi + b^2 \cos^2 \varphi\right)^{\frac{3}{2}}}, \qquad r_2 = \frac{a^2}{\left(a^2 \sin^2 \varphi + b^2 \cos^2 \varphi\right)^{\frac{1}{2}}}, \tag{14.8}$$

where φ denotes the location on the meridian.

Alternatively, the local curvatures can be calculated numerically by fitting an analytic function (e.g., a lower order bivariate polynomial or a spherical surface) to a small region about a point of interest, then evaluating the principal curvatures r_1 and r_2 using the first and second fundamental forms (Tonge et al. 2013). The accuracy of these analytical approaches depends directly on the accuracy of the radius of curvature calculations and the assumption of an axisymmetric surface. The assumptions of the analytical membrane model for the stress components prohibit its application in regions adjacent to the boundaries, regions with large material discontinuities, and regions of negative curvature, where the effects of bending stresses cannot be neglected. For the sclera, this precludes the analytical stress calculation in the peripapillary region adjacent to the optic nerve head. A computational model of the specimen under inflation is needed to describe the stress response and determine the material properties from inflation testing for these regions. This requires developing constitutive models for the stress response and specimen-specific finite element models of the cornea and sclera.

14.3 Structural Constitutive Models

The cornea and sclera exhibit a complex hierarchical structure ranging over many length scales as described in Sect. 14.1. Each structural level exhibits distinct deformation mechanisms that dominate under different stress states and time scales. The multiscale nature of soft tissues presents formidable challenges to the development of predictive models and simulation tools, which is needed for a wide range of biomedical applications from surgical planning to development of collagen crosslinking therapies, tonometric and elastographic diagnostic techniques, and engineered biomaterials. Early modeling efforts treated the corneal and scleral tissues as a homogeneous, isotropic, linear elastic, or hyperelastic materials (Bryant et al. 1991; Asejczyk-Widlicka 2007; Hanna et al. 1989; Yan et al. 1994; Dongqi and Zeqin 1999; Bellezza et al. 2000; Sigal et al. 2004, 2005). Development of constitutive models for the nonlinear, anisotropic, and viscoelastic behavior remains active areas of research. Phenomenological models for soft tissues, such as those based on the structure tensor approach (Spencer 1971), often contain a large number of parameters to be determined experimentally (e.g., Pioletti et al. 1998; Limbert and Middleton 2004). Thus, most recently developed anisotropic models for the cornea and sclera have followed a structural approach, which base the description of the mechanical anisotropy of the tissue stress response on the anisotropic collagen structure. This reduces the number of parameters and extends the predictive capability of the model beyond the range of available experimental data. Structural models can be categorized into discrete fiber models, which idealize the collagen structure as having discrete numbers of fiber orientations (Alastrué et al. 2006; Pandolfi and Manganiello 2006; Pandolfi et al. 2009) (e.g., two in the central region for the human cornea), and distributed fiber models, which represent the collagen structure using a continuous orientation distribution (Pinsky et al. 2005; Pandolfi and Holzapfel 2008; Nguyen et al. 2008; Girard et al. 2009c; Studer et al. 2010; Nguyen and Boyce 2011; Pandolfi and Vasta 2012). For the cornea and sclera, discrete fiber models do not consider that the collagen fibers are arranged in all possible orientations despite having preferred orientations. When applied to simulate the response of the tissues to intraocular pressure, discrete fiber models can produce non-physiological bulging zones at physiological pressures (Pandolfi and Holzapfel 2008).

Distributed fiber models can be further split into fully integrated and pre-integrated models. Fully integrated models assume a strain energy density at the fiber-level rather than at the tissue level. These are integrated over all possible fiber orientations weighted by a continuous probability density function (PDF) of the fiber orientation to describe the anisotropic behavior of the tissue. The PDF describes the preferred fiber orientation as well as the dispersion about the preferred orientation. A fully integrated model reduces to a structure tensor model when there is no dispersion in the fiber orientation. This modeling approach has its origins in the work of Lanir (1983), and has become the standard approach in modeling the cornea (Pinsky et al. 2005; Nguyen et al. 2008; Studer et al. 2010; Nguyen and Boyce 2011), sclera (Girard et al. 2009c; Coudrillier et al. 2013), and many other collagenous tissues (Sacks 2003).

The integral formulation of distributed fiber models cannot in general be evaluated analytically except for special forms of the PDF and fiber potential (Raghupathy and Barocas 2009). In a finite element framework, numerical evaluation of the integral formulation at each integration point for each iteration is computationally expensive. Gasser et al. (2006) introduced an alternative model formulation in which the continuous PDF is pre-integrated in the reference configuration to give an equivalent generalized structure tensor. The generalized structure tensor is a linear mixture of an isotropic tensor representing the population of fibers equally distributed in all orientations and a transversely isotropic structure tensor for the dominant fiber orientation. The strain energy density for the tissue then can be expressed as an isotropic function of the deformation tensor and generalized structure tensor. Pre-integrated distributed fiber models have been developed for the cornea (Pandolfi and Holzapfel 2008) and the sclera (Grytz et al. 2011). Though rooted in the same structural descriptions, the fully integrated and pre-integrated models are equivalent only for planar distributions under equibiaxial stretch. They demonstrate divergent results for cases with a low degree of fiber alignment (Federico and Herzog 2008; Cortes et al. 2010), which is characteristic of the cornea and sclera.

The subsections below will present in detail the development of the fully integrated distributed fiber modeling approach for the large deformation mechanical behavior of the cornea and sclera. The following subsection will present an extension of the model developed by the author to describe the nonlinear anisotropic behavior of the cornea.

14.3.1 Fully Integrated Distributed Fiber Model

Distributed fiber model is developed from the following assumptions. First, the stroma can be idealized as a continuum mixture of fibers arranged in an isotropic matrix. The fiber orientation is represented in the reference configuration by an orientation vector $\mathbf{N}$. For a 2D fiber arrangement, e.g., distributed in the plane of the tissue, $\mathbf{N} = (\cos\gamma, \sin\gamma)$, where γ is an angle defined in the plane of the tissue. The anisotropic collagen structure can be described by a continuous PDF $D(\gamma, \mathbf{X})$ that can vary with the material position. Pinsky et al. (2005) and Studer et al. (2010) used trigonometric functions to represent angular dependence of the normalized scattering intensity data of the WAXS measurements. Girard et al. (2009c) and the author and coworkers (Nguyen and Boyce 2011) have used a von Mises distribution function

$$D(\gamma, \mathbf{X}) = \frac{\exp\left(k(\mathbf{X})\cos\left[2\left(\gamma - \gamma_p(\mathbf{X})\right)\right]\right)}{2\pi I_0(k(\mathbf{X}))}, \tag{14.9}$$

where k is the dispersion parameter for the degree of fiber alignment and γ_p is the preferred fiber orientation. Girard and coworkers (Girard et al. 2009b, 2011) applied IFEA to fit both parameters to the inflation response, while Coudrillier et al. (2013), used the normalized scattering intensity measured independently by WAXS directly to describe $D(\gamma, \mathbf{X})$ without fitting to an analytic function.

Second, the fibers and matrix deform affinely with the tissue, such that their deformation can be described by the macroscopic deformation gradient

$$\mathbf{F} = \frac{\partial \chi(\mathbf{X}, t)}{\partial \mathbf{X}}, \tag{14.10}$$

where $\mathbf{x} = \chi(\mathbf{X}, t)$ is the motion that maps material points $\mathbf{X}$ in the reference configuration to spatial points $\mathbf{x}$ in the deformed configuration at time t. The deformation gradient maps material lines from the reference configuration to spatial lines in the deformed configuration. From $\mathbf{F}$, we can define the symmetric left and right deformation tensors $\mathbf{C} = \mathbf{F}^{\mathrm{T}}\mathbf{F}$ and $\mathbf{b} = \mathbf{F}\mathbf{F}^{\mathrm{T}}$ for the reference and deformed configurations. The fiber stretch then can be computed as,

$$\lambda(\gamma) = (\mathbf{N}(\gamma) \cdot \mathbf{C}\mathbf{N}(\gamma))^{1/2} \tag{14.11}$$

and the deformed fiber orientation is given as $\mathbf{n} = \lambda^{-1}\mathbf{F}\mathbf{N}$.

Finally, the fibers and matrix behave independently from one another, i.e., they deform in parallel. Together, these assumptions lead to a model for the strain energy density of the tissue that is mixture of an isotropic part for the proteoglycan matrix and anisotropic part from the collagen fibers,

$$W(\mathbf{C}, \mathbf{X}) = W_m(I_1, I_2, I_3) + \int_{-\pi}^{\pi} W_f(\lambda(\gamma))\, D(\gamma, \mathbf{X})\, \mathrm{d}\gamma, \tag{14.12}$$

where $I_1 = \mathbf{C} : \mathbf{1}$, $I_2 = \frac{1}{2}\left(I_1^2 - \mathbf{C} : \mathbf{C}\right)$, and $I_3 = \det[\mathbf{C}]$ are the first and third invariants of the deformation tensor. In Eq. (14.12), the anisotropic component is represented by the integrated contributions of the free energy density of the fibers W_f weighted by the PDF of the fiber orientation.

A compressible Neo-Hookean model is commonly applied to represent the compliant isotropic matrix of the free energy density, e.g.,

$$W_m(I_1, I_3) = \frac{\mu}{2}\left(\bar{I}_1 - 3\right) + \frac{\kappa}{4}\left(I_3 - \ln I_3 - 1\right), \tag{14.13}$$

where $\overline{I_1} = I_3^{-\frac{1}{3}} I_1$ is the distortional part of I_1 and μ and κ are the shear modulus bulk moduli in the reference configuration. A nearly incompressible response can be achieved by choosing $\kappa \gg \mu$. The strain-stiffening stress response observed for the cornea and sclera is widely described using an exponential function such as Nguyen and Boyce (2011)

$$W_f(\lambda_f) = \frac{\alpha}{\beta}\left[\exp\left(\beta\left(\lambda_f^2 - 1\right)\right) - \beta\lambda_f^2\right]. \tag{14.14}$$

The model has two parameters, α and β where β is related to the strain at the onset of stiffening. The expression $4\alpha\beta$ signifies the small-strain stiffness of the fiber family and β is the strain-stiffening parameter. Elastica models for the strain-stiffening stress response of the collagen fiber has also been developed. Grytz et al. (2009) used a helical coiled beam to represent the crimped collagen fiber and calculated the axial response of the coiled beam to determine the stress response of the fiber.

The second Piola–Kirchhoff stress tensor can be calculated from hyperelasticity theory as $\mathbf{S} = 2\frac{\partial W}{\partial \mathbf{C}}$ to give

$$\begin{aligned} \mathbf{S}(\mathbf{C}) = {} & \mu I_3^{-\frac{1}{3}}\left(\mathbf{1} - \frac{1}{3}I_1\mathbf{C}^{-1}\right) + \frac{\kappa}{2}(I_3 - 1)\,\mathbf{C}^{-1} \\ & + \int_{-\pi}^{\pi} S_f\left(\lambda_f(\theta), t\right)\mathbf{N}(\theta)\otimes\mathbf{N}(\theta)\; D(\theta, \mathbf{X})\,\mathrm{d}\theta, \end{aligned} \tag{14.15}$$

where the fiber stress, S_f, is given by

$$S_f = \alpha\left[\exp\left(\beta\left(\lambda_f^2 - 1\right)\right) - 1\right]. \tag{14.16}$$

The Cauchy stress is determined by applying the Piola transformation $\boldsymbol{\sigma} = \frac{1}{\sqrt{I_3}}\mathbf{F}\mathbf{S}\mathbf{F}^{\mathrm{T}}$ to give

$$\begin{aligned} \boldsymbol{\sigma} = {} & \frac{\mu}{\sqrt{I_3}}\left(\overline{\mathbf{b}} - \frac{1}{3}\overline{I}_1\mathbf{1}\right) + \frac{\kappa}{2\sqrt{I_3}}(I_3 - 1)\,\mathbf{1} \\ & + \frac{1}{\sqrt{I_3}}\int_{-\pi}^{\pi} s_f\left(\lambda_f(\theta)\right)\mathbf{n}(\theta)\otimes\mathbf{n}(\theta)\; D(\theta, \mathbf{X})\,\mathrm{d}\theta, \end{aligned} \tag{14.17}$$

where $\overline{\mathbf{b}} = I_3^{-\frac{1}{3}}\mathbf{F}\mathbf{F}^{\mathrm{T}}$ is the distortional component of the left deformation tensor and $s_f = \lambda_f^2 S_f$.

Inverse Finite Element Analysis for Parameter Determination

The distributed fiber model contains two sets of parameters, one for the collagen fiber and matrix (α, β, μ) and another for the PDF of the collagen orientation. Girard and coworkers (Girard et al. 2009a, 2011) used IFEA to fit both sets to the 3D displacement field measured for inflation testing of the monkey sclera using ESPI at different pressure steps. Finite element models were created for each specimen from digital surface reconstruction and ultrasonic pachymeter measurements of the spatially varying thickness. The models were then segmented into eight regions, four sectors in the peripapillary and four in the midposterior region. The von Mises PDF in Eq. (14.9) was used to describe the collagen structure of each region. The parameters of the collagen fibers and matrix were assumed to be uniform over the entire region. For the parameters of the PDF, each sector was assumed to exhibit a different preferred orientation γ_p, while the peripapillary and midposterior regions were assigned different dispersion parameters for fiber orientation. This model attributed all spatial variations in material response to the collagen structure, and the assumption that the two regions had a different degree of fiber alignment was justified from histology. An evolutionary algorithm was used to find the set of 13 parameters that globally minimized the Euclidian error norm between the ESPI measured and FEA calculated displacement field over the different pressure steps. A good fit was obtained to the experimental data, but the resulting preferred fiber orientation showed large differences between different sectors, though they were on average circumferential over all the regions (Girard et al. 2009a). This method was extended recently by Grytz et al. (2013) to determine the parameters of the human posterior sclera. This work used an elastica model to describe the strain-stiffening response of the collagen fibers, a smooth interpolation of the parameters of the von Mises PDF between material regions, and a compliant spring support to describe the boundary conditions at the clamps. The inverse analysis fits the spring constant of the compliant support and parameters describing the location of the material regions, as well as those of the elastica model and von Mises PDF, resulting in a total of 21 fitting parameters. In a different approach, the author and coworkers used the specimen-specific WAXS measurements of the collagen structure directly to represent the PDF of the collagen orientation (Coudrillier et al. 2013). For each measurement point, the angular variation of the X-ray scattering intensity was normalized to obtain the PDF of the fiber orientation. Consequently, only three parameters (α, β, μ) for the collagen fibril and proteoglycan matrix needed to be determined from the DIC measured inflation response. This highlights the tremendous advantage of the structural modeling approach. Allowing structural measurements to inform the models greatly reduces the number of parameters needed to be determined from mechanical experiments.

The parameters obtained from inverse analysis were applied in a computational study of the effects of the heterogeneous collagen fiber structure on the anisotropic mechanical behavior of the sclera (Coudrillier et al. 2013). Micromechanical models were created that applied equibiaxial loading to virtual sections of the sclera to calculate the anisotropic strain response. Different size sections were examined, including 0.5 mm^2 squares containing one WAXS measurement, 2.0 mm^2 squares equal to the width of the peripapillary region, and 4.5 mm^2 squares similar in size used in the biaxial experiments of Eilaghi et al. (2010). The simulations applied normal tractions consistent with a pressure loading of 30 mmHg in the preferred fiber and perpendicular orientations for the 0.5 mm^2 sections and in the circumferential and meridional directions for all other sections. Figures 14.8 and 14.9 plot the resulting maps of the anisotropic ratio of the maximum over

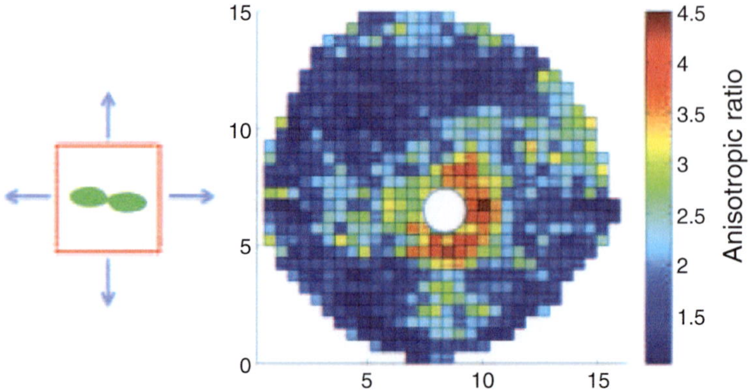

Fig. 14.8 Map of anisotropic ratio comparing strain response of fiber and perpendicular orientation to equibiaxial loading of 0.5 mm^2 regions containing a single WAXS measurement (Coudrillier et al. 2013)

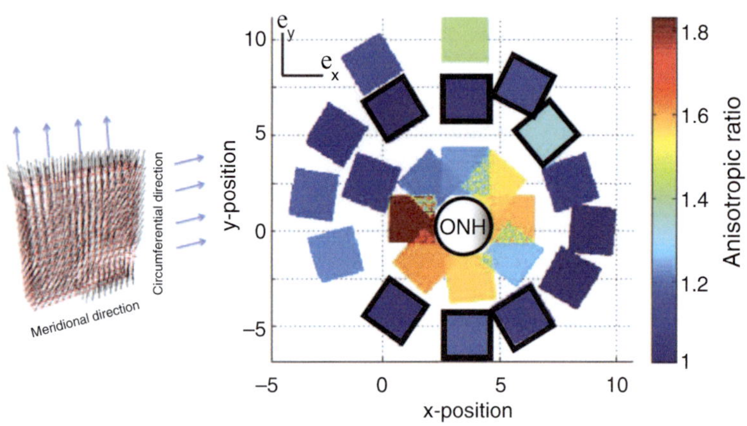

Fig. 14.9 Map of anisotropic ratio comparing strain response of circumferential and meridional orientation to equibiaxial loading of 2.0 mm^2 regions containing 16 WAXS measurements. Reproduced from Coudrillier et al. 2013 with permission of copyright holder

minimum strain response for the different length scales. The peripapillary region is more anisotropic than the midposterior region for all length scales. For the 0.5 mm^2 sections, the anisotropic ratio was on average 2.0 (fiber orientation is twice as stiff as perpendicular orientation) in the peripapillary region and 1.4 in the midposterior region. Moreover, there was a large spatial variation in the anisotropic ratio in the peripapillary region. For the 2.0 mm^2 sections, the anisotropic ratio of both regions decreased; but while the peripapillary region remained anisotropic with a ratio of 1.76 (e.g. the circumferential orientation is 1.76 times stiffer than the meridional direction), the midposterior region became nearly isotropic with an anisotropic ratio of 1.2. The large spatial variations in the preferred fiber orientation and low degree of fiber alignment lead to an isotropic behavior at larger length scales.

14.3.2 Viscoelastic Distributed Fiber Model

We developed an anisotropic viscoelastic distributed fiber model for the nonlinear creep response of the cornea (Nguyen et al. 2007, 2008; Nguyen 2010) based on the results of uniaxial strip tests (Boyce et al. 2007). The experiments showed extensive creep deformation and a strong dependence of the normalized creep rate on the applied stress. The model attributes the large creep strains to the viscoelastic deformation of the collagen fibers and the nonlinear creep behavior to multiple stress-activated viscoelastic processes. The viscoelastic constitutive relation is specified at the fiber level then homogenized to obtain the viscoelastic constitutive relation for the stroma. As a result, the anisotropy of the viscoelastic behavior is described entirely by the anisotropic collagen structure, which can be measured independently from WAXS or similar histological methods. The remaining parameters characterize the stiffness and viscosity of the fibers.

The model was developed using the nonlinear viscoelastic framework of Reese and Govindjee (1998). The kinematics of each viscoelastic process is described by the multiplicative split of the deformation gradient $\mathbf{F}$ into N elastic and viscous parts

$$\mathbf{F} = \mathbf{F}_k^{\mathrm{e}}\mathbf{F}_k^{\mathrm{v}}, \quad k = 1 \ldots N, \tag{14.18}$$

where $\mathbf{F}_k^{\mathrm{v}}$ is the viscous component of the deformation gradient and $\mathbf{F}_k^{\mathrm{e}}$ is the complementary elastic component. The deformation tensors and rates of deformation tensors associated with the elastic and viscous deformation gradient are defined as

$$\begin{aligned} &\mathbf{C}_k^{\mathrm{v}} = \mathbf{F}_k^{\mathrm{v}^{\mathrm{T}}}\mathbf{F}_k^{\mathrm{v}}, \quad \mathbf{b}_k^{\mathrm{v}} = \mathbf{F}_k^{\mathrm{v}}\mathbf{F}_k^{\mathrm{v}^{\mathrm{T}}}, \quad \mathbf{C}_k^{\mathrm{e}} = \mathbf{F}_k^{\mathrm{e}^{\mathrm{T}}}\mathbf{F}_k^{\mathrm{e}}, \quad \mathbf{b}_k^{\mathrm{e}} = \mathbf{F}_k^{\mathrm{e}}\mathbf{F}_k^{\mathrm{e}^{\mathrm{T}}}, \\ &\mathbf{D}_k^{\mathrm{v}} = \frac{1}{2}\left(\dot{\mathbf{F}}_k^{\mathrm{v}}\mathbf{F}_k^{\mathrm{v}^{-1}} + \mathbf{F}_k^{\mathrm{v}^{-T}}\dot{\mathbf{F}}_k^{\mathrm{v}^{\mathrm{T}}}\right), \quad \mathscr{L}_v\mathbf{b}^{\mathrm{e}} = \mathbf{F}_k^{\mathrm{e}}\mathbf{D}_k^{\mathrm{v}}\mathbf{F}_k^{\mathrm{e}^{T}}, \end{aligned} \tag{14.19}$$

where $\mathbf{D}_k^{\mathrm{v}}$ is the viscous rate of deformation tensor and $\mathscr{L}_v\mathbf{b}^{\mathrm{e}}$ is the Lie time derivative, an objective rate, of the elastic left deformation tensor. The viscous and elastic components of the fiber stretch for the kth relaxation mechanism are defined as

$$\lambda_k^{\mathrm{v}} = \sqrt{\mathbf{N}\cdot\mathbf{C}_k^{\mathrm{v}}\mathbf{N}}, \quad \lambda_k^{\mathrm{e}} = \frac{\lambda}{\lambda_k^{\mathrm{v}}}. \tag{14.20}$$

The theoretical developments for the stretch components are detailed in Nguyen et al. (2007).

To model the viscoelastic behavior of the fibers, the free energy density of the collagen fiber is split into an equilibrium component dependent on the total fiber stretch and N time-dependent nonequilibrium parts dependent on the elastic fiber stretch. This results in the following formulation for the free energy density

$$W\left(\mathbf{C},\mathbf{C}_k^{\mathrm{v}},\mathbf{X}\right) = W_m\left(I_1,I_2,I_3\right) + \int_{-\pi}^{\pi}\left(W_f^{\mathrm{eq}}\left(\lambda\left(\gamma\right)\right) + \sum_k W_{fk}^{\mathrm{neq}}\left(\lambda_k^{\mathrm{e}}\left(\gamma\right)\right)\right)D\left(\gamma,\mathbf{X}\right)\mathrm{d}\gamma. \quad (14.21)$$

We assumed that the equilibrium and nonequilibrium components of the free energy density can be described using a phenomenological exponential function:

$$W_f^{\mathrm{eq}}\left(\lambda\right) = \alpha^{\mathrm{eq}}\left[\exp\left(\beta\left(\lambda^2-1\right)\right) - \frac{\beta}{\lambda^2}\right],$$
$$W_{fk}^{\mathrm{neq}}\left(\lambda\right) = \alpha^{\mathrm{neq}}\left[\exp\left(\beta\left(\lambda_k^{\mathrm{e}2}-1\right)\right) - \frac{\beta}{\lambda_k^{\mathrm{e}2}}\right], \quad (14.22)$$

where the same β was used for both to reduce the number of parameters.

The normal stress tensor can be determined from the Coleman-Noll (Coleman and Gurtin 1967) theory of internal variables as, $\mathbf{S} = 2\frac{\partial W}{\partial \mathbf{C}}$. Evaluating the derivatives and applying the Piola transformation gives an expression for the stress response.

$$\boldsymbol{\sigma} = \boldsymbol{\sigma}_m\left(\mathbf{b}\right) + \frac{1}{\sqrt{I_3}}\int_{-\pi}^{-\pi} s_f^{\mathrm{eq}}\left(\lambda\right)\mathbf{n}\otimes\mathbf{n}D\left(\gamma,\mathbf{X}\right)\mathrm{d}\gamma + \sum_k^N \underbrace{\frac{1}{\sqrt{I_3}}\int_{-\pi}^{-\pi} s_{fk}^{\mathrm{neq}}\left(\lambda_k^{\mathrm{e}}\right)\mathbf{n}\otimes\mathbf{n}D\left(\gamma,\mathbf{X}\right)\mathrm{d}\gamma}_{\boldsymbol{\sigma}_{fk}^{\mathrm{neq}}}, \quad (14.23)$$

where $s_f^{\mathrm{eq}} = \frac{\partial W_{\mathrm{f}}^{\mathrm{eq}}}{\partial \lambda}\lambda$ and $s_{fk}^{\mathrm{neq}} = \frac{\partial W_{fk}^{\mathrm{neq}}}{\partial \lambda_k^{\mathrm{e}}}\lambda_k^{\mathrm{e}}$ are the equilibrium and nonequilibrium axial stress response of the collagen fiber, and $\boldsymbol{\sigma}_m$ is the isotropic matrix stress response.

To complete the model, an evolution equation must be specified for the elastic fiber stretches λ_k^{e}. An Eyring model, which describes stress activated viscous flow, was used to describe the nonlinear time-dependent response of the fiber. The evolution equation can be expressed in terms of the viscous stretch rate as follows:

$$\frac{\dot{\lambda}_k^{\mathrm{v}}}{\lambda_k^{\mathrm{v}}} = \frac{s_{0_k}}{\eta_{0_k}}\sinh\left(\frac{s_{fk}^{\mathrm{neq}}}{s_{0_k}}\right), \quad (14.24)$$

where η_{0_k} is the characteristic viscosity and s_{0_k} is the activation stress of the fiber for viscous flow. An effective evolution equation for the viscous deformation of the stroma can be developed from the fiber evolution equation (14.24) as

$$\boldsymbol{\sigma}_{f_k}^{\text{neq}} = \mathscr{V}_{f_k} : \mathbf{b}_k^{\text{e}^{-1}} \left(-\frac{1}{2} \mathscr{L}_v \mathbf{b}_k^{\text{e}} \right) \mathbf{b}_k^{\text{e}^{-1}},$$

$$\mathscr{V}_{f_k} = \left[\int_{-\pi}^{-\pi} \eta_{0_k} \left[\frac{s_{0_k}}{s_{f_k}^{\text{neq}}} \sinh \left(\frac{s_{f_k}^{\text{neq}}}{s_{0_k}} \right) \right]^{-1} \lambda_k^{\text{e}^2} \mathbf{n} \otimes \mathbf{n} \otimes \mathbf{n} \otimes \mathbf{n} D(\theta, \mathbf{X}) \, \text{d}\theta \right], \tag{14.25}$$

where $\boldsymbol{\sigma}_{f_k}^{\text{neq}}$ is the nonequilibrium stress tensor for the kth nonequilibrium process in Eq. (14.23). The tensor $\mathscr{V}_k$ is the effective viscosity for the kth nonequilibrium process that relates the nonequilibrium stress of the stroma to the viscous rate of deformation $\mathbf{b}_k^{\text{e}^{-1}} \left(-\frac{1}{2} \mathscr{L}_v \mathbf{b}_k^{\text{e}} \right) \mathbf{b}_k^{\text{e}^{-1}}$ of the tissue. The anisotropy of the viscosity tensor is derived entirely from the PDF of the collagen fiber orientation.

The model was applied to reproduce the nonlinear creep response of the cornea measured by uniaxial tension tests at three different stress levels (Fig. 14.10a). Three viscoelastic processes ($N = 3$) were used to fit the experimental data, and the procedure to determine the parameters is described in detail in Nguyen et al. (2008). By including stress-activated flow mechanisms, the model successfully reproduced the dramatic increase in the creep rate with the applied stress. Figure 14.10b plots the reduced creep and stress relaxation functions evaluated from the model for the preferred fiber orientation and the perpendicular orientation. The reduced creep function was evaluated by normalizing the uniaxial tension creep response with the instantaneous strain response to the step increase in the applied stress. Similarly, the reduced relaxation response was normalized by the instantaneous stress response to the applied step strain. The simulations used a PDF determined from a WAXS measurement for a location in the limbus (Boote et al. 2006; Pinsky et al. 2005). The reduced relaxation and creep functions were identical for the stiffest and most compliant orientations. The time dependent response of the model is independent of orientation because the viscoelastic behavior is specified at the fiber level and each fiber has the same viscoelastic properties.

14.4 Mechanical Function of the Collagen Fiber Anisotropy

14.4.1 The Anisotropic Limbus

The limbus is the narrow band of the cornea that joins with the sclera. It has a collagen structure that is strongly aligned in the circumferential direction. By comparison, the human central cornea displays a nearly transversely isotropic collagen structure with weak preferential alignment in the orthogonal nasal-temporal and

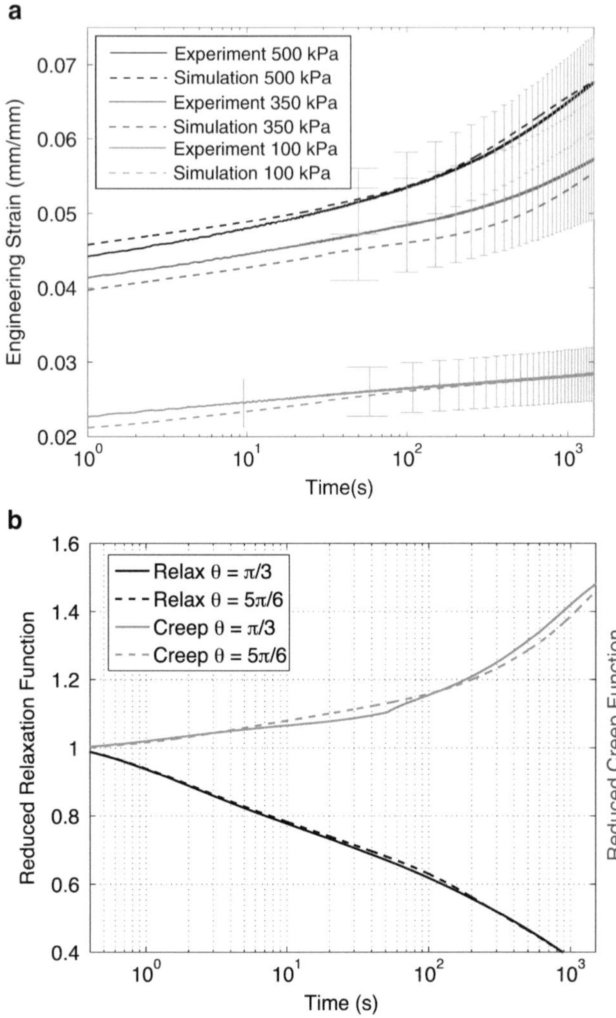

Fig. 14.10 Time-dependent behavior of viscoelastic distributed fiber model. (**a**) The uniaxial tensile creep response at different applied stresses comparing simulations and experiments for bovine cornea. Reproduced from Nguyen et al. (2008) with permission from copyright holder. The model parameters were fit to data for 100 and 500 kPa and used to predict the creep response to 300 kPa. (**b**) The reduced relaxation function for $\lambda = 1.04$ and reduced creep function for 500 kPa for the preferred fiber and perpendicular orientation. Reproduced from Nguyen et al. (2008) with permission from copyright holder

inferior–superior directions. The author developed a finite element modeling study of the effect of the anisotropic collagen structure on the deformation response of bovine cornea in response to pressure changes (Nguyen and Boyce 2011). The distributed fiber model in Sect. 14.3.1 was used to describe the anisotropic behavior of bovine corneal tissue. The cornea was segmented into two regions, the central cornea and limbus, each with a different preferential fiber orientation obtained from WAXS (Hayes et al. 2007) and dispersion parameter of the von Mises PDF for the degree of fiber alignment. Inverse finite element analysis was used to fit the material properties of the fibers and matrix as well as the dispersion parameters for the two regions to the DIC displacement field measured at an inflation pressure of 53 mmHg. Results confirmed a high degree of fiber alignment in the limbus region with a dispersion parameter $k = 6$, but returned a transversely isotropic central region with a dispersion parameter $k = 0$. The WAXS measurements showed a weak preferential alignment in the inferior–superior direction for the bovine cornea. It was possible the resulting mechanical anisotropy was too subtle to detect by the inflation tests and optimization method.

To determine the effect of the collagen anisotropy, a parameter study compared the deformed shapes of model corneas with a strongly anisotropic limbus and different anisotropic fiber structures in the central region, including a transversely isotropic structure, orthogonal structure, circumferential structure, and two additional structures with a preferred orientation in either the nasal-temporal or inferior–superior directions. Except for the transversely isotropic case, the same dispersion parameter $n = 6$ was applied for the von Mises PDF for both the central cornea and limbus to evaluate the effect of a strongly anisotropic central cornea. The results in Fig. 14.11 calculated from a pressure loading of 53 mmHg showed that deformation for the case with a transversely isotropic central region was localized in the strongly anisotropic limbus. The limbus exhibited the largest displacement gradients, while the displacement field in the central region was nearly uniform, indicate little change in the curvature. The strong circumferential fiber reinforcement caused the tissue to be more compliant meridionally in the limbus than the central cornea. The same was obtained for the case with a strongly anisotropic orthogonal structure in the central cornea. Significant shape changes were computed for the three other highly anisotropic fiber structure. The case with a highly aligned circumferential structure in the central cornea deformed into a conical shape. The two cases with a high degree of fiber alignment in either the nasal-temporal or inferior–superior directions developed a distorted deformed shape. The central cornea is the primary refractive component of the eye, and the shape of the central cornea determines 75 % of the unaccommodated optical power of the eye. The particular combination of a strongly aligned circumferential collagen structure in the limbus and a nearly transversely isotropic collagen structure in the central region may serve to preserve the shape of the central cornea for refraction during pressure changes. The results suggest that one function of the corneal collagen structure is to provide the cornea with a robust design for refraction.

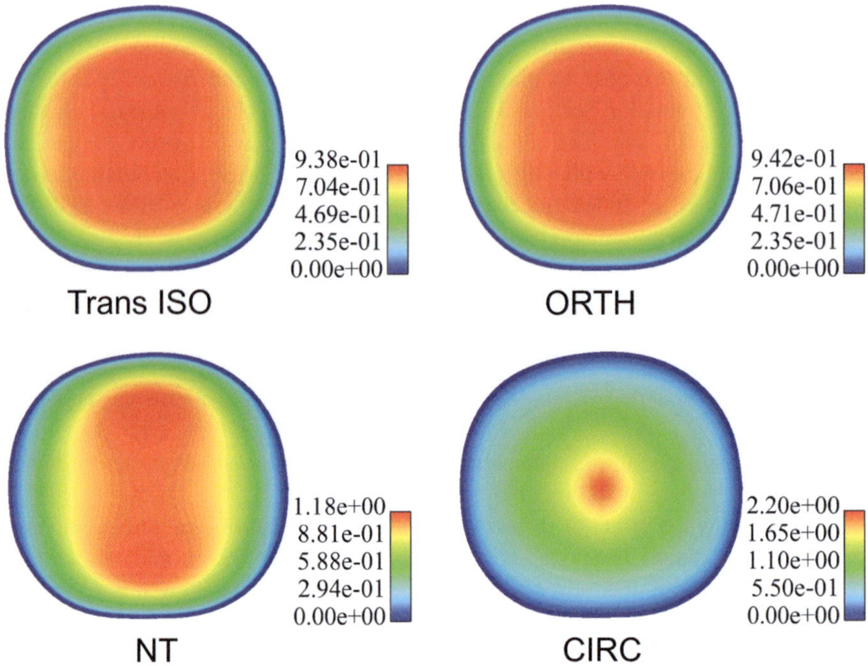

Fig. 14.11 Contours of the displacement magnitude calculated at 53 mmHg for different model collagen structures in the central bovine cornea. In all cases, the collagen fibers are highly oriented in the circumferential orientation and along the model preferred orientations of the central cornea. The model fiber structures are transversely isotropic (Trans ISO), orthogonal (ORTH), nasal-temporal (NT), inferior–superior (IS), and circumferential (CIRC). Reproduced from Nguyen and Boyce 2011 with permission of copyright holder

14.4.2 The Anisotropic Peripapillary Sclera

The peripapillary region of the sclera lies immediately adjacent to the optic nerve head. It provides the structural support for the lamina cribrosa, which is a thin fenestrated membrane that guides the optic nerve axons away from the eye. WAXS measurements show that the peripapillary sclera has a high degree of collagen alignment in the circumferential direction, which provides mechanical reinforcements against the stress concentration caused by the presence of the more compliant optic nerve head (Pijanka et al. 2012). The region displays a large variation in the degree of fiber alignment. The highest degree of fiber alignment in the peripapillary sclera of normal human eyes occurs in the superior-temporal quadrant, while the lowest degree of fiber alignment occurs in the superior-nasal quadrant. The degree of fiber alignment decreases dramatically away from the peripapillary sclera and the preferred fiber orientation also becomes more spatially heterogeneous in the midposterior region. To study the effects of the anisotropic structure on the deformation of the optic nerve head, the author and coworkers

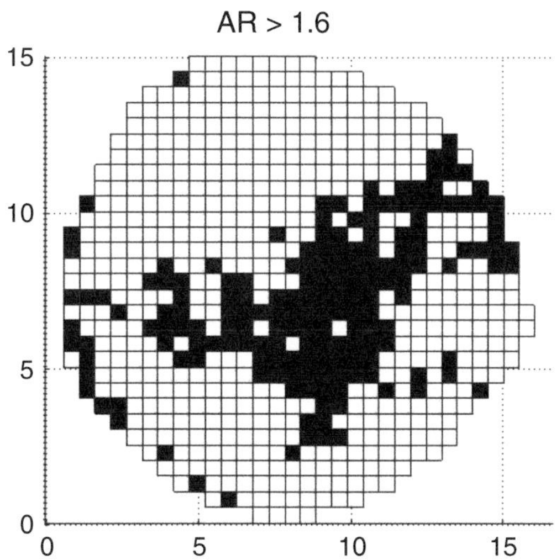

Fig. 14.12 Model fiber structure for a threshold value of the anisotropic ratio $AR = 1.6$. The *black squares* indicate regions where the PDF was obtained directly from WAXS measurements, while the *white squares* were regions with an isotropic PDF. Reproduced from Coudrillier et al. (2013) with permission from copyright holder

developed finite element models for specimens of the posterior sclera characterized by inflation and WAXS experiments (Coudrillier et al. 2013). The distributed fiber model in Sect. 14.3.1 was used to describe the mechanical behavior of the tissue, where the WAXS data was used directly to represent the PDF of the collagen orientation. The remaining model parameters for the collagen fiber and matrix were determined using IFEA to fit to DIC measured displacement field of the inflation experiments.

The effects of the spatially heterogeneous degree of alignment were examined by comparing the optic nerve head deformations calculated for different model representations of the collagen structure to those calculated for the physiological collagen structure measured by WAXS. Specifically, the simulations calculated the scleral canal expansion, posterior bowing of the lamina cribrosa, and maximum tensile principal and shear strain in the lamina cribrosa. The model representations of the collagen structure were created by progressively filtering out the less anisotropic features of the WAXS measurements based on a threshold criterion for the anisotropic ratio in Fig. 14.8. Increasing the anisotropic threshold from 1.0 (WAXS fiber structure) to 1.6 converted 71 % of the tissues of the posterior sclera to a transversely isotropic material (Fig. 14.12); however, this did not noticeably alter the deformation response of the optic nerve head (Figs. 14.13 and 14.14). Increasing the filtering threshold for the anisotropic ratio beyond 1.6 began to convert the peripapillary region to a transversely isotropic material. This caused a significant decrease in the posterior bowing of the lamina cribrosa (Fig. 14.13), a significant increase in the scleral canal expansion Fig. 14.14, and significant increases in the maximum principal and shear strains of the lamina cribrosa (Coudrillier et al. 2013). The results confirmed the conclusions of the previous micromechanical studies for the midposterior sclera (Sect. 14.3.1) that large variations in the fiber dispersion and preferred fiber orientation at the 0.5 mm length scale of the WAXS measurements

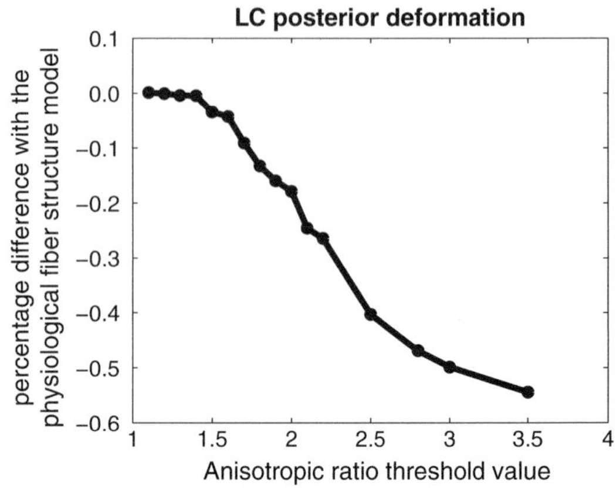

Fig. 14.13 The posterior deformation of the lamina cribrosa for different threshold values of the anisotropic ratio. Reproduced from Coudrillier et al. (2013) with permission from copyright holder

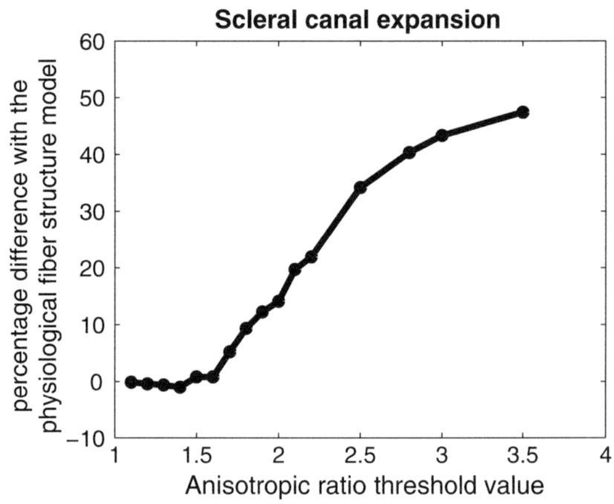

Fig. 14.14 The scleral canal expansion for different threshold values of the anisotropic ratio. Reproduced from Coudrillier et al. (2013) with permission from copyright holder

translate to a transversely isotropic behavior at the tissue length scale. In contrast, the anisotropy of the peripapillary sclera plays a major role in the mechanical behavior of the tissues of the optic nerve head. Without it, the tissues of the optic nerve head would experience larger tensile and shear strains from the increased hoop expansion of the scleral canal. Furthermore, spatial variation in the degree of fiber alignment in the peripapillary sclera has a significant effect on the deformation response of the ONH. The sclera of glaucoma eyes have an altered collagen structure (Pijanka et al. 2012), where the maximum fiber anisotropy no longer consistently resides in the superior-temporal sector. These alterations in the fiber structure of the peripapillary sclera may significantly alter the biomechanical environment of the optic nerve head, elevating the susceptibility to the development of glaucoma damage and spurring the progression of the disease.

14.5 Future Outlook

Tremendous advancements in experimental methods have been made in the past decade that has enabled spatial mapping of the collagen structure and mechanical properties of the cornea and sclera. These have in turn propelled advancements in computational modeling of the structure–function relationship of the tissues. The state of the art in computational models for the sclera now includes specimen-specific geometries, maps of the fiber structure, and material properties. Despite these recent advances, many challenges remain in experimental characterization and computational modeling of the tissues of the eye wall. Current understanding of the tissue structure and material behavior remains essentially 2D because experimental methods such as ESPI, stereo DIC, and WAXS provide either surface or through-thickness averaged measurements. The trend is moving towards 3D volume characterization methods for the tissue structure and mechanical properties. WAXS measurement of the sectioned specimens of the central cornea shows a gradient in the degree of fiber alignment through the thickness (Abahussain et al. 2009), while multiphoton laser scanning microscopy using second harmonic generated (SHG) signals reveals significant interweaving of the collagen fibers in the anterior third of the central cornea (Winkler et al. 2011) and posteriorly in the peripapillary sclera (Pijanka et al. 2012). The degree of interweaving remains unknown and current experiments provide more qualitative than quantitative information regarding the 3D fiber structure. Lamellar interweaving enhances the out-of-plane tensile and shear stiffness as well as the toughness of the tissues. These feature may be important to the mechanical response and function of the tissues. Similar efforts are underway to characterize the depth-dependent mechanical properties and deformation response of the tissues to mechanical loading. Petsche et al. (2012) measured the shear modulus of sectioned specimens of the central cornea and found a gradient in the shear modulus through the thickness that was consistent with the SHG measurements of Winkler et al. (2011) of the 3D collagen structure. New methods such as optical coherence elastography (Ford et al. 2011), ultrasonic elastography (Tang and Liu 2012), and digital volume correlation (Girard 2012) hold great promise to provide 3D maps of the deformation response in the tissue volume to inflation testing. The ability to characterize through-thickness variation in structure and deformation response is needed to accurately determine the material properties of the tissues of the eye wall in critical regions such as the peripapillary sclera, that display large and irregular three-dimensional variations in anatomic and microstructural features.

The development of inflation methods has focused on obtaining higher resolution and more accurate strain mapping. High resolution and accurate methods are also needed to map the local stress response and determine the variations in material properties. The available methods currently are IFEA and analytical models for thin membranes subjected to uniform pressure loading. The former is computationally intensive and unfeasible for a large number of parameters. For highly nonlinear problems, IFEA may not return a unique set of parameters. Moreover, the method requires accurate description of the tissue structure. Inaccuracies in the specimen geometry will manifest in the converged parameters returned by the IFEA. Analytical membrane models allow evaluation of a large number specimens for statistical comparison. The application of membrane models is limited to regions away from the boundaries, which in the sclera precludes analysis of the peripapillary region, and the accuracy of the models is determined by the accuracy of the local curvature calculations. Calculating the principal curvatures requires double differentiation of the displacement data, which amplifies the uncertainties and noise in the displacement measurements.

Microstructural approaches have been applied successfully to model the mechanical anisotropy that originates from the collagen fiber structure. Similar approaches are needed to model the effect of elastin reinforcement in the sclera. Elastin is heavily concentrated and highly oriented in the peripapillary region, yet its influence on the stress response is not well understood. Most models for the collagen anisotropy continue to rely on phenomenological exponential models for the mechanical behavior of the collagen fibers for expediency. The physical significance of the model parameters is unclear, which makes it difficult to interpret the significance of alterations in the material properties with aging and disease development. Both the cornea and sclera exhibit viscoelastic behavior, but few viscoelastic models have been developed for the tissues. More fundamentally, the deformation mechanisms underlying the strain-stiffening stress response and viscoelastic behavior are poorly understood. Advances in structural models require development of experimental methods to characterize the deformation of the collagen fibril and proteoglycans relative to the tissue under applied mechanical loads, as well as the fibril-fibril and fibril–proteoglycan interactions.

Acknowledgements The author would like to thank Dr. Michael Sacks for his invitation to write this review. In addition, the author would like to acknowledge the support of the NEI Public Health Service Research Grants EY021500 and the National Glaucoma Research program of the BrightFocus Foundation.

References

Abahussain M, Hayes S, Cartwright NK, Kamma-Lorger CS, Khan JMY, Meek KM. Invest Ophthalmol Vis Sci. 2009;50:5159.
Aghamohammadzadeh H, Newton RH, Meek KM. Structure. 2005;12:249.
Alastrué V, Calvo B, Pena E, Doblaré M. J Biomech Eng. 2006;128:150.
Andreassen TT, Simonsen AH, Oxlund H. Exp Eye Res. 1980;31:435.
Asejczyk-Widlicka M, Śródka DW, Kasprzak H, Pierscionek BK. Eye. 2007;21:1087.
Bellezza AJ, Hart RT, Burgoyne CF. Invest Ophthalmol Vis Sci. 2000;41:2991.
Bisplinghoff JA, McNally C, Manoogian SJ, Duma SM. J Biomech. 2009;42:1493.
Boote C, Dennis S, Newton RH, Puri H, Meek KM. Invest Ophthalmol Vis Sci. 2003;44:2941. doi:10.1167/iovs.03-0131.
Boote C, Dennis S, Meek KM. J Struct Biol. 2004;146:359.
Boote C, Hayes S, Abahussain M, Meek KM. Invest Ophthalmol Vis Sci. 2006;47:901.
Boote C, Hayes S, Jones S, Quantock AJ, Hocking PM, Ingleheran CF, Ali M, Meek KM. J Struct Biol. 2008;161:1.

Boyce BL, Jones RE, Nguyen TD, Grazier JM. J Biomech. 2007;40:2367. doi:10.1016/j.jbiomech.2006.12.001.
Boyce BL, Jones RE, Nguyen TD, Grazier JM. Biomaterials. 2008;28:3896.
Bryant MR, Velinsky SA, Plesha ME, Clarke GP. CLO J. 1991;13:238.
Coleman BD, Gurtin ME. J Chem Phys. 1967;47:597.
Cortes DH, Lake SP, Kadlowec JA, Soslowsky LJ, Elliott DM. Biomech Model Mechanobiol. 2010;9:651.
Coudrillier B, Tian J, Alexander S, Myers K, Quigley HA, Nguyen TD. Invest Ophthalmol Vis Sci. 2012;53(4):1714. doi:10.1167/iovs.11-8009.
Coudrillier B, Boote C, Quigley H, Nguyen T. Scleral anisotropy and its effects on the mechanical response of the optic nerve head. In press 2013. doi:10.1007/s10237-012-0455-y.
Curtin BJ. Trans Am Ophthalmol Soc. 1969;75:417.
Dongqi H, Zeqin R. J Biomech. 1999;32:579.
Downs JC, Thomas KA, Hart RT, Burgoyne CF. Invest Ophthalmol Vis Sci. 2001;42(Suppl):S831.
Downs JC, Suh JK, Thomas KA, Belleza AJ, Burgoyne CF. J Biomech Eng. 2003;125:124.
Downs JC, Suh JKF, Thomas KA, Belleza AJ, Hart RT, Burgoyne CF. Invest Ophthalmol Vis Sci. 2005;46:540.
Eilaghi A, Flanagan JG, Tertinegg I, Simmons CA, Brodland GW, Ethier CR. J Biomech. 2010;43:1696.
Elsheikh A, Anderson K. Am J Ophthalmol. 2006;141:868.
Elsheikh A, Wang D, Brown M, Rama P, Campanelli M, Pye D. Curr Eye Res. 2007;32:11.
Elsheikh A, Wang D, Rama P, Campanelli M, Garway D. Curr Eye Res. 2008;33:205.
Elsheikh A, Geraghty B, Alhasso D, Knappett J, Campanelli M, Rama P. Exp Eye Res. 2010a;90:624.
Elsheikh A, Geraghty B, Rama R, Campanelli M, Meek KM. J Soc Interface. 2010b;51:1475.
Fazio MA, Grytz R, Bruno L, Girard MJA, Gardiner S, Girkin CA, Downs JC. Invest Ophthalmol Vis Sci. 2012;53:5326.
Federico S, Herzog W. J Biomech. 2008;41:3309.
Ford MR, Dupps WJ, Rollins AM, Roy AS, Hu Z. J Biomed Opt. 2011;16:016005.
Friberg TR, Lace JW. Exp Eye Res. 1988;47:429.
Fung YC. Biomechanics: mechanical properties of living tissues. New York, NY: Springer; 1993.
Gasser TC, Ogden RW, Holzapfel GA. J R Soc Interface. 2006;3:15.
Gelman S, Cone FE, Pease ME, Nguyen TD, Myers K, Quigley HA. Exp Eye Res. 2010;90(2):210. doi:10.1016/j.exer.2009.10.007.
Girard MJA. ASME summer bioengineering conference; 2012.
Girard M, Suh JKF, Hart RT, Burgoyne CF, Downs JC. Curr Eye Res. 2007;32:465470.
Girard MJA, Downs JC, Burgoyne CF, Suh JF. J Biomech Eng. 2009a;131:051012.
Girard MJA, Suh JKF, Bottlang M, Burgoyne CF, Downs JC. Invest Ophthalmol Vis Sci. 2009b;50:5226. doi:10.1167/iovs.08-3363.
Girard MJA, Downs JC, Burgoyne CF, Suh JF. J Biomech Eng. 2009c;131:051011.
Girard MJA, Suh JKF, Bottlang M, Burgoyne CF, Downs JC. Invest Ophthalmol Vis Sci. 2011;52:5656.
Grytz R, Meschke G. J Mech Behav Biomed Mater. 2009;2:522.
Grytz R, Meschke G, Jonas JB. Biomech Model Mechanobiol. 2011;10:371.
Grytz R, Fazio MA, Girard MJ, Libertiaux V, Bruno L, Gardiner S, Girkin CA, Downs JC. J Mech Behav Biomed Mater. In press 2013. doi:10.1016/j.jmbbm.2013.03.027.
Hanna KD, Jouve FE, Waring GO. Arch Ophthalmol. 1989;107:911.
Hayes S, Boote C, Lewis J, Sheppard J, Abahussain M, Quantock AJ, Purslow C, Votruba M, Meek KM. Anat Rec. 2007;290:1542.
Hjortdal JO. J Biomech. 1996;29:931.
Hoeltzel DA, Altman P, Buzard K, Choe KI. J Biomech Eng. 1992;114:202.
Hommer A, Fuchsjäger-Mayrl G, Resch H, Vass C, Garhofer G, Schmetterer L. Invest Ophthalmol Vis Sci. 2008;49:4046.
Jue W, Maurice DM. J Biomech. 1991;24:907.

Kamma-Lorger CS, Boote C, Hayes S, Moger J, Burghammer M, Knupp C, Quantock AJ, Sorensen T, Di Cola E, White N, Young RD, Meek KM. J Struct Biol. 2010;169:424.
Kampmeier J, Radt B, Birngruber R, Brinkmann R. Cornea. 2000;19:355.
Kaufman PL, Alm A. Adler's physiology of the eye: clinical applications. 10th ed. St. Louis, MO: Mosby; 2003.
Ke XD, Schreier HW, Sutton MA, Wang YQ. Exp Mech. 2011;51:423.
Lanir Y. J Biomech. 1983;16:1.
Lari D, Schultz D, Wang A, Lee O, Stewart J. Exp Eye Res. 2012;94:128.
Limbert G, Middleton J. Int J Solids Struct. 2004;2004:4237.
Maurice DM. J Physiol. 1957;136:263.
McBrien NA, Gentle A. Prog Ret Eye Res. 2003;22:307.
McBrien NA, Lawlor P, Gentle A. Invest Ophthalmol Vis Sci. 2001;42:2179.
Meek KM, Boote C. Prog Retin Eye Res. 2009;28:369.
Meek KM, Tuft SJ, Huang Y, Gill PS, Hayes S, Newton RH, Bron AJ. Invest Ophthalmol Vis Sci. 2005;46:1948.
Myers KM, Cone FE, Quigley HA, Gelma S, Pease ME, Nguyen TD. Exp Eye Res. 2010a;91(6):866. doi:10.1016/j.exer.2010.09.009.
Myers K, Coudrillier B, Boyce BL, Nguyen TD. Acta Biomater. 2010b;6(11):4327. doi:10.1016/j.actbio.2010.06.007.
Nguyen TD. In: Garikipati K, Arruda EM, editors. IUTAM symposium on cellular, molecular and tissue mechanics. Volume 16. IUTAM book series. Netherlands: Springer; 2010. p. 19–29. doi:10.1007/978-90-481-3348-2_2.
Nguyen TD, Boyce BL. Biomech Model Mechanobiol. 2011;10(3):323. doi:10.1007/s10237-010-0237-3.
Nguyen TD, Jones RE, Boyce BL. Int J Solids Struct. 2007;44:8366. doi:10.1016/j.ijsolstr.2007.06.020.
Nguyen TD, Jones RE, Boyce BL. J Biomech Eng. 2008;130(4):041020. doi:10.1115/1.2947399.
Nguyen C, Cone FE, Nguyen TD, Coudrillier B, Pease ME, Steinhart MR, Oglesby EN, Jefferys JL, Quigley HA. Invest Ophthalmol Vis Sci. 2013;54(3):1767. doi:10.1167/iovs.12-10952.
Nyquist GW. Exp Eye Res. 1968;7:183.
Pandolfi A, Holzapfel G. J Biomech Eng. 2008;130:061006.
Pandolfi A, Manganiello F. Biomech Model Mechanobiol. 2006;5:237.
Pandolfi A, Vasta M. Mech Mater. 2012;44:151.
Pandolfi A, Fotia G, Manganiello F. Eng Comput. 2009;25:15.
Phillips JR, McBrien NA. Ophthalmic Physiol Opt. 1995;15:357.
Phillips JR, Khalaj M, McBrien NA. Invest Ophthalmol Vis Sci. 2000;41:2028–34.
Pijanka JK, Coudrillier B, Ziegler K, Sorensen T, Meek KM, Nguyen TD, Quigley HA, Boote C. Invest Ophthalmol Vis Sci. 2012;53(9):5258. doi:10.1167/iovs.12-9705.
Pinsky PM, van der Heide D, Chernyak D. J Cataract Refract Surg. 2005;31:136.
Pioletti DP, Rakotomanana LR, Benvnuti JF, Leyvraz PF. J Biomech. 1998;31:753.
Petsche SJ, Chernyak D, Martiz J, Levenston ME, Pinksy PM. Invest Ophthalmol Vis Sci. 2012;53:873.
Quigley HA, Dorman-Pease M, Brown AE. Curr Eye Res. 1991;10:877.
Quigley HA, Brown A, Dorman-Pease ME. Br J Ophthalmol. 1996;75:552.
Raghupathy R, Barocas V. J Biomech. 2009;42:1424.
Reese S, Govindjee S. Int J Solids Struct. 1998;35:3455.
Ruberti JW, Roy AS, Roberts CJ. Annu Rev Biomed Eng. 2011;13:269. doi:10.1146/annurev-bioeng-070909-105243.
Sacks MS. J Biomech Eng. 2003;125:280.
Shin TJ, Vito RP, Johnson L, McCarey BE. J Biomech. 1997;30:497.
Siegwart JT, Norton TT. Vision Res. 1999;39:387.
Sigal IA, Flanagan JG, Tertinegg I, Ethier CR. Invest Ophthalmol Vis Sci. 2004;45:4378.
Sigal IA, Flanagan JG, Tertinegg I, Ethier CR. Technol Health Care. 2005;13:313.

Spencer AJM . In: Eringen A, editor. Volume I. Continuum physics. New York: Academic; 1971. p. 239–53.
Steinhart MR, Cone FE, Nguyen C, Nguyen TD, Pease ME, Puk O, Graw J, Ogelsby EN, Quigley HA. Mol Vision. 2012;18:1093.
Studer H, Larrea X, Riedwyl H, Buchler P. J Biomech. 2010;43:836.
Sutton MA. Image correlation for shape, motion, and deformation measurements: basic concepts, theory and application. Berlin: Springer; 2009
Sutton MA, Ke X, Lessner SM, Goldbach M, Yost M, Zhao F, Schreier HW. J Biomed Mater Res Part A. 2007;84A:178.
Tang J, Liu J. J Biomech Eng. 2012;134:091007.
Thornton IL, Dupps WJ, Roy AS, Krueger RR. Invest Ophthalmol Vis Sci. 2009;50:1227.
Tomoshenko SP. Theory of plates and shells. New York: McGraw-Hill; 1959.
Tonge TK, Voo LM, Nguyen TD. Acta Biomater. 2013;9(4):5926. doi:10.1016/j.actbio.2012.11.034.
Watson PG, Young RD. Exp Eye Res. 2004;78:609. doi:10.1016/S0014-4835(03)00212-4.
Winkler M, Chai D, Krilling S, Nien CJ, Brown DJ, Jester B, Juhasz T, Jester JV. Invest Ophthalmol Vis Sci. 2011;52:8818.
Wollensak G, Spoerl E, Seiler T. J Cataract Refract Surg. 2003;29:1780.
Woo SLY, Kobayashi AS, Schlegel WA, Lawrence C. Exp Eye Res. 1972;14:29.
Yan DB, Coloma FM, Metheetrairut A, Trope GE, Heathcote JG, Ethier CR. Br J Ophthalmol. 1994;78:643.
Young RD. J Cell Sci. 1985;74:95.

Chapter 15
Mechanical Modeling of Skin

Cees Oomens and Gerrit Peters

Abstract The chapter describes the work that was performed in the soft tissue biomechanics laboratory at Eindhoven University of Technology on the biomechanics of skin. A rationale is given for the changes from standard testing methods to inverse methods, from in vitro to in vivo and back to in vitro testing and for the more detailed studies on individual skin layers of the last decade. The chapter tries to explain how our vision on testing methods and modeling changed over the years and why the pursuit towards a complete constitutive model is still ongoing.

15.1 Introduction

When Yoram Lanir did his postdoc in California in the lab of professor Y. C. Fung, he wrote two seminal papers about biaxial tests on rabbit skin (Lanir and Fung 1974a, b). The first paper described the equipment. The second paper extensively described the tests on rabbit skin. Their conclusion was that biaxial testing was a valuable technique, but that results had to be interpreted with caution. Repeatability was an issue, swelling and the difficulty to reach an equilibrium in relaxation tests also caused problems, but it appeared that the skin could be seen as an orthotropic, highly nonlinear material that could be described reasonably well with Fung's pseudo-elastic exponential strain energy equation. At that time, they were probably already thinking about developing microstructural models to create constitutive equations, eventually leading to Yoram's most cited paper from 1983 in the journal of biomechanics on constitutive equations for fibrous connective tissues. This was the first comprehensive paper on microstructural models for biological materials, including the distribution of collagen fibers as well as undulation. Many papers followed after that, with applications to other tissues like blood vessels and the brain. Yoram Lanir kept exploring microstructural models with an increasing level of sophistication, extending models to mixtures of solids and fluids and non-affine deformation of fibers. When we were asked to write a chapter for this book, we decided not to review the microstructural fibers models, because other authors have

C. Oomens (✉) • G. Peters
Eindhoven University of Technology, Eindhoven, The Netherlands
e-mail: c.w.j.oomens@tue.nl

G.S. Kassab, M.S. Sacks (eds.), *Structure-Based Mechanics of Tissues and Organs*,
DOI 10.1007/978-1-4899-7630-7_15

more experience on this, but we decided to focus on skin, the first biological material Yoram Lanir has worked on, and focus on methods that were used over the years to measure properties. In addition, we focus on one aspect of the microstructure; more specific the different top layers of the skin, i.e., the epidermis with a thickness of 100–150 μm and the stratum corneum (SC) with a thickness of 10 μm. Apart from a series of publications on SC in the 1960s and 1970s of the last century most studies on skin between 1970 and 2000 were on the tissue as a whole, which is dominated by the behavior of the dermis. In the last decennium, the interest is more focused on the very thin top layers and attempts are being made to determine the individual properties of those layers. The reason for this is an increased interest in the interaction of personal care products and medical instruments with the top layer of skin, but also because of new techniques for transepidermal drug delivery.

The chapter is not a comprehensive review of the literature, but rather a chronicle of the work that was done in our own lab in the last 30 years, including the change in vision over the years on how mechanical testing of biological materials has to be done, but we will give credits to work of others whenever that is opportune.

The biaxial test as performed by Yoram Lanir in the 1970s is an example of a standard mechanical test. A displacement or velocity field is enforced, which is independent of the material that is being tested and usually uniform in some part of the tissue specimen. This requires homogeneous properties for the sample (i.e., usually small samples) and a very specific shape of the sample. The force(s) required to obtain this field are measured. So force and displacement are known and the material behavior can be derived from this.

There were several reasons why in the 1980s researchers started with what is now known as inverse or numerical/experimental methods. The displacement field was no longer uniform or known but had to be measured too. This required a numerical model to describe the experiment and an iterative loop to fit the model on the experiment. The advantage was that there were less restrictions to make the samples and the method had the potential to be used in vivo. The disadvantage was the long computing times and difficulties to interpret the outcome of experiments. We started 25 years ago with these inverse methods and attempts to measure properties of skin and other biological tissues in vivo, returned to ex vivo studies and more or less standard testing on small samples in the last 10 years.

15.2 Inverse Methods to Characterize Materials

The incentive for the use of inverse methods in our lab came from studies that were done by Peters (1987) on collagenous connective tissues from the elbow region. Peters used an imaging technique to measure the strain distribution at long slender samples of connective tissue in a uniaxial stress test. It appeared that the strain field was far from homogeneous and the behavior was very much affected by a destruction of the internal coherence of the structure and the extreme anisotropy of the tissue (Fig. 15.1).

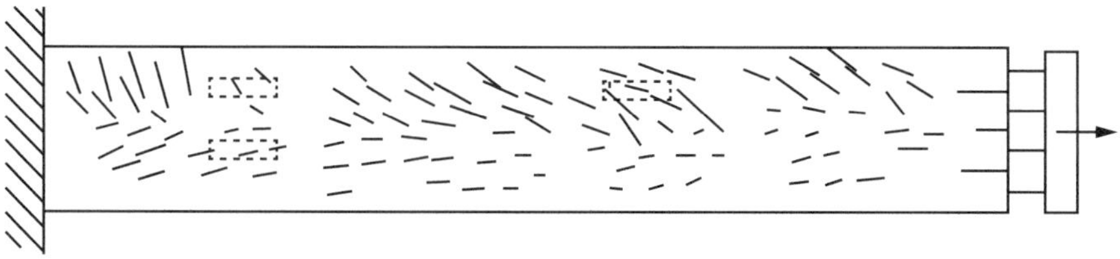

Fig. 15.1 Measured strain distribution on a specimen obtained from the overlaying fascia of the extensor digitorum muscle. Only the positive principle strains are shown (Peters 1987)

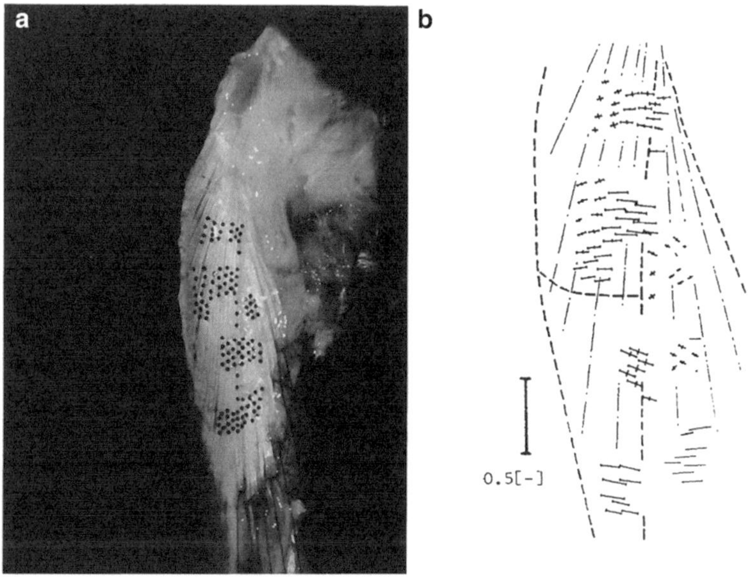

Fig. 15.2 (a) Image of a prepared arm with exposed *m.* extensor digitorum septum. The black dots are markers used to measure displacement. (b) Strain distributions derived from mechanically loaded tissue in in situ test (Peters 1987)

The solution to this problem was to use much larger samples so the internal coherence was maintained, and to measure the heterogeneous strain field and boundary load to gain as much information as possible so that even when the properties changed with position they could be measured (identified). This could be extended to in situ tests in which the internal coherence of the structure was preserved. An example of an in situ test with a structure that is left intact as much as possible is shown in Fig. 15.2. Figure 15.2a shows a prepared arm. Exposed is the septum extensor digitorum, which primarily consists of collagenous connective tissue. The black dots are markers that were used to measure a displacement field during loading by means of a two-camera system. From the marker displacements

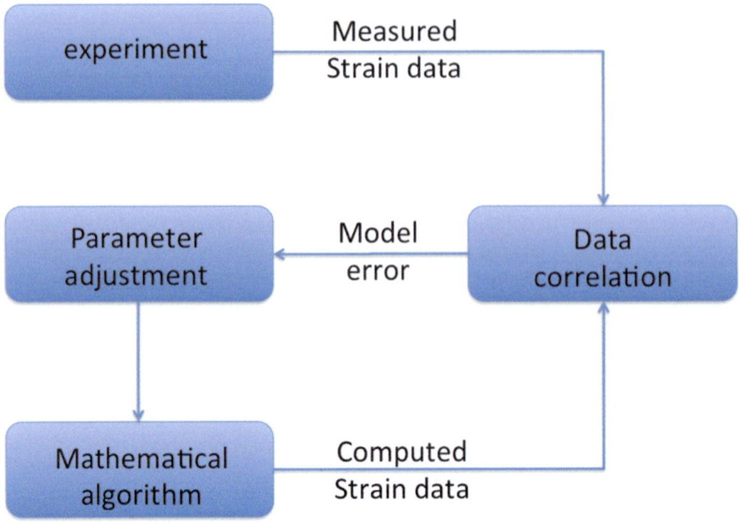

Fig. 15.3 Diagram for the inverse method to determine material parameters

strains can be derived. The principal strain distribution is given in Fig. 15.2b. The length of the bars corresponds to the magnitude of the principal strain in the direction of the bar.

By combining the measured displacement fields with detailed dedicated finite element models, it should be possible to determine the material properties. The idea of using displacement fields in this way was proposed earlier (Pister 1974; Distefano 1974; Yettram and Vinson 1979; Kavanagh 1972), but the required computer power and means to measure displacement fields with sufficient accuracy on biological materials was lacking for these pioneers.

Figure 15.3 shows a schematic of the method. An in vivo or in vitro experiment is set up and the strain field, resulting from some loading condition, is measured.

Options to measure displacement fields were Moiré, laser-speckle interferometry for small strains and methods using a grid of dots in combination with imaging techniques for larger displacements. Later Digital Imaging Correlation (Sutton et al. 1986) was the better option and for internal strain measurements Magnetic Resonance Imaging, using tagging (Alistair and Young 1996).

The measured strain field and boundary load was compared to the calculated strain field and boundary load based on an assumed constitutive model for the material that was tested and initial assumptions for the parameters. These initial assumptions were often determined from standard tests. The difference between the computed and measured strains and forces was used to update the parameters in an iterative loop until convergence was reached. Parameter adjustment was usually done by means of a weighted least squares algorithm and a Newtonian or Levenberg–Marquardt technique to find the best fit (Oomens et al. 1993).

Although the method was used successfully in a number of applications in our group (Meijer et al. 1999; Op den Camp et al. 1999; Hendriks et al. 2003, 2006; Cox et al. 2008; Dam 2007) and is still being used quite extensively, it did not live up completely to the original expectations. The hypothesis was that by adding more information to the experimental setup in the form of a full displacement field this would give indications about the (constitutive) model error of the system and clues on how to improve the model, more or less inspired by similar methods that are used to identify and control dynamic systems. However, a real understanding of how material behaves has to be obtained from a thorough study of the microstructure of the material and the physics of the interaction between the components. Unless the constitutive model is a proper description of these microstructural interactions, the numerical/experimental method does not add much to more standard techniques.

15.3 Mechanical Properties of the Stratum Corneum and Epidermis

Most studies on skin mechanics until 2000 were on total skin. It was assumed that the relatively thick dermis (compared to epidermis and stratum corneum) with its collagenous structure would dominate the mechanical behavior of the skin. Apart from a series of papers in the 1970s of the last century on stratum corneum the mechanical properties of the top layers of the skin were hardly studied separately. This situation changed because applications in cosmetics, personal care equipment, and clinical applications like transepidermal drug delivery required more knowledge on mechanical stiffness and strength and diffusion properties. Especially, the top 10–150 μm of the skin became an area of interest.

We attempted first to study these properties in vivo with noninvasive means and making use of a numerical experimental procedure. The technique used was the suction test. A suction device with different aperture sizes was used in combination with ultrasound and optical coherence tomography to measure the mechanical properties of the volar forearm skin of healthy volunteers. The shape and thickness changes of the top layers were measured while a negative pressure was applied to the skin by means of the suction device. The idea was that the aperture size of the suction device could be varied and, depending on the diameter, different layers would dominate the mechanical behavior.

The suction experiments combined with ultrasound allowed measuring the properties of the full skin (Hendriks et al. 2003).

By using optical coherence tomography, it became possible to measure the thickness change of the papillary layer of the skin. A two-layer finite element model of the experimental setup was developed using Neo-Hookean material for the upper layer (combined papillary layer and stratum corneum) and the reticular dermis, resulting in a very high difference in stiffness. The strain energy density function for a Neo-Hookean model is:

$$W = C_{10}\,(I_1 - 3) \tag{15.1}$$

with I_1 is the first invariant of the left Cauchy-Green tensor $\boldsymbol{B}$.

For the reticular dermis, $C_{10} = 0.16$ MPa was found and for the upper layers $C_{10} = 0.11$ kPa (Hendriks et al. 2006).

Next to this, Hendriks also developed a setup to measure internal strain fields in the epidermis by combining Digital Image Correlation (DIC) with in vivo confocal imaging, an attempt to design a noninvasive technique to characterize the epidermis (Hendriks 2005). Figure 15.3 shows a result with a reconstruction of displacements at different depths.

The severe limitations and difficulties of in vivo testing and better knowledge on skin preparation and preservation as a result of experiences with tissue engineering formed the incentive to design novel in vitro experiments to determine properties of stratum corneum and epidermis.

Geerligs et al. (2011a, b) developed two experimental setups: (1) micro-indentation of very thin skin samples and (2) a technique to measure shear properties of epidermis and stratum corneum (Fig. 15.4).

Indentation tests were carried out on ex vivo human skin. Epidermal sheets were obtained using a dermatome. The dermatomed slices of 100 μm thickness were cut into pieces of approximately 1 cm^2. To obtain stratum corneum (SC) samples, dermatomed skin slices of 200 μm were immersed in a solution of 0.1 % trypsin in an incubator at 37 °C for 2–3 h. After that, the sheets were rinsed in PBS and also cut into pieces of approximately 1 cm^2.

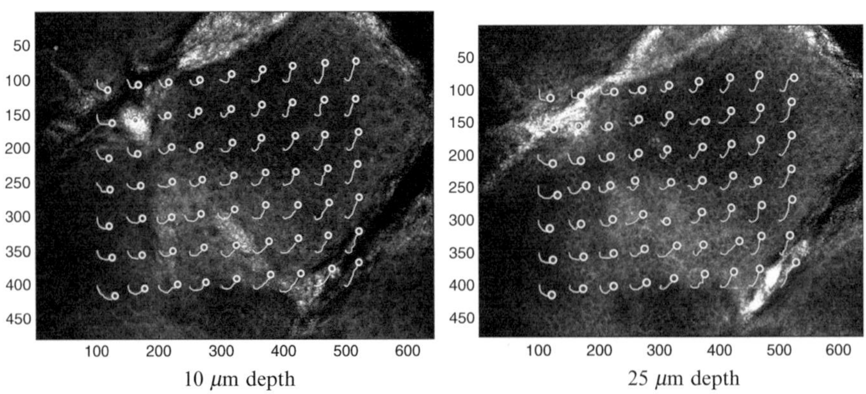

Fig. 15.4 Calculated displacements at different depths under the skin surface during shear loading. *Circles* represent the final position of a point and the lines represent the followed track. The pad left of the images was moved downwards, the pad right of the images was moved upwards. Pixel numbers are shown along the axes. Image size is 0.6–0.4 mm (Hendriks 2005)

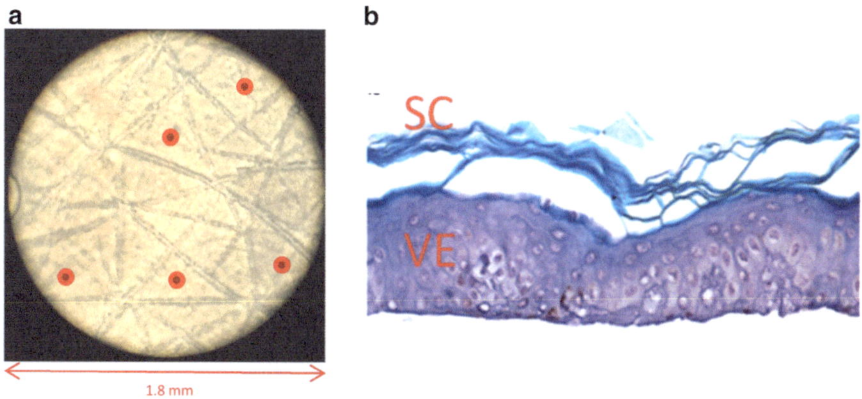

Fig. 15.5 (**a**) The top center of the triangles, highlighted by the large *red* points, formed by the glyphics was chosen as indentation location on the skin samples. (**b**) Dermatomed skin with a set thickness of 100 μm consisting of the epidermal layer only (Geerligs 2010)

Micro-indentation experiments were applied, using a spherical tip with a diameter of 1 mm. The indentation distance and required force were measured with resolutions of 1 nm and 1 nN, respectively. A ramp load was applied of 0.2 mN for the SC samples and 1 mN for the epidermis. The loading/unloading rate was 0.01 mN/s with a hold phase of 30 s.

The load/displacement data were used to determine the Young's moduli by using an analytical solution by Oliver and Pharr (1992) and fine tuned by comparison with a numerical model of the indentation test. No significant differences in stiffness between the stratum corneum and viable epidermis were observed. Estimates for the Young's moduli were found to be 2.6 ± 0.6 and $1.1 + 0.2$ MPa for the stratum corneum and epidermis, respectively. The value for the epidermis was much higher than those found in vivo by Hendriks et al. (2006) (Fig. 15.5).

The shear tests were performed with a parallel plate rheometer on samples of stratum corneum and the viable epidermis. The method differs from standard testing methods. To increase the torque on the upper plate of the rheometer, the sample is placed in eccentric position (see Fig. 15.6), a technique that was first applied by van Turnhout et al. (2005) on skeletal muscle samples and later with success repeated on brain tissue (Hrapko et al. 2006) and aortic aneurism thrombus (Van Dam et al. 2006).

A setup was built around the sample that enabled control of the temperature and the relative humidity. Preliminary tests were performed to determine the linear viscoelastic range, the effect of normal loading on the sample and the time to reach equilibrium after changes of temperature and relative humidity. The study shows that reproducible results can be obtained for the shear properties of epidermis in an in vitro setup. The dynamic shear modulus for stratum corneum ranged from about 4–12 kPa, decreasing with increasing relative humidity. The values were considerably lower than the shear modulus value based on tensile Young's moduli

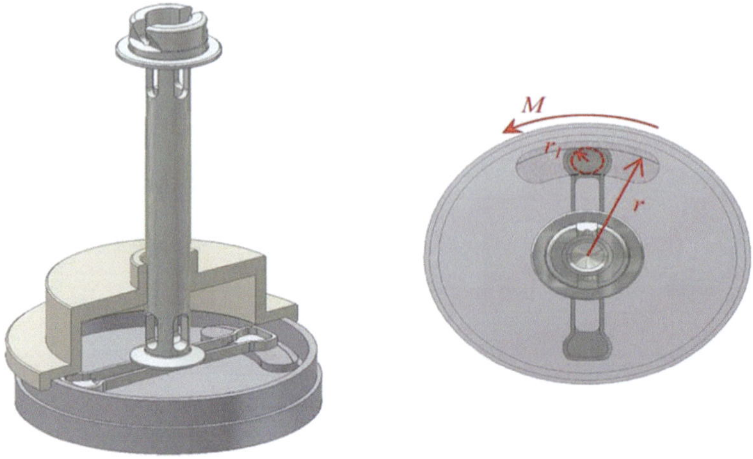

Fig. 15.6 Eccentric configuration for rotational shear experiments. A sample with radius *r*1 is rotated at a radius *r* with a torque *M*. The groove following the perimeter facilitated the positioning of the samples

in the literature, indicating a highly anisotropic material behavior. Results for the epidermis were of the same order of magnitude, but were less consistent possibly due to a less well-defined tissue composition. But these shear moduli were of the same order of magnitude as those found by Hendriks et al. (2006) in vivo.

Two years ago, we decided to extend the shear work to high deformations. Still interested in the properties of the individual layers we decided to work with a combination of the displacement distribution-based method that we developed and used in the 1990s and the more standard rheometer testing in the last decade. Porcine full thickness skin slices were used (1.2–1.6) to develop the method. Like in earlier experiments, the samples were eccentrically placed in a plate-plate rheometer setup and subjected to oscillatory shear strains up to 10 %. A random grey value pattern was sprayed on the side face of the sample by means of a high performance air brush. With a microscope and a high speed camera system, the deformation distribution of the side face of the skin was determined using DIC. To prevent slippage, the samples were glued to the plates.

Figure 15.7a shows an image of undeformed skin in the setup after it is sprayed with ink. Figure 15.7b shows the skin with an undeformed and deformed grid on top, as was reconstructed by means of DIC. From the reconstructed displacements the local shear strain was determined along a vertical line in the middle of the sample.

The result is given in Fig. 15.8. It is clear that the shear modulus in the epidermis is higher than in the dermis.

These are just preliminary data for a few samples of porcine skin, but it shows that in principle the method, which is a combination of the numerical/experimental technique we developed 20 years ago and the well-established standard rheometry, works fine. Currently, we are improving the method, and we apply it to human skin.

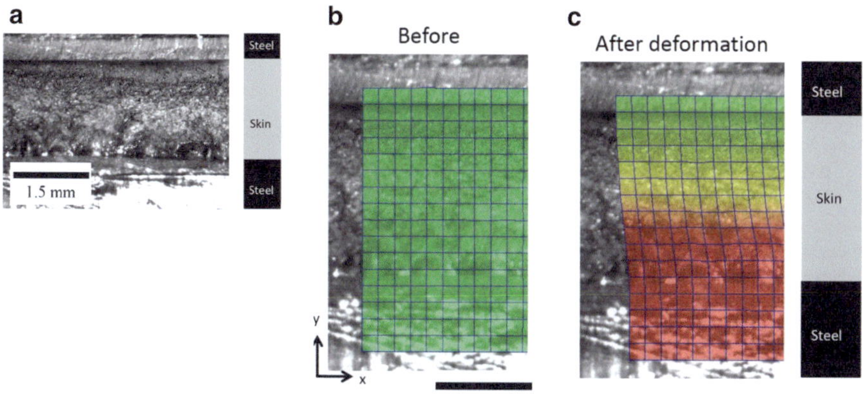

Fig. 15.7 (**a**) Undeformed configuration of porcine skin sample, (**b**) undeformed skin with a grid projected on the skin, and (**c**) deformed grid with a deformed grid based on digital image correlation projected upon the skin (adapted from Gerhardt et al. 2012)

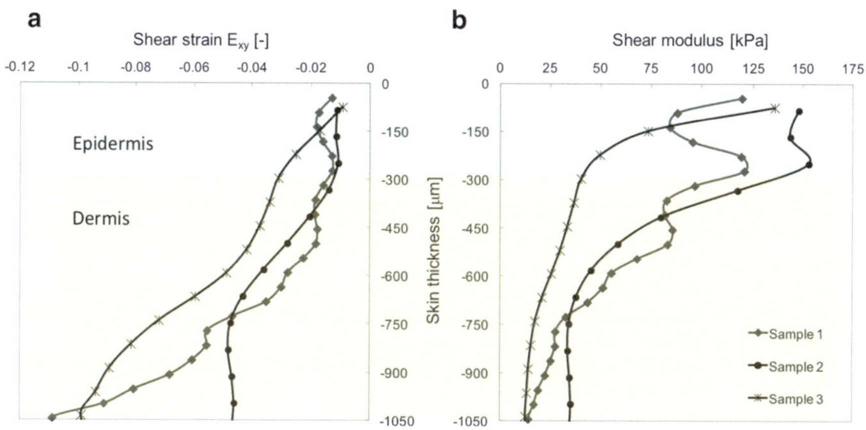

Fig. 15.8 (**a**) The shear strain for three different skin samples as derived from the images by means of Digital Image Correlation. (**b**) The local shear modulus for the three skin samples

15.4 Conclusion

Over the last year, our vision on how to characterize skin has evaluated from standard testing, via in vivo testing using numerical/experimental or inverse methods to a kind of mixture between standard tests and inverse methods, but with using well defined, small samples in in vitro setups. It has become clear that the top layers of the skin behave very different from the dermal layer and the behavior is highly anisotropic. Shear tests lead to orders of magnitude different results than tensile and indentation tests. An explanation for this has to be found in studying the microstructure and the interaction between different components.

References

Alistair A, Young ZAF. Right ventricular midwall surface motion and deformation using magnetic resonance tagging. Am J Physiol. 1996;270:281.

Cox MAJ, Gawlitta D, Driessen NJB, Oomens CWJ, Baaijens FPT. The non-linear mechanical properties of soft engineered biological tissues determined by finite spherical indentation. Comput Methods Biomech Biomed Engin. 2008;11(5):585–92.

Distefano N. Nonlinear processes in engineering. New York: Academic; 1974.

Geerligs M. Skin layer mechanics (Ph.D. thesis). Eindhoven: T. U. Eindhoven, 2010.

Geerligs M, van Breemen LCA, Peters GWM, Ackermans PAJ, Baaijens FPT, Oomens CWJ. In vitro indentation to determine the mechanical properties of epidermis. J Biomech. 2011a;44:1176–81.

Geerligs M, Oomens CWJ, Ackermans PAJ, Baaijens FPT, Peters GWM. Linear shear response of the upper skin layers. Biorheology. 2011b;48(3–4):229–45.

Gerhardt LC, Schmidt J, Sanz-Herrera JA, Baaijens FP, Ansari T, Peters GW, Oomens CW. A novel method for visualising and quantifying through-plane skin layer deformations. J Mech Behav Biomed Mater. 2012;14:199–207.

Hendriks FM. Mechanical behaviour of human epidermal and dermal layers (PhD-thesis). Eindhoven: Eindhoven University of Technology, 2005.

Hendriks FM, Brokken D, van Eemeren J, Oomens CWJ, Baaijens FPT, Horsten JBAM. A numerical-experimental method to characterize the non-linear mechanical behaviour of human skin. Skin Res Technol. 2003;9:274–83.

Hendriks FM, Brokken D, Oomens CWJ, Bader DL, Baaijens FPT. The relative contributions of different skin layers to the mechanical behaviour of human skin in vivo using suction experiments. Med Eng Phys. 2006;28(3):259–66.

Hrapko M, van Dommelen JA, Peters GW, Wismans JS. The mechanical behavior of brain tissue: large strain response and constitutive modeling. Biorheology. 2006;43:623–36.

Kavanagh KT. Extension of classical experimental techniques for characterizing composite material behavior. Exp Tech. 1972;12:55–6.

Lanir Y. Constitutive equations for fibrous connective tissues. J Biomech. 1983;16(1):1–12.

Lanir Y, Fung YC. Two-dimensional mechanical properties of rabbit skin—I. Experimental system. J Biomech. 1974a;7:29–34.

Lanir Y, Fung YC. Two-dimensional mechanical properties of rabbit skin II: experimental results. J Biomech. 1974b;7:171–82.

Meijer R, Douven LFA, Oomens CWJ. Characterisation of anisotropic and non-linear behaviour in human skin in-vivo. Comput Methods Biomech Biomed Engin. 1999;1:13–27.

Oliver WC, Pharr GM. An improved technique for determining hardness and elastic modulus using load and displacement sensing indentation experiments. J Mater Res. 1992;7:1564–83.

Oomens CWJ, Ratingen MRV, Janssen JD, Kok JJ, Hendriks MAN. A numerical-experimental method for a mechanical characterization of biological materials. J Biomech. 1993;26(4/5):617–21.

Op den Camp OMGC, Oomens CWJ, Veldpaus FE, Janssen JD. An efficient algorithm to estimate material parameters of biphasic mixtures. Int J Numer Methods Eng. 1999;45:1315–31.

Peters GWM. Tools for the measurement of stress and strain fields in soft tissue (PhD-thesis) Limburg: University of Limburg, 1987.

Pister KS. Constitutive modeling and numerical solution of field problems. Nucl Eng Des. 1974;28:137–46.

Sutton MA, Cheng M, Peters WH, Chao YJ, McNeill SR. Application of an optimized digital correlation method to planar deformation analysis. Image Vis Comput. 1986;4:143–50.

Van Dam EA. Morphology and mechanical properties of abdominal aortic aneurysms (PhD-thesis). Eindhoven: T. U. Eindhoven, 2007.

Van Dam EA, Dams SD, Peters GW, Rutten MC, Schurink GW, Buth J, van de Vosse FN. Determination of linear viscoelastic behavior of abdominal aortic magnitude physiologically relevant shear. Biorheology. 2006;43:695–707.
van Turnhout M, Peters G, Stekelenburg A, Oomens C. Passive transverse mechanical properties as a function of temperature of rat skeletal muscle in vivo. Biorheology. 2005;42:193–207.
Yettram AL, Vinson CA. Orthoptropic elastic moduli for left ventricular mechanical behavior. J Biomech. 1979;12:841–50.

Chapter 16
Undesirable Anisotropy in a Discrete Fiber Bundle Model of Fibrous Tissues

Cormac Flynn and M.B. Rubin

Abstract Lanir (J Biomech. 16(1):1–12, 1983) proposed a structural model for the anisotropic response of fibrous tissues with fiber bundles oriented in space by a continuous orientation distribution. Each fiber bundle was assumed to have the same undulation distribution that characterizes its nonlinear elastic response. Recently, a discrete fiber icosahedron model for fibrous soft tissues has been introduced, which is based on fiber bundles parallel to the six lines that connect opposing vertices of a regular icosahedron. Although the parameters in the icosahedron model can be determined to match experimental data for the anisotropic response of various tissues, the icosahedron model predicts anisotropic response when the weights of the six fiber bundles are equal. This chapter quantifies this undesirable anisotropic response and refers to a new icosahedron model based on a generalized invariant which also matches experimental data and analytically reduces to an isotropic form when the weights of the fiber bundles are equal.

16.1 Introduction

Lanir (1983) proposed a structural model for the anisotropic elastic response of fibrous tissues which was based on the idea that the tissue is a collection of fiber bundles that are characterized by continuous orientation and undulation distribution functions. More specifically, it was assumed that each fiber bundle is a collection of coiled or undulated fibers and that an individual fiber does not resist compression or extension when it is undulated. Consequently, it resists extension only when it is straight. Thus, the undulation distribution characterizes the nonlinear response of

C. Flynn (✉)
Centre for Engineering and Industrial Design, Wintec-Waikato Institute of Technology, Tristram St, Whitiora, Hamilton 3200, New Zealand
e-mail: Cormac.Flynn@wintec.ac.nz

M.B. Rubin
Faculty of Mechanical Engineering, Technion—Israel Institute of Technology, Haifa 32000, Israel
e-mail: mbrubin@tx.technion.ac.il

G.S. Kassab, M.S. Sacks (eds.), *Structure-Based Mechanics of Tissues and Organs*,
DOI 10.1007/978-1-4899-7630-7_16

the fiber bundle to stretching. Moreover, it was assumed that undulation distribution is independent of orientation with each fiber bundle exhibiting the same response to extension.

Within the context of this type of structural model the strain energy function is expressed as a double integral over the orientation and undulation distributions. Due to nonlinearity induced by general undulation distributions, it is usually not possible to evaluate these integrals analytically. A number of procedures for numerical integration over a sphere have been discussed in Bazant and Oh (1986), Ehret et al. (2010), and Itskov et al. (2010) which evaluate the integrand at a finite set of specific orientations and which cause varying degrees of unphysical anisotropy due to discretization.

Structural models with a finite collection of fibers have been used to study the response of low-density materials with open cells and fiber-dominated matrix composites (Christensen 1986; Christensen 1987). Models of this type that are based on orientations determined by opposing vertices of a regular icosahedron and of a dodecahedron (ten fibers) have been studied in Elata and Rubin (1994, 1995). Also, an icosahedron model for anisotropic response of fibrous soft tissue using six discrete fiber bundles oriented in the directions of opposing vertices of a regular icosahedron was recently considered in Flynn et al. (2011). Specifically, in Flynn et al. (2011) the strain energy function for each fiber bundle was assumed to be the same function of the stretch of the fiber bundle and the strain energy of the entire tissue was taken to be a weighted sum of the strain energies of each of the six fiber bundles in the discrete icosahedron model. Moreover, the strain energy function was determined by simple undulation distributions which ensure that the fiber bundle cannot be compressed. It was shown in Flynn et al. (2011) that the weights and the material parameters of the undulation distribution can be determined to match large deformation experimental data for the anisotropic response of various tissues. However, it was also noted in Flynn et al. (2011) that for the proposed undulation distributions, the tissue response was not isotropic even when the weights of the strain energy of each fiber bundle are the same. This means that unequal weights cannot be interpreted as the sole contribution to anisotropy.

The objective of this chapter is to analyze this undesirable anisotropy induced by fiber undulation distributions in a discrete icosahedron model similar to the one discussed in Flynn et al. (2011). An outline of this chapter is as follows. Section 16.2 describes a simplified icosahedron model for which the strain energy function of each fiber bundle is taken to be a function of the Lagrangian strain of the fiber bundle and not its stretch. As in Flynn et al. (2011), the strain energy of the entire tissue is a weighted sum of the strain energies of the specified fiber bundles. Section 16.3 uses the response to isochoric extension to quantify undesirable anisotropy caused by the nonlinearity of simple undulation distributions. The undesirable anisotropy caused by discreteness of the icosahedron model is also analyzed in Sect. 16.4 using a refined icosahedron model for which the strain energy function is an average of the strain energy function for N icosahedron models with different fiber orientations. Section 16.5 introduces a randomly oriented fiber model and Sect. 16.6 presents conclusions.

16.2 An Icosahedron Model of the Fiber Distribution

For an icosahedron model of the fiber distribution the six unit vectors $\mathbf{N}_i$ ($i = 1, 2, \ldots, 6$) that are parallel to the six lines connecting opposing vertices of a regular icosahedron are specified relative to the rectangular Cartesian base vectors $\mathbf{e}_i$ ($i = 1, 2, 3$) by the expressions (see Fig. 16.1)

$$\mathbf{N}_1 = \frac{2}{\sqrt{5}}\mathbf{e}_1 + \frac{1}{\sqrt{5}}\mathbf{e}_3, \mathbf{N}_2 = \frac{1}{2}\left(1 - \frac{1}{\sqrt{5}}\right)\mathbf{e}_1 + \sqrt{\frac{1}{2}\left(1 + \frac{1}{\sqrt{5}}\right)}\ \mathbf{e}_2 + \frac{1}{\sqrt{5}}\mathbf{e}_3,$$

$$\mathbf{N}_3 = -\frac{1}{2}\left(1 + \frac{1}{\sqrt{5}}\right)\mathbf{e}_1 + \sqrt{\frac{1}{2}\left(1 - \frac{1}{\sqrt{5}}\right)}\ \mathbf{e}_2 + \frac{1}{\sqrt{5}}\mathbf{e}_3,$$

$$\mathbf{N}_4 = -\frac{1}{2}\left(1 + \frac{1}{\sqrt{5}}\right)\mathbf{e}_1 - \sqrt{\frac{1}{2}\left(1 - \frac{1}{\sqrt{5}}\right)}\ \mathbf{e}_2 + \frac{1}{\sqrt{5}}\mathbf{e}_3,$$

$$\mathbf{N}_5 = \frac{1}{2}\left(1 - \frac{1}{\sqrt{5}}\right)\mathbf{e}_1 - \sqrt{\frac{1}{2}\left(1 + \frac{1}{\sqrt{5}}\right)}\ \mathbf{e}_2 + \frac{1}{\sqrt{5}}\mathbf{e}_3, \quad \mathbf{N}_6 = \mathbf{e}_3. \tag{16.1}$$

Moreover, it is convenient to define the symmetric structural tensors $\mathbf{B}_i$, such that

$$\mathbf{B}_i = \mathbf{N}_i \otimes \mathbf{N}_i \,(\text{no sum on } i = 1, 2, \ldots, 6), \tag{16.2}$$

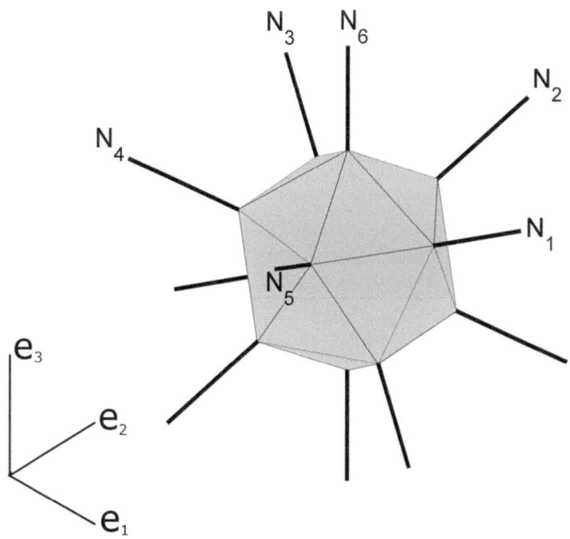

Fig. 16.1 Sketch of a regular icosahedron showing the vectors $\mathbf{N}_i$

where $\otimes$ denotes the tensor product operator. Then, using the work in Elata and Rubin (1994) it can be shown that for an arbitrary second order tensor $\mathbf{E}$

$$\sum_{i=1}^{6} \mathbf{B}_i = 2\,\mathbf{I}, \quad \sum_{i=1}^{6} \mathbf{B}_i \cdot \mathbf{E} = 2\,\mathbf{E} \cdot \mathbf{I}, \tag{16.3a, b}$$

$$\sum_{i=1}^{6} (\mathbf{B}_i \otimes \mathbf{B}_i) \cdot (\mathbf{E} \otimes \mathbf{E}) = \sum_{i=1}^{6} (\mathbf{B}_i \cdot \mathbf{E})^2 = \frac{2}{5}\left[(\mathbf{E} \cdot \mathbf{I})^2 + 2\,(\mathbf{E} \cdot \mathbf{E})\right], \tag{16.3c}$$

where $\mathbf{I}$ is the second order unity tensor, $\mathbf{A} \cdot \mathbf{B} = \mathrm{tr}\left(\mathbf{A}\mathbf{B}^{\mathrm{T}}\right)$ denotes the inner product between two second order tensors $\{\mathbf{A}, \mathbf{B}\}$ and (16.3c) generalizes the inner product operator for fourth order tensors.

Next, recall that a material point located by $\mathbf{X}$ in the reference configuration is deformed to the position $\mathbf{x}$ in the present configuration at time t. The mapping from the reference to present configurations, the deformation gradient $\mathbf{F}$, dilatation J, and Lagrangian strain $\mathbf{E}$ are given by

$$\mathbf{x} = \mathbf{x}\,(\mathbf{X},\ t)\,, \quad \mathbf{F} = \partial \mathbf{x}/\partial \mathbf{X}, \quad J = \det\,(\mathbf{F}) > 0, \quad \mathbf{E} = \frac{1}{2}\,(\mathbf{C} - \mathbf{I})\,, \quad \mathbf{C} = \mathbf{F}^{\mathrm{T}}\mathbf{F}. \tag{16.4}$$

Moreover, the Lagrangian strains E_i ($i = 1, 2, \ldots, 6$) of the material fibers in the $\mathbf{N}_i$ directions are defined by

$$E_i = \mathbf{E} \cdot \mathbf{B}_i\ (i = 1,\ 2, \ldots,\ 6)\,. \tag{16.5}$$

Now, for a compressible hyperelastic material the strain energy function W per unit mass for the icosahedron model is specified by

$$\rho_0 W = \sum_{i=1}^{6} w_i f\,(E_i), \quad w_i \geq 0, \quad \sum_{i=1}^{6} w_i = 1, \tag{16.6}$$

where ρ_0 is the reference mass density, the strain energy function f of each fiber bundle has the same form, and w_i are nonnegative weighting functions. Using the usual arguments it follows that the symmetric Piola–Kirchhoff stress $\mathbf{S}$ and the Cauchy stress $\mathbf{T}$ associated with (16.6) are given by

$$\mathbf{S} = \sum_{i=1}^{6} w_i \frac{df\,(E_i)}{dE_i} \mathbf{B}_i, \quad \mathbf{T} = J^{-1}\mathbf{F}\mathbf{S}\mathbf{F}^{\mathrm{T}}. \tag{16.7a, b}$$

For the simple case of a single fiber bundle it follows that

$$w_1 = 1 \quad \text{all other} \quad w_i = 0, \quad S_1 = \mathbf{S} \cdot \mathbf{B}_1 = \frac{df\,(E_1)}{dE_1}, \tag{16.8}$$

so that the stiffness K of the fiber bundle is given by

$$K = \frac{dS_1}{dE_1} = \frac{d^2 f(E_1)}{dE_1^2}. \tag{16.9}$$

As discussed by Lanir (1983), the collagen fiber bundles in soft connective tissues are typically coiled in the stress-free reference configuration and the stress response of each fiber bundle is characterized by an undulation distribution $D(x)$ which is normalized so that

$$\int_0^\infty D(x)\ dx = 1, \tag{16.10}$$

where the fraction of fibers that are straight at the strain E is given by

$$\int_0^E D(x)\ dx. \tag{16.11}$$

Furthermore, assuming that the stiffness of each collagen fiber in the bundle is constant E_c when the fiber is straight, the function f in (16.9) is determined by integrating the expression

$$\frac{d^2 f(E)}{dE^2} = E_c \int_0^E D(x)\ dx \quad \text{for} \quad E \geq 0. \tag{16.12}$$

In this expression it is tacitly assumed that the fiber in the bundle is coiled when it is compressed and that it makes no contribution to the stress when it is not straight ($E \leq 0$).

To investigate undesirable anisotropy caused by nonlinearity of the undulation distribution in the discrete icosahedron model, the weights are taken to be equal

$$w_i = \frac{1}{6}\, (i = 1,\ 2, \ldots,\ 6)\,, \tag{16.13}$$

and the strain energy function of the tissue is given by

$$\rho_0 W = \frac{1}{6} \sum_{i=1}^{6} f\,(E_i). \tag{16.14}$$

16.2.1 *Anisotropic Response Case I*

The simplest distribution considered in Flynn et al. (2011) is a step distribution that vanishes for $x \leq x_1$ and $x > x_2$ and is constant in the interval $x_1 \leq x \leq x_2$, such that

$$\begin{aligned} D(x) &= 0 \qquad \text{for } x < x_1 \text{ and } x > x_2, \\ D(x) &= \frac{1}{x_2 - x_1} \text{for } 0 \leq x_1 \leq x \leq x_2, \end{aligned} \tag{16.15}$$

where x_1 is a nonnegative constant that characterizes the strain when the first fiber in the fiber bundle becomes straight. It then follows that the solution $f_{\rm I}$ of (16.12) is given by

$$\begin{aligned} f(E) = f_{\rm I}(E) &= \frac{E_{\rm c}\langle E - x_1\rangle^3}{6\,(x_2 - x_1)} && \text{for } E \le x_2, \\ f(E) = f_{\rm I}(E) &= \frac{E_{\rm c}(x_2 - x_1)^2}{6} + \frac{E_{\rm c}\,(E - x_1)\ (E - x_2)}{2} && \text{for } E > x_2, \end{aligned} \tag{16.16}$$

where the Macaulay brackets $\langle x\rangle$ are defined by

$$\langle x\rangle = \frac{1}{2}\,(x + |x|)\,. \tag{16.17}$$

16.2.2 Isotropic Response

For the simple case when the stiffness of the collagen fiber bundle is constant $E_{\rm c}$ and the fiber is allowed to resist compression, the strain energy function f in (16.9) is given by

$$f(E) = \frac{E_{\rm c}E^2}{2}. \tag{16.18}$$

It follows from (16.3c), (16.5), and (16.14) that the associated strain energy function for the tissue is an isotropic function of the strain $\mathbf{E}$ given by

$$\rho_0 W = \frac{E_{\rm c}}{12}\sum_{\rm i=1}^{6} E_i^2 = \frac{E_{\rm c}}{30}\left[(\mathbf{E}\cdot\mathbf{I})^2 + 2\,(\mathbf{E}\cdot\mathbf{E})\right]. \tag{16.19}$$

16.2.3 Anisotropic Response Case II

In order to analyze the influence of the assumption that the fibers cannot support compression when they are coiled, the strain energy function (16.19) is modified to take the form

$$f(E) = f_{\rm II}(E) = \frac{E_{\rm c}}{2}\langle E - x_1\rangle^2, \quad \rho_0 W = \frac{E_{\rm c}}{12}\sum_{i=1}^{6}\langle E_i - x_1\rangle^2. \tag{16.20}$$

Figure 16.2 plots the functions $f_{\rm I}$ and $f_{\rm II}$ that characterize the strain energy of the fiber bundles associated with (16.16 and 16.20), respectively.

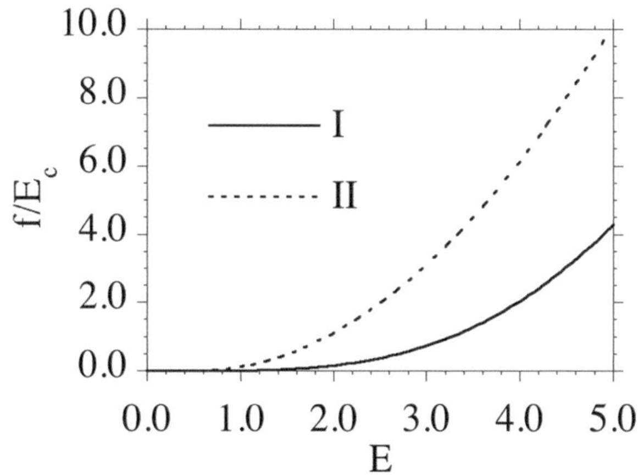

Fig. 16.2 Functions f_I and f_{II} characterizing the strain energy of fiber bundles

16.3 An Example of Isochoric Extension

In order to prove that a strain energy function characterizes isotropic response it is necessary to prove analytically that it depends on the strain $\mathbf{E}$ only through its invariants. In contrast, it is sufficient to consider a single numerical simulation to prove that a strain energy function exhibits undesirable anisotropy. To this end, it is convenient to define the right-handed orthonormal triad $\mathbf{a}_i$ in the reference configuration with $\{\mathbf{a}_1, \mathbf{a}_2\}$ being in the plane of the vectors $\{\mathbf{N}_1, \mathbf{N}_2 + \mathbf{N}_6\}$, such that

$$\mathbf{a}_1 = \mathbf{N}_1, \quad \mathbf{a}_2 = \mathbf{a}_3 \times \mathbf{a}_1, \quad \mathbf{a}_3 = \frac{\mathbf{a}_1 \times (\mathbf{N}_2 + \mathbf{N}_6)}{|\mathbf{a}_1 \times (\mathbf{N}_2 + \mathbf{N}_6)|}. \tag{16.21}$$

Moreover, it can be shown that the angle β between $\mathbf{a}_1$ and the vector $(\mathbf{N}_2 + \mathbf{N}_6)$ is given by

$$\beta = \cos^{-1}\left(\frac{\sqrt{2}}{\sqrt{5+\sqrt{5}}}\right) > 0. \tag{16.22}$$

Then, it is possible to define another right-handed orthonormal triad of vectors $\mathbf{A}_i$ parametrically in terms of the parameter α, such that

$$\begin{aligned} \mathbf{A}_1 &= \cos(\alpha\beta)\ \mathbf{a}_1 + \sin(\alpha\beta)\ \mathbf{a}_2, \quad \mathbf{A}_2 = -\sin(\alpha\beta)\ \mathbf{a}_1 + \cos(\alpha\beta)\ \mathbf{a}_2, \\ \mathbf{A}_3 &= \mathbf{a}_3, \qquad 0 \le \alpha < 1. \end{aligned} \tag{16.23}$$

Specifically, this causes $\mathbf{A}_1$ to rotate about the $\mathbf{A}_3$ axis from the orientation $\mathbf{a}_1$ to the vector $\mathbf{a}_2$ that is parallel to $(\mathbf{N}_2 + \mathbf{N}_6)$ as α ranges from zero to unity (see Fig. 16.3).

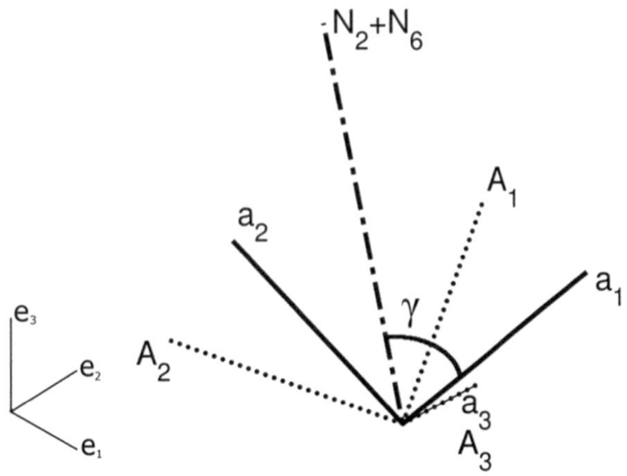

Fig. 16.3 Sketch of the angle γ and triads $\mathbf{a}_i$ and $\mathbf{A}_i$ of vectors characterizing the orientation of the sample of material that is being loaded in isochoric extension

Next, consider isochoric extension relative to $\mathbf{A}_i$ and specify $\mathbf{F}$ in the form

$$\mathbf{F} = a\,\mathbf{A}_1 \otimes \mathbf{A}_1 + \frac{1}{\sqrt{a}} \left(\mathbf{A}_2 \otimes \mathbf{A}_2 + \mathbf{A}_3 \otimes \mathbf{A}_3\right), \quad a > 0, \tag{16.24}$$

where a is the stretch of a material fiber in the $\mathbf{A}_1$ direction. It follows that this deformation field can be used to examine the response of samples of the material with different orientations in the reference configuration (characterized by the value of α) to the same isochoric extension (characterized by the value of a).

Using the deformation (16.24), it is possible to calculate the value of the strain energy as a function of $\{a, \alpha\}$. In particular, when the weights w_i are equal (16.13), the strain energy function (16.14) takes the values $\{W_{\mathrm{I}}, W_{\mathrm{II}}\}$, respectively, for the specifications (16.16) and (16.20) with

$$\rho_0 W_{\mathrm{I}}(a,\alpha) = \frac{1}{6}\sum_{i=1}^{6} f_{\mathrm{I}}(E_i), \quad \rho_0 W_{\mathrm{II}}(a,\alpha) = \frac{1}{6}\sum_{i=1}^{6} f_{\mathrm{II}}(E_i). \tag{16.25}$$

Therefore, the relative errors $\{ER_{\mathrm{I}}, ER_{\mathrm{II}}\}$ in these strain energy functions for a specific value of a and varying values of α can be defined by

$$ER_{\mathrm{I}}(\alpha) = \frac{W_{\mathrm{I}}(a,\alpha)}{W_{\mathrm{I}}(a,0)} - 1, \quad ER_{\mathrm{II}}(\alpha) = \frac{W_{\mathrm{II}}(a,\alpha)}{W_{\mathrm{II}}(a,0)} - 1. \tag{16.26a, b}$$

In the following example it will be shown that both of the models (16.16 and 16.20) predict undesirable anisotropy even though the weights w_i have been taken to be equal (16.13). Specifically, for the example specify

$$x_1 = 1, \quad x_2 = 4, \quad a = 3. \tag{16.27}$$

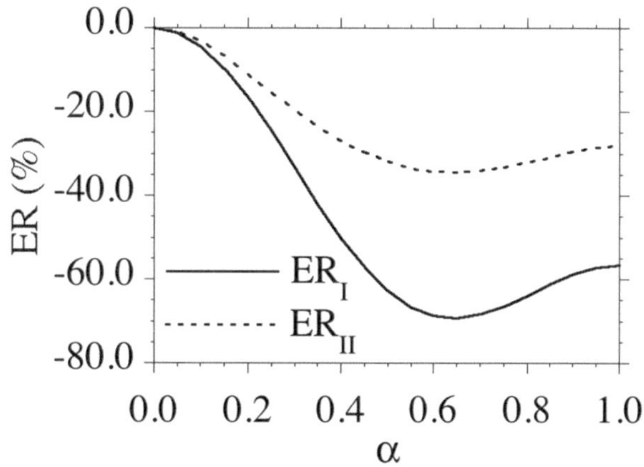

Fig. 16.4 Errors $\{ER_{\mathrm{I}}, ER_{\mathrm{II}}\}$ quantifying the undesirable anisotropy for the two icosahedron models characterized by the undulation distributions, (16.16) and (16.20), respectively

Figure 16.4 plots the errors $\{ER_{\mathrm{I}}, ER_{\mathrm{II}}\}$ as functions of α. The maximum magnitudes of these errors are about $\{69\,\%, 34\,\%\}$, respectively, for $\{ER_{\mathrm{I}}, ER_{\mathrm{II}}\}$. Since the values of these errors are nonzero, it follows that nonlinearity of the undulation distribution causes undesirable anisotropy in the discrete icosahedron model. Moreover, comparison of (16.16) with the isotropic strain energy (16.18) indicates that anisotropy predicted by (16.16) {i.e., ER_{I}} is due to both the cubic dependence on strain and the presence of the cutoff strain $E = x_1$. In contrast, comparison of (16.20) with the isotropic strain energy (16.18) indicates that anisotropy predicted by (16.20) {i.e., ER_{II}} is due solely to the cutoff strain $E = x_1$.

16.4 A Refined Icosahedron Model

In order to further analyze the influence of discreteness of the icosahedron model on undesirable anisotropy, it is convenient to define a refined icosahedron model. Within the context of the icosahedron model described in Sect. 16.2, it is necessary to evaluate the strain energy function for only six directions defined by the vectors $\mathbf{N}_i$ in (16.1). A refined icosahedron model can be obtained by defining the strain energy function as an average of $N = 6\{5(4)^{J-1} + 1\}$ icosahedron models with N structural tensors $\mathbf{B}_i$ associated with N fiber orientations. The value of J (=1, 2 ...) determines the level of refinement as discussed presently.

To this end, it is noted that vectors $\mathbf{N}_i$ in (16.1) can be used to define five equilateral triangles with the following triads of vectors locating the vertices of the triangles

$$\{\mathbf{N}_I,\ \mathbf{N}_{I+1},\ \mathbf{N}_6\}\ \text{for}\ I = 1,\ 2,\ldots,\ 4, \quad \{\mathbf{N}_5,\ \mathbf{N}_1,\ \mathbf{N}_6\}\ \text{for}\ I = 5, \tag{16.28}$$

each of which has the same vertex located by $\mathbf{N}_6$. These five triangles can be tessellated into 4^{J-1} equilateral triangles ($J = 1,\ 2,\ \ldots$) as shown in Fig. 16.5.

Fig. 16.5 Sketch of tessellation of an equilateral triangle into 4^{J-1} equilateral triangles

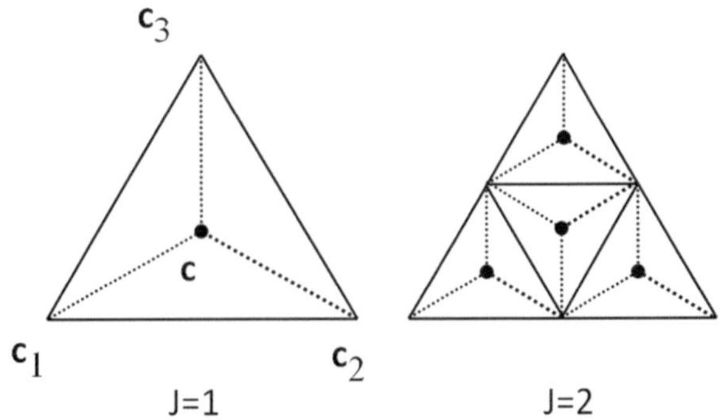

For example, a typical triangle has vertices located by the three unit vectors $\mathbf{c}_i$ ($i = 1$, 2, 3) and its centroid is located by the unit vectors $\mathbf{c}$ defined by

$$\mathbf{c} = \frac{\mathbf{c}_1 + \mathbf{c}_2 + \mathbf{c}_3}{|\mathbf{c}_1 + \mathbf{c}_2 + \mathbf{c}_3|}. \tag{16.29}$$

Next, the proper orthogonal rotation tensor $\mathbf{R}(\mathbf{c})$ is defined which rotates the vector $\mathbf{N}_6$ to the vector $\mathbf{c}$ counterclockwise by the angle δ about the unit direction $\mathbf{n}_3$, which is normal to the $\mathbf{N}_6$–$\mathbf{c}$ plane. Specifically, define the right-handed triad of vectors $\mathbf{n}_i$ by the expressions

$$\mathbf{n}_1 = \mathbf{N}_6, \quad \mathbf{n}_2 = \frac{\mathbf{c} - (\mathbf{c} \cdot \mathbf{n}_1)\,\mathbf{n}_1}{|\mathbf{c} - (\mathbf{c} \cdot \mathbf{n}_1)\,\mathbf{n}_1|}, \quad \mathbf{n}_3 = \mathbf{n}_1 \times \mathbf{n}_2. \tag{16.30}$$

Then, $\mathbf{R}(\mathbf{c})$ can be written in the form

$$\mathbf{R}(\mathbf{c}) = (\cos\delta\mathbf{n}_1 + \sin\delta\mathbf{n}_2) \otimes \mathbf{n}_1 + (-\sin\delta\mathbf{n}_1 + \cos\delta\mathbf{n}_2) \otimes \mathbf{n}_2 + \mathbf{n}_3 \otimes \mathbf{n}_3, \tag{16.31}$$

where the acute angle δ between the vectors $\mathbf{n}_1 = \mathbf{N}_6$ and $\mathbf{c}$ is given by

$$\delta = \cos^{-1}(\mathbf{n}_1 \cdot \mathbf{c}). \tag{16.32}$$

For each value of $\mathbf{R}(\mathbf{c})$ an additional icosahedron model is generated using the six structural tensors $\mathbf{B}_i$ ($i = 1, 2, \ldots, 6$) in (16.2) to obtain the following six additional structural tensors

$$\mathbf{R}(\mathbf{c})\mathbf{B}_i\mathbf{R}^T(\mathbf{c}) \quad \text{for } i = 1,\ 2, \ldots,\ 6. \tag{16.33}$$

The resulting refined model has N structural tensors $\mathbf{B}_i$, associated with N fiber orientations and the strain energy function is specified by

$$\rho_0 W = \rho_0 W_I(a,\ \alpha) = \frac{1}{N}\sum_{i=1}^{N} f_I(E_i), \quad E_i = \mathbf{E} \cdot \mathbf{B}_i \quad (i = 1,\ 2, \ldots,\ N). \tag{16.34}$$

In these expressions, E_i is the component of the Lagrangian strain of a fiber bundle in the direction $\mathbf{B}_i$, the function f_{I} associated with the undulation distribution of the fiber bundle is specified by (16.16), and the parameters $\{\alpha, a\}$ in (16.23) and (16.24) characterize the loading. This model has N fibers, which are equally weighted. In this regard, it should be emphasized that the definitions of the fiber orientations in the refined icosahedron model are different conceptually from orientations used to obtain numerical approximations of integrals over the unit sphere, which are weighted unequally in order to increase accuracy of integrating specific functional forms (e.g., Ehret et al. 2010; Bazant and Oh 1985). Moreover, it is noted that since $\mathbf{B}_i$ $(i = 1, 2, \ldots, 6)$ satisfy (16.3a) it follows that the refined icosahedron model has the symmetry that

$$\sum_{i=1}^{N} \mathbf{B}_i = \frac{3}{N}\mathbf{I}, \tag{16.35}$$

for any level of refinement J.

As an example, use is made of the specifications (16.27) and the error ER_{I} is defined by (16.6). Figure 16.6 shows predictions of the error ER_{I} for different values of refinement. Specifically, for $N = 6$ the results correspond to the simple icosahedron model described in Sect. 16.2; and the other predictions correspond to the refined icosahedron model with $N = 36$ for $J = 1$, $N = 126$ for $J = 2$, and $N = 486$ for $J = 3$. Also, Fig. 16.7 focuses attention on the results for the higher values of N. These results indicate that even for a relatively simple strain energy function a large number of fibers are needed to obtain nearly isotropic material response. At this point it is not clear why the error predicted by the refined icosahedron model does not reduce monotonically with increased refinement.

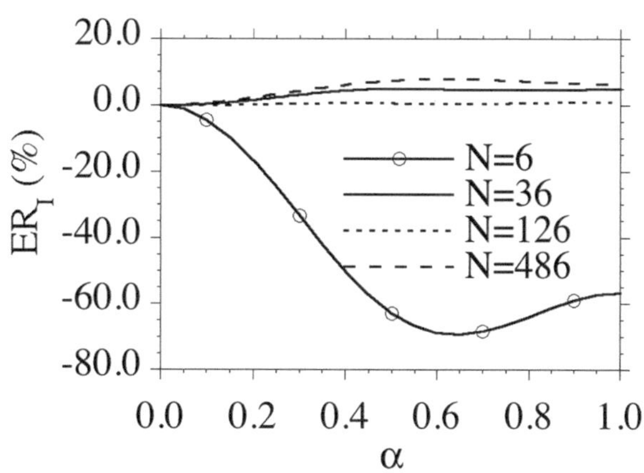

Fig. 16.6 Reduction in the error ER_{I} predicted by the refined icosahedron model for the values $N = 6$, 36, 126, 486 of the number of fiber bundles in the model

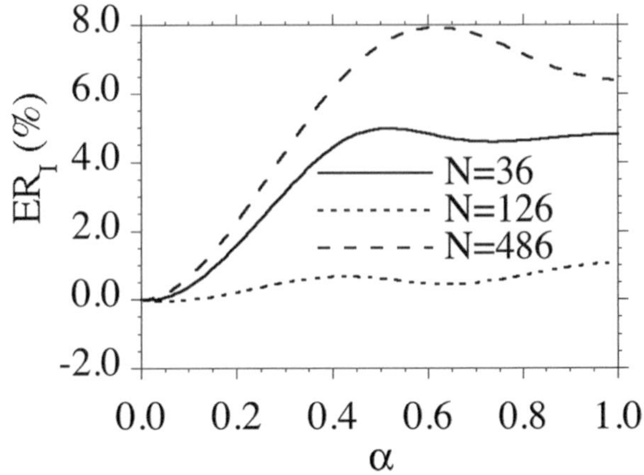

Fig. 16.7 Reduction in the error ER_I predicted by the refined icosahedron model for the higher values $N = 36$, 126, 486 of the number of fiber bundles in the model

16.5 Equal Area Model

It is well known that the regular polyhedron (Platonic solid) with the greatest number of faces is the regular icosahedron with 20 faces. Consequently, with regard to numerical integration schemes over the unit sphere, Bazant and Oh (1985) state that “we cannot have, for a hemisphere, a numerical integration formula with more than $N = 10$ regularly spaced points ....” Nevertheless, in this section it is of interest to consider a model based on N oriented fibers which locate the centroids of patches of a hemisphere that have equal areas. In particular, this model is used in conjunction with the refined icosahedron model of the Sect. 16.4 to help quantify the number of fibers needed to reduce the error due to unphysical anisotropy.

An approximate uniform distribution of fibers can be developed by dividing the surface area of a hemisphere into patches that have the same areas. Specifically, consider the unit vector $\mathbf{N}$ defined in terms of the spherical polar angles $\{\theta, \phi\}$ by

$$\mathbf{N} = \mathbf{N}(\theta, \phi) = \sin(\phi)\,[\cos(\theta)\,\mathbf{e}_1 + \sin(\theta)\,\mathbf{e}_2] + \cos(\phi)\,\mathbf{e}_3. \tag{16.36}$$

It follows that the upper surface of hemisphere is characterized by the ranges

$$0 \le \theta \le 2\pi, \quad 0 \le \cos(\phi) \le 1. \tag{16.37}$$

Moreover, the area A of a patch on a hemisphere with unit radius for $\{\theta, \phi\}$ in the ranges $\Theta_i \le \theta \le \Theta_{i+1}$ and $\Phi_j \le \phi \le \Phi_{j+1}$ is given by

$$A = (\Theta_{i+1} - \Theta_i)\left[\cos(\Phi_j) - \cos(\Phi_{j+1})\right]. \tag{16.38}$$

Consequently, the area of this hemisphere can be divided into $N = K^2$ equal areas by specifying $\{\Theta_i, \Phi_j\}$ in the forms

$$\Theta_i = \frac{2\pi\,(i-1)}{K} \quad \text{for } i = 1,\ 2, \ldots,\ K+1,$$

$$\Phi_j = \cos^{-1}\left(\frac{K+1-j}{K}\right) \quad \text{for } j = 1,\ 2, \ldots,\ K+1. \qquad (16.39)$$

Then, the values $\{\theta_i, \phi_j\}$ of $\{\theta, \phi\}$, which locate the centroids of these regions can be defined by

$$\theta_i = \frac{1}{2}\left(\Theta_i + \Theta_{i+1}\right) \quad \text{for } i = 1,\ 2, \ldots,\ K,$$

$$\phi_j = \cos^{-1}\left[\frac{1}{2}\left\{\cos\left(\Phi_j\right) + \cos\left(\Phi_{j+1}\right)\right\}\right] \quad \text{for } j = 1,\ 2, \ldots,\ K. \qquad (16.40)$$

Now, using these values $\{\theta_i, \phi_j\}$ the fibers are oriented in the N directions $\mathbf{N}_i$ defined by (16.36) with θ taking the K values θ_i for each of the K values of ϕ_j. Also, the associated structural tensors $\mathbf{B}_i$ are given by

$$\mathbf{B}_i = \mathbf{N}_i \otimes \mathbf{N}_i \,(\text{no sum on } i = 1,\ 2, \ldots,\ N)\,. \qquad (16.41)$$

Next, it is convenient to define the average structural tensor $\overline{\mathbf{B}}$ by the expression

$$\overline{\mathbf{B}} = \frac{1}{N}\sum_{i=1}^{N} \mathbf{B}_i. \qquad (16.42)$$

For fibers uniformly distributed over the hemisphere, it would be expected that this average structural tensor would be a scalar times the unity tensor $\mathbf{I}$. Consequently, an error measure of uniformity can be defined in terms of the relative magnitude of the deviatoric part $\overline{\mathbf{B}}'$ of $\overline{\mathbf{B}}$ defined by

$$ER_{\mathrm{B}} = \sqrt{\frac{9\overline{\mathbf{B}}' \cdot \overline{\mathbf{B}}'}{\left(\overline{\mathbf{B}} \cdot \mathbf{I}\right)^2}}, \quad \overline{\mathbf{B}}' = \overline{\mathbf{B}} - \frac{1}{3}\left(\overline{\mathbf{B}} \cdot \mathbf{I}\right)\mathbf{I} \qquad (16.43)$$

Due to the result (16.35), the refined icosahedron model will predict that ER_{B} vanishes for all levels of refinement.

For the equal area model the strain energy function $W_{\mathrm{I}}(a, \alpha)$ is specified by (16.34) and the error $ER_{\mathrm{I}}(\alpha)$ is specified by (16.26a) using the values (16.27) and the deformation (16.24). Figure 16.8 shows the predictions of $ER_{\mathrm{I}}(\alpha)$ for different values of the number N of fibers and Table 16.1 records the associated values of the error ER_{B} in (16.43). These results indicate that for $N = 49$ the equal area model is less accurate than the refined icosahedron model for $N = 36$. However, the monotonic error reduction with increasing values of N predicted by the equal area model suggests that the fibers are more uniformly distributed in the model than

Fig. 16.8 Reduction in the error ER_I predicted by the equal area model for different values of the number N of fibers

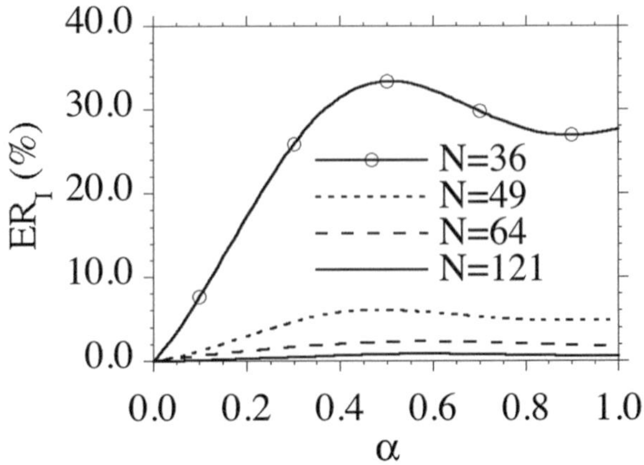

Table 16.1 Values of the error ER_B in (16.40) predicted by the equal area model

N	ER_B (%)
36	0.85
49	0.62
64	0.48
121	0.25

in the refined icosahedron model. Moreover, it is noted that the equal area model predicts the error due to unphysical anisotropy to be less than 3 % for $N = 64$. Even for this simple strain energy function, the computational effort required to evaluate the equal area model for $N = 64$ is significant since the constitutive equation must be evaluated at each Gauss point in a finite element program. An alternative model that significantly reduces the computational effort is discussed in the next section.

16.6 Conclusions

The structural model for anisotropic elastic response of fibrous connective tissue proposed by Lanir (1983) has the simplicity that the undulation distribution of fibers in each fiber bundle is independent of the orientation distribution. This suggests that the orientation distribution can be correlated to histological observations of fiber orientations. In particular, a random orientation of fibers should lead to isotropic response of the tissue. However, from a computational point of view it is necessary to discretize the evaluation of the integral over the orientation region. This discretization yields a finite number N of nonlinear strain energy functions (characterizing fiber bundles in specified orientations) that need to be evaluated for each strain at each material point.

The example in Sect. 16.4 considered a refined icosahedron model and the example in Sect. 16.5 considered an equal area model. For each of these models

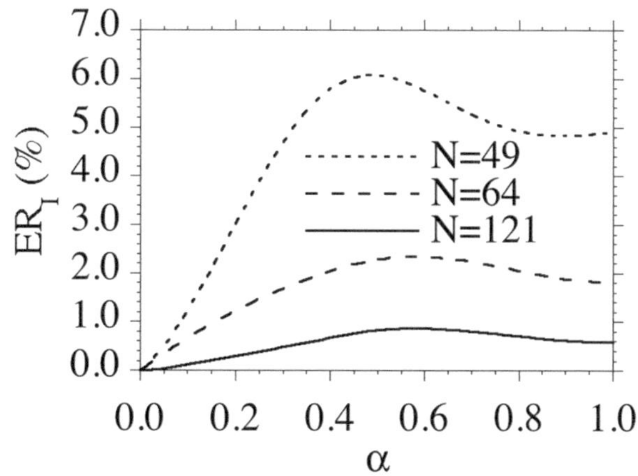

Fig. 16.9 Reduction in the error ER_{I} predicted by the equal area model for the higher values $N = 49, 64, 121$ of the number of fiber bundles in the model

the strain energy function needs to be evaluated for each of the N fibers. Also, the Cauchy stress is expressed in terms of a weighted sum of N symmetric tensors $\mathbf{B}_{\mathrm{i}}$ defined in (16.33). It was shown in Figs. 16.7 and 16.9 that the error ER_{I} due to unphysical anisotropy for the refined icosahedron model is less than 5 % for $N = 36$ and for the uniform area model is less than 7 % for $N = 49$. Of course, the magnitude of the unphysical anisotropy depends on the specific loadings considered. Moreover, these levels of refinement cause considerable increased complexity, which may not be justified by the accuracy and availability of experimental data.

Within the context of the icosahedron model proposed in Flynn et al. (2011) the tissue is modeled by only six fiber bundles, each of which has the same undulation distribution. This model can be thought of as a specific discretization of the model in Lanir (1983). It was shown in Flynn et al. (2011) that this icosahedron model can successfully reproduce experimental data exhibiting anisotropic response. However, it was shown in Sect. 16.3 here that this icosahedron model exhibits significant undesirable anisotropy when the weighting functions w_i in (16.6) are equal (16.13). This means that anisotropy in the model is influenced by both the nonlinearity of the undulation distribution and differences in the values of the weights. Consequently, the weights w_i are not pure measures of anisotropy of the histological orientations of fibers in the tissue.

Itskov and Ehret (2009) have proposed an alternative model of tissues which is based on a generalized invariant of deformation determined by a weighted average of different structural tensors. This idea has been used in Flynn and Rubin (2012) to develop a generalized icosahedron model. Specifically, a generalized structural tensor $\mathbf{W}$ is defined in terms of the weights w_i and the structural tensors $\mathbf{B}_i$ in (16.2) of the icosahedron model by

$$\mathbf{W} = \sum_{i=1}^{6} w_i \ \mathbf{B}_i, \quad w_i \geq 0, \quad \mathbf{W} \cdot \mathbf{I} = \sum_{i=1}^{6} w_i = 1. \tag{16.44}$$

Also, the generalized strain invariant γ is defined by

$$\gamma = \left(\mathbf{C} + \mathbf{C}^{-1}\right) \cdot \mathbf{W} - 2 \geq 0, \tag{16.45}$$

where $\mathbf{C}$ is the right Cauchy–Green deformation tensor (16.4). Then, the strain energy is taken to be a nonlinear function of γ. In view of the property (16.3a) of $\mathbf{B}_i$ it follows that for equal weights w_i (16.13), γ becomes an isotropic invariant of $\mathbf{C}$ given by

$$\gamma = \frac{1}{3}\left(\mathbf{C} + \mathbf{C}^{-1}\right) \cdot \mathbf{I} - 2 \geq 0. \tag{16.46}$$

In particular, the response of the tissue is analytically isotropic for any nonlinear dependence of the strain energy on γ in (16.46). Moreover, it was shown in Flynn and Rubin (2012) that when the strain energy is a simple polynomial function of γ, the coefficients of the polynomial and the weights w_i can be determined to match large deformation experimental data for the anisotropic response of various tissues.

The advantages of this generalized icosahedron model are that the number of material constants is small, it can be simplified to produce isotropic response exactly and γ depends on only a single structural tensor $\mathbf{W}$ so the constitutive response is simple to evaluate numerically. Specifically, this is a phenomenological model that characterizes a coupled network of fiber bundles. In this regard, the generalized icosahedron model has the disadvantage that the nonlinear elastic response of the model is not simply connected to an undulation distribution of each fiber bundle, as proposed in the structural model of Lanir (1983).

Acknowledgments This research was partially supported by MB Rubin's Gerard Swope Chair in Mechanics. The authors would also like to acknowledge helpful discussions with A Rubin about the equal area model.

References

Bazant Z, Oh BH. Microplane model for progressive fracture of concrete and rock. J Eng Mech ASCE. 1985;111(4):559–82.

Bazant ZP, Oh BH. Efficient numerical integration on the surface of a sphere. Z Angew Math Mech. 1986;66(1):37–49.

Christensen RM. Mechanics of low density materials. J Mech Phys Solids. 1986;34(6):563–78.

Christensen RM. Sufficient symmetry conditions for isotropy of the elastic moduli tensor. J Appl Mech. 1987;54(4):772–7.

Ehret AE, Itskov M, Schmid H. Numerical integration on the sphere and its effect on the material symmetry of constitutive equations—a comparative study. Int J Numer Methods Eng. 2010;81(2):189–206.

Elata D, Rubin MB. Isotropy of strain energy functions which depend only on a finite number of directional strain measures. J Appl Mech. 1994;61(2):284–9.

Elata D, Rubin MB. A new representation for the strain energy of anisotropic elastic materials with application to damage evolution in brittle materials. Mech Mater. 1995;19(2–3):171–92.

Flynn C, Rubin MB. An anisotropic discrete fibre model based on a generalised strain invariant with application to soft biological tissues. Int J Eng Sci. 2012;60:66–76.

Flynn C, Rubin MB, Nielsen P. A model for the anisotropic response of fibrous soft tissues using six discrete fibre bundles. Int J Numer Methods Biomed Eng. 2011;27(11):1793–811.

Itskov M, Ehret AE. A universal model for the elastic, inelastic and active behaviour of soft biological tissues. GAMM-Mitteilungen. 2009;32(2):221–36.

Itskov M, Ehret AE, Dargazany R. A full-network rubber elasticity model based on analytical integration. Math Mech Solids. 2010;15:655–71.

Lanir Y. Constitutive equations for fibrous connective tissues. J Biomech. 1983;16(1):1–12.

Chapter 17
Finite Element Implementation of Structural Constitutive Models

Michael S. Sacks

Abstract It is well established that the highly nonlinear and anisotropic mechanical behaviors of soft tissues are an emergent behavior of the underlying tissue microstructure. Numerical solutions form the cornerstone in the application of constitutive models in contemporary biomechanics. Herein, a structural constitutive model into a finite element framework specialized for membrane tissues. Multiple deformation modes were simulated, including strip biaxial, planar biaxial with two attachment methods, and membrane inflation. Detailed comparisons with experimental data were undertaken to insure faithful simulations of both the macro-level stress–strain insights into adaptations of the fiber architecture under stress, such as fiber reorientation and fiber recruitment. Results indicated a high degree of fidelity and demonstrated interesting microstructural adaptions to stress and the important role of the underlying tissue matrix.

17.1 Introduction

Traditionally, soft tissues are modeled as pseudo-hyperelastic materials using either phenomenological or structural approaches (Criscione et al. 2003; Holzapfel and Ogden 2009; Sacks 2000). A common phenomenological model is the Fung-type (Fung 1993; Tong and Fung 1976), in which the strain energy function is a quadratic exponential function of the Green-Lagrange strain tensor. The original form was based on the observed linear relation between tissue stiffness and stress under uniaxial conditions (Fung 1993). However, phenomenological models lack physical interpretation and cannot, in general, be used for simulations beyond the strain range utilized in parameter estimation. This effect has been shown to be the case even when the strain magnitudes did not exceed the maximum values measured but where substantially far from the available experimental data (Sun et al. 2003). While the

M.S. Sacks (✉)
Department of Biomedical Engineering, Institute for Computational Engineering and Sciences, The University of Texas at Austin, 201 East 24th Street, ACES 5.438, 1 University Station, C0200, Austin, TX 78712-0027, USA
e-mail: msacks@ices.utexas.edu

G.S. Kassab, M.S. Sacks (eds.), *Structure-Based Mechanics of Tissues and Organs*,
DOI 10.1007/978-1-4899-7630-7_17

underlying reasons for this still need to be elucidated, models which possess greater links to the underlying physical mechanisms appear to be the next step.

Like any biological or synthetic biomaterial, the complex mechanical behavior of soft tissues results from the deformations and interactions of the constituent phases. For most soft tissues, these include collagen, elastin, muscular, and related matrix components such as glycosaminoglycans and proteoglycans. The idea of accounting for tissue structure into mechanical models of soft tissues goes back to at least the work on leather mechanics in 1945 by Mitton. More contemporary work on structural approaches followed, with growing popularity in the 1970s (Beskos and Jenkiins 1975), with the concept of stochastic constituent fiber recruitment developed about the same time (Soong and Huang 1973) based on related structural studies (Kenedi et al. 1965). In part a result of the availability of the first planar biaxial data for soft tissues, Lanir developed the first comprehensive, multidimensional structural constitutive model formulation (Lanir 1979). With various modifications, Lanir et al. applied this approach to many soft tissues such as lung (Lanir 1983) and myocardium (Horowitz et al. 1988).

By linking tissue deformation at macroscopic scale and microscopic (fiber) scale through affine deformation assumption, the structural constitutive model can be considered a statistical multi-scale approach. Above all, structural constitutive modeling approaches can, in principle, provide valuable insight into tissue function. For example, Billiar and Sacks (2000a, b) demonstrated for aortic valve leaflets that, using a simplified leaflet structure, angular rotation of the fibers account for such important features such as pronounced mechanical anisotropy, axial coupling, and very large strains (>80 %) even though the tissue is composed of collagen fibers that fail at less than $\sim$12 % strain. Later, Sacks demonstrated that with the use of only an equi-biaxial test and the experimentally measured fiber orientation distribution, the complete in-plane biaxial response could be simulated (Sacks 2003). More recently, structural approaches have been used for a wide range of native and engineered tissue applications, such as elastomeric tissue engineering scaffolds (Courtney et al. 2006), urinary bladder wall (Wognum et al. 2009), and many others (Fata et al. 2014; Hansen et al. 2009; Hollander et al. 2011; Kao et al. 2011).

Due to the need to solve soft tissue problems that involve complex anatomical geometries and boundary conditions, many constitutive models for soft tissues in various forms have been implemented into a computational framework (Driessen et al. 2007; Holzapfel et al. 1996; Prot et al. 2007; Sun and Sacks 2005; Hariton et al. 2007). Yet, robust evaluation and rigorous validation of structural constitutive models remain quite limited. Moreover, studies on structural model have mainly focused on either material parameter estimation (Jor et al. 2011) or comparison with different constitutive models (Bischoff 2006; Cortes et al. 2010; Tonge et al. 2013). Structural models that incorporate fiber recruitment have rarely been used, largely due to computational demands of the additional integration. Recent ability to get detailed fiber recruitment data (e.g., Chen et al. 2011; Fata et al. 2014; Hansen et al. 2009) makes this approach all the more relevant. The deep insights that structural models can provide, such as the role of fiber structure and kinematics, still have yet to be fully explored by simulation.

Since many soft tissues are relatively thin, they can be modeled using shell or membrane elements in FE analysis, greatly speeding up the simulations. In the present study, we implemented a planar structural constitutive model into the commercial finite element (FE) package ABAQUS. By numerical simulation of one single element subjected to uniaxial tension, we first revealed that matrix must be present to prevent unrealistic tissue deformations. Flexural simulations were utilized to estimate the matrix modulus, since the underlying collagen fibers remain undulated due to the small extensional strains, and thus have little effect on tissue stress development. Strip biaxial strain and equi-biaxial tension simulations were also performed and compared with experimental collagen fiber measurements to demonstrate the effects of initial fiber orientation distribution on fiber reorientation. Simulation of membrane inflation tests were also applied to further test the structural model. In addition to prediction of macroscopic mechanical response of soft tissues, we demonstrate how the structural model can provide insights into tissue micro-structural events.

17.2 Methods

17.2.1 Theoretical Formulation

Soft biological tissues primarily have two major load-bearing components: the fibrous network and the nonfibrous (i.e., amorphous) ground matrix. Based on Fung (1993), we idealize the elastic behavior of soft tissues as pseudo-hyperelastic composite materials. Thus, the total strain energy function $\varPsi$ of soft tissue at a represent volume element (RVE) is defined using

$$\Psi\left(\tilde{C}\right) = \sum_{i=1}^{n} \phi_{\mathrm{f}}^{i}\Psi_{\mathrm{f}}^{i} + \phi_{\mathrm{m}}\Psi_{\mathrm{m}} + p\left(J-1\right), \tag{17.1}$$

where $\varPsi_{\mathrm{f}}$ and $\varPsi_{\mathrm{m}}$ are the strain energy functions for the fiber and matrix phases, respectively, ϕ_{f}^{i} and ϕ_{m} are the volume fraction of fiber and matrix, respectively, with $\sum_{i=1}^{n}\phi_{\mathrm{f}}^{i} + \phi_{\mathrm{m}} = 1$, $J = \det\left(\tilde{\mathbf{F}}\right)$, and p the Lagrange multiplier to enforce incompressibility due to soft tissue's high water content. The contributions of the nonfibrous components and fluid phases are assumed to be responsible for the incompressibility of the tissue. Based on previous results (Buchanan and Sacks 2013), we model the matrix phase (which compromises all nonfibrous components) as a single isotropic hyperelastic Neo-Hookean material with the strain energy function $\Psi_{\mathrm{m}} = C_1\left(I_{\mathrm{C}} - 3\right)$. Here, I_{C} is the first invariant of the right Cauchy-Green tensor $\tilde{\mathbf{C}} = \tilde{\mathbf{F}}^{\mathrm{T}}\tilde{\mathbf{F}}$, and to be consistent with linear elasticity $C_1 = \mu/2$, where μ is the shear modulus. Using $\mathbf{S} = \partial\Psi/\partial\mathbf{E}$, the resulting second-Piola Kirchhoff stress is given by

$$\mathbf{S}_{\rm m} = \phi_{\rm m}\frac{C_1}{2}(I_1 - 3) - p\tilde{\mathbf{C}}^{-1}. \tag{17.2}$$

Next, without loss of generality, we focus on a single, undulated Type I collagen fiber type with planar structures and a plane stress state. As in related work on collagenous tissues (Lanir 1983; Sacks 2003), we assume a linear $S_{\rm f} = \eta E_{\rm f}$ relationship for the individual collagen fibers

$$\Psi_{\rm f}(E_{\rm f}) = \frac{\eta}{2}E_{\rm f}^2, \tag{17.3}$$

where η is the elastic modulus of individual straight collagen fibers (Fig. 17.1a). Due to their crimped structure, we express individual fiber's true fiber strain using $E_{\rm t} = \frac{E_{\rm f}-E_{\rm s}}{1+2E_{\rm s}}$ where $E_{\rm s}$ is the fiber slack strain (Fig. 17.1a). The resulting *individual* collagen fiber strain energy is thus

$$\Psi_{\rm f}(E_{\rm t}) = \frac{\eta}{2}E_{\rm t}^2 = \frac{\eta}{2}\left(\frac{E_{\rm f} - E_{\rm s}}{1 + 2E_{\rm s}}\right)^2. \tag{17.4}$$

To develop the first level homogenization, we define a fiber ensemble as the collection of all fibers within the RVE with a common direction $\widehat{\mathbf{N}} = [\cos(\theta)\ \sin(\theta)\ 0]$ (Fig. 17.1b). The collective mechanical contribution from the

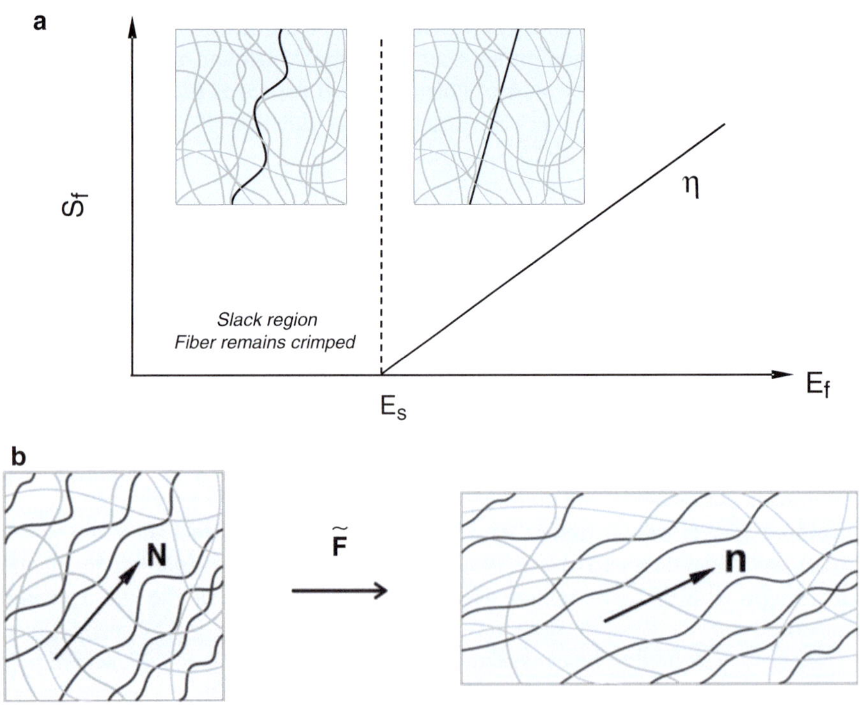

Fig. 17.1 (**a**) Assumed stress–strain response of a single undulated fiber. Graphical depiction of a (**b**) fiber ensemble

ensemble is represented by its strain energy Ψ_{ens}. Assuming affine deformation (Lanir 1983; Sacks 2003), the fiber ensemble strain E_{ens} in the direction $\widehat{\mathbf{N}}$ is related to the macroscopic tissue-level Green-Lagrange strain tensor $\tilde{\mathbf{E}} = \frac{1}{2}\left(\tilde{\mathbf{C}} - \tilde{\mathbf{I}}\right)$ by

$$E_{ens}(\theta) = \widehat{\mathbf{N}}^{\mathbf{T}}\,\tilde{\mathbf{E}}\,\widehat{\mathbf{N}}. \tag{17.5}$$

We make the distinction between fiber and ensemble strains here since individual collagen fibers will have a different strain levels due to their undulations. Note that the nonlinearity of the tissue evolves from the gradual recruitment of the linearly elastic collagen fibers (Lanir 1983), and is thus a structural as opposed to a material property.

To stochastically account for the gradual recruitment of the collagen fiber in each fiber ensemble with strain, we define the function $D(E_s)$ over the ensemble strain range $E_{ens} \in [E_{lb}, E_{ub}]$. Here, E_{lb} and E_{ub} represent the lower and upper bounds of collagen fiber ensemble recruitment strain levels, with $E_{ub} > E_{lb} > 0$ and $\int_{E_{lb}}^{E_{ub}} D(x)dx = 1$. The ensuing fiber ensemble strain energy and stress–strain relation are then described as the sum of individual fiber strain energies of the ensemble weighted by the distribution of slack strains D, so that

$$\Psi_{ens} = \frac{\eta}{2}\int_0^{E_{ens}} D(x)\left(\frac{E_{ens} - x}{1 + 2x}\right)^2 dx \qquad S_{ens} = \eta\int_0^{E_{ens}} D(x)\frac{E_{ens} - x}{(1 + 2x)^2}dx. \tag{17.6}$$

D was represented as a Beta distribution B defined over $E_s \in [E_{lb}, E_{ub}]$,

$$D(x) = \begin{cases} \frac{x^{\alpha-1}(1-x)^{\beta-1}}{B(\alpha,\beta)(E_{ub}-E_{lb})}, & \text{for } x \in [0, 1] \\ 0, & \text{otherwise} \end{cases}, \quad x = (E_{ens} - E_{lb}) \,/\, (E_{ub} - E_{lb}), \tag{17.7}$$

where α and β are the shape factors. Note that for simplicity we chose $E_{lb} = 0$, although generally it is not (Fata et al. 2014).

In situations where computational demands are very high, we also present an alternative formulation for the ensemble stress–strain relation using simplified exponential form that emulates the recruitment behavior at both low and high strains. The novel aspect here is that the terminal stiffness of the fiber ensemble is reproduced for ensemble strains above E_{ub}. This simple but important modification helps avoid unrealistically high fiber stresses at high strains when using an exponential model alone. For an exponential model, this becomes

$$S_{ens}(E_{ens}) = \begin{cases} \mathrm{A}\left(e^{B\,E_{ens}} - 1\right), \; for\; E_{ens} \le E_{ub} \\ \mathrm{A}\left(e^{B\,E_{ub}} - 1\right) + \mathrm{AB}e^{B\,E_{ub}}\left(E_{ens} - E_{ub}\right), \; \text{for } E_{ens} > E_{ub} \end{cases}, \tag{17.8}$$

where A and B are material constants. Note that the tangent modulus is continuous at $E = E_{ub}$.

In the final step, we homogenize the ensemble response to the tissue level by defining the tissue strain energy as the sum of the strain energy of fiber ensembles, weighted by the orientation distribution function (ODF) $\Gamma(\theta)$. Thus, we have

$$\Psi_c = \int_{-\pi/2}^{\pi/2} \Gamma(\theta)\, \Psi_{\text{ens}}(E_{\text{ens}})\, d\theta, \tag{17.9}$$

with the normalization constraint $\int_{-\pi/2}^{\pi/2} \Gamma(\theta)\, d\theta = 1$. In summary, the total strain energy function of soft tissue in the RVE is expressed as

$$\Psi = \frac{\phi_{\text{f}}\eta}{2} \int_{-\pi/2}^{\pi/2} \Gamma(\theta) \left[\int_0^{E_{\text{ens}}} D(x) \left(\frac{E_{\text{ens}} - x}{1 + 2x} \right)^2 dx \right] d\theta + \frac{\phi_{\text{m}}\mu_{\text{m}}}{2} (I_1 - 3) + p\,(J - 1). \tag{17.10}$$

For the plane stress case, the out-of-plane stress component $S_{33} = 2\partial\Psi/\partial C_{33} = 0$, so that the Lagrange multiplier p can be determined directly from

$$p = -\phi_{\text{m}}\mu_{\text{m}} C_{33}. \tag{17.11}$$

The total second Piola-Kirchhoff stress can be thus written as

$$\tilde{\mathbf{S}} = \phi_{\text{f}}\eta_{\text{f}} \int_{-\pi/2}^{\pi/2} \Gamma(\theta) \left[\int_0^{E_{\text{ens}}} D(x) \frac{E_{\text{ens}} - x}{(1 + 2x)^2} dx \right] \left(\widehat{\mathbf{N}} \otimes \widehat{\mathbf{N}} \right) d\theta + \phi_{\text{m}}\mu_{\text{m}} \left(\tilde{\mathbf{I}} - \mathrm{C}_{33} \tilde{\mathbf{C}}^{-1} \right). \tag{17.12}$$

for the recruitment model, and

$$\tilde{\mathbf{S}} = \int_{-\pi/2}^{\pi/2} \Gamma(\theta)\, S_{\text{ens}}\left[E_{\text{ens}}(\theta)\right] \left(\widehat{\mathbf{N}} \otimes \widehat{\mathbf{N}} \right) d\theta + \phi_{\text{m}}\mu_{\text{m}} \left(\tilde{\mathbf{I}} - \mathrm{C}_{33} \tilde{\mathbf{C}}^{-1} \right). \tag{17.13}$$

for the simplified model (17.8).

17.2.2 Finite Element Implementation

The structural model was implemented into commercial finite element software ABAQUS/Standard (Dassault Systèmes Simulia Corp., Providence, RI) via user-defined material subroutine UMAT. The stress tensor components utilized in UMAT is defined in a co-rotational coordinate system in which the local material axes defined in the initial configuration rotates with the material (ABAQUS 2011). Using polar decomposition theorem (Marsden and Hughes 1983) we have $\tilde{\mathbf{F}} = \tilde{\mathbf{R}}\tilde{\mathbf{U}}$, where $\tilde{\mathbf{R}}$ is the rigid body rotation tensor and $\tilde{\mathbf{u}}$ is the right symmetric stretch tensor. Also, the rotated Cauchy stress can be determined using $\tilde{\mathbf{t}} = \mathrm{J}^{-1}\tilde{\mathbf{U}}\tilde{\mathbf{S}}\tilde{\mathbf{U}}$ and the fourth-rank material elasticity tensor $\mathbb{C}^{\mathbf{SE}}$ are updated in the UMAT code. In the actual implementation, $\tilde{\mathbf{S}}$ and $\mathbb{C}^{\mathrm{SE}}$ require integration over $\theta \in [-\pi/2, \pi/2]$ as well as $E_{\mathrm{ens}} \in [E_{\mathrm{lb}}, E_{\mathrm{ub}}]$. Since a closed form solutions are not available in general, a numerical integration scheme was used as follows. During implementation, the angle domain and the fiber strain domain were separated into twenty segments with equal size. In each segment, Gaussian quadrature integration rule (Hughes 2000) was performed with five integration points.

17.2.3 Further Model Modifications and Material Parameter Estimation

For the present work we merged (without loss of generality) the material parameters ϕ_{m} and μ_{m} for matrix component into μ_{m}, as well as ϕ_{f} and η for the fiber component were also combined into η. Due to its high collagen Type I content, generally planar tissue architecture, well-characterized structure and mechanical properties, and previous use in structural models (Sacks 2003) made native bovine pericardium natural choice for the representative tissue for simulations. To obtain the value for μ_{m}, we utilized flexural data from native bovine pericardium (Mirnajafi et al. 2005). In that study, a nearly linear moment-curvature relation has been observed. This suggested that the collagen fibers have little effect in flexure, which is consistent with the very low strains that occur in this deformation mode (so that the collagen fibers remain fully undulated and only the matrix contributes). We thus obtained μ_{m} by fitting the moment-curvature curve (Mirnajafi et al. 2005), using methods described in the next section.

The total fiber angular distribution function is expressed as a linear combination of Gaussian distribution and uniform distribution

$$\Gamma(\theta) = d\left[\frac{\exp\left(-\frac{\theta^2}{2\sigma^2}\right)}{\operatorname{erf}\left(\frac{\pi}{2\sqrt{2}\sigma}\right)\sqrt{2\pi}\sigma}\right] + \frac{(1-d)}{\pi}. \tag{17.14}$$

Equation (17.14) was chosen to allow graduations in aligned and isotropic fiber distributions to be simulated easily. Here σ denotes the standard deviation of the Gaussian distribution function, and the error function erf() is introduced so that the integration of the Gaussian distribution function over angle domain $\theta \in [-\pi/2, \pi/2]$ is equal to unity. The fiber angular distribution function was obtained previous measurements and fitting the experimental data with $d = 1$ (Billiar and Sacks 1997).

One way to evaluate the robustness and accuracy of the FE implementation is to examine applications where very large strains are known to occur, which induce large fiber rotations and stretches. Previous experimental results have revealed that the mechanical behavior of soft collagenous tissue is strongly dependent on gripping methods (Waldman and Michael Lee 2002). In particular, we noted in that study that clamps induced large rotations in the corner regions between the clamps. Thus, the material parameters from both model forms were obtained by fitting stress–strain curve from the equi-biaxial loading stress–strain data with suturing arrangement from Waldman and Michael Lee (2002). This allowed us to directly compare the FE results to the experimental findings from that study.

17.2.4 Finite Element Simulations

We start with a basic simulation of a single element under uniaxial tension to investigate the effects of matrix. For this example, a square element was subjected to uniaxial strain in X_1 direction (Fig. 17.2a). Nodes 1 and 2 were constrained in X_2 direction, and nodes 1 and 4 were constrained in X_1 direction, with uniform displacements applied to nodes 2 and 3 in the X_1 direction. The preferred fiber orientation coincided with the X_1 direction. Next, to verify minimal fiber recruitment occurred during flexure, simulation of a bending test was performed. The "The length and width of the specimen..." length and width of the specimen used for bending simulation is 20.0 and 3.0 mm, respectively, with the thickness of the tissue as 0.4 mm and the span 16.0 mm (Fig. 17.2b). The loading was applied at the center of the tissue through the middle post, with the three posts being considered as rigid bodies. The friction coefficient was assumed to be zero for the tissue in contact with the left and right posts.

To investigate the effects of both boundary conditions (as both localized point and distributed loads) on fiber reorientation, we simulated native bovine pericardium using sutures under strip biaxial tension using data from Billiar and Sacks (1997) and equi-biaxial tension using clamped boundary conditions using data from Waldman et al. (2002). To simulate both high and low orientations, we utilized two levels of d ($d = 1.0$ and $d = 0.25$). For the first test, the dimensions of the specimen were 19.2 mm × 19.2 mm and the thickness was 0.4 mm. As in the original experiment, uniform displacements were applied on the seven suture attachment points along each side the specimen, with the initial preferred fiber orientation set to 27° from the X_1 axis (Fig. 17.3). Two loading cases were considered; 30 % along X_1 direction/0 % for the X_2, and 30 % along the X_2 direction/0 % along the X_1. For the clamped equi-biaxial tension test, the dimensions of the specimen were 22.0 mm × 22.0 mm (Fig. 17.4) and the thickness 0.4 mm. The tissue was stretched 10 % in X_1 and X_2 directions. The initial fiber orientation was assumed to be the X_1 direction (Fig. 17.3d).

As a final test, we simulated a fetal membrane (FM) inflation test using data from Joyce et al. (2009). Uniform pressure was applied on the top surface of a circular membrane and only half of the tissue was modeled due to symmetry. The radius of the circular membrane was 21.0 mm and the thickness 0.228 mm. The tube was modeled as rigid body with an inner radius of 15.0 mm and the edge of the tissue fixed. SALS measurements of the intact FM (Joyce et al. 2009) revealed that the tissue contains no preferred collagen direction, therefore a uniform fiber angular distribution function $\Gamma(\theta) = 1/\pi$ was utilized. The friction coefficient was

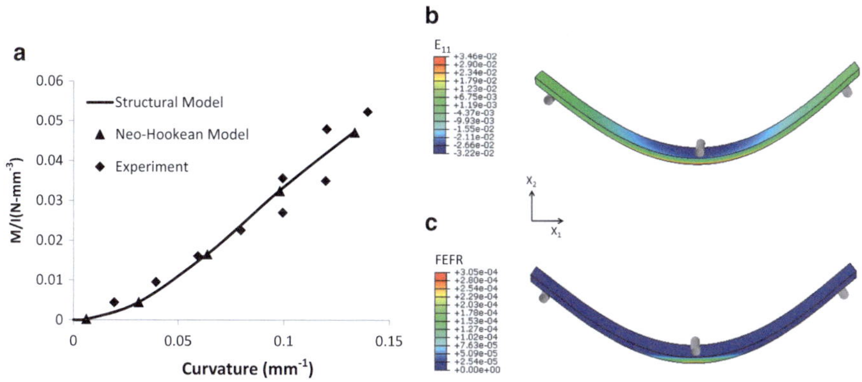

Fig. 17.2 Effective deformations of fibers on soft tissues under three-point bending showing (**a**) the moment-curvature curve relation obtained using a Neo-Hookean model with matrix only and structural model with fiber and matrix, (**b**) Green Strain E_{11} contour, (**c**) Fractional ensemble fiber recruitment contour in X_1 direction

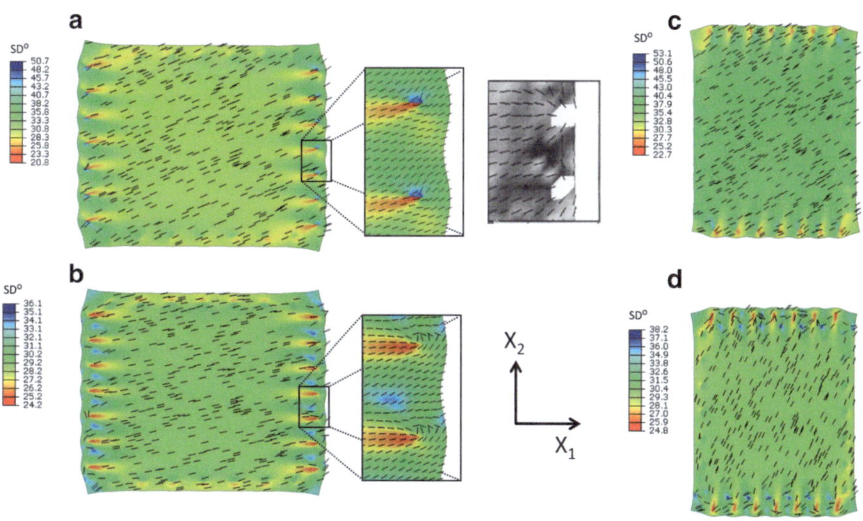

Fig. 17.3 Preferred fiber reorientation and standard deviation contour of fiber angular distribution function from simulation under strip biaxial stretch (**a**) in X_1 direction with $d = 1.0$, (**b**) in X_1 direction with $d = 0.25$, (**c**) in X_2 direction with $d = 1.0$, (**d**) in X_2 direction with $d = 0.25$. *Inset*—measured SALS data from the Billiar and Sacks 1997 study, showing very good agreement

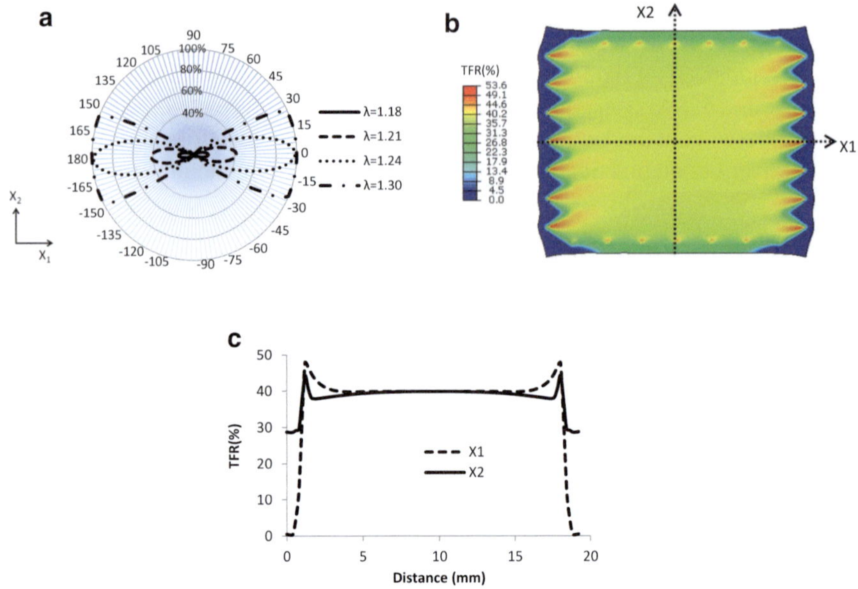

Fig. 17.4 (**a**) Azimuthal plot of the fractional ensemble fiber recruitment under different stretch ratios, (**b**) total percentage of fiber recruitment contour after stretch in X_1 direction, (**c**) total percentage of fiber recruitment along the line X_1 and X_2 in (**b**)

assumed to be zero for the contact interaction of tissue with the rigid tube. Note that for flexural simulations, four-node quadrilateral shell elements were used, and for all the other simulations four-node quadrilateral membrane elements were used.

17.2.5 Simulation Post-processing

To provide insights into the deformations of soft tissue microstructure under strain, we implemented the following post-processing procedures. The following two-dimensional form

$$\Gamma_t(\beta) = \Gamma(\theta)\frac{\lambda^2}{J_{2D}} = \Gamma(\theta)\frac{\widehat{\mathbf{N}}\cdot\tilde{\mathbf{C}}\cdot\widehat{\mathbf{N}}}{\det\left[\tilde{\mathbf{F}}_{2D}\right]}, \tag{17.15}$$

where J_{2D} is the determinate of the in-plane deformation gradient. Note that $\Gamma_t(\beta)$ should be plotted against the deformed fiber angle $\beta = \tan^{-1}\left(\frac{F_{21}\cos(\theta)+F_{22}\sin(\theta)}{F_{11}\cos(\theta)+F_{12}\sin(\theta)}\right)$ since it refers to the deformed (convected) fiber direction. In addition to its ability to provide high fidelity simulations of the stress–strain behaviors, the structural model also provides a great deal of information structural adaptation to within the RVE. These are usually overlooked in the literature. Thus, in the present study we defined the *fractional ensemble fiber recruitment* (FEFR) structural metric for a given direction as

$$\mathrm{FEFR}\,(\theta) = \int_0^{E_{\mathrm{ens}}(\theta)} D(x)dx. \tag{17.16}$$

In addition, the *total fiber recruitment* (TFR) structural metric for a given RVE was defined as

$$\mathrm{TFR} = \int_{-\pi/2}^{\pi/2} \Gamma\,(\theta) \left[\int_0^{E_{\mathrm{ens}}(\theta)} D(x)dx \right] d\theta. \tag{17.17}$$

Note that both metrics are expressed using a percentile scale.

17.3 Results

17.3.1 Uniaxial Tension Simulation

Overall, we determined that the matrix had a signification effect on the simulated deformation of soft tissue under uniaxial tension. When the tissue is modeled with fibers only, the lateral deformation of the tissue may be unrealistic. Specifically, since under uniaxial tension the stress component in the X_2 direction is $S_{22} = \int_{-\pi/2}^{\pi/2} \Gamma\,(\theta)\, S_{\mathrm{ens}}\,(E_{\mathrm{ens}})\sin^2(\theta)\,d\theta$, with $S_{\mathrm{ens}} = D(x) = 0$ in this direction since the fibers cannot carry any load when compressed. Since $\Gamma\,(\theta) \geq 0$ and $\sin^2(\theta) \geq 0$, S_{22} must be greater than zero, yet under uniaxial tension $S_{22} = 0$, so that an equilibrium state cannot be achieved. For example, when the element is stretched 6 % (Fig. 17.3a) the Green-Lagrange strain $E_{11} = 0.618$ and $E_{22} = -0.0434$ with a ratio of $-E_{22}/E_{11}$ of 0.702. When the element is stretched 12 % (Fig. 17.3b), the strain ratio $-E_{22}/E_{11}$ increased to 2.049. The deformation in X_2 direction is larger than that in X_1 direction, and the element collapses to a single line when the stretch ratio is greater than 15 %. However, when the matrix is included, the strain ratio $-E_{22}/E_{11}$ is 0.459 and 0.424 as the element is stretched 6 % and 12 %, respectively, so that the deformation of the element is acceptable. Thus for the present model a matrix should be present and in sufficient quantity in the structural model to prevent unphysical characterization of the mechanical behavior of soft tissues under uniaxial tension.

17.3.2 Flexural Simulations

The moment-curvature curves from FE simulation using only the isotropic neo-Hookean model were in good agreement with the published experimental results (Mirnajafi et al. 2005) (Fig. 17.2a). The simulation results also confirmed that collagen fiber contributions were negligible; the moment-curvature curves from FE simulations are the same using Neo-Hookean model and the structural model with both fiber and matrix (Fig. 17.2). The maximum tensile strain (Green-Lagrange strain) under bending was 0.0346 located at the bottom surface. Virtually all (>99 %) of the fibers were still undulated in this loading configuration (Fig. 17.2), supporting our use of these studies to determine ground matrix mechanical behavior.

17.3.3 Biaxial Test Simulations: Sutured Boundary Conditions

For the strip biaxial test stretched in the X_1 direction, fiber orientation from SALS measurements (Billiar and Sacks 1997) revealed that the overall preferred fiber direction was reoriented towards the direction of stretch and the degree of fiber alignment was increased. When $d = 1.0$ in the fiber angular density function, the preferred fiber direction rotated only about 4° towards the stretch direction (Fig. 17.3a). However, for $d = 0.25$, the preferred fiber direction rotated about 15° towards the stretch direction (Fig. 17.3b), in agreement with SALS measurements (Billiar and Sacks 1997). Around the suture points, the simulation results with $d = 0.25$ (Fig. 17.3b) demonstrated that the preferred fiber direction rotated towards the suture points, also in agreement with the SALS data (Billiar and Sacks 1997). When the tissue was stretched in the X_2 direction with $d = 1.0$, the overall preferred fiber direction reoriented only about 4° towards the stretch direction (Fig. 17.3c). While for $d = 0.25$, the overall preferred fiber direction reoriented about 35.0° towards the stretch direction (Fig. 17.3d).

When the tissue was stretched in the X_1 (preferred) direction, more fibers are recruited in the X_1 direction than those in the X_2 direction. For $d = 0.25$, the polar FEFR plot (Fig. 17.4a) under different stretch ratios revealed the increasing FEFR in all directions. At $\lambda = 1.18$ the FEFR in all directions was less than 20 %. As the stretch increased to 1.21, more than 40 % of fibers were recruited in the X_1 direction. All the fibers in the X_1 direction were straightened when the strain in X_1 direction just reaches the upper bound strain at $\lambda = 1.24$, with all fibers within 24° from the X_1 direction straightened by a stretch of 1.4. The TFR is 40 % uniformly distributed in the center region of the soft tissue (Figs. 17.4c and 17.6b). The maximum TFR occurs at the suture points (Fig. 17.4b, c) due to stress concentration.

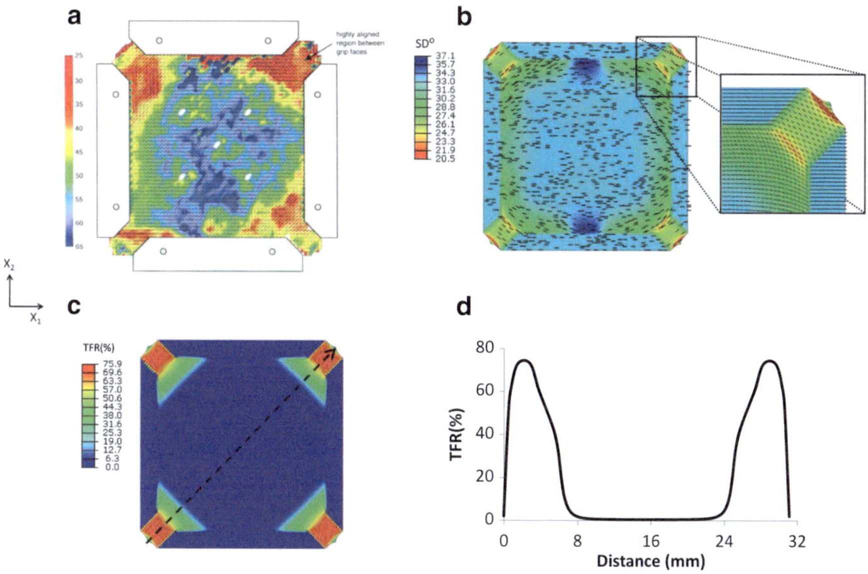

Fig. 17.5 (**a**) SALS data (Waldman et al. 2002), (**b**) preferred fiber orientation and standard deviation contour of fiber angular distribution function from simulation with $d = 0.25$, (**c**) total percentage of fiber recruitment contour, (**d**) total percentage of fiber recruitment along the diagonal line in (**c**)

17.3.4 Biaxial Test Simulations: Clamped Boundary Conditions

For the tissue with clamped boundary conditions, SALS measurements (Waldman et al. 2002) indicated that fibers between grip faces were highly aligned due to shearing (Fig. 17.5a). The preferred fiber orientation rotated about 37.0° at the corner regions from simulation with $d = 0.25$. The distribution of the standard deviation of fiber density from simulation is similar to the distribution of orientation index (OI) from SALS measurements. The structural model provided deep insights into the deformation of fibers. All the fibers initially undulated were straightened gradually with increasing load. Less than 1 % of fibers were recruited in the center region of the tissue specimen and more than 70 % of fibers in the corner regions were recruited. Along the diagonal line, the TFR is uniformly distributed around the center region of the tissue specimen. The TFR increased dramatically at the corner regions due to large shear strain.

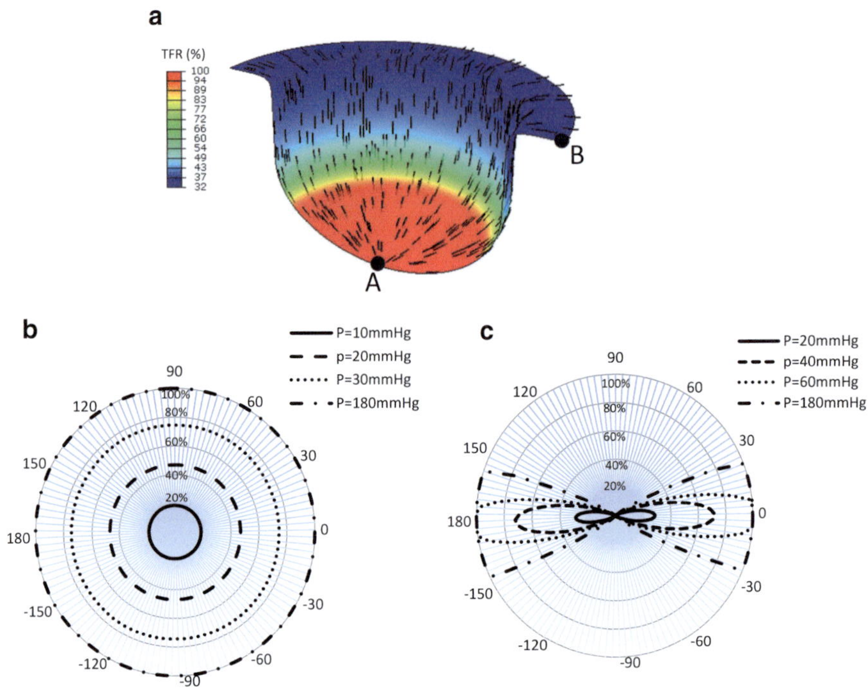

Fig. 17.6 Results from the membrane inflation simulations showing (**a**) preferred fiber orientation and total percentage of fiber recruitment contour from simulation, and the azimuthal plot of the fractional ensemble fiber recruitment of the elements located at (**b**) element A and (**c**) element B. Note here that while increased pressure loading predictably increased the total amount of fiber recruitment at both locations, substantial angular variations with pressure was observed at element B

17.3.5 Membrane Inflation Simulations

Simulation results revealed that all fibers around the center dome region were straightened (Fig. 17.6a). The preferred fiber direction in the deformed shape was rotated to the radial direction of the tissue (Fig. 17.6a). The TFR decreased gradually from 100 % in the center region to 32 % at the edge (Fig. 17.6a). For element A (Fig. 17.6a) at the center of tissue, the strain is the same in each direction. Therefore the fractional ensemble fiber recruitment is equal in each direction (Fig. 17.6b). However for location B (Fig. 17.6a) at the tissue edge, the strain in the radial direction is larger than that in the circumferential direction. More fibers were recruited in the radial direction (Fig. 17.6c).

17.4 Discussion

A framework for the implementation of a structural constitutive model for soft tissues into a finite element framework was developed and validated. Simulation of a single element under uniaxial tension revealed that when using fiber systems with angular dispersion, a matrix phase must be present to prevent nonphysiological deformations. This finding may shed insight into how the non-collagenous components of soft tissues play an important role guiding the overall mechanical responses. For example, Lake and Barocas (2011) studied the effects of simulated non-fibrillar matrix using an agarose analog on the behavior of a collagen-agarose co-gel in uniaxial tension. They reported that the Poisson's ratio of co-gel decreased from a range of 1.5–3.0 (with a large volume decrease) with no agarose to $\sim$0.5 (i.e., nearly incompressible) with high concentration of agarose. Thus, both the experimental results and the present simulation results suggest that matrix phase may have significant effects on the mechanical behavior of soft tissues. Biaxial tension simulations demonstrated that the presence of an isotropically oriented fiber phase can significantly affect the overall fiber orientation in the deformed configuration. For $d = 1.0$, the preferred fiber orientations were far from the SALS measurement for both strip biaxial tests and equi-biaxial tests. However, simulation results with $d = 0.25$ are in good agreement with the SALS measurement. Not surprisingly, this observation revealed that accurate measurement of the fiber ODF is critical to the structural model. By incorporating fiber orientation distribution and fiber recruitment distribution into the strain energy function, the structural model can predict not only the mechanical behavior of soft tissues at the macroscopic scale, but also fiber deformations patterns (in a statistical sense) at the microscopic scale. The new indices introduced herein may be particularly useful in understanding structural adaptations and could easily be used in structural optimization studies in engineered tissue design. Moreover, the present results underscore the importance of architecture in tissue modeling development and application.

Acknowledgements Funding for this work was supported by FDA contract HHSF223201111595P and NIH/NHLBI Grant NHLBI R01 HL108330 and R01 HL119297-01.
Conflict of Interests The authors have no conflict of interests to report in this work.

References

ABAQUS. Abaqus user subroutines reference manual. 2011.
Beskos DE, Jenkiins JT. A mechanical model for mammalian tendon. J Appl Mech. 1975;42:755.
Billiar KL, Sacks MS. A method to quantify the fiber kinematics of planar tissues under biaxial stretch. J Biomech. 1997;30:753–6.
Billiar KL, Sacks MS. Biaxial mechanical properties of the native and glutaraldehyde-treated aortic valve cusp: Part II–A structural constitutive model. J Biomech Eng. 2000a;122:327–35.
Billiar KL, Sacks MS. Biaxial mechanical properties of the natural and glutaraldehyde treated aortic valve cusp–Part I: experimental results. J Biomech Eng. 2000b;122:23–30.

Bischoff JE. Continuous versus discrete (invariant) representations of fibrous structure for modeling non-linear anisotropic soft tissue behavior. Int J Non Linear Mech. 2006;41:167–79.
Buchanan RM, Sacks MS. Interlayer micromechanics of the aortic heart valve leaflet. Biomech Model Mechanobiol. 2013;3(4):1–4.
Chen H, Liu Y, Slipchenko MN, Zhao X, Cheng JX, Kassab GS. The layered structure of coronary adventitia under mechanical load. Biophys J. 2011;101:2555–62.
Cortes DH, Lake SP, Kadlowec JA, Soslowsky LJ, Elliott DM. Characterizing the mechanical contribution of fiber angular distribution in connective tissue: comparison of two modeling approaches. Biomech Model Mechanobiol. 2010;9:651–8.
Courtney T, Sacks MS, Stankus J, Guan J, Wagner WR. Design and analysis of tissue engineering scaffolds that mimic soft tissue mechanical anisotropy. Biomaterials. 2006;27:3631–8.
Criscione JC, Sacks MS, Hunter WC. Experimentally tractable, pseudo-elastic constitutive law for biomembranes: I. Theory. J Biomech Eng. 2003;125:94–9.
Driessen NJ, Mol A, Bouten CV, Baaijens FP. Modeling the mechanics of tissue-engineered human heart valve leaflets. J Biomech. 2007;40:325–34.
Fata B, Zhang W, Amini R, Sacks MS. Insights into regional adaptations in the growing pulmonary artery using a meso-scale structural model: effects of ascending aorta impingement. J Biomech Eng. 2014;136:021009.
Fung YC. Biomechanics: mechanical properties of living tissues. 2nd ed. New York: Springer; 1993.
Hansen L, Wan W, Gleason RL. Microstructurally motivated constitutive modeling of mouse arteries cultured under altered axial stretch. J Biomech Eng. 2009;131:101015.
Hariton I, de Botton G, Gasser TC, Holzapfel GA. Stress-driven collagen fiber remodeling in arterial walls. Biomech Model Mechanobiol. 2007;6:163–75.
Hollander Y, Durban D, Lu X, Kassab GS, Lanir Y. Experimentally validated microstructural 3D constitutive model of coronary arterial media. J Biomech Eng. 2011;133:031007.
Holzapfel GA, Eberlein R, Wriggers P, Weizascker HW. Large strain analysis of soft biological membranes: formulatin and finite element analysis. Comput Methods Appl Mech Eng. 1996;132:45–61.
Holzapfel GA, Ogden RW. Constitutive modelling of passive myocardium: a structurally based framework for material characterization. Philos Transact A Math Phys Eng Sci. 2009;367:3445–75.
Horowitz A, Lanir Y, Yin FC, Perl M, Sheinman I, Strumpf RK. Structural three-dimensional constitutive law for the passive myocardium. J Biomech Eng. 1988;110:200–7.
Hughes TJR. The finite element method: linear static and dynamic finite element analysis. New York: Dover; 2000.
Jor JW, Nash MP, Nielsen PM, Hunter PJ. Estimating material parameters of a structurally based constitutive relation for skin mechanics. Biomech Model Mechanobiol. 2011;10:767–78.
Joyce EM, Moore JJ, Sacks MS. Biomechanics of the fetal membrane prior to mechanical failure: review and implications. Eur J Obstet Gynecol Reprod Biol. 2009;144 Suppl 1:S121–7.
Kao PH, Lammers S, Tian L, Hunter K, Stenmark KR, Shandas R, Qi HJ. A microstructurally-driven model for pulmonary artery tissue. J Biomech Eng. 2011;133:051002.
Kenedi RM, Gibson T, Daly CH. Biomechanics and related bio-engineering topics. In: Kenedi RM, editor. Bioengineering studies of human skin. Oxford: Pergamon; 1965. p. 147–58.
Lake SP, Barocas VH. Mechanical and structural contribution of non-fibrillar matrix in uniaxial tension: a collagen-agarose co-gel model. Ann Biomed Eng. 2011;39:1891–903.
Lanir Y. A structural theory for the homogeneous biaxial stress-strain relationships in flat collagenous tissues. J Biomech. 1979;12:423–36.
Lanir Y. Constitutive equations for fibrous connective tissues. J Biomech. 1983;16:1–12.
Marsden JE, Hughes TJR. Mathematical foundations of elasticity. Don Mills: Dover; 1983.
Mirnajafi A, Raymer J, Scott MJ, Sacks MS. The effects of collagen fiber orientation on the flexural properties of pericardial heterograft biomaterials. Biomaterials. 2005;26:795–804.
Mitton R. Mechanical properties of leather fibers. J Soc Leather Trades' Chem. 1945;29:169–94.

Prot V, Skallerud B, Holzapfel G. Transversely isotropic membrane shells with application to mitral valve mechanics. Constitutive modelling and finite element implementation. Int J Numer Methods Eng. 2007;71:987–1008.

Sacks M. Biaxial mechanical evaluation of planar biological materials. J Elast. 2000;61:199–246.

Sacks MS. Incorporation of experimentally-derived fiber orientation into a structural constitutive model for planar collagenous tissues. J Biomech Eng. 2003;125:280–7.

Soong TT, Huang WN. A stochastic model for biological tissue elasticity in simple elongation. J Biomech. 1973;6:451–8.

Sun W, Sacks MS. Finite element implementation of a generalized Fung-elastic constitutive model for planar soft tissues. Biomech Model Mechanobiol. 2005;4(2-3):190–9.

Sun W, Sacks MS, Sellaro TL, Slaughter WS, Scott MJ. Biaxial mechanical response of bioprosthetic heart valve biomaterials to high in-plane shear. J Biomech Eng. 2003;125:372–80.

Tong P, Fung YC. The stress-strain relationship for the skin. J Biomech. 1976;9:649–57.

Tonge TK, Voo LM, Nguyen TD. Full-field bulge test for planar anisotropic tissues: part II–a thin shell method for determining material parameters and comparison of two distributed fiber modeling approaches. Acta Biomater. 2013;9:5926–42.

Waldman SD, Michael Lee J. Boundary conditions during biaxial testing of planar connective tissues. Part 1: dynamic behavior. J Mater Sci Mater Med. 2002;13:933–8.

Waldman SD, Sacks MS, Lee JM. Boundary conditions during biaxial testing of planar connective tissues: Part II: Fiber orientation. J Mater Sci Lett. 2002;21:1215–21.

Wognum S, Schmidt DE, Sacks MS. On the mechanical role of de novo synthesized elastin in the urinary bladder wall. J Biomech Eng. 2009;131:101018.

Chapter 18
Fibers to Organs: How Collagen Fiber Properties Modulate the Closing Behavior of the Mitral Valve

Chung-Hao Lee and Michael S. Sacks

Abstract We developed a micro- and macro anatomically accurate MV finite element model by incorporating actual fiber microstructural architecture and a realistic structure-based constitutive model. Comparative and parametric studies were conducted to identify essential model fidelity and information for achieving desirable accuracy. More importantly, for the first time, the interrelationship between the local fiber ensemble behavior and the organ-level MV closing behavior was investigated using a computational simulation. These novel results indicated not only the appropriate parameter ranges, but also the importance of the microstructural tuning (i.e., straightening and reorientation) of the collagen/elastin fiber networks at the microscopic tissue level for facilitating the proper coaptation and natural functioning of the MV apparatus under physiological loading at the organ level. The proposed computational model would serve as a logical first step toward our long-term modeling goal—facilitating simulation-guided design of optimal surgical repair strategies for treating diseased MVs with significantly enhanced durability.

18.1 Introduction

The mitral valve (MV) is one of the four heart valves located between the left atrium and left ventricle, and it regulates the flow between these two respective chambers. The MV is considered an "apparatus" (Komeda et al. 1997) with four primary components: the anterior and posterior leaflets (MVAL and MVPL), the papillary muscles (PMs) that project from the left ventricular wall, the chordae tendineae that provide connections between the papillary muscles and the MV leaflets and prevent the leaflets from prolapse during MV functioning, and the annulus that is part of the conceptual transition between the MV leaflets and the left atrium. The

---

C.-H. Lee • M.S. Sacks, Ph.D. (✉)
Center for Cardiovascular Simulation, Department of Biomedical Engineering, Institute for Computational Engineering and Sciences (ICES), The University of Texas at Austin, 201 East 24th Street, 1 University Station C0200, POB 5.236, Austin, TX 78712, USA
e-mail: msacks@ices.utexas.edu

G.S. Kassab, M.S. Sacks (eds.), *Structure-Based Mechanics of Tissues and Organs*, DOI 10.1007/978-1-4899-7630-7_18

opening of the MV allows blood flow from the left atrium to left ventricle in diastole of the cardiac cycle. In clinical practice, MV repair and replacement are the two options for treating MV diseases, such as mitral regurgitation (MR) presumably caused by MV prolapse (Adams et al. 2010; Gillinov et al. 2008) and ischemic mitral regurgitation (IMR) due to post-infarction ventricular remodeling (Gorman and Gorman 2006). After two decades of emphasis on valve replacement, cardiac surgeons have gradually turned to MV surgical repair (Shuhaiber and Anderson 2007; Vassileva et al. 2011) to treat valvular dysfunctions and disease. Promising MV repair concepts include *leaflet augmentation* that restores leaflet mobility (Robb et al. 2011; Kincaid et al. 2004; Jassar et al. 2012), *saddle-shaped annuloplasty* that reinstates normal annular shape (Mahmood et al. 2010; Jensen et al. 2011), *leaflet resection* for repairing leaflet prolapse (Carpentier et al. 1980; Carpentier 1983), and *chordal replacement* for ruptured or inadequately functioning native chordae (Frater et al. 1990; David 1999). However, recent long-term studies showed an unsatisfactory recurrence rate of severe MR 3–5 years after surgical repair (Gillinov et al. 2008; Braunberger et al. 2001; Flameng et al. 2003, 2008). It has been suggested that excessive tissue stress and the resulting strain-induced tissue failure are possible etiological factors controlling the success of MV surgical repair (David et al. 2005; Schoen and Levy 2005). The resulting surgery-induced excessive tissue stresses will then lead to changes in MV interstitial cell (MVIC) metabolism and protein biosynthesis, which are essential in understanding the mechanobiological responses at the organ, tissue, and cellular levels (Grande-Allen et al. 2005; Rabkin-Aikawa et al. 2004; Dal-Bianco et al. 2009).

Based on these observations, we hypothesized that restoration of MV leaflet tissue stresses in MV repair techniques that most closely approximate the normal range would ultimately lead to improved repair durability. This would occur through the restoration of normal MVIC biosynthetic responses and homeostatic state. Although extant MV models represent an important step toward developing physiologically realistic MV computational models, few in vitro or in vivo validations have been thoroughly performed. Moreover, organ-level computational modeling only gives us basic information about the deformed geometry and overall pseudo-elastic responses. *The ability to reproduce the native valve function is only the first step, and understanding why the MV is designed, in its natural and functional way, is essential to provide insights into the MV apparatus considering normal, pathological, and optimally repaired scenarios.* Therefore, a multi-scale, biomechanical computational modeling framework could thus provide a means for accomplishing this by connecting cellular transduction to adaptions of tissue structure and further to organ-level mechanical responses. As a feasible first step toward our long-term modeling goals, we aimed, in the present study, at developing an anatomically and microstructurally accurate MV finite element model in conjunction with the direct use of realistic MV leaflet tissue microstructure and realistic structurally-driven constitutive model to investigate the effects of the MV leaflet microstructure and mechanical properties on the MV closing behavior.

18.2 Methods

We developed a comprehensive framework for modeling the MV (Fig. 18.1), which consisted of four key components: (1) image-based organ-level geometry, (2) incorporation of the fiber morphology, (3) realistic tissue-level mechanical behavior characterized from available biomechanical testing data, and (4) applicable boundary and loading conditions simulating MV closing mechanism. This computational framework was then implemented, by integrating an anatomically accurate FE model based on high-resolution micro-CT images, structurally-driven constitutive model for describing MV leaflet mechanical behaviors, and mapping of the fiber microstructural architecture via affine fiber kinematics, into a standard nonlinear, large-deformation FE analysis procedure. The details of the proposed model were provided in the subsequent sections.

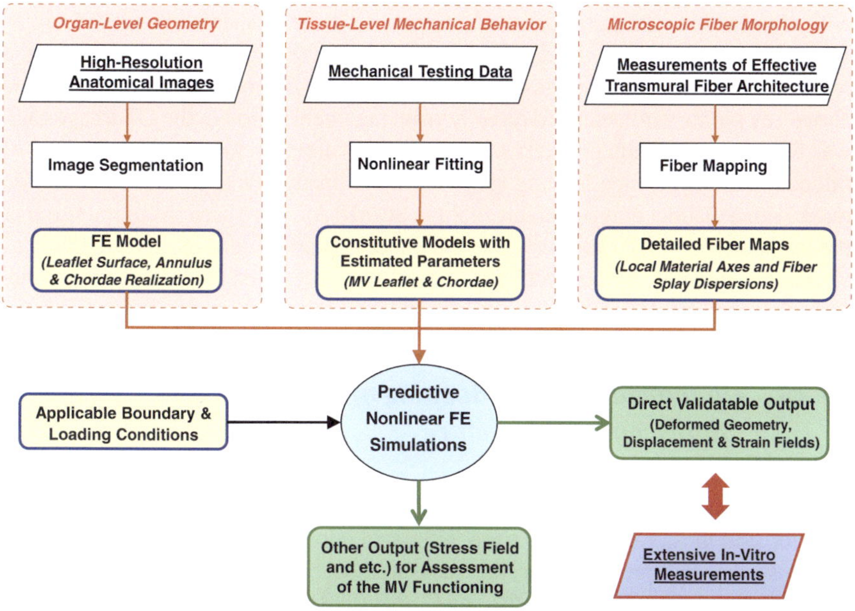

Fig. 18.1 A schematic diagram of the proposed FE computational framework for modeling the MV with the following four key ingredients: (1) anatomically accurate organ-level geometry obtained from high-resolution images for construction of the computationally tractable MV FE model, (2) constitutive models for MV components based on available mechanical testing data, (3) microscopic fiber morphological architecture incorporated with the FE model via the mapping based on affine fiber kinematics, and (4) applicable boundary and loading conditions

18.2.1 Acquisition of the In Vitro MV Geometry

A fresh ovine heart from a 35 kg sheep was acquired from a local USDA approved abattoir and imaged in a micro-CT system. The stack of 3D micro-CT images were segmented using ScanIP (Simpleware Ltd., United Kingdom) to obtain anatomically accurate geometry of the MV apparatus at the stress-free, 30 and 70 mmHg pressure-loaded states. The atrial and ventricular leaflet surfaces were identified from the segmented MV geometry at each state, and a median MV leaflet surface was reconstructed in Geomagic Studio (Morrisville, NC, USA) based on these two leaflet surfaces. Local MV leaflet thicknesses were determined using the distance between the spline-parametrized atrial and ventricular surfaces and were part of the FE input data. Then, the FE mesh at the fully-open (reference) configuration associated with this MV leaflet median surface was generated in HyperMesh. For idealization of the MV chordae tendineae using truss elements, 3D locations of the landmark points, such as MV papillary muscle attaching points, points associated with chordae branching, transitional points with distinguishable changes in cross-sectional area, and MV leaflet attaching points, were quantified based on the reconstructed micro-CT images. Note that the cross-sectional area of each of the above key points was measured directly from the voxel region in the CT images and was used for the section property of each 3D truss element. Finally, 3D positions of fiducial markers at three loading states were obtained via a separate segmentation mask with a distinct gray-scale threshold in ScanIP and were used in computation of the displacement errors and served as the material points in the strain calculation as well as the mapping of fiber architecture as will be introduced in the next subsection.

18.2.2 Quantification and Mapping of the Fiber Microscopic Architecture

After acquisition of the micro-CT image data at various configurations, the MV anterior and posterior leaflets were separated along the anterior and posterior commissures and prepared for the measurement of the gross (effective) fiber microscopic architecture using the small angle light scattering (SALS) technique developed previously (Sacks et al. 1997). In brief, each excised, flatten MV leaflet was placed in glycerol for dehydration and scanned in the SALS device. The light scattering patterns were measured at 254-μm increments over the entire MV leaflet specimen, and the measurements were analyzed to quantify the effective fiber orientation distribution function (ODF) $\Gamma(\theta)$ with the local preferred fiber direction μ_f and the degree of fiber splay σ_f derived from the orientation index (OI) by assuming a Gaussian distribution. Note that the OI value was defined as the angular width in which 50 % of the total number of fibers occurs, and the fiber splay alignment can be computed via $\mathrm{erf}\left(\frac{OI}{2\sqrt{2}\sigma_f}\right) - \frac{1}{2}\mathrm{erf}\left(\frac{180^\circ}{2\sqrt{2}\sigma_f}\right) = 0$ (Sacks et al. 1997; Sacks 2003).

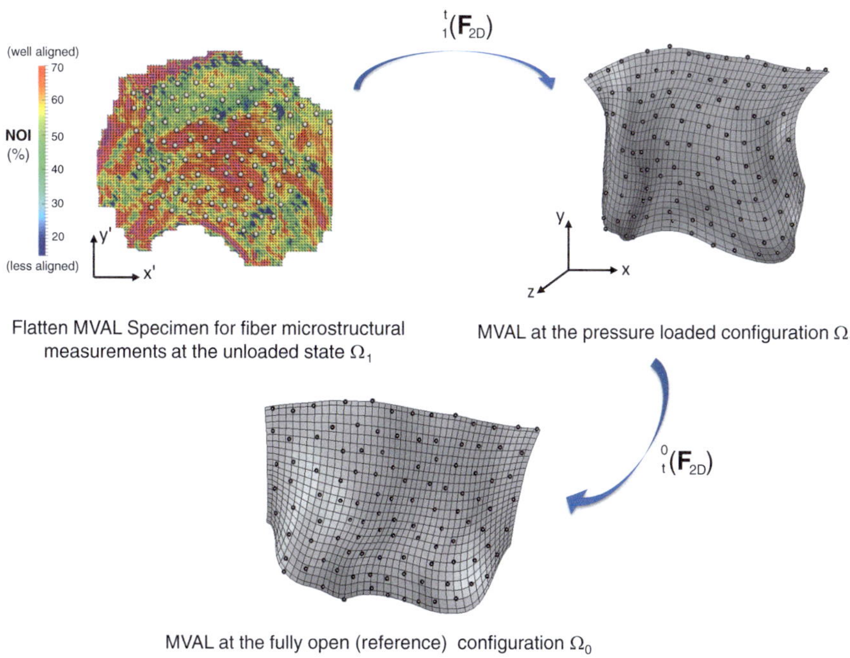

Fig. 18.2 A schematic diagram of the proposed technique for mapping the collagen fiber architecture onto the 3D FE mesh (MVAL as an illustration). Fiber splay microstructural morphology was measured via the SALS technique for the flatten MV leaflet tissues at the unloaded state Ω_1, showing both the preferred fiber directions and strengths of fiber splay alignment. The measured fiber architecture was first mapped onto the generated MVAL FE mesh at the pressure loaded state Ω_t via the mapping based on affine fiber kinematics and using the deformation gradient ${}^{t}_{1}(\mathsf{F}_{2D})$ computed from the fiducial marker positions between states Ω_1 and Ω_t. The corresponding fiber microstructural architecture was then transformed onto the MVAL FE mesh at the reference state Ω_0 using ${}^{0}_{t}(\mathsf{F}_{2D})$ to obtain element local material axes and the degree of fiber splay for the simplified structural constitutive model

To incorporate the measured fiber microstructural architecture with the constructed MV FE model, we developed the following two-stage mapping algorithm (Fig. 18.2) based on *affine fiber kinematics* (Lee et al. 2015; Lee et al. n.d.). Briefly, the fundamental property of affine fiber kinematics states that the total number of fibers remains constant within an infinitesimal angular element undergoing given in-plane deformation $\mathbf{F}_{2D}$, and the fiber ODF at the deformed configuration $\Gamma(\beta)$ can then be related to the ODF at the reference state $\Gamma(\theta)$ by

$$\Gamma(\beta) = \Gamma(\theta)\,\frac{\mathsf{N}(\theta)\cdot[\mathsf{C}_{2D}\mathsf{N}(\theta)]}{J_{2D}} = \Gamma(\theta)\,\frac{\lambda_{\mathrm{N}}^{2}}{J_{2D}}, \qquad (18.1)$$

where $\mathbf{N}(\theta)$ denotes the unit vector of the local fiber axis at the undeformed configuration, $\mathbf{C}_{2D} = (\mathbf{F}_{2D})^{\mathrm{T}}(\mathbf{F}_{2D})$ is the right Cauchy–Green deformation tensor,

J_{2D} is the determinant of the in-plane deformation gradient, and $\lambda_{\mathbf{N}}$ is the stretch ratio along **N**. The fiber angle at the current configuration can be computed by

$$\beta = \tan^{-1}\left[F_{21}\cos(\theta) + F_{22}\sin(\theta)/F_{11}\cos(\theta) + F_{12}\sin(\theta)\right]. \tag{18.2}$$

This rendered the general framework for describing the fiber microstructural architecture with tissue-level deformations. The MV leaflets at the unloaded, full pressure loaded (where the SALS measurements were made), and fully-open state are Ω_1, Ω_{t}, and Ω_0, respectively, with the corresponding local material axes $\left\{\mathbf{e}_1^{(1)} \equiv [1, 0], \mathbf{e}_2^{(1)} \equiv [0, 1]\right\}$, $\{\mathbf{e}_1^{\mathrm{t}}, \mathbf{e}_2^{\mathrm{t}}\}$ and $\{\mathbf{e}_1^{0}, \mathbf{e}_2^{0}\}$, respectively. Since the fiber architectural measurements were made after the MV leaflets were fixed under 70 mmHg hydrostatic pressure, the local material axes at state Ω_{t} were determined by using the in-plane deformation gradient ${}_1^{\mathrm{t}}(\mathrm{F}_{2D})$ computed based on the fiducial marker positions between state Ω_{t} and state Ω_1

$$\mathbf{e}_1^{\mathrm{t}} = {}_1^{\mathrm{t}}(\mathrm{F}_{2D}) \cdot \mathbf{e}_1^{(1)} \text{ and } \mathbf{e}_2^{\mathrm{t}} = {}_1^{\mathrm{t}}(\mathrm{F}_{2D}) \cdot \mathbf{e}_2^{(1)}. \tag{18.3}$$

Therefore, we reached the fiber dispersion and local material axes for both leaflets at the pressure-loaded state Ω_{t}. Following the similar procedure, the corresponding fiber ODF and fiber angle at the reference state Ω_0, which can be incorporated with a structure-based constitutive model as introduced in the subsequent section and served as part of the FE input, were calculated via the reverse affine mapping in the second step of our mapping algorithm using the in-plane deformation gradient ${}_{\mathrm{t}}^{0}(\mathrm{F}_{2D})$ computed from the fiducial marker positions at states Ω_{t} and Ω_0.

18.2.3 Constitutive Models for the MV Apparatus

We employed an incompressible, transversely isotropic simplified structural model (SSM) in this work to simulate the in vitro passive mechanical behaviors of the MV leaflet tissues by assuming a homogeneous pseudo-hyperelastic, collagenous fiber-reinforced composite material (Sacks 2003; Fung 1993; Lanir 1983; Fan and Sacks 2014). The adopted SSM enables the integration of the tissue composition and structure into the function and mechanics of the MV leaflet tissues and allows the full in-plane coupling responses within physiological loading range and the direct use of the mapped fiber microstructural architecture. In brief, the MV leaflet tissues are assumed to consist of two major load-bearing components, a ground matrix material with nonfibrous substances and water and an effective fibrous material model that homogenized the elastin and type-I collagen fibers, and the corresponding total strain energy function (SEF) Ψ is expressed as

$$\Psi^{\text{leaflet}} = \Psi_{\text{m}} + \Psi_{\text{ens}} = \int_{-\pi/2}^{\pi/2} \Gamma(\theta)\, \Psi_{\text{f}}\left[E_{\text{f}}(\theta)\right] d\theta + \frac{\mu_{\text{m}}}{2}(I_1 - 3) + p(J-1), \tag{18.4}$$

where Ψ_{m} is the SEF associated with the matrix component using a neo-Hookean material with the neo-Hookean constant μ_{m} responsible for the low-strain response and the incompressibility of the planar tissues, Ψ_{ens} is the SEF associated with the effective fiber networks as the sum of the SEF of each individual fiber Ψ_{f}, $\Gamma(\theta)$ is the ODF of the fiber ensembles, $E_{\text{f}}(\theta) = \mathsf{N}^{\mathsf{T}}(\theta)\,\mathsf{E}\mathsf{N}(\theta)$ is the effective fiber strain, $\mathsf{N}(\theta) = [\cos(\theta), \sin(\theta), 0]^{\mathsf{T}}$, $\mathsf{E} = (\mathsf{C} - \mathsf{I})/2$ is the Green-Lagrange strain tensor, $\mathsf{C} = \mathsf{F}^{\mathsf{T}}\mathsf{F}$ is the right Cauchy–Green deformation tensor, F is the deformation gradient tensor, I is the identity tensor, $I_1 = \mathsf{trace}(\mathsf{C})$, $J = \det(\mathsf{F})$, and p is the Lagrange multiplier to enforce the incompressibility of the planar tissues. Based on the pseudo-hyperelastic formation (Fung 1993), the second Piola–Kirchhoff stress tensor $\mathsf{S}^{\text{leaflet}}$ can be derived by

$$\mathsf{S}^{\text{leaflet}} = \frac{\partial \Psi^{\text{leaflet}}}{\partial \mathsf{E}} = \int_{-\pi/2}^{\pi/2} \Gamma(\theta)\, S_{\text{f}}\left[E_{\text{f}}(\theta)\right] \mathsf{N}(\theta) \otimes \mathsf{N}(\theta)\, d\theta + \mu_{\text{m}}\left(\mathsf{I} - C_{33}\mathsf{C}^{-1}\right). \tag{18.5}$$

Here, $C_{33} = 1/\left(C_{11}C_{22} - C_{12}^2\right)$ is the consequence of the incompressibility condition, $p = -\mu_{\text{m}}C_{33}$ is derived from the plane–stress condition ($\mathsf{S}_{33}^{\text{leaflet}} = 0$), and $\otimes$ denotes the dyadic tensor product. In this work, we adopted an exponential model with a terminal stiffness for highly nonlinear fiber stress–strain behavior (Fan and Sacks 2014)

$$S_{\text{f}}(E_{\text{f}}) = \begin{cases} c_0\left[\exp(c_1 E_{\text{f}}) - 1\right] & \text{for } E_{\text{f}} \le E_{\text{ub}} \\ c_0\left[\exp(c_1 E_{\text{ub}}) - 1\right] + c_0 c_1 \exp(c_1 E_{\text{ub}})(E_{\text{f}} - E_{\text{ub}}) & \text{for } E_{\text{f}} > E_{\text{ub}}. \end{cases} \tag{18.6}$$

Herein, S_{f} is the fiber stress, c_0 and c_1 are material constants, and E_{ub} is the cutoff fiber strain above which a linear fiber tangent modulus is considered. The effective fiber ODF was expressed by a Gaussian distribution function:

$$\Gamma(\theta) = \frac{\exp\left[-\frac{(\theta-\mu)^2}{2\sigma^2}\right]}{\operatorname{erf}\left(\frac{\pi}{2\sqrt{2}\sigma}\right)\sqrt{2\pi}\sigma}, \tag{18.7}$$

where $\theta \in [-\pi/2, +\pi/2]$, and $\int_{-\pi/2}^{\pi/2} \Gamma(\theta)\, d\theta = 1$. The Cauchy stress tensor can then be obtained by the push-forward operation $\boldsymbol{\sigma}^{\text{leaflet}} = \mathbf{F}\mathbf{S}^{\text{leaflet}}\mathbf{F}^{\text{T}}$.

For modeling the MV chordae tendineae, we adopted an incompressible, isotropic hyperelastic material with the following stress–strain relationship

$$\mathrm{S}^{\text{chordae}} = \frac{\partial \Psi^{\text{chordae}}}{\partial E_{11}} = C_{10}\left[\exp\left(C_{01}E_{11}\right) - 1\right], E_{11} > 0, \tag{18.8}$$

where $E_{11} = \frac{1}{2}\left(F_{11}^2 - 1\right)$ is the uniaxial strain of the chordae tendineae, and C_{10} and C_{01} are the material constants.

18.2.4 Parameter Estimation Based on the Inverse Modeling Approach

For characterization of the planar mechanical behavior of the ovine MV leaflet tissues, we employed load-controlled biaxial testing protocols previously developed (Grashow et al. 2006). In brief, square central regions (20 mm x 20 mm) of ovine MV anterior and posterior leaflets were dissected, submerged in isotonic saline, and then connected to the biaxial testing equipment with the tissue-preferred fiber (circumferential, *C*) and cross-preferred fiber (radial, *R*) directions approximated aligned with the device axes. The stress and strain behaviors in the two directions under equi-biaxial loading were recorded and used for the following parameter estimation. Note that the shear strains in this study were negligible.

To estimate the model parameters associated with the SSM, we adopted an inverse modeling approach based on the nonlinear fit of the stress–strain curves between the simulated results and the biaxial testing data. Briefly, the SSM was first implemented in FE software ABAQUS (SIMULIA, Dassault Systèmes, Providence, RI, USA) via the user-defined material subroutine (cf. details in Fan and Sacks 2014). The equi-biaxial loading protocol was then modeled, and the simulated stress–strain behaviors in both circumferential and radial directions were compared to the experimental data for characterizing the material parameters by iteratively minimizing the errors of the stress–strain curves between the simulation results and experimental data. Similarly, the uniaxial stress–strain behavior of the chordae tendineae was simulated by three-dimensional truss elements (T3D2). Prescribed force and fixed boundary condition were applied on the opposite ends of the truss element, respectively. The simulated stress–strain curves of both the basal and marginal chordae were compared to available uniaxial testing data (Ritchie et al. 2006), and parameters C_{10} and C_{01} were then characterized.

18.2.5 Finite Element Simulations of the MV Closing Behavior

We first assumed that the interaction between blood flow and the MV apparatus is idealized by transvalvular surface pressure loads acting on the MV leaflets, and the effect of the left ventricular contraction on the MV mechanics is mimicked by prescribed boundary motions of the MV annulus and PMs. Alterations of the MV stress and strain fields in response to external loading were of the interest in the current study, and simulations of MV closure were performed in ABAQUS. Nodal displacements and rotations, and elemental strain and stress fields were the primary output from the simulations and were post-processed via a Python script to compute the predicted fiducial marker 3D positions as well as the in-surface maximum and minimum principal stretches compared with the in vitro data at 30 and 70 mmHg transvalvular pressures. The following FE modeling specifications were considered:

1. Explicit dynamics was utilized as a more general computational framework for future extensions, such as in vivo modeling and surgical simulations. Proper mass scaling ($dt = 1 \text{ x } 10^{-6}$) of the inertial effects was employed to simulate the quasi-static behavior associated with the in vitro experimental condition.
2. Shell (S4) and 3D uniaxial truss (T3D2) elements were used for the MV leaflets and chordae, respectively.
3. Spatially varied, element-based thicknesses determined from anatomically accurate micro-CT images were supplied as part of input using DISTRIBUTION TABLE (length).
4. Element-based local material coordinate obtained from the mapped preferred fiber directions was defined using ORIENTATION feature along with DISTRIBUTION TABLE (coor3d).
5. User-defined subroutine VUSDFLD was adopted to specify element-based fiber splay dispersion σ.
6. Constitutive models of the MV apparatus were implemented in user-defined subroutine VUMAT.
7. AMPLITUDE feature in conjunction with DSLOAD was used to prescribe the transvalvular pressure loading.
8. Although clamped boundary conditions were considered in the in vitro experiments, subroutine VDISP was implemented for prescribing displacement boundary conditions of the MV annulus and PM tips.
9. General self-contact algorithm was adopted for handling the coaptation and interactions of the MV leaflets during MV closing process, and a separate element set consisting of all leaflet elements is defined for treating both leaflets as a contiguous entity and their faces toward the left atrium as potential contacting surfaces. A surface interaction/behavior with a sliding-friction coefficient of 0 (frictionless) and a linear pressure-overclosure coefficient of 0.5.

18.2.6 Comparative and Parametric Studies

In addition to thorough validations with extensive in vitro data, we further investigated the effect of the MV leaflet structural property on the numerically predicted closing behavior of the MV through simulations of the following four perturbations with an increasing level of modeling fidelity: level I—an exponential-type isotropic material ($\sigma = 180°$), level II—a transversely isotropic material with *uniformly curvilinear fiber directions*[1] and *identical fiber dispersions*, level III—a transversely isotropic material with *uniformly curvilinear fiber directions* and *mapped fiber dispersions* based on (18.1), and level IV—a transversely isotropic material with *mapped fiber directions* and *mapped fiber dispersions* based on the proposed mapping algorithms. Comparisons of the numerically predicted displacements and in-plane principal stretches were made among these four cases to examine whether a higher model fidelity yielded more accurate solutions. Moreover, we conducted a series of parametric studies on how the micromechanical ensemble fiber stress–strain behavior (Fig. 18.5) affects the predicted MV deformations (Fig. 18.5), von Mises stress field, and the deformed fiber architecture (represented by NOI values) subject to transvalvular pressure loading up to *100 mmHg*.

18.3 Results

18.3.1 FE Model Information

The final FE model constructed in this study for the native ovine MV was composed of 6836 nodes (6720 nodes corresponding to the MV leaflets; 192 nodes representing the MV annulus; and 16, 39, 48, and 13 nodes associated with the chordae branching points, chordae transitional points, leaflet attachment points, and papillary muscle tips, respectively), 6528 leaflet elements (2176 and 4352 four-node shell elements for the MVAL and MVPL, respectively), and 297 chordae tendineae 3D truss elements. For the anatomical information about the MV chordae tendineae, 4 and 2 intermediary chordae attaching to the leaflet belly regions were identified for the MVAL and MVPL, respectively, and 6 and 2 basal chordae were reconstructed for the MVPL and two commissure regions, respectively. Each of the above intermediary and basal chordae contained various numbers of marginal chordae branches attaching to the free-edge of the MV leaflets. Note that 189 nodes on the MV leaflet shell elements were connected with the above-mentioned leaflet attachment points *in a fork fashion* (Fig. 18.5) to avoid the

---

[1] A convective curvilinear cylindrical coordinate was adopted to describe the contiguous MV leaflet entity, and fiber directions were assumed to be all uniformly aligned with the circumferential direction in this coordinate system for the case study associated with *uniformly curvilinear fiber directions*.

stress concentration via the single-node attachment of the chordae truss elements to the leaflet elements. Element-based thicknesses determined from the median leaflet surfaces were 1.11 ± 0.29 and 0.93 ± 0.22 mm for the MVAL and MVPL, respectively.

18.3.2 Fiber Microstructural Mapping

The measured fiber microstructural information, including the preferred fiber direction μ_f and the OI value, was mapped onto the FE mesh of both leaflets (Fig. 18.3., top panel), accounting for the transformation and deformation between the excised state $\boldsymbol{\Omega}_1$ and the in vitro pressure-loaded state $\boldsymbol{\Omega}_t$. Note that the gross fibers were fairly continuous for the MVAL and MVPL, which provide smooth local material axes for contiguous leaflet elements, with higher aligned fibers observed in the central regions of both leaflets which sustain the greater amount of stretching under pressure loading. The fiber architecture associated with the FE mesh at the reference configuration (state $\boldsymbol{\Omega}_0$) was determined by the affine-transformation reverse mapping (Fig. 18.3, bottom panel). Clearly, the gross fibers were less aligned at state $\boldsymbol{\Omega}_0$, whereas the fibers at state $\boldsymbol{\Omega}_t$ were re-oriented due to stretches in both circumferential and radial direction, and became better aligned during the MV closing process.

18.3.3 Effects of Local Ensemble Fiber Mechanical Behavior on the MV Closure

We found that increasing curvature of the microscopic fiber stress–strain relationship led to *earlier* contact interaction and coaptation of both anterior and posterior leaflets at lower transvalvular pressure loading which corresponds to the low stress–strain region primarily governed by material parameter μ_m, and *less* leaflet deformation toward the atrial chamber at higher pressure loading which is predominantly influenced by the progressive engagement of the straightening of collagen and elastin fiber networks (Fig. 18.4). This finding underscores the importance of adopting a microstructurally-informed constitutive model, which is able to capture both the long toe-region and the rapid growth region in material nonlinearity of the MV leaflet tissue's overall stress–strain behavior, for simulating physiological deformed profiles of the functioning MV. The predicted von Mises stress field and the NOI values at 100 mmHg transvalvular pressure were compared among these models (Fig. 18.4), and the results showed that a linear ensemble fiber stress–strain model (red in Fig. 18.4) yielded *lower tissue stresses* and *better aligned fibers* typically in the central regions of both leaflets compared to the current model (blue in Fig. 18.4) and a fiber stress–strain model with much larger curvature (green in Fig. 18.4), *suggesting an important balancing between microscopic straightening and realignment of collagen fiber network and the overall tissue-level stresses for maintaining proper functioning of the MV apparatus*. These results highlight the essential role of the microscopic fiber mechanical properties played in the MV closing behavior as well as in preserving the realistic deformed geometry of the MV leaflets under physiological loading.

18.4 Discussion

18.4.1 Overall Findings and Implications

In this study, we developed a finite element computational model for analyses of the MV closure, which comprised of an anatomically accurate organ-level geometry reconstructed from high-resolution micro-CT images, the incorporation of the measured microscopic fiber morphology, and a realistic structure-based

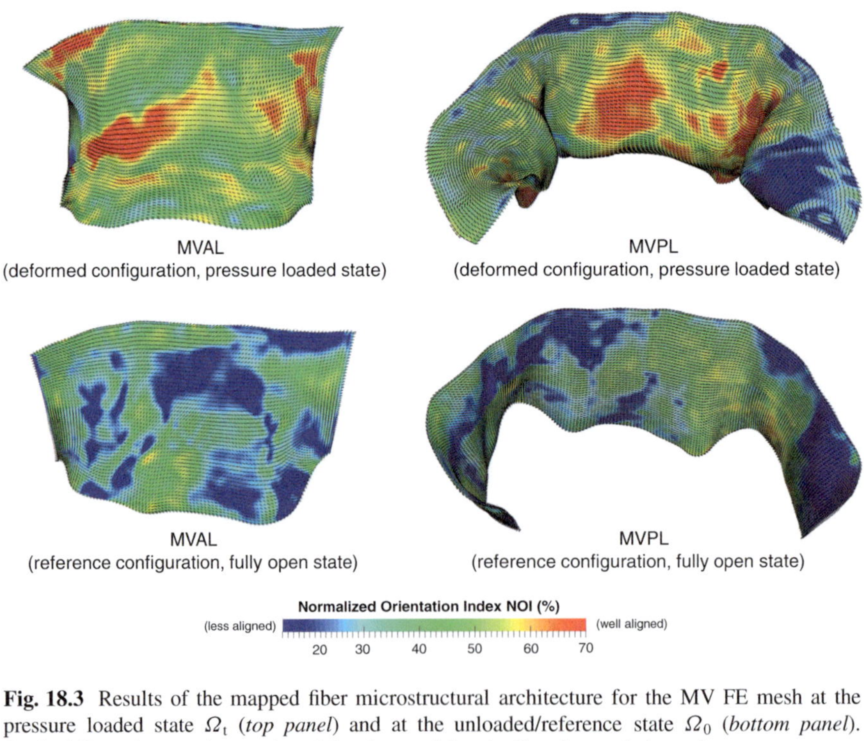

Fig. 18.3 Results of the mapped fiber microstructural architecture for the MV FE mesh at the pressure loaded state Ω_t (*top panel*) and at the unloaded/reference state Ω_0 (*bottom panel*). *Dashed lines* denote element-based local material axis direction and the *color contour* represents the strength of fiber splay. Note that the normalized orientation index is computed by NOI = (90° − OI)/90°

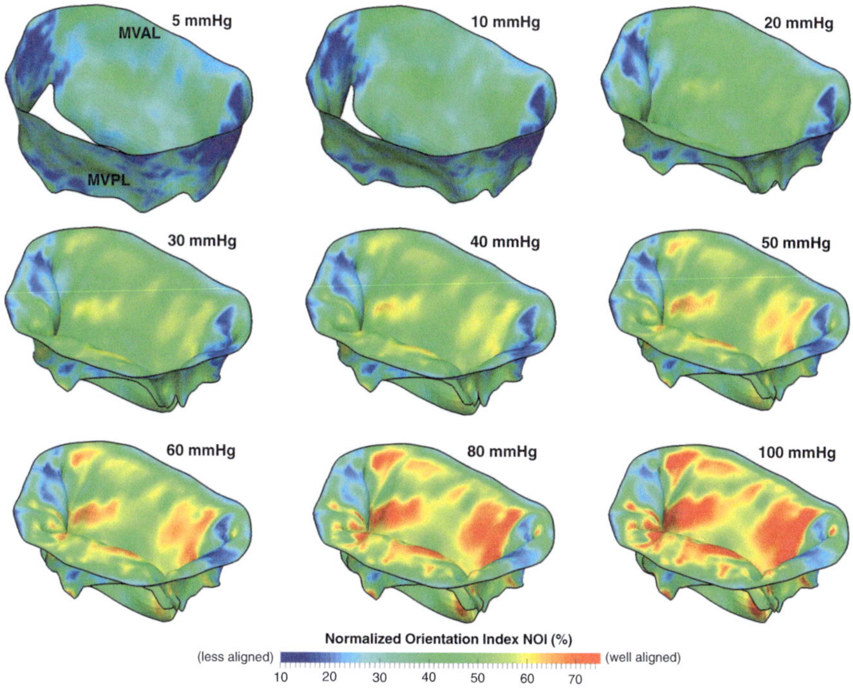

Fig. 18.4 Progressive predicted NOI values of the MV leaflets at various transvalvular pressure levels

constitutive model for direct employment of the fiber orientation and fiber splay dispersion. The proposed computational model addressed some of the challenges in the state-of-the-art MV modeling field (Choi et al. 2014; Wang and Sun 2013; Mansi et al. 2012; Stevanella et al. 2011; Votta et al. 2008) with our substantial improvements, such as the capability of reconstructing all the main components of the MV apparatus, a more detailed and precise description of the MV leaflet thicknesses, a better idealization of the chordae tendineae structures with realistic key points and cross-section areas identified, and the detailed mapping of actual fiber microstructure. More importantly, for the first time, we investigated the interrelationship between the local fiber ensemble mechanical behavior and the MV closing behavior (Fig. 18.4) via *computer simulations*. These novel results indicated not only the appropriate parameter ranges but also the importance of the microstructural tuning (like straightening and reorientation) of the collagen/elastin fiber networks at the microscopic tissue level for facilitating the proper coaptation and natural functioning of the MV apparatus under physiological loading at the organ level. Understanding the important underlying mechanisms of the normal MV allows us to reiterate and design the surgical repair procedures, and, in the meantime, maintain the above physiological loading scenario. Furthermore, the proposed computational framework with reasonable future extensions, as will be discussed

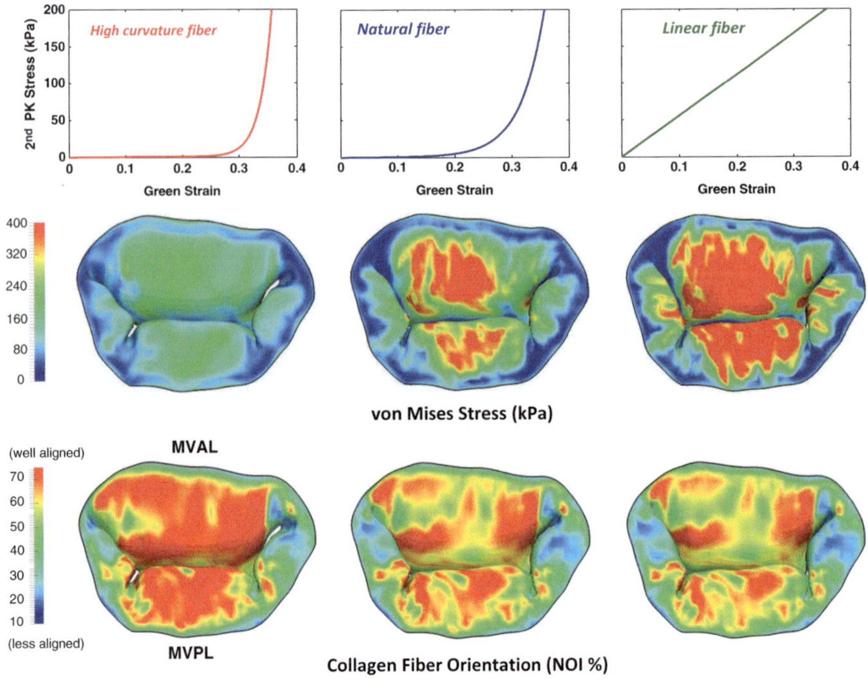

Fig. 18.5 Comparison of the predicted von Mises stress and predicted normalized orientation index (NOI) of the MV leaflets considering different curvatures of the ensemble fiber stress-strain relationship at 100 mmHg transvalvular pressure

in the subsequent section, will serve as the basis for utilizing simulations (1) to better understand the MV biomechanics and physiological functionality, including how the MV respond to disease- or repair-induced stress alteration particularly associated with the hypo- and hyper-physiological states, (2) to gain better insight into the remodeling of the MV due to disease progression and surgical intervention, and (3) to ultimately provide guideline for improving treatment strategies and surgery planning. Any such modeling effort would essentially depend on currently unavailable quantitative data for determination of the local stresses at the cellular, tissue, and organ levels within complex functioning physiological systems.

18.4.2 Conclusions

In summary, we have developed a computational framework for modeling of the functioning MV and successfully implemented a computational model into the nonlinear FE simulation procedure, which included micro- and meso scale anatomically accurate geometry of the MV apparatus, mapped fiber microstructural architecture, and a structure-based constitutive model directly integrated with tissue microstructure. This computational model, to our knowledge, is the first of this kind that has been thoroughly validated with extensive in vitro data for its model accuracy and robustness. This study underscored in parametric studies appropriate parameter ranges and micromechanical fiber responses for representing the proper coaptation and natural functioning behaviors of the MV apparatus under physiological loading. Understanding the underlying mechanisms of the normal MV could help in reiterating and designing surgical repair procedures for maintaining the proper physiological loading scenario. Finally, the computational model developed in this work serves as an important first step toward our long-term modeling goals, which are to employ computer simulations for assessing the performance of existing MV repair techniques and to provide insightful guidance in the design of optimal surgical repair strategies for treating diseased MVs with better restored functionality and improved long-term durability.

Acknowledgments Support from the National Institutes of Health (NIH) grants R01 HL119297 is greatly acknowledged. Dr. Chung-Hao Lee was supported in part by the American Heart Association (AHA) Postdoctoral Fellowship (14POST18160013) and a UT Austin ICES Postdoctoral Fellowship.

Conflict of Interest: None of the authors have a conflict of interest with the present work.

References

Adams DH, Rosenhek R, Falk V. Degenerative mitral valve regurgitation: best practice revolution. Eur Heart J. 2010;31(16):1958–66. doi:10.1093/eurheartj/ehq222. pii: ehq222. PubMed PMID: 20624767; PubMed Central PMCID: PMC2921508. Epub 2010/07/14.

Braunberger E, Deloche A, Berrebi A, Abdallah F, Celestin JA, Meimoun P, et al. Very long-term results (more than 20 years) of valve repair with carpentier's techniques in nonrheumatic mitral valve insufficiency. Circulation. 2001;104(12 Suppl 1):I8–11. PubMed PMID: 11568021.

Carpentier A. Cardiac valve surgery–the "French correction∞". J Thorac Cardiovasc Surg. 1983;86(3):323–37. PubMed PMID: 6887954.

Carpentier A, Chauvaud S, Fabiani JN, Deloche A, Relland J, Lessana A, et al. Reconstructive surgery of mitral valve incompetence: ten-year appraisal. J Thorac Cardiovasc Surg. 1980;79(3):338–48. PubMed PMID: 7354634.

Choi A, Rim Y, Mun JS, Kim H. A novel finite element-based patient-specific mitral valve repair: virtual ring annuloplasty. Bio-Med Mater Eng. 2014;24(1):341–7.

Dal-Bianco JP, Aikawa E, Bischoff J, Guerrero JL, Handschumacher MD, Sullivan S, et al. Active adaptation of the tethered mitral valve: insights into a compensatory mechanism for functional mitral regurgitation. Circulation. 2009;120(4):334–42. doi:10.1161/CIRCULATIONAHA.108.846782. Epub 2009/07/15. PubMed PMID: 19597052; PubMed Central PMCID: PMC2752046.

David TE. The Toronto SPV bioprosthesis: clinical and hemodynamic results at 6 years. Ann Thorac Surg. 1999;68(3 Suppl):S9–13.

David TE, Ivanov J, Armstrong S, Christie D, Rakowski H. A comparison of outcomes of mitral valve repair for degenerative disease with posterior, anterior, and bileaflet prolapse. J Thorac Cardiovasc Surg. 2005;130(5):1242–9.

Fan R, Sacks MS. Simulation of planar soft tissues using a structural constitutive model: finite element implementation and validation. J Biomech. 2014;47:2043–54.

Flameng W, Herijgers P, Bogaerts K. Recurrence of mitral valve regurgitation after mitral valve repair in degenerative valve disease. Circulation. 2003;107(12):1609–13. PubMed PMID: 12668494.

Flameng W, Meuris B, Herijgers P, Herregods M-C. Durability of mitral valve repair in Barlow disease versus fibroelastic deficiency. J Thorac Cardiovas Surg. 2008;135(2):274–82.

Frater R, Vetter H, Zussa C, Dahm M. Chordal replacement in mitral valve repair. Circulation. 1990;82(5 Suppl):IV125–30.

Fung YC. Biomechanics: mechanical properties of living tissues. 2nd ed. New York: Springer; 1993. p. 568.

Gillinov AM, Blackstone EH, Nowicki ER, Slisatkorn W, Al-Dossari G, Johnston DR, et al. Valve repair versus valve replacement for degenerative mitral valve disease. J Thorac Cardiovas Surg. 2008;135(4):885–93.e2.

Gorman III JH, Gorman RC. Mitral valve surgery for heart failure: a failed innovation? Semin Thorac Cardiovasc Surg. 2006;18(2):135–8. doi:10.1053/j.semtcvs.2006.07.003. Epub 2006/12/13.

Grande-Allen KJ, Borowski AG, Troughton RW, Houghtaling PL, Dipaola NR, Moravec CS, et al. Apparently normal mitral valves in patients with heart failure demonstrate biochemical and structural derangements: an extracellular matrix and echocardiographic study. J Am Coll Cardiol. 2005;45(1):54–61. PubMed PMID: 15629373.

Grashow JS, Yoganathan AP, Sacks MS. Biaxial stress-stretch behavior of the mitral valve anterior leaflet at physiologic strain rates. Ann Biomed Eng. 2006;34(2):315–25. PubMed PMID: 16450193.

Jassar AS, Minakawa M, Shuto T, Robb JD, Koomalsingh KJ, Levack MM, et al. Posterior leaflet augmentation in ischemic mitral regurgitation increases leaflet coaptation and mobility. Ann Thorac Surg. 2012;94(5):1438–45.

Jensen MO, Jensen H, Levine RA, Yoganathan AP, Andersen NT, Nygaard H, et al. Saddle-shaped mitral valve annuloplasty rings improve leaflet coaptation geometry. J Thorac Cardiovasc Surg. 2011;142(3):697–703. doi:10.1016/j.jtcvs.2011.01.022. Epub 2011/02/19. PubMed PMID: 21329946; PubMed Central PMCID: PMC3224846.

Kincaid EH, Riley RD, Hines MH, Hammon JW, Kon ND. Anterior leaflet augmentation for ischemic mitral regurgitation. Ann Thorac Surg. 2004;78(2):564–8. doi:10.1016/j.athoracsur.2004.02.040; discussion 8. Epub 2004/07/28. PubMed PMID: 15276520.

Komeda M, Glasson JR, Bolger AF, Daughters II G, MacIsaac A, Oesterle S, et al. Geometric determinants of ischemic mitral regurgitation. Circulation. 1997;96(9 Suppl):II-128–33.

Lanir Y. Constitutive equations for fibrous connective tissues. J Biomech. 1983;16:1–12.

Lee CH, Carruthers CA, Ayoub S, Gorman RC, Gorman 3rd JH, Sacks MS. J Theor Biol. 2015;373:26–9. doi:10.1016/j.jtbi.2015.03.004.

Lee CH, Amini R, Yusuke S, Carruthers CA, Ankush A, Gorman RC, et al. Mitral valves: a computational framework. In: Suvranu D, Kuhl E, Hwang W, editors. Multiscale modeling in biomechanics and mechanobiology. London: Springer; 2015.

Mahmood F, Gorman III JH, Subramaniam B, Gorman RC, Panzica PJ, Hagberg RC, et al. Changes in mitral valve annular geometry after repair: saddle-shaped versus flat annuloplasty rings. Ann Thorac Surg. 2010;90(4):1212–20. doi:10.1016/j.athoracsur.2010.03.119. Epub 2010/09/28. pii: S0003-4975(10)00938-0. PubMed.

Mansi T, Voigt I, Georgescu B, Zheng X, Mengue EA, Hackl M, et al. An integrated framework for finite-element modeling of mitral valve biomechanics from medical images: application to MitralClip intervention planning. Med Image Anal. 2012;16(7):1330–46.

Rabkin-Aikawa E, Farber M, Aikawa M, Schoen FJ. Dynamic and reversible changes of interstitial cell phenotype during remodeling of cardiac valves. J Heart Valve Dis. 2004;13(5):841–7. PubMed PMID: 15473488.

Ritchie J, Jimenez J, He Z, Sacks MS, Yoganathan AP. The material properties of the native porcine mitral valve chordae tendineae: An in vitro investigation. J Biomech. 2006;39(6):1129–35. PubMed PMID: 16549101.

Robb JD, Minakawa M, Koomalsingh KJ, Shuto T, Jassar AS, Ratcliffe SJ, et al. Posterior leaflet augmentation improves leaflet tethering in repair of ischemic mitral regurgitation. Eur J Cardiothorac Surg. 2011;40(6):1501–7. doi:10.1016/j.ejcts.2011.02.079. PubMed PMID: 21546260: Epub 2011/05/07.

Sacks MS. Incorporation of experimentally-derived fiber orientation into a structural constitutive model for planar collagenous tissues. J Biomech Eng. 2003;125(2):280–7. PubMed PMID: 12751291.

Sacks MS, Smith DB, Hiester ED. A small angle light scattering device for planar connective tissue microstructural analysis. Ann Biomed Eng. 1997;25(4):678–89.

Schoen FJ, Levy RJ. Calcification of tissue heart valve substitutes: progress toward understanding and prevention. Ann Thorac Surg. 2005;79(3):1072–80. PubMed PMID: 15734452.

Shuhaiber J, Anderson RJ. Meta-analysis of clinical outcomes following surgical mitral valve repair or replacement. Eur J Cardiothorac Surg. 2007;31(2):267–75.

Stevanella M, Maffessanti F, Conti CA, Votta E, Arnoldi A, Lombardi M, et al. Mitral valve patient-specific finite element modeling from cardiac MRI: application to an annuloplasty procedure. Cardiovas Eng Tech. 2011;2(2):66–76.

Vassileva CM, Boley T, Markwell S, Hazelrigg S. Meta-analysis of short-term and long-term survival following repair versus replacement for ischemic mitral regurgitation. Eur J Cardiothorac Surg. 2011;39(3):295–303.

Votta E, Caiani E, Veronesi F, Soncini M, Montevecchi FM, Redaelli A. Mitral valve finite-element modelling from ultrasound data: a pilot study for a new approach to understand mitral function and clinical scenarios. Philos Transact A Math Phys Eng Sci. 2008;366(1879): 3411–34. doi:10.1098/rsta.2008.0095. Epub 2008/07/08. PubMed PMID: 18603525.

Wang Q, Sun W. Finite element modeling of mitral valve dynamic deformation using patient-specific multi-slices computed tomography scans. Ann Biomed Eng. 2013;41(1):142–53. doi:10.1007/s10439-012-0620-6.

Chapter 19
Mesoscale Structural Models in the Growing Pulmonary Artery

Bahar Fata, Will Zhang, Rouzbeh Amini, and Michael S. Sacks

Abstract We utilized the extensive experimental measurements of the growing ovine PA from our previous study (J. Biomech Eng. 2013 Jul 1;135(7):71010–12) to develop a structural constitutive model for the PA wall tissue. Novel to the present approach was the treatment of the elastin network as a distributed fiber network rather than a continuum phase. We then utilized this model to delineate structure–function differences in the PA wall at the juvenile and adult stages. Overall, the predicted elastin exhibited minor regional differences moduli remained largely unchanged with age and region (in the range of 150–200 kPa). Similarly, the predicted collagen moduli ranged from ~1600 to 2700 kPa in the four regions studied in the juvenile state. Interestingly, we found for the medial region that the elastin and collagen fiber splay underwent opposite changes (collagen standard deviation juvenile = 17° to adult = 28°, elastin standard deviation juvenile = 35° to adult = 27°), along with a trend toward more rapid collagen fiber strain recruitment with age, along with a drop in collagen fiber moduli, which went from 2700 kPa for the juvenile stage to 746 kPa in the adult. These changes were likely due to the previously observed impingement of relatively stiff ascending aorta on the growing PA medial region. Intuitively, the effects of the local impingement would be to lower the local wall stress, consistent with the observed decrease in collagen modulus. This result suggests that during the postnatal somatic growth period local stresses can substantially modulate regional tissue microstructure and mechanical behaviors in the PA. We further underscore that our previous studies indicated an increase in effective PA wall stress with postnatal maturation. When taken together with the fact that the observed changes in mechanical behavior and structure in the growing PA wall were modest in the other three regions studied, our collective results suggest that the majority of the growing PA wall is subjected to increasing stress levels with age without undergoing major structural adaptations. This observation is contrary to the accepted theory of maintenance of homeostatic stress levels in the regulation of vascular function, and suggests alternative mechanisms might regulate postnatal

B. Fata • W. Zhang • R. Amini • M.S. Sacks, Ph.D. (✉)
Center for Cardiovascular Simulation, Department of Biomedical Engineering, Institute for Computational Engineering and Science, University of Texas, 201 East 24th Street, ACES 5.438, 1 University Station, C0200, Austin, TX 78712-0027, USA
e-mail: msacks@ices.utexas.edu

G.S. Kassab, M.S. Sacks (eds.), *Structure-Based Mechanics of Tissues and Organs*, DOI 10.1007/978-1-4899-7630-7_19

somatic growth. Understanding the underlying mechanisms will help to improve our understanding of congenital defects of the PA and lay the basis for functional duplication in their repair and replacement.

19.1 Introduction

Congenital abnormalities of pulmonary artery often necessitate surgical repair or the use of autologous tissue and synthetic biomaterials as vascular grafts (Kogon et al. 2009; Ono et al. 2007; Rosenberg et al. 1987). The patency of synthetic conduit replacements remains limited, often requiring further surgical re-interventions due to lack of adaptation to the normal growth of the child and/or functional failure of the graft (Mayer 1995). The autologous conduit replacements are limited in supply and may not adjust to different flow environment of the graft site. Above all, an optimal vascular replacement should be able to accommodate somatic growth and closely mimic the structure, function, and physiologic function of the native vessel. In recent years, there has been a growing interest in the development of a living, autologous tissue graft that could address the critical need for growing substitutes for the repair of congenital cardiac defects (Cho et al. 2009; Hoerstrup et al. 2006; Shinoka et al. 1998; Mol et al. 2009), especially the pulmonary valve and artery (PA). The engineering foundation of such novel approaches must thus rest on an understanding of changes in the structure–function relationship that occur during postnatal maturation. Moreover, the distensibility of great arteries are important determinants of ventricular afterload and eventual dysfunction in the pulmonary hypertension, as well as many congenital defects (Wang and Chesler 2011). Yet, relatively little is known of the postnatal somatic growth characteristics of the PA.

In general, the altered mechanical properties of the great arteries are primarily associated with remodeling of the collagen and elastin fiber networks. For example, biochemical studies in animals have shown a significant upsurge in the collagen and elastin synthesis and mass, as well as reorganization in hypertensive pulmonary arteries (Ooi et al. 2010; Poiani et al. 1990). The perinatal period is associated with significant elastin and collagen accumulation in the pulmonary trunk and aorta in preparation for a marked postnatal increase in arterial pressure (Leung et al. 1977; Langille et al. 1990). It is well known that newborn animals develop more severe pulmonary hypertension than adults with dramatic vascular changes (Tucker et al. 1984), possibly due to the elastin and collagen synthesis being particularly sensitive to modulation by hypoxia during this time of rapid growth. Lammers et al. (2008) have delineated the prominent role of elastin in the alteration of pulmonary artery mechanics in hypertensive calves. Moreover, structural and degradative alterations of medial elastin are found to be a major contributing factor in physiological phenomena such as aging, and the initiation and development of cardiovascular disease, such as aortic aneurysms (Mceniery et al. 2007; Schwartz et al. 1991).

We have recently demonstrated complex patterns of spatial growth in the growing ovine PA (Fata et al. 2013a; Gottlieb et al. 2013). Our results indicated that the spatial and temporal surface growth deformation patterns of both arteries were heterogeneous, including an increase in taper in both arteries and increase in cross-sectional ellipticity of the PA. Interestingly, contact between the PA and AA resulted in increasing spatial heterogeneity in postnatal growth, with the PA demonstrating the greatest changes. Results of this study clearly underscored the fact that functional growth of the PA during postnatal maturation involves complex geometric adaptations. In a parallel study, we quantified the structural and biomechanical properties over the same age period (Fata et al. 2013b). Here, the PA wall demonstrated significant mechanical anisotropy, except in the posterior region where it was nearly isotropic, and overall modest changes in regional mechanical properties with growth. Perhaps our most interesting finding was that we found that the PA wall thickness was maintained over the entire growth period in spite of the substantial increase in vessel diameter. This suggests that the PA grows by in-plane tissue accumulation only, resulting in a 40 % average increase in hoop stress over the growth period. Therefore, unlike the arterial wall remodeling due to hypertension (Ooi et al. 2010; Poiani et al. 1990; Tozzi et al. 1994), there is not strictly held homeostatic maintenance of wall stress during the postnatal growth period. This rather surprising result opens the door for many questions, such as if there are alterations in the effective moduli of the collagen and elastin networks during the growth period.

To begin to address these questions, one can utilize constitutive models which incorporate several important aspects of the underlying microstructure. Structurally based models can help elucidate the mechanisms governing the structure–function relationship of biological tissues and elucidate what happens during the remodeling period. Such approaches have been utilized for arterial tissue remodeling (Hollander et al. 2011; Hansen et al. 2009), as well as the PA (Hunter et al. 2006; Zhang et al. 2007; Kao et al. 2011). However, the models developed so far have not addressed the mechanical behavior of normal growing vessels from the early juvenile to the adult states, either as a contiguous growth model or as a set of quasi-static steps. Moreover, the reliability of such models considerably depends on the accurate quantification of organization and load-bearing behavior of fibrous components of the tissue. In the case of the PA, these are collagen, elastin, and to a lesser extent, smooth muscle as the mechanically significant structural components. Thus, the elastin and collagen structure–function relationship of the normal PA and in connection to growth changes need to be utilized in such models.

In the present study, we utilized the extensive experimental measurements of the growing PA to develop a structural model for the PA in the juvenile and adult states. Novel to this approach was the explicit treatment of the elastin phase as a population of fibers as opposed to single material phase as in other studies of the PA. The developed constitutive model also took account of contributions of the ground matrix, consisting of smooth muscle cells and other noncellular materials. When applied to the available data, we demonstrate that we were able to delineate the structure–function relationship of PA wall in the postnatal growth period in the juvenile and adult states.

19.2 Methods

19.2.1 Mechanical and Structural Data

Extensive structural and biaxial mechanical properties and collagen and elastin fiber architecture measurements from four regions of the PA wall were taken from Fata et al. (2013b) (Fig. 19.1a). These data included elastin, collagen, and smooth muscle mass fractions obtained histologically (Fig. 19.1b). In addition to these data, collagen recruitment of the medial region (Fig. 19.1b) was available from Fata et al. (2013b). Details of the exact available data and its pre- and post-processing are given in the following.

19.2.2 Considerations and Assumptions

While the PA is a multilayered structure (Fig. 19.1b) undergoing continuous growth, to make the present problem tractable we model the PA wall as a single homogenized layer at two quasi-static time points of postnatal growth: juvenile and adult. This is in part based on our previous observations that no changes in relative

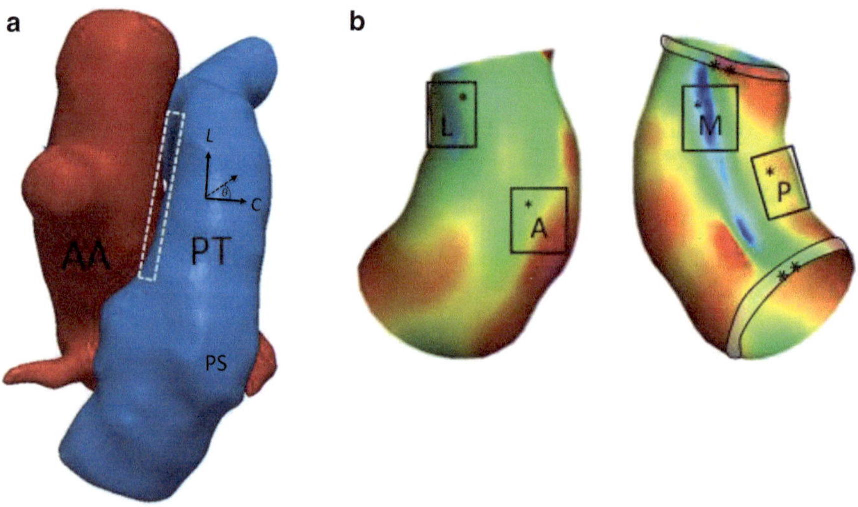

Fig. 19.1 (**a**) 3D reconstruction of the ovine ascending aorta (AA) and the pulmonary artery showing the pulmonary trunk (PT) view, with the white dashed box indicating area of contact between the two great vessels. Also shown are the posterior, anterior, lateral, and medial regions of the PA. (**b**) Transmural micrograph of the ovine PA and volume fraction results of the PA at the juvenile and adult states. Note that all regions demonstrated an increase in collagen mass with age, with the medial region indicating the largest mass fraction in the adult state

thickness of individual layers and total wall thickness were observed in the growing PA (Fata et al. 2013b). Next, the following comprehensive set of assumptions were observed.

1. The general form of the constitutive model was assumed to be maintained during the postnatal maturation period. This assumption is partially supported by the fact that the stress–strain curve shapes were conserved with growth (Fata et al. 2013b).
2. The collagen and elastin fiber networks are considered to be the mechanically dominant components compared to the inactive smooth muscle cells (Zulliger et al. 2004).
3. Both elastin and collagen fibers bear load only along their fiber axes and have negligible resistance to compressive forces.
4. The elastin–collagen mechanical interactions were ignored and the net tissue response is considered to be the sum of the individual constituent responses only.
5. The contribution of both fiber systems is in the plane of the wall, with the radial mechanical behavior of the artery governed by the incompressible muscle and ground matrix constituents.
6. The load required to straighten the collagen fibers is negligible compared to the load transmitted by elastin or stretched collagen. Thus, collagen bears load only when completely straight.
7. Based on the finding of Zulliger et al. (Lanir 1983), the contribution of inactive smooth muscle tone to load bearing structures was assumed to be very small compared to the fibers, and are thus modeled as a single material phase assuming an isotropic response.

19.2.3 *Tissue Level Strain Energy Function*

The tissue level pseudo-hyperelastic strain energy density Ψ (Fung 1993) of a representative volume element (RVE), which is assumed to be small enough so that the deformation gradient tensor **F** is homogenous (i.e., does not locally depend on position), yet large enough to allow local averaging of the constituent microstructure. We further assume the mechanical contributions of elastin (e), collagen (c), and smooth muscle (m) are weighted by their respective volume fractions φ

$$\Psi(\mathbf{C}) = \phi_c \Psi_c(\mathbf{C}) + \phi_e \Psi_e(\mathbf{C}) + \phi_m \Psi_m(\mathbf{C}), \tag{19.1}$$

where $\mathbf{C} = \mathbf{F} \cdot \mathbf{F}^T$ is the left Cauchy–Green stretch tensor. The tissue level response in terms of the second Piola–Kirchhoff stress tensor **S** is derived using

$$\mathbf{S} = \frac{\partial \Psi}{\partial \mathbf{E}} - p \cdot \mathbf{C}^{-1} = \phi_c \frac{\partial \Psi_c}{\partial \mathbf{E}} + \phi_e \frac{\partial \Psi_e}{\partial \mathbf{E}} + \phi_m \frac{\partial \Psi_m}{\partial \mathbf{E}} - p\mathbf{C}^{-1}, \tag{19.2}$$

where $\mathbf{E} = (\mathbf{C}\text{-}\mathbf{I})/2$ is the Green–Lagrange strain tensor and $\mathbf{I}$ the identity tensor, along with the Lagrange multiplier p to enforce incompressibility.

We develop the strain energy function for each tissue phase as follows. First, based on assumption 7 above, we model the muscle component as a Neo-Hookean material with small shear modulus. This is consistent with the inactivated state of the tissue investigated. Using a shear modulus of $\kappa_m = 1$ kPa (divided by two to be consistent with linear elasticity), the second Piola–Kirchhoff stress for the nonviable smooth muscle and nonfibrous ground matrix phases and water is given by

$$\mathbf{S}_m = \phi_m \frac{\kappa_m}{2}(I_1 - 3) - p\mathbf{C}^{-1}. \tag{19.3}$$

The contributions of the non-fibrous components and fluid phases are assumed to be responsible for the incompressibility of the tissue. Details for the fibrous components are given in the following.

19.2.4 Constitutive Model for Elastin

Novel to this study is the approach to model the elastin network as discrete fibers oriented in the plane of the tissue, rather than a material phase (e.g., Kao et al. 2011). This has been made possible by the quantification of the elastin structure in Fata et al. (2013b). The results from Fata et al. (Fata et al. 2013b) and related studies (Hunter et al. 2006; Zhang et al. 2007; Kao et al. 2011) suggest that a linear **S**–**E** relation is observed in the low stress region. We assume that elastin dominates the low stress. This assumption was further supported in our previous study that collagen recruitment does not occur until strain range of the low stress region was exceeded (Fata et al. 2013b). Thus we utilize the following elastin fiber strain energy

$$\Psi_e^f(\mathrm{E_f}) = \frac{\kappa_e}{2} E_f^2, \tag{19.4}$$

where κ_e is the effective elastin fiber modulus. Assuming affine transformation for the RVE, the elastin fiber strain E_f was derived from the bulk tissue strain using $E_f = \mathbf{N}^T\ \mathbf{E}\ \mathbf{N}$, where $\mathbf{N}(\theta) = \cos(\theta)\widehat{X}_1 + \sin(\theta)\widehat{X}_2$ with θ measured from the circumferential axis (Fig. 19.1a). Next, measured elastin orientation distributions from the multiphoton imaging data from Fata et al. (2013b) were normalized to unit area to obtain $\Gamma'_e(\theta)$ (Fig. 19.2). Next, since elastin fibers to not appear to undergo any type of recruitment the initial expression for the total elastin stress is given by

$$\mathbf{S}_e(\mathbf{E}) = \phi_e \frac{\partial \Psi_e}{\partial \mathrm{E_f}} \frac{\partial \mathrm{E_f}}{\partial \mathbf{E}} = \phi_e \int_\theta \Gamma'_e(\theta) S_e^f(\mathrm{E_f}) \frac{\partial E_f}{\partial \mathbf{E}} d\theta = \phi_e \kappa_e \int_\theta \Gamma'_e(\theta)\, \mathrm{E_f}\ \mathbf{N} \otimes \mathbf{N} d\theta. \tag{19.5}$$

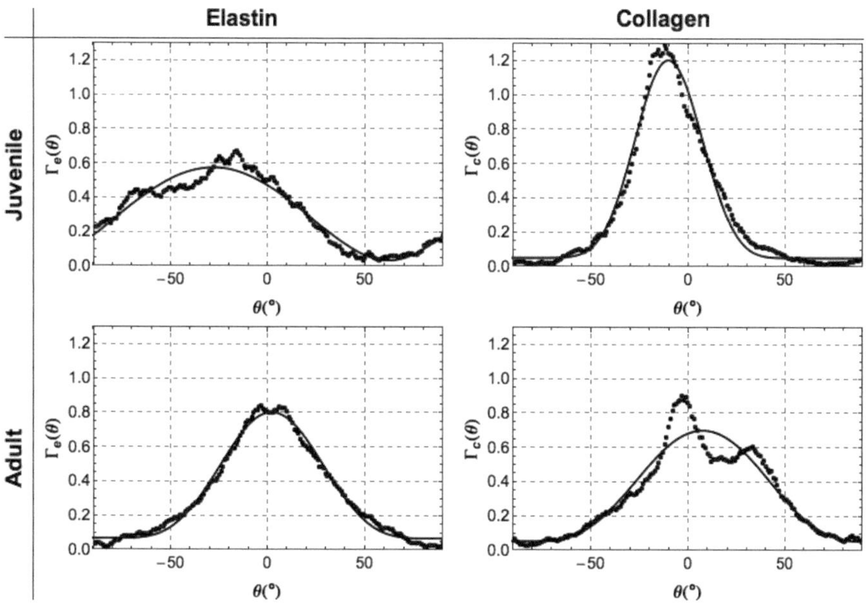

Fig. 19.2 Final average elastin and collagen measured orientation distributions and the Modified Beta distribution probability distribution fits in the juvenile and adult states

However, in the multiphoton images of the PA wall, elastin appeared both as oriented fibrous and sheet-like isotropic structures. These contiguous sheet-like structures, although visualized, were not reliably quantifiable in terms of mass fraction. Supportive evidence of their mechanical contribution came from the low stress regions where we observed significant stress development in the longitudinal direction, even though there were no measurable elastin fibers aligned in the longitudinal direction in the juvenile and adult states (Fig. 19.2). Therefore, we modified (19.5) to model the elastin fiber network as the sum of an oriented phase of a measured orientation function $\Gamma'_e(\theta)$ and an isotropic, randomly distributed fiber phase. Thus, the total fiber orientation distribution function is

$$\Gamma_e(\mathsf{d}_e, \theta) = \mathsf{d}_e \cdot \Gamma'_e(\theta) + (1 - \mathsf{d}_e)/\pi. \tag{19.6}$$

The resulting final expression for the elastin phase stress is given by

$$\mathbf{S}_e = \phi_e \kappa_e \mathsf{d}_e \int_\theta \Gamma_e(\mu_e, \sigma_e, \mathsf{d}_e, \theta) E_f(\mathbf{E}, \theta) \mathbf{N} \otimes \mathbf{N} d\theta, \tag{19.7}$$

with ϕ_e and Γ'_e taken from experimental measurements (Figs. 19.1 and 19.2, respectively) and κ_e and d_e estimated from the biaxial mechanical data, as detailed in the following sections.

19.2.5 Constitutive Model for Collagen

Following Lanir (1983), we treat collagen fibers with a common orientation as a fiber ensemble that undergoes a uniaxial strain $E_{\mathrm{ens}} = \mathbf{N}^{\mathrm{T}}\ \mathbf{E}\ \mathbf{N}$. As in related work on collagenous tissues (Lanir 1983; Sacks 2003), we assume a linear fiber stress–strain relationship for individual collagen fibers

$$\Psi_{\mathrm{c}}^{\mathrm{f}}\left(\mathsf{E}_{\mathrm{f}}\right) = \frac{\kappa_{\mathrm{c}}}{2}\mathsf{E}_{\mathrm{f}}^{2}, \tag{19.8}$$

where κ_{c} is the elastic modulus of individual straight collagen fibers. Due to their crimped structure, we express individual collagen fiber's true fiber strain using $\mathsf{E}_{\mathrm{t}} = \frac{\mathsf{E}_{\mathrm{ens}} - \mathsf{E}_{\mathrm{s}}}{1+2\mathsf{E}_{\mathrm{s}}}$ where E_{s} is the collagen fiber slack strain. The resulting *individual* collagen fiber strain energy is then

$$\Psi_{\mathrm{c}}^{\mathrm{f}}\left(\mathsf{E}_{\mathrm{t}}\right) = \frac{\kappa_{\mathrm{c}}}{2}\mathsf{E}_{\mathrm{t}}^{2}. \tag{19.9}$$

While it is evident that $E_{\mathrm{ens}} = E_{\mathrm{f}}$, we make the distinction between fiber and ensemble strains here since individual collagen fibers will have a different strain levels due to their undulations.

Next, we account for the gradual recruitment of the collagen fiber in each fiber ensemble with strain stochastically using the function $D = D(E_{\mathrm{s}})$, defined over the ensemble strain range $\mathsf{E}_{\mathrm{ens}} \in [\mathsf{E}_{\mathrm{lb}}, \mathsf{E}_{\mathrm{ub}}]$, where E_{lb} and E_{ub} represent the lower and upper bounds of collagen fiber ensemble recruitment strain levels, with $E_{\mathrm{ub}} > E_{\mathrm{lb}} > 0$. The ensuing fiber ensemble strain energy and stress–strain relation are then described as the sum of individual fiber strain energies of the ensemble weighted by the distribution of slack strains D, as

$$\Psi_{\mathrm{c}}^{\mathrm{ens}} = \frac{\kappa_{\mathrm{c}}}{2}\int_{0}^{E_{\mathrm{ens}}} D(x)\left(\frac{E_{\mathrm{ens}} - x}{1+2x}\right)^{2} dx, \quad S_{\mathrm{c}}^{\mathrm{ens}} = \kappa_{\mathrm{c}}\int_{0}^{E_{\mathrm{ens}}} D(x)\frac{E_{\mathrm{ens}} - x}{(2x+1)^{2}}dx, \tag{19.10}$$

where $\int_{E_{\mathrm{lb}}}^{E_{\mathrm{ub}}} D(x)dx = 1$. A scaled Beta distribution $B(\alpha,\beta)$ was used for D as follows:

$$D(x) = \begin{cases} \frac{x^{\alpha-1}(1-x)^{\beta-1}}{B(\alpha,\beta)(E_{\mathrm{ub}} - E_{\mathrm{lb}})}, & \text{for } x \in]0,1[\\ 0, & \text{otherwise} \end{cases}$$

$$x = \left(E_{\mathrm{ens}} - E_{\mathrm{lb}}\right) / \left(E_{\mathrm{ub}} - E_{\mathrm{lb}}\right), \tag{19.11}$$

where α and β are the shape constants with mean μ_{r} and standard deviation σ_{r} determined using

$$\mu_r' = (\mu_r - E_{lb}) / (E_{ub} - E_{lb}), \qquad \sigma_r' = \sigma_r / (E_{ub} - E_{lb}),$$
$$\alpha = \frac{(\mu_r')^2 - (\mu_r')^3 - (\sigma_r')^2 \mu_r'}{(\sigma_r')^2}, \qquad \beta = \alpha \frac{1 - \mu_r'}{\mu_r'}. \tag{19.12}$$

The resulting total collagen fiber phase second Piola–Kirchhoff stress is the sum of individual fiber second Piola-Kirchhoff stress weighted by the angular probability function Γ_c

$$\mathbf{S}_c = \phi_c \kappa_c \int_\theta \Gamma_c(\theta) \left[\int_0^{E_{ens}} D(\mu_r, \sigma_r, x) \frac{E_{ens} - x}{(2x+1)^2} dx \right] \mathbf{N} \otimes \mathbf{N} \, d\theta. \tag{19.13}$$

19.2.6 Complete Constitutive Model Form

Since complete structural information was only available from the medial region, some final modifications were necessary to provide maximum fitting flexibility and to facilitate numerical computations. As for the treatment for elastin we simulated the measured collagen fiber orientation distribution using

$$\Gamma_c(\mu_c, \sigma_c, d_c) = d_c \cdot \Gamma_c'(\mu_c, \sigma_c) + \frac{(1 - d_c)}{\pi}, \tag{19.14}$$

where the oriented component Γ'_c is modeled using a Beta probability distribution function with mean (μ_c) and standard deviation (σ_c), along with a weighting term (d_c). Note that when applied to the measured medial region data the collagen baseline response given by d_c serves only to improve the model by quantifying randomly oriented collagen fibers observed in the actual measured data (Fata et al. 2013b, Fig. 19.2). Thusly, by combining (19.2, 19.3, 19.6, and 19.12), we have the final expression for the total second Piola–Kirchhoff stress $\mathbf{S}$ for the PA wall

$$\begin{aligned} \mathbf{S} = {} & \phi_c \kappa_c \int_\theta \Gamma_c(\mu_c, \sigma_c, d_c, \theta) \left[\int_0^{E_{ens}} D(\mu_r, \sigma_r, x) \frac{E_{ens} - x}{(2x+1)^2} dx \right] \mathbf{N} \otimes \mathbf{N} \, d\theta \\ & + \phi_e \kappa_e d_e \int_\theta \Gamma_e(\mu_e, \sigma_e, d_e, \theta) \, E_f(\mathbf{E}, \theta) \, \mathbf{N} \otimes \mathbf{N} \, d\theta \\ & + \phi_m \frac{\kappa_m}{2} (I_1 - 3) - p \cdot \mathbf{C}^{-1}. \end{aligned} \tag{19.15}$$

The final model has a total of 12 parameters.

19.2.7 Model Parameter Estimation

Direct parameter estimation for such a highly nonlinear model would be ill-advised with respect to uniqueness and confidence of a global minimum. Rather, we developed the following systematic means to obtain optimal parameter values by dividing the total model parameter estimation effort into the determination of the effective fiber properties, followed by the complete in-plane behavior. For the medial region, due to the availability of the requisite data, we only had to fit the collagen and elastin moduli and d_e. Moreover, while important regional differences were observed in the measured biomechanical responses (Fata et al. 2013b), the overall qualitative shapes of the stress–strain curves were similar. We thus utilized both the predicted and measured responses of the medial region as initial estimates for the remaining three regions. Details of the fitting approach are provided in the following.

As presented in our work on pericardium (Sacks 2003), when utilizing a structural model intrinsic *fiber ensemble* responses (i.e., the S_{ens}–E_{ens} relation) can be successfully determined directly from an equi-biaxial strain loading path (that is, $S_{ens} = S_{11} + S_{22}$, $E_{ens} = E_{11} = E_{22}$, $E_{12} = 0$) since there are no fiber rotations in this kinematic mode. This simple method allows one to directly fit fiber moduli and recruitment model parameters independent of fiber splay (and its associated parameters). However, the available mechanical testing data (Fata et al. 2013b) was performed under stress control with no equi-biaxial strain data available (Fig. 19.3a). To obtain these data, we interpolated the equi-biaxial response from the experimentally acquired multi-protocol data using bicubic Hermite elements to fit the second Piola–Kirchhoff stress surfaces. Smoothness of the surface was

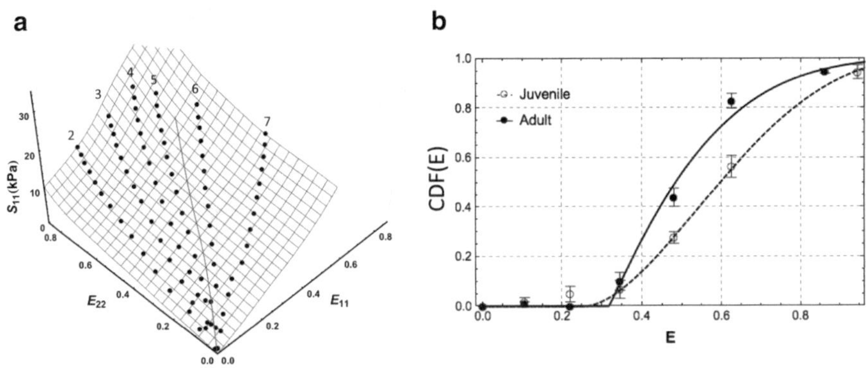

Fig. 19.3 (**a**) Results of the bicubic Hermite surface interpolation of the second Piola–Kirchhof stress biaxial test responses to allow interpolation of an equi-biaxial strain path, shown here in *red*. (**b**) Fit of the medial region collagen fiber recruitment using a Beta cumulative distribution function for both juvenile and adult states, revealing both an excellent fit to the data along with a trend toward more rapid recruitment with strain in the adult stage

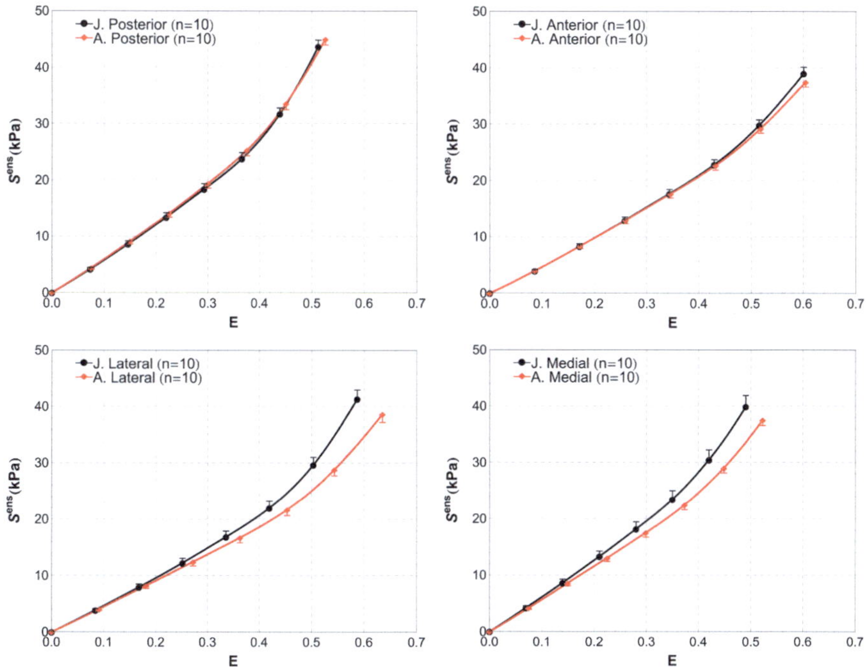

Fig. 19.4 Fiber ensemble stress–strain results for the interpolated equi-biaxial strain path responses for all four regions at both age time points

enforced in the Sobolev norm (Smith et al. 2000). An interpolated protocol where $E_{11} = E_{22}$ from the resulting surface was used to simulate the equi-biaxial response (Fig. 19.3a).

Initial inspections of the S_{ens}–E_{ens} responses revealed a noticeable and consistent increase in stiffness for $S_{\mathrm{ens}} \geq 20$ kPa, corresponding to $E_{\mathrm{ens}} \cong 0.30$ (Figs. 19.4 and 19.5). When compared to the available collagen fiber recruitment data for the medial region (Fig. 19.3b), it was observed that collagen engagement also initiated at E $\cong$ 0.30. We thus attributed the change in total fiber ensemble stiffness to the initiation of collagen fiber recruitment, so that E_{lb} could be directly determined from the experimental data for each specimen, removing this parameter from the fitting procedures. Moreover, this finding indicated that for fiber ensemble strains below E_{lb}, (i.e., $E_{\mathrm{ens}} \in [0, E_{\mathrm{lb}}[$), the S_{ens}–E_{ens} responses are completely dominated by elastin only (i.e., no collagen contribution). By taking the difference between the circumferential (S_{11}) and the longitudinal (S_{22}) components S_{Δ} from (19.15), with the collagen phase removed since it is assumed that it does not bear load under equi-biaxial strain, we obtain

$$S_{\Delta} = S_{11} - S_{22} = d_{\mathrm{e}}\phi_{\mathrm{e}}\kappa_{\mathrm{e}} \int_{\theta} \Gamma'_{\mathrm{e}}\left(\mu_{\mathrm{c}}, \sigma_{\mathrm{c}}\right) E_{\mathrm{f}} \left(\cos^2\theta - \sin^2\theta\right) d\theta, \qquad (19.16)$$

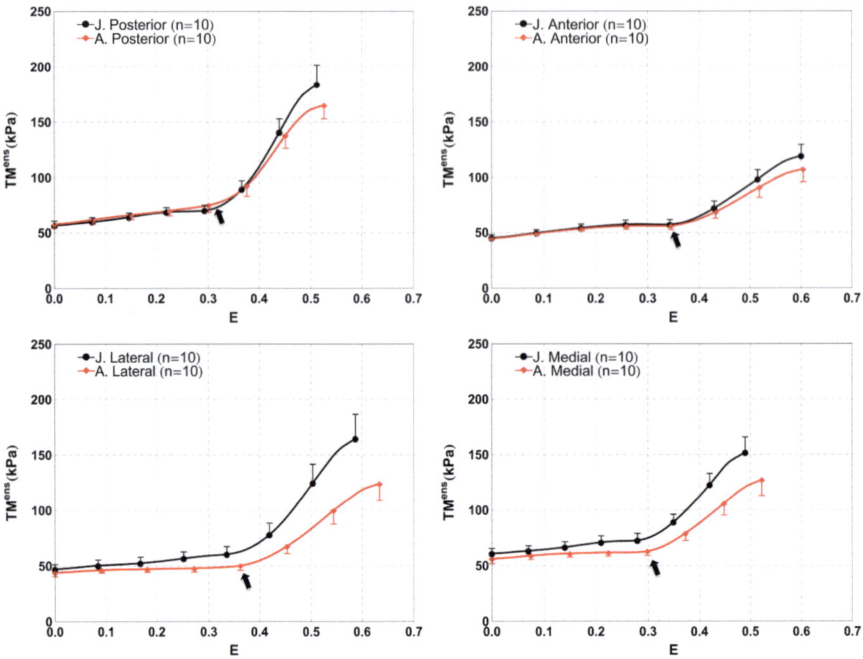

Fig. 19.5 Fiber ensemble tangent modulus–strain results for the interpolated equi-biaxial strain path responses for all four regions at both age time points. Note here the sharp increase in stiffness due to collagen fiber engagement (*arrows*). Note too the decreases in stiffness with age in the lateral and medial regions

following a similar procedure developed in Sacks (2000). The corresponding cost function for fitting the S_{ens}–E_{ens} for $E_{\text{ens}} \in [0, E_{\text{lb}}[$ is simply

$$\text{SSE} = \sum_i \left(S_\Delta - \left(\widehat{S}_{11,i} - \widehat{S}_{22,i}\right)\right)^2 + \left(\left(S_{11}^{\text{aniso}} + S_{11}^{\text{iso}}\right) - \widehat{S}_{11,i}\right)^2 + \left(\left(S_{22}^{\text{aniso}} + S_{22}^{\text{iso}}\right) - \widehat{S}_{22,i}\right)^2. \tag{19.17}$$

This procedure allowed the parameters d_e and κ_e to be estimated directly from the interpolated equi-biaxial data, rather than in the full model.

Next, to further facilitate fitting of the medial region data only, we approximated the measured average collagen fiber recruitment probability distribution from Fata et al. (2013b) using (19.10) to determine μ_r, σ_r, E_{lb}, and E_{ub} directly (Fig. 19.3b). The resulting parameters were shifted to match E_{lb} as estimated by the equi-biaxial strain interpolation described above on a per specimen basis, while keeping the span of the distribution constant. The respective parameters are then used as constants for the model. For the remaining 3 regions, E_{lb} was estimated directly from the equi-biaxial data using the distinct increase in tangent modulus (TM) observed when the collagen recruitment initiated as described above. The remaining recruitment parameters from the medial region were used as the initial guess during optimization with E_{lb} fixed, since it was determined directly.

A custom written Mathematica program (Wolfram Research Corp.) was developed and utilized to fit the complete model (19.15) to the multi-protocol mechanical stress–strain data (Fig. 19.3a). A stepwise approach was used to reduce computational cost and increase the rate of convergence as follows. First, using the parameters obtained from the equi-biaxial strain data over the full measured strain region, we used three protocols (numbers 3–5, Fig. 19.3a), then five protocols (numbers 2–6, Fig. 19.3a) were fitted sequentially, with the best fit parameter from the prior step used as the initial guess for the current optimization. This approach also improved convergence to the global minimum using a gradient approach without the need of time-intensive genetic algorithms. The interior point method was chosen for its self-concordant barrier function, allowing us to easily constraint the parameters under the Karush–Kuhn–Tucker conditions. The predictive quality of the fit was examined by comparing the predicted response to the data for the remaining two protocols (numbers 1 and 7, Fig. 19.3a).

19.3 Results

19.3.1 Post-processed Structural and Mechanical Data

The averaged fiber orientation distribution data of elastin and collagen fibers were well fit by their respective modified Beta probability distribution functions (19.6 and 19.14, respectively), with $r^2 > 0.94$ (Fig. 19.2). Next, (19.11) (the modified Beta density function for collagen recruitment) provided an accurate representation of the mean measured collagen recruitment data of juvenile ($r^2 = 0.97$) and adult ($r^2 = 0.98$) specimens, respectively (Fig. 19.3b). This last result provided important evidence to support the use of (19.11) to represent collagen fiber recruitment with ensemble strain for the ovine PA at both the juvenile and adult stages.

The interpolated equi-biaxial responses demonstrated excellent consistency within each region and age group, for both the stress (Fig. 19.4) and tangent moduli (Fig. 19.5). In all four regions and two ages, the initiation of collagen recruitment was clearly marked (Fig. 19.5) at $E_{ens} = \sim 0.30$. Consistent with our previous study (Fata et al. 2013b), the anterior and posterior regions demonstrated minimal changes in tangent moduli with age, whereas the lateral and medial regions both showed a decrease in maximum stiffness (Fig. 19.5). Since no fiber rotations can occur for equi-biaxial strain loading path, the observed decrease in regional tissue stiffness must be due to changes in either fiber moduli or collagen fiber recruitment. Next, from these data the corresponding E_{lb} were obtained as transition points in the tangent modulus–strain relations (Fig. 19.5), which demonstrated no changes with age and only significant differences between the lateral and medial groups.

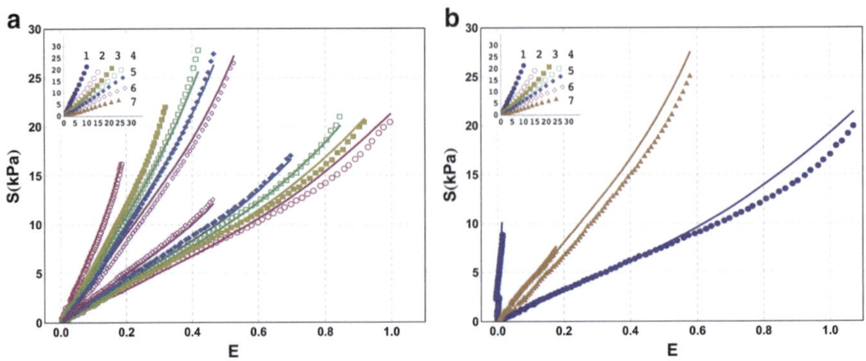

Fig. 19.6 (**a**) Constitutive model fit to the average five-protocol biaxial stress–stretch data (protocols 2–6, see *inset*) of juvenile and adult medial PA wall specimens. (**b**) Predicted fit results to protocols 1 and 7 (see *inset*), showing excellent agreement

19.3.2 Overall Fit Results

When applied to the medial region data, the complete model (19.15) demonstrated an excellent quality of fit to all protocols used ($r^2 \geq 0.99$, protocols 2–6, Fig. 19.6). Moreover, when applied to measured strains from the "outer" protocols (numbers 1 and 7, Fig. 19.6), the excellent predictive capabilities were apparent (Fig. 19.6b). Interestingly, no apparent differences in fit quality or values for the k_e and k_c were observed when either the actual measured or interpolated fiber splay data (Fig. 19.2) were used. This last result suggested that the Beta distribution approximation to the fiber splay is sufficient to capture the tissue response. Finally, equivalent quality fits were obtained for the remaining three regions for all mechanical data measured.

19.3.3 Regional Differences

In the juvenile state, key model parameters exhibited differences between all four regions of the PA. For the predicted fiber splay responses, the posterior region had the lowest elastin d_e value of ~0.5, followed by increasing values for the remaining three regions (~0.5 to ~0.75). In contrast, the collagen d_c values did not appear to exhibit marked regional differences (all about 0.7–0.8, Fig. 19.7b). Similar trends were found for the actual dispersion σ_e and σ_c with only the medial region exhibiting substantial differences from the other regions, especially for elastin. It is also interesting to note that the collagen and elastin fiber splays were neither similar in value nor exhibited similar regional differences, and thus appear to be structurally decoupled. Collagen fiber recruitment followed similar trends, with the posterior region's mean recruitment being the smallest, followed by increasing values in order of anterior, lateral, and medial regions. In contrast, the recruitment

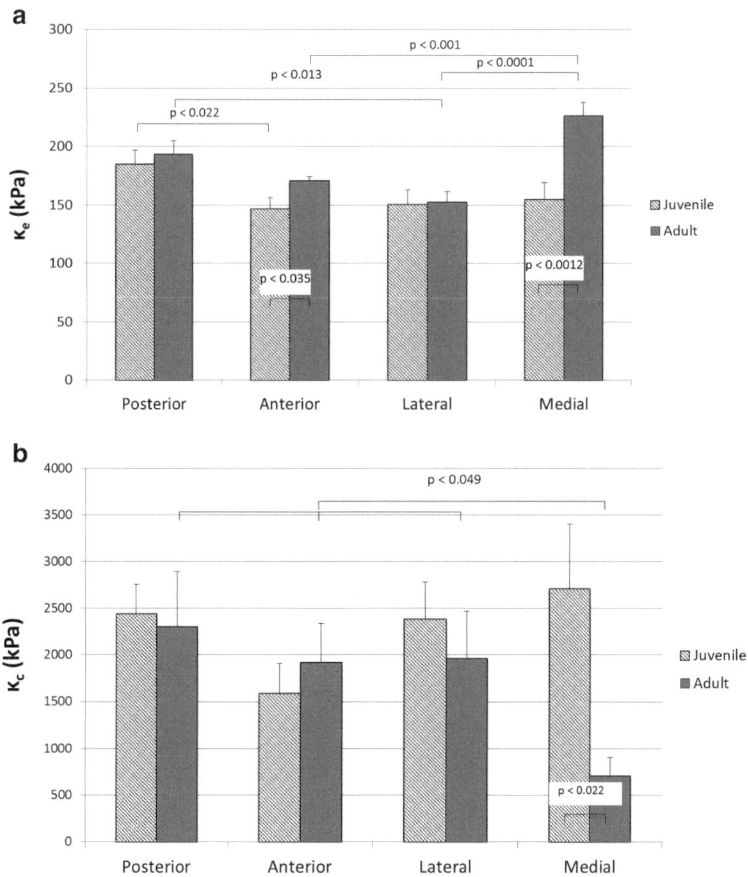

Fig. 19.7 Predicted collagen (κ_c) and elastin (κ_e) fiber moduli regions at both age time points. The most drastic changes were for the elastin modulus k_e increasing by ~50 % and the collagen modulus k_c decreasing to only ~25 % of the juvenile value in the medial region

standard deviation demonstrated that the medial region was larger and thus recruited over a larger strain range. Regional differences in collagen and elastin moduli in the juvenile state were generally similar, with elastin in the range of 150–200 kPa and collagen from ~1600 to 2700 kPa (Fig. 19.7).

19.3.4 Effects of Age

No major differences were found between the juvenile and adult PA in the posterior, anterior, and lateral regions for most model parameters. Some difference is found between elastin modulus for the anterior region ($p < 0.035$). However, this is only a 14 % drop in stiffness than the adult anterior region, which does not produce significant difference in the tissue level biomechanics (Fig. 19.5). Slight differences are seen in the equi-biaxial strain data for the lateral region. This is due to the mass fraction difference (Fig. 10) and difference in the amount of fibrous elastin; the difference fails to be statistically significant ($p < 0.11$). For the adult PA, we found the regional differences to be generally similar to the juvenile PA. While some trends were apparent in d_e and d_c, these were not marked. For the fiber dispersion σ_e and σ_c, only the lateral and medial σ_c demonstrated regional differences in the adult state. Similar responses were found for the recruitment parameters. The anterior region had a highest μ_r for the adult largely due to a slight increase with age ($p < 0.12$) and the drop in the medial region with age. No other major regional differences were found in the adult comparing to the juvenile.

The primary finding for age-related changes was for the collagen and elastin moduli in the medial region (Fig. 19.7). While the elastin modulus for this region only modestly increased from ~150 to 225 kPa, the collagen modulus dropped from 2711 kPa for the juvenile to slightly more than one quarter of its value at 747 kPa for the adult ($p < 0.022$, Fig. 19.7). Corresponding histological and structural measurements also demonstrated (1) a 47 % increase in volume fraction of collagen (Fig. 19.1c) and (2) a significant drop in elastin splay ($p < 0.00002$) a 60 % increase in the percentage of collagen splay (Fig. 10). Interestingly as well, both predicted μ_r and σ_r decreased with age in the medial region, in agreement with the observed more rapid measured collagen fiber recruitment for this region (Fig. 19.3b). Collectively, the modeling results suggest substantial tissue remodeling in the medial region with age, whereas the other three measured regions were largely stable with age.

19.4 Discussion

19.4.1 Overall Findings

Overall, the predicted elastin moduli remained largely unchanged with age and compared well with between the four regions studied, lying within a relatively narrow range of 150–200 kPa. Similarly, the predicted collagen moduli ranged from 1600 to 2700 kPa in the four regions studied in the juvenile state. However, major age effects were found in the medial aspect of the PA wall (Fig. 19.1). In this region, we observed that the elastin and collagen fiber splay underwent opposite changes, with the collagen standard deviation juvenile = 17° to adult = 28°, elastin standard deviation juvenile = 35° to adult = 27°. In parallel, we observed more rapid collagen fiber strain recruitment with age in this similar to that observed experimentally (Fata et al. 2013b), but these were not statistically significant.

Clearly, the most profound changes in this region were observed for the collagen fiber moduli, which went from 2700 kPa for the juvenile stage to 746 kPa in the adult (Fig. 19.7). These changes were likely due to the impingement of relatively stiff ascending aorta on the growing PA medial region, and were apparently a local effect. Intuitively, the effects of the local impingement would be to lower the local wall stress, consistent with the observed decrease in collagen modulus. While this change was in part driven by the observed increase in total tissue elastin volume fraction relative to the collagen volume fraction (Fig. 19.1c), these changes were also a direct result of the changes in local tissue structure.

The results of the present study extend these studies by utilizing a detailed mesoscale (i.e., bulk fiber level) structural model, coupled with extensive experimental measures, to obtain further insight into the PA remodeling process. This approach was taken since altered mechanical properties of arteries are critically associated with microstructural remodeling (Humphrey 2009; Ambrosi et al. 2011). The mechanical loading–deformation relation of elastin and collagen fibers is fundamental to understanding the underlying microstructural mechanisms of arterial tissue behavior. Structurally motivated models, such as used herein, incorporate significant mechanical aspects of the underlying microstructure to better predict the mechanical behavior and understand the mechanisms governing the structure–function relationship of biological tissues. The present constitutive model takes account of contributions of the collagen component, elastin or elastic fibers, ground matrix consisting of smooth muscle cells and other noncellular materials, along with detailed in-plane biaxial mechanical behaviors (Fata et al. 2013b) (Figs. 19.1, 19.2, 19.3, 19.4, and 19.5).

As observed in our geometric studies (Fata et al. 2013a; Gottlieb et al. 2013), the focal changes observed in the medial region are likely due to the impingement of relatively stiff ascending aorta on the growing PA. This effect is likely more pronounced since these two great vessels are held together within a connective tissue sheath (J.E. Mayer, Jr., private communication). Intuitively, the effects of the local impingement would be to lower the local wall stress. If so, then this would be consistent with the observed decrease in collagen modulus. However, it does not explain the concomitant increase in elastin modulus. Clearly, some type of compensatory mechanism is in play here, but the nature of which remains unknown. What can be said with some certainty is that, during the postnatal somatic growth, local stresses can substantially modulate the development of regional tissue microstructure and mechanical behaviors in the PA.

In our parallel study on the corresponding mechanical properties (Fata et al. 2013b) the medial and lateral locations experienced local, moderate increases in anisotropy. Moreover, the PA thickness remained constant with growth. When this fact is combined with the observed stable overall mechanical behavior and increase in vessel diameter with growth, a simple Laplace Law wall stress estimate suggests an increase in effective PA wall stress with postnatal maturation. This observation is contrary to the accepted theory of maintenance of homeostatic stress levels in the regulation of vascular function, and suggests alternative mechanisms regulate postnatal somatic growth.

The present study extends our previous investigations into postnatal PA growth. As observed in our geometric studies (Fata et al. 2013a; Gottlieb et al. 2013), the focal changes observed in the medial region are likely due to the impingement of relatively stiff ascending aorta on the growing PA. Intuitively, the effects of the local impingement would be to lower the local wall stress. If so, then this would be consistent with the observed decrease in collagen modulus. However, it does not explain the concomitant increase in elastin modulus. Clearly, some type of compensatory mechanism is in play here, but the nature of which remains unknown. What can be said with some certainty is that, during the postnatal somatic growth, local stresses can substantially modulate the development of regional tissue microstructure and mechanical behavior in the PA. However, this finding should be taken in light that we observed relatively stable overall mechanical behavior over most of the PA with growth. As the PA thickness remained constant with growth, a simple Laplace Law wall stress estimate suggests an increase in effective PA wall stress with postnatal maturation. This observation is contrary to the accepted theory of maintenance of homeostatic stress levels in the regulation of vascular function, and suggests alternative mechanisms regulate postnatal somatic growth. Understanding the underlying mechanisms, including incorporating important structural features during growth, will help to improve our understanding of congenital defects of the PA and lay the basis for functional duplication in their repair and replacement.

Acknowledgment This research was supported by NIH grant R01 HL-089750.

References

Ambrosi D, Ateshian GA, Arruda EM, Cowin SC, Dumais J, Goriely A, Holzapfel GA, Humphrey JD, Kemkemer R, Kuhl E, Olberding JE, Taber LA, Garikipati K. Perspectives on biological growth and remodeling. J Mech Phys Solids. 2011;59(4):863–83.

Cho SW, Kim IK, Kang JM, Song KW, Kim HS, Park CH, Yoo KJ, Kim BS. Evidence for in vivo growth potential and vascular remodeling of tissue-engineered artery. Tissue Eng Part A. 2009;15(4):901–12.

Fata B, Gottlieb D, Mayer JE, Sacks MS. Estimated in vivo postnatal surface growth patterns of the ovine main pulmonary artery and ascending aorta. J Biomech Eng. 2013a;135(7):71010–2.

Fata B, Carruthers CA, Gibson G, Watkins SC, Gottlieb D, Mayer JE, Sacks MS. Regional structural and biomechanical alterations of the ovine main pulmonary artery during postnatal growth. J Biomech Eng. 2013b;135(2):021022.

Fung YC. Biomechanics: mechanical properties of living tissues. New York: Springer; 1993.

Gottlieb D, Fata B, Powell AJ, Cois CA, Annese D, Tandon K, Stetten G, Mayer Jr JE, Sacks MS. Pulmonary artery conduit in vivo dimensional requirements in a growing ovine model: comparisons with the ascending aorta. J Heart Valve Dis. 2013;22(2):195–203.

Hansen L, Wan W, Gleason RL. Microstructurally motivated constitutive modeling of mouse arteries cultured under altered axial stretch. J Biomech Eng. 2009;131(10):101015.

Hoerstrup SP, Cummings Mrcs I, Lachat M, Schoen FJ, Jenni R, Leschka S, Neuenschwander S, Schmidt D, Mol A, Gunter C, Gossi M, Genoni M, Zund G. Functional growth in tissue-engineered living, vascular grafts: follow-up at 100 weeks in a large animal model. Circulation. 2006;114(1 Suppl):I159–66.

Hollander Y, Durban D, Lu X, Kassab GS, Lanir Y. Constitutive modeling of coronary arterial media—comparison of three model classes. J Biomech Eng. 2011;133(6):061008.
Humphrey JD. Vascular mechanics, mechanobiology, and remodeling. J Mech Med Biol. 2009;9(2):243–57.
Hunter KS, Lanning CJ, Chen SY, Zhang Y, Garg R, Ivy DD, Shandas R. Simulations of congenital septal defect closure and reactivity testing in patient-specific models of the pediatric pulmonary vasculature: a 3D numerical study with fluid-structure interaction. J Biomech Eng. 2006;128(4):564–72.
Kao PH, Lammers S, Tian L, Hunter K, Stenmark KR, Shandas R, Qi HJ. A microstructurally-driven model for pulmonary artery tissue. J Biomech Eng. 2011;133(5):051002.
Kogon BE, Patel M, Pernetz M, Mcconnell M, Book W. Late pulmonary valve replacement in congenital heart disease patients without original congenital pulmonary valve pathology. Pediatr Cardiol. 2009;31(1):74–9.
Lammers SR, Kao PH, Qi HJ, Hunter K, Lanning C, Albietz J, Hofmeister S, Mecham R, Stenmark KR, Shandas R. Changes in the structure-function relationship of elastin and its impact on the proximal pulmonary arterial mechanics of hypertensive calves. Am J Physiol Heart Circ Physiol. 2008;295(4):H1451–9.
Langille BL, Brownlee RD, Adamson SL. Perinatal aortic growth in lambs: relation to blood flow changes at birth. Am J Physiol. 1990;259(28):H1247–53.
Lanir Y. Constitutive equations for fibrous connective tissues. J Biomech. 1983;16:1–12.
Leung DYM, Glagov S, Mathews MB. Elastin and collagen accumulation in rabbit ascending aorta and pulmonary trunk during postnatal growth. Correlation of cellular synthetic response with medial tension. Circ Res. 1977;41(3):316–23.
Mayer Jr JE. Uses of homograft conduits for right ventricle to pulmonary artery connections in the neonatal period. Semin Thorac Cardiovasc Surg. 1995;7(3):130–2.
Mceniery CM, Wilkinson IB, Avolio AP. Age, hypertension and arterial function. Clin Exp Pharmacol Physiol. 2007;34(7):665–71.
Mol A, Smits AI, Bouten CV, Baaijens FP. Tissue engineering of heart valves: advances and current challenges. Expert Rev Med Devices. 2009;6(3):259–75.
Ono M, Goerler H, Kallenbach K, Boethig D, Westhoff-Bleck M, Breymann T. Aortic valve-sparing reimplantation for dilatation of the ascending aorta and aortic regurgitation late after repair of congenital heart disease. J Thorac Cardiovasc Surg. 2007;133(4):876–9.
Ooi CY, Wang Z, Tabima DM, Eickhoff JC, Chesler NC. The role of collagen in extralobar pulmonary artery stiffening in response to hypoxia-induced pulmonary hypertension. Am J Physiol Heart Circ Physiol. 2010;299(6):H1823–31.
Poiani GJ, Tozzi CA, Yohn SE, Pierce RA, Belsky SA, Berg RA, Yu SY, Deak SB, Riley DJ. Collagen and elastin metabolism in hypertensive pulmonary arteries of rats. Circ Res. 1990;66(4):968–78.
Rosenberg HG, Williams WG, Trusler GA, Higa T, Rabinovitch M. Structural composition of central pulmonary arteries. Growth potential after surgical shunts. J Thorac Cardiovasc Surg. 1987;94(4):498–503.
Sacks MS. A Structural constitutive model for chemically treated planar connective tissues under biaxial loading. Comput Mech. 2000;26(3):243–9.
Sacks MS. Incorporation of experimentally-derived fiber orientation into a structural constitutive model for planar collagenous tissues. J Biomech Eng. 2003;125(2):280–7.
Schwartz CJ, Valente AJ, Sprague EA, Kelley JL, Nerem RM. The pathogenesis of atherosclerosis: an overview. Clin Cardiol. 1991;14(2 Suppl 1):I1–16.
Shinoka T, Shum-Tim D, Ma PX, Tanel RE, Isogai N, Langer R, Vacanti JP, Mayer Jr JE. Creation of viable pulmonary artery autografts through tissue engineering. J Thorac Cardiovasc Surg. 1998;115(3):536–45. discussion 545–6.
Smith DB, Sacks MS, Vorp DA, Thornton M. Surface geometric analysis of anatomic structures using biquintic finite element interpolation. Ann Biomed Eng. 2000;28(6):598–611.
Tozzi CA, Christiansen DL, Poiani GJ, Riley DJ. Excess collagen in hypertensive pulmonary arteries decreases vascular distensibility. Am J Respir Crit Care Med. 1994;149(5):1317–26.

Tucker A, Migally N, Wright ML, Greenlees KJ. Pulmonary vascular changes in young and aging rats exposed to 5,486 m altitude. Respiration. 1984;46(3):246–57.
Wang Z, Chesler NC. Pulmonary vascular wall stiffness: an important contributor to the increased right ventricular afterload with pulmonary hypertension. Pulm Circ. 2011;1(2):212–23.
Zhang Y, Dunn ML, Hunter KS, Lanning C, Ivy DD, Claussen L, Chen SJ, Shandas R. Application of a microstructural constitutive model of the pulmonary artery to patient-specific studies: validation and effect of orthotropy. J Biomech Eng. 2007;129(2):193–201.
Zulliger MA, Rachev A, Stergiopulos N. A constitutive formulation of arterial mechanics including vascular smooth muscle tone. Am J Physiol Heart Circ Physiol. 2004;287(3):H1335–43.

Chapter 20
A Microvascular Model in Skeletal Muscle Fascia

Frank G. Jacobitz, Niki L. Yamamura, Adam M. Jones, and Geert W. Schmid-Schönbein

Abstract This study considers the microcirculation in skeletal muscle fascia. Simulations are performed using a comprehensive approach to the problem with realistic reconstruction of the microvasculature, blood rheology and vessel wall properties, and Stokes flow in the microvessels. The simulation results provide detailed network displays of basic hemodynamic parameters. For example, an approximately normal distribution was found for the hematocrit. High hematocrit values are observed in areas with low blood perfusion, e.g., in the peripheral regions of the network. A range of velocity values was found in the capillary vessels of the network, in contrast to experimental observations which suggest a relative narrow distribution of capillary velocities. This finding points to the need of an improved treatment of mechanisms for the control of vessel diameter. A local mechanism based on the shear stress is proposed for future studies of the microcirculation.

20.1 Introduction

Since the introduction of morphological and rheological models in the 1970s, biomechanical analysis of blood flow in microvascular networks has made major progress. Using realistic reconstructions of the network topology and models of blood properties in narrow blood vessels, an increasing level of detail is analyzed and predictions are critically tested by experiments. The analysis is organ specific and serves as the first level of a more comprehensive quantitative description of mass transport, metabolism, and specific physiological functions. Microvascular blood rheology (Pries and Secomb 2005), pressure-dependent response, shear stress response, metabolic regulation, hematocrit distribution in microvessels and at bifurcations (Pries et al. 1990), the endothelial surface layer (glycocalyx) (Pries and

F.G. Jacobitz • N.L. Yamamura • A.M. Jones
Mechanical Engineering Program, University of San Diego, San Diego, CA, USA

G.W. Schmid-Schönbein (✉)
Department of Bioengineering, University of California, San Diego, La Jolla, CA, USA
e-mail: gwss@bioeng.ucsd.edu

G.S. Kassab, M.S. Sacks (eds.), *Structure-Based Mechanics of Tissues and Organs*,
DOI 10.1007/978-1-4899-7630-7_20

Secomb 2005), acute adaptive mechanisms by pressure and wall shear stress (Pries et al. 2001) have been subject to analysis. Several details of the static and dynamic aspects of hemodynamics in skeletal muscle have been studied (Schmid-Schönbein 1988; Lee et al. 1994).

Skeletal muscle is surrounded by a fascia which consists of a connective tissue with a collagen network and fibers that integrate into the extracellular matrix of the skeletal muscle fibers. The fascia in many muscles has its own microvascular network that is positioned parallel (without direct connections) to the microvasculature of the skeletal muscle although the feeding arterioles and draining venules of the fascia connect to those of skeletal muscle microvasculature. The fascia microcirculation contains the major vessel classes encountered in skeletal muscle, e.g., arcade arterioles and venules, terminal arterioles and collecting venules, but a relatively sparse capillary network compared to that of skeletal muscle (Stokke 1999). The details are distinctly different between the two tissues.

In this study of the microcirculation, a computational approach is used in the fascia of rat spinotrapezius muscle with account for the shear- and hematocrit-dependent rheology of blood. The simulations are based on a realistic network reconstructions derived from microscopic images of the fascia. The model includes fundamental aspects about passive properties of microvessels (Skalak and Schmid-Schönbein 1986a, b) as well as active myogenic constriction in arterioles (Lee and Schmid-Schönbein 1996). We also present the development of a software tool to perform a spatial analysis of the hemodynamic results. This tool provides local information about the flow properties within a spatial microvessel network map in order to optimize the comparison with experiments. We conclude with a discussion of the flow in capillaries, taking into account the unusual influence of white blood cells and the pseudopod formation in endothelium on the hemodynamics in a capillary network.

20.2 Simulation of Blood Flow in Fascia Microcirculation

In the following section, we present the methods of analysis, including a description of the rat spinotrapezius muscle fascia network, the models for vessel elasticity and blood rheology, the numerical method to compute the flow in the network, and the post-processing tool to display spatial information in a network map.

20.2.1 Rat Spinotrapezius Muscle Fascia

The microcirculation in adult muscle fascia, similar to muscle per se, contains a network structure of distinct vascular segments that include the arcade arterioles, transverse arterioles, capillaries, collecting venules, and arcade venules, similar to the one found in skeletal muscle (Stokke 1999). Contrary to muscle, the capillary vessels have no preferred direction in the muscle fascia, but form a mesh-like structure. We reconstructed these networks in several specimen and developed topological parameters that represent ensemble averages (Stokke 1999).

In the following, we will focus on a single capillary network in the fascia. It is supplied by a single terminal (transverse) arteriole but drained by multiple collecting venules, typical for capillary networks. The particular capillary bundle considered in this study consists of 286 vessels interconnected at 190 nodes. The capillary vessels are linked periodically to avoid any regions with reduced blood flow.

20.2.2 Vessel Elasticity

Blood vessels have passive viscoelastic properties and their diameter depends nonlinearly on transmural pressure. In the present simulations, we focus on the elastic nonlinear vessel expansion as a function of blood pressure (Skalak and Schmid-Schönbein 1986a, b). In addition, arterioles contain vascular smooth muscle and actively contract in response to blood pressure (Davis 1963; Meininger and Davis 1992). Here, a myogenic constriction for pressures above a threshold value of 40 mmHg is assumed (Lee and Schmid-Schönbein 1996).

20.2.3 Blood Rheology

Blood is a complex non-Newtonian fluid and its apparent viscosity is determined by vessel diameter, hematocrit, and shear. The vessel diameter and hematocrit dependence of the apparent viscosity of blood is captured by a model developed by Pries et al. (1990), referred to as the Pries model in the following. This model determines the apparent viscosity of blood μ_{app} from the plasma viscosity μ_{plasma} and an empirical factor η_{P} used to describe diameter and hematocrit dependence:

$$\mu_{\mathrm{app}} = \mu_{\mathrm{plasma}}\ \eta_{\mathrm{P}} \tag{20.1}$$

where

$$\eta_{\mathrm{P}} = 1 + \left(\mathrm{e}^{H_{\mathrm{d}}\,\alpha} - 1\right) / \left(\mathrm{e}^{0.45\,\alpha} - 1\right)\left(110\,\mathrm{e}^{-1.424\,D} + 3 - 3.45\mathrm{e}^{-0.035\,D}\right) \tag{20.2}$$

and

$$\alpha = 4/\left(1 + \mathrm{e}^{-0.593(D-6.74)}\right) \tag{20.3}$$

Here H_{d} is the discharge hematocrit. The tube hematocrit H_{t} is related to the discharge hematocrit through the following relationship:

$$H_{\mathrm{t}} = H_{\mathrm{d}}\big(H_{\mathrm{d}} + (1 - H_{\mathrm{d}})\left(1 + 1.7\mathrm{e}^{-0.415\,D} - 0.6\mathrm{e}^{-0.011D}\right)\big) \tag{20.4}$$

The model uses the following expression for velocity of blood:

$$v = (0.4 \times D - 1.9) \times 1000 \tag{20.5}$$

The Pries model does not consider the shear thinning of blood. An improved model for the apparent viscosity of blood to include the shear thinning of blood was developed by Jacobitz et al. (2009). A comprehensive set of measurements on the shear rate dependence of apparent viscosity of blood was provided by Lipowsky et al. (1980) from in vivo measurements in the mesentery of the cat. Using these results, the apparent viscosity of blood, μ_{app}, as a function of shear rate

$$\gamma = (8 \times v)/D \tag{20.6}$$

was determined in form of the following expression (referred to as *shear model*):

$$\mu_{\text{app}} = \mu_{\text{plasma}} \eta_{\text{P}}\ \eta_{\text{shear}} \tag{20.7}$$

η_{shear} is a shear factor describing the shear thinning effect of blood as a function of the shear rate:

$$\eta_{\text{shear}} = A_0 + A_1 \gamma^{A2} \tag{20.8}$$

where γ is the shear rate (20.6). The coefficients A_0, A_1, and A_2 and their dependence on the hematocrit are

$$A_0 = 1 - 0.29688\, H_t^2 \tag{20.9}$$

$$A_1 = 19.7182\, H_t \tag{20.10}$$

$$A_2 = -0.68131 \tag{20.11}$$

by optimal fit to Lipowsky's experimental data. The results of the model were validated by comparison with the shear dependence of the apparent viscosity of blood observed in viscometer measurements (Chien et al. 1970).

In the simulations, the hematocrit is found from the cell-plasma phase-separation at each bifurcation. The fraction of red blood cell flow is higher for larger daughter vessels, leading to a decreased hematocrit in the smaller daughter vessel (Schmid-Schönbein et al. 1980; Pries et al. 1990).

20.2.4 Network Simulation

At the level of the muscle fascia we have low Reynolds Number Stokes flow in each microvessel. The velocity profile is assumed to be parabolic but with apparent viscosity given by the shear model (20.7). The application of Kirchhoff's Law (mass conservation) at each bifurcation of microvessels results in a system of linear equations, which is solved using a sparse matrix solver. Nonlinear effects introduced by blood rheology and vessel elasticity are accounted for by an iterative approach (Jacobitz et al. 2009). Typically, a convergent solution is obtained after just a few iterations (about 5 iterations to obtain 3 significant digits in the results).

20.2.5 Implementation of Spatial Microhemodynamic Display

A black-and-white micrograph of the blood vessel network was digitized (JPEG format) and converted into a matrix (Matlab), which contained the node positions. This matrix was then exported (Excel format) and used to create a second matrix which contained the spatial microvessel positions. A data file (Excel) containing start and end node coordinates was used to number the blood vessels (1–286). The pixels that made up each blood vessel were designed to contain the blood vessel assignment number. Code was then written (Matlab) that used the blood vessel matrix and the vessel data file, which also contained other information about the vessel network including mean pressure, velocity, diameter, hematocrit, shear, and viscosity from the simulations to create the spatial representation for the entire vessel network.

The sample code diagram (Fig. 20.1) shows how the spatial representation of pressure in the vessel network was created. The code serves to read in the information from the data file for pressure (using the vessel network matrix as the

```
A=network1;          //store network1 in a temporary matrix, so all work is done on A instead of network1
temp=A<1;            //the location of all the zeros in A are saved to another temporary matrix
A(temp)=255;         //the values of the zeros in the temporary matrix are all changed to 255
i=1;                 //the value of the incremental variable i is initially set to 1
matrixPressure=A;    //a new matrix called matrixPressure is create and is identical to A
while i<255          //beginning of a while loop, the loop runs as long as the incremental variable i is less than 255
    temp=A<i+1;      //this saves the location of all the places in A that have values less than i+1 to the temp matrix
    matrixPressure(temp)=pressure(i);      //this inserts values from the pressure data table into the pressure matrix
    A(temp)=255;     //this changes all the values in A for which pressure data has already been inserted to 255
    i=i+1;           //this increments the variable i by 1
end                  //end of while loop
A=network2;          //this starts the same process over again with the remaining elements 255-286
temp=A<1;            //same as line 2
A(temp)=255;         //same as line 3
while i<287          //same as line 4
  temp=A<i-253;      //same as line 5
  matrixPressure(temp)=pressure(i);        //same as line 6
  A(temp)=255;       //same as line 7
  i=i+1;             //same as line 8
end                  //same as line 9
temp=border>0;       //this saves the locations of the border from the border matrix to a temporary matrix
matrixPressure(temp)=0;                    //this enters the value zero in the pressure matrix in all locations saved in temp
```

Fig. 20.1 Sample code for the spatial representation of pressure map for location of each vessel) and enters each value of the pressure data file into the appropriate location in a final pressure matrix. This pressure matrix was then converted into an image and a specific color map (called "Jet") was used to help visualize the pressure changes across the network.

20.3 Results

In this section, first the results from simulations of the microcirculation in rat spinotrapezius muscle fascia are presented. Then, the velocity distribution in a capillary network is discussed and a model for local velocity control is described. The model is based on wall shear stress and justified by the study of endothelial pseudopods. Finally, a rheological control of capillary flow is briefly addressed.

20.3.1 Hemodynamics in Fascia Microcirculation

Figure 20.2 shows the mean pressure distribution in a vessel network in muscle fascia with an arteriolar pressure of 100 mmHg and a venular pressure of 20 mmHg. The mean pressure in each vessel is shown as the average of inflow and exit pressures. The majority of the mean pressure values are close to 24 mmHg (left histogram of the Fig. 20.1). The vessel map (right part of Fig. 20.1) shows that the highest pressures are concentrated at the center of the vessel network, which is at the location of the transverse arteriole. The pressure decreases as the blood flows further from the transverse arteriole and this can be seen by the colors in the spatial representation changing from dark red at the center to lighter red, orange, yellow, green, light blue, and finally to dark blue. The lowest pressures are observed at the top and bottom edges of the network, where the collecting venules are located.

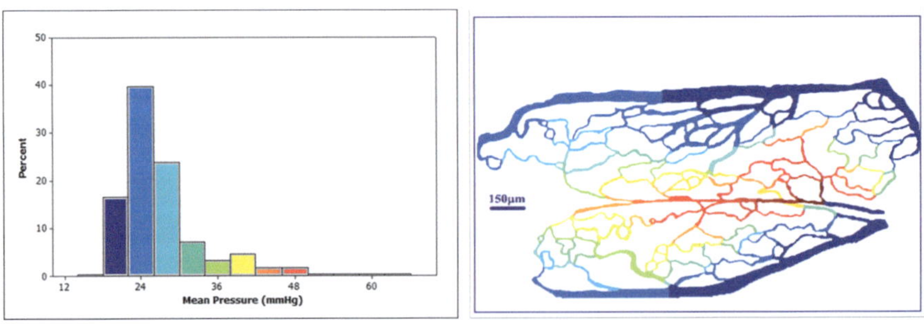

Fig. 20.2 Mean pressure (mmHg) histogram (*left*) and structure (*right*) for the microvessels of the capillary bundle

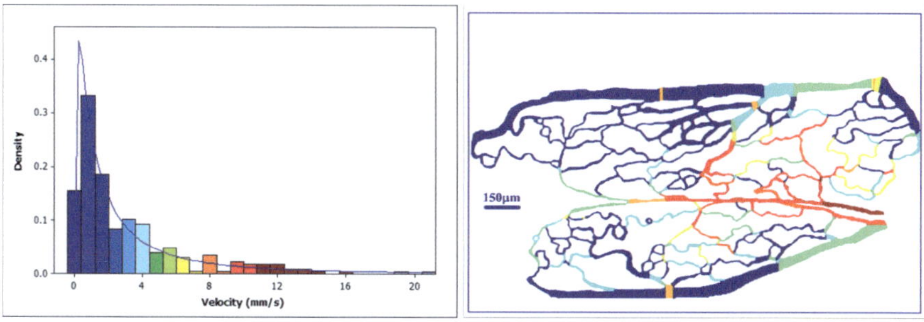

Fig. 20.3 Velocity (mm/s) histogram (*left*) and structure (*right*) for the microvessels of the capillary bundle

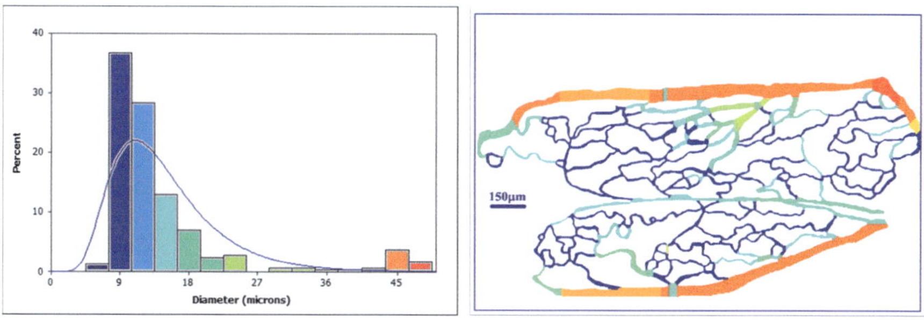

Fig. 20.4 Diameter (microns) histogram (*left*) and structure (*right*) for the microvessels in the capillary bundle

Figure 20.3 presents the mean velocity distribution in the vessel network. The histogram (left) indicates a log-normal distribution of velocities in the network. The vessel map (right) indicates that the highest velocities are at the center, where the blood enters the capillary bundle through the transverse arteriole. The velocity slows as it travels further through the bundle. Low velocities are obtained in the collecting venule as well as in capillaries away from the inflow and outflow vessels of the bundle.

Figure 20.4 shows the distribution of vessel diameter in the muscle fascia network. The histogram (left) indicates that capillaries with a vessel diameter of about 9 μm are the most frequently observed vessels in the bundle. A log-normal fit to the histogram is given for comparison. The vessel map (right) clearly indicates the location of the transverse arteriole in the center of the bundle with vessel diameters ranging from about 15 to 21 μm. The collecting venule has vessel diameters larger than 30 μm. The muscle fascia shows a larger range of diameter values due to an extensive collecting venular vessel system characterized by microvascular loops, a feature found in the fascia but less in the skeletal muscle per se.

Figure 20.5 shows the hematocrit distribution in the capillary bundle. The histogram (left) indicates an approximately normal distribution of hematocrit values in the network. The simulations were performed with an inflow hematocrit of 0.35

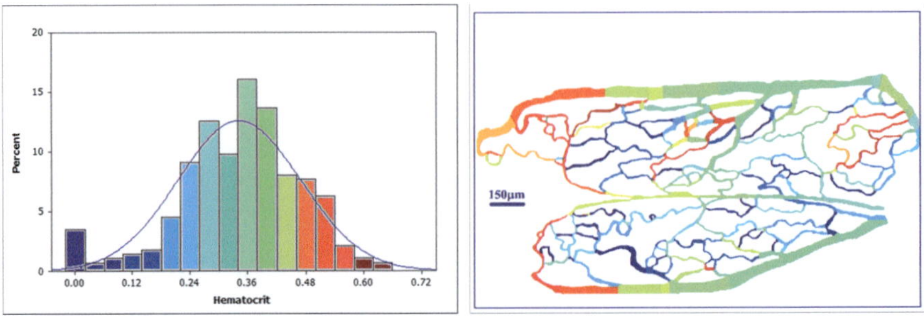

Fig. 20.5 Hematocrit histogram (*left*) and structure (*right*) for the microvessels in the capillary bundle

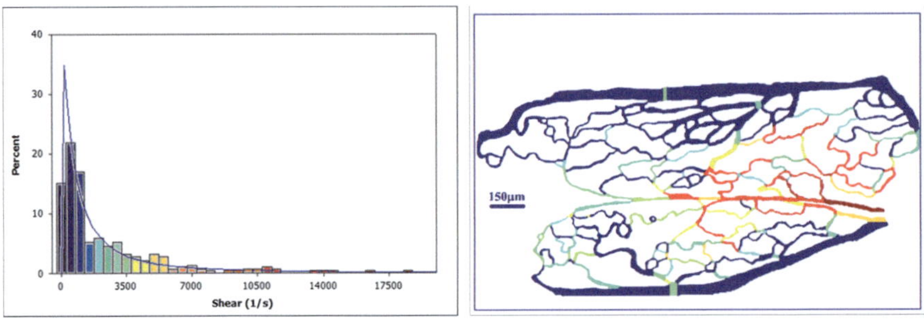

Fig. 20.6 Shear rate (1/s) histogram (*left*) and structure (*right*) for the microvessels in the capillary bundle

into the transverse arteriole and a large number of vessels reflect this value. The vessel map (right) generally shows high hematocrit values in regions with low velocities and low hematocrit values in regions with high velocities.

Figure 20.6 shows the shear rate distribution in the capillary bundle. A log-normal fit is added to the histogram (left). High shear rates are generally found in arterioles and capillary vessels with high velocity. Lower shear rates are found in venules, venule loop vessels, and capillaries away from the inflow and outflow vessels of the bundle.

Figure 20.7 presents the apparent viscosity in the capillary bundle. The computational values in the histogram are closely fit by a log-normal distribution (left). The blood plasma has a viscosity of 1.4 cP and the apparent viscosity is computed from vessel diameter, hematocrit, and shear rate using the shear model presented above. A few capillaries show apparent viscosity values of about 1.4 cP and these vessels correspond to those with low hematocrit. Generally, the apparent viscosity is highest in venule loop vessels as well as in the collecting venule. The vessels are characterized by large diameter and low shear rate. Despite its larger vessel diameter, the high shear rate in the transverse arteriole results in apparent viscosity values similar to those observed in capillaries. In the transverse arteriole, the shear thinning effect counteracts the apparent viscosity increase caused by the larger diameter.

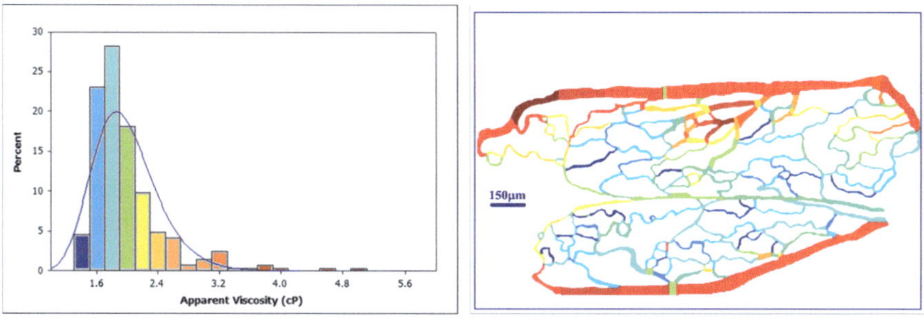

Fig. 20.7 Apparent viscosity (cP) histogram (*left*) and structure (*right*) for the microvessels in the capillary bundle

20.3.2 The Velocity Distribution in Capillary Network

The current analysis brings to light that the velocity in individual vessels of a network of capillaries varies considerably from one vessel to the next. The velocities in these narrow vessels are exquisitely sensitive to the diameter one obtains from analysis of the network images, largely due to the low Reynolds number characteristics in such vessels with a pressure-flow relationship that has a strong diameter dependence (e.g., the fourth order dependence in Poiseuille's Law).

Yet, measurements of capillary velocities in a healthy microvascular network show a narrow distribution with lower variance as compared to the capillary velocity distribution in networks obtained from image reconstructions (Scheidler 2007). A major reason for this discrepancy needs to be looked for in the measurement accuracy to determine capillary diameters. Fractions of a micrometer make a difference in the velocity inside capillaries, especially if they fall into the large class of vessels below 10 μm. In fact the narrower a capillary, the more accurate diameters need to be measured to obtain a realistic picture of the hemodynamics in such a vessel. Several issues surrounding the limitations of current technology to make such measurements have been discussed (Ellis et al. 1983).

Direct observation of a capillary network, for example in skeletal muscle with many capillaries arranged in parallel to skeletal muscle fibers, shows that in healthy muscle the red cell velocity is remarkably uniform from one capillary to the next (Tyml and Cheng 1995), with network variances (standard deviation/mean velocity) in the range of 25–30 % (Scheidler 2007). This relatively low variance is preserved during vasomotor cycles or during acute application of an arteriolar vasodilator (papaverine) indicating that the control mechanisms are inherent to the capillary network and independent of the controls of blood flow in the arterioles or the venules.

The variance in capillary velocity is significantly increased in the capillary network of an animal with chronic disease (e.g., hypertension and metabolic syndrome; Scheidler 2007) or during aging (Tyml et al. 1992). The variance in capillary red cell velocity is also increased when nitric oxide is inhibited pharmacologically.

These observations suggest that there may be a mechanotransduction mechanism that serves to adjust red cell velocity in capillaries per se. The mechanisms need to have a "fine control" over diameters to achieve the remarkable uniformity of the velocity in a network irrespective of a particular branching pattern.

While being influenced by many factors (e.g., glycocalyx, membrane ion transporters, membrane receptors) the mechanotransduction mechanisms fundamentally depend on shear stress on the endothelium or on local pressure. Whatever the mechanisms, they are operational in different regions of a capillary network at higher and lower hydrostatic pressures (e.g., closer to the arteriolar inflows versus the venular outflows, respectively). This evidence suggests that less the pressure and may be more the fluid shear stress is involved.

Wall shear stress controls the thickness of the endothelium in vitro (Barbee et al. 1994; Sato et al. 2000). Thus, there is a possibility that the radius of a capillary, a, may depend on fluid shear stress on the capillary endothelium. An increase in fluid shear stress over a limited physiological range causes a reduction of the lumen diameter and vice versa. Thus the fluid shear stress on the endothelium in a capillary, τ_W, is inversely proportional to the radius a such that

$$\tau_W \sim 1/a^\alpha \tag{20.12}$$

For example, for $\alpha = 1$, we hypothesize there exists an optimal wall shear stress, τ_O, and an optimal lumen radius, a_O, such that

$$\tau_W = a_O\ \tau_O/a \tag{20.13}$$

If we note that for a Newtonian fluid with viscosity μ, the mean velocity for a volumetric flow rate Q, $v_M = Q/\pi\ a^2$, is given by

$$v_M = -\left(a^2/8\mu\right)\ (\partial p/\partial z) \tag{20.14}$$

where $\partial p/\partial z$ is the pressure gradient and the wall shear stress is

$$\tau_W = -(a/2)\ (\partial p/\partial z) = 4\ \mu\ v_M/a \tag{20.15}$$

Then in such a vessel the mean velocity assumes an optimal value

$$v_{MO} = a_O\ t_O/8\mu \tag{20.16}$$

This optimal mean velocity v_{MO} depends only on the optimal shear stress τ_O and the associated optimal lumen radius a_O without dependence on the local pressure gradient $\partial p/\partial z$. Thus it is possible that all capillaries in a network of microvessels that adjust their diameter according to the condition outlined in (20.13) reach the same mean velocity and thus similar red cell velocities. The capillary radius a_O

and an optimal shear stress τ_O are independently adjusted quantities, but their combination serves to adjust capillary velocities to an optimal configuration. The molecular mechanism is currently unknown, but the phenomenon of a capillary velocity with small statistical variance suggests it does exist.

20.3.3 Capillary Endothelium Pseudopod Formation

A suggestion that fluid shear stress controls the morphology of capillaries comes from the study of endothelial pseudopods. While pseudopods generated by actin polymerization on endothelial cells have relatively minor influence on the hemodynamics in vessels outside the microcirculation, in capillaries with single file blood cells their presence has a dramatic effect on the hemodynamics (Lee and Schmid-Schönbein 1995). Endothelial pseudopods reaching into the lumen of a capillary reduce the plasma velocity to zero inside the narrow capillary lumen, dramatically raising the hemodynamic resistance. Cessation of fluid shear stress leads to formation of capillary pseudopods, especially at the junctions between endothelial cells (Hueck et al. 2008). Shear stress on the endothelium serves to attenuate formation of pseudopod projections as well as to reduce the length of existing pseudopods. The process directly involves the endothelial cytoskeleton and the ability to regulate actin polymerization via membrane receptor sensors (Moazzam et al. 1997; Makino et al. 2006). In the capillary network this important regulation mechanism is unexplored.

20.3.4 Rheological Control of Capillary Flow

Under normal physiological conditions the rheological properties of blood in large vessels are dominated by the high volume fraction of red cells. Removal of white cells or platelets during a viscosity measurement using a flow field that has large dimensions compared to those of red cells or white cells has a negligible influence on apparent viscosity.

But white cells cannot be neglected in capillaries with single file flow of blood cells. Even the relatively low number of white blood cells (compared to red cells) leads to a major effect on the hemodynamic resistance if blood is flowing through capillaries. This is due to the effect that white cells have on the motion of red cells, less the white cells by themselves (Helmke et al. 1997). The concentric position of red cells in a capillary, determined by the fluid film between its membrane and the endothelium, is disturbed by the presence of white cells that due to their larger size and higher cell stiffness move with a lower axial velocity than the red cells (Thompson et al. 1989). The consequence of this inequality in red and white cell motion is accumulation and tumbling of red cells upstream of white cells in capillaries, and an increase in their apparent viscosity. The effect of this form of cell interaction in capillaries is strong and causes a major change of the hemodynamic resistance (Helmke et al. 1997), especially if the white cells are activated and form pseudopods, as is seen in inflammatory states.

20.4 Summary and Outlook

Modeling of microvascular blood flow has made significant progress. Yet, we have only seen the beginning of an analysis that will eventually include a large number of cell control mechanisms both on the side of the blood cells and on the side of the endothelium and the pericytes that form the capillary wall. There is a bright future to this field of investigation, especially if applied to disease. Many diseases manifest themselves first at the capillary level, be it from reduced capillary perfusion to a dramatic and complete loss of capillaries, as seen in rarefaction (Tran et al. 2011).

References

Barbee KA, Davies PF, Lal R. Shear stress-induced reorganization of the surface topography of living endothelial cells imaged by atomic force microscopy. Circ Res. 1994;74(1):163–71.

Chien S, Usami S, Bertles JF. Abnormal rheology of oxygenated blood in sickle cell anemia. J Clin Invest. 1970;49(4):623–34.

Davis DL. Small blood vessel responses to sympathetic stimulation. Am J Physiol. 1963;205(3):579–84.

Ellis CG, Safranyos RG, Groom AC. Television-computer method for in vivo measurement of capillary diameter, based on the passage of red cells. Microvasc Res. 1983;26(2):139–50.

Helmke BP, Bremner SN, Zweifach BW, Skalak R, Schmid-Schönbein GW. Mechanisms for increased blood flow resistance due to leukocytes. Am J Physiol. 1997;273(6 Pt 2):H2884–90.

Hueck IS, Rossiter K, Artmann GM, Schmid-Schönbein GW. Fluid shear attenuates endothelial pseudopodia formation into the capillary lumen. Microcirculation. 2008;15(6):531–42.

Jacobitz FG, Anderson S, Babila E, Gabbard M, Weiss C. A comparison of the microcirculation in rat spinotrapezius muscle and muscle fascia. Marseille: Congrès Français de Méchanique; 2009.

Lee J, Schmid-Schönbein GW. Biomechanics of skeletal muscle capillaries: hemodynamic resistance, endothelial distensibility, and pseudopod formation. Ann Biomed Eng. 1995;23(3): 226–46.

Lee S-Y, Schmid-Schönbein GW. Biomechanical model for the myogenic response in the microcirculation: part I. Formulation and initial testing. J Biomech Eng. 1996;118(2):145–51.

Lee S-Y, Sutton D, Fenster M, Schmid-Schönbein GW. Biomechanical model for skeletal muscle microcirculation with reference to red and white blood cell perfusion and autoregulation. In: Mow VC, Guilak F, Tran-Son-Tay R, Hochmuth RM, editors. Cell mechanics and cellular engineering. New York: Springer; (1994) ch. 29, pp. 534–64.

Lipowsky HH, Usami S, Chien S. In vivo measurements of "apparent viscosity" and microvessel hematocrit in the mesentery of the cat. Microvasc Res. 1980;19(3):297–319.

Makino A, Prossnitz ER, Bünemann M, Wang JM, Yao W, Schmid-Schönbein GW. G Protein-coupled receptors serve as mechanosensors for fluid shear stress in neutrophils. Am J Physiol Cell Physiol. 2006;290(6):C1633–9.

Meininger GA, Davis MJ. Cellular mechanisms involved in the vascular myogenic response. Am J Physiol. 1992;263(3 Pt 2):H647–59.
Moazzam F, DeLano FA, Zweifach BW, Schmid-Schönbein GW. The leukocyte response to fluid stress. Proc Natl Acad Sci U S A. 1997;94(1):5338–43.
Pries AR, Secomb TW. Microvascular blood viscosity in vivo and the endothelial surface layer. Am J Physiol Heart Circ Physiol. 2005;289(6):H2657–64.
Pries AR, Secomb TW, Gaehtgens P, Gross JF. Blood flow in microvascular networks. Experiments and simulation. Circ Res. 1990;67(4):826–34.
Pries AR, Reglin B, Secomb TW. Structural adaptation of microvascular networks: functional roles of adaptive responses. Am J Physiol Heart Circ Physiol. 2001;281(3):H1015–25.
Sato M, Nagayama K, Kataoka N, Sasaki M, Hane K. Local mechanical properties measured by atomic force microscopy for cultured bovine endothelial cells exposed to shear stress. J Biomech. 2000;33(1):127–35.
Scheidler NM. A proposed mechanism for control of capillary network perfusion: the role of endothelial mechanotransduction through nitric oxide signaling and the glycocalyx. Dissertation, University of California, San Diego; 2007.
Schmid-Schönbein GW. A theory of blood flow in skeletal muscle. J Biomech Eng. 1988;110(1):20–6.
Schmid-Schönbein GW, Skalak R, Usami S, Chien S. Cell distribution in capillary networks. Microvasc Res. 1980;19(1):18–44.
Skalak TC, Schmid-Schönbein GW. The micro-vasculature in skeletal muscle. IV. A model of the capillary network. Microvasc Res. 1986a;32(3):333–47.
Skalak TC, Schmid-Schönbein GW. Viscoelastic properties of microvessels in rat spinotrapezius muscle. J Biomech Eng. 1986b;108(3):193–200.
Stokke KE. An analysis of the microvasculature and blood flow in rat spinotrapezius muscle fascia. M.S. Thesis, University of California, San Diego; 1999.
Thompson TN, La Celle PL, Cokelet GR. Perturbation of red blood cell flow in small tubes by white blood cells. Pflugers Arch. 1989;413(4):372–7.
Tran ED, Yang M, Chen A, DeLano FA, Murfee WL, Schmid-Schönbein GW. Matrix metalloproteinase activity causes VEGFR-2 cleavage and microvascular rarefaction in rat mesentery. Microcirculation. 2011;18(3):228–37.
Tyml K, Cheng L. Heterogeneity of red blood cell velocity in skeletal muscle decreases with increased flow. Microcirculation. 1995;2(2):181–93.
Tyml K, Mathieu-Costello O, Budreau CH. Distribution of red blood cell velocity in capillary network, and endothelial ultrastructure, in aged rat skeletal muscle. Microvasc Res. 1992;44(1):1–13.

Chapter 21
Network Approaches to the Mechanical Failure of Soft Tissues: Implications for Disease and Tissue Engineering

Béla Suki and Arnab Majumdar

Abstract Damage and mechanical failure of soft tissues can lead to diseases and even life threatening conditions including the failure of prosthetic heart valves, capillary stress failure, tissue destruction in pulmonary emphysema, and vessel wall aneurysms. Microscale damage occurs when mechanical forces in the tissue are sufficiently high to rupture intercellular connections or enzymatically weakened extracellular matrix (ECM) elements. When the microscale damage reaches a critical amount, tissue or organ failure can happen. In this chapter, we first briefly review the failure of the main constituents of tissues including molecules, cells, elastin, collagen, and proteoglycans. We then discuss failure of tissues modeled as complex networks of these constituents. A key concept here is percolation, a process where network elements reach from one boundary of the network to another. We show that when a certain type of fiber percolates the tissue, the rupture process of the network is primarily governed by the failure properties of the individual fibers. When the main fiber component does not percolate, the failure stress is still dominated by the stiffness of the fiber, but the failure strain emerges as a network phenomenon. Finally, we conclude by proposing some general concepts of how to potentially minimize the risk of failure and best repair a damaged network.

21.1 Introduction

With the realization that "from molecules to organisms, everything must obey the laws of mechanics" (Fung 1993), biomechanics has evaded virtually all of life sciences including physiology, biology, and pathology. Perhaps the most important concept in biomechanics is that the biomechanical properties of biological tissues play fundamental roles in the normal functioning of virtually all connective tissue, organ, and organism. These biomechanical properties are critical determinants of how mechanical interactions of the body with the environment produce physical

B. Suki (✉) • A. Majumdar
Department of Biomedical Engineering, Boston University, Boston, MA 02215, USA
e-mail: bsuki@bu.edu

G.S. Kassab, M.S. Sacks (eds.), *Structure-Based Mechanics of Tissues and Organs*,
DOI 10.1007/978-1-4899-7630-7_21

forces at the fiber, cell, and single molecule levels. Consequently, many studies have explored and modeled the mechanical properties of various tissues and cells at multiple length scales (Fung 1993, 1997; Humphrey and Delange 2004).

Characterization of the macroscopic mechanical properties of living tissues and organs requires establishing mathematical relations, called the constitutive equations, linking the mechanical response such as stress to a change in the size and/or shape of a body usually given in terms of strain. Realistic constitutive equations are nonlinear and time dependent giving rise to viscoelastic behavior. A major advance in biomechanics occurred when Yoram Lanir introduced structure–function relations which formulate the constitutive equations in terms of the microstructure and the properties of the constitutive elements (Lanir 1983). Because of the complexity in the microstructure of tissues, often statistical descriptions are invoked in deriving constitutive equations (Lokshin and Lanir 2009) which can be based on microscopic imaging of the structure and measurement of the stress–strain relations of the constituents in isolation.

An important area of biomechanics that has received considerably less attention is related to damage and mechanical failure of tissues. Damage usually appears first at the microscale and can occur under a variety of normal conditions such as sport injuries or labor. When the microscale damage exceeds a critical amount, the consequences can reach the macroscopic level eventually leading to failure. In diseases, microdamage can also percolate to the macroscopic level and result in abnormal function or total failure of a tissue or organ culminating in life threatening conditions. For example, capillary stress failure in the lung can occur during severe exercise in athletes or in animals bred for high aerobic activity (West 2000). Capillary failure can also be induced by mechanical ventilation of patients suffering from acute lung injury (Bland et al. 1989). During normal pregnancy, the membrane surrounding the fetus has a protective role, but premature membrane rupture occurs in 5–10 % of pregnancies with an increased perinatal morbidity and mortality (Joyce et al. 2009). In diseases such as vascular aneurysm and pulmonary emphysema, the extracellular matrix (ECM) undergoes proteolytic digestion and because both the vessels and the lung operate under a preexisting mechanical stress condition, also called prestress, the weakened fibrils and fibers can rupture leading to severe conditions including death (Suki et al. 2003; Vorp 2007). Implanted tissue-engineered constructs such as prosthetic heart valves can also undergo digestion and subsequent failure (Lee et al. 2001). Similarly, atherosclerotic plaques are exposed to high stresses, and following enzymatic digestion, plaque rupture can occur suddenly triggering acute coronary syndromes such as angina or myocardial infarction (Arroyo and Lee 1999).

Despite the obvious importance of damage and failure in both healthy and diseased tissue, the corresponding processes at the microscale are much less understood than those during large but continuous deformations of tissues. While various continuum approaches to modeling damage have been proposed (Christian Gasser 2011; Ito et al. 2010; Volokh 2008), the ECM is heterogeneous with the cells and fibers forming a network embedded in a proteoglycan (PG) gel. Therefore, a better understanding of the network organization of the ECM, and how its components contribute to the failure process should provide insight into the progression of diseases. This chapter focuses on network approaches to mechanical failure of soft fibrous tissues. First, we will describe the general structure and failure properties of the main constituents of soft tissues including molecules, elastin, collagen, PGs, and cells. Next, we will consider a process called percolation that has relevance to the mechanics of failure. Several examples of failure will then be described in light of networks and percolation both numerically and analytically including published as well as new results. Finally, we will propose some general concepts of how one could potentially minimize the risk of failure and best repair a damaged network.

21.2 Failure of Molecules, ECM Components, Cells, and Tissues

Tissues are composed of biomolecules which organize themselves to form larger ECM constructs such as fibrils and fibers as well as cells. The main components that form the complex structural network of the ECM are elastin, collagen, and PGs. Both cells and ECM components are necessary for tissues to function properly. Additionally, tissues can also contain other smaller molecules such as fibronectin, laminin, and various microfibrils. However, from the point of view of mechanics, elastin and collagen are the key fibrous proteins capable of carrying loads and resisting tension (Kielty et al. 2002; Silver et al. 2003). The PGs make up the gel in which elastin and collagen fibers as well as cells are embedded. The PGs are necessary for the stability of the ECM and for functional cell–ECM interactions (Carey 1991). Varying the structure and composition of the ECM has major effects on its mechanical (Angele et al. 2004; Black et al. 2008; Marque et al. 2001; Suki et al. 2005; Tanaka et al. 2001) as well as failure properties (Billiar et al. 2005; Black et al. 2008; Donovan et al. 1990; Ito et al. 2005; Jesudason et al. 2007; Joyce et al. 2009; Kononov et al. 2001; Ritter et al. 2009; Vorp et al. 2003; Wilson et al. 1998).

21.2.1 Failure of Molecules

At the molecular level, the physical process of rupture of a single bond has only been recently elucidated (Chen and Chu 2005; Dias et al. 2005). The essence of these theories is that two particles (atoms or molecules) are held together via a specific bond characterized by for example the Lennard–Jones potential. When the molecule is held at a given strain, rupture of bonds can occur as a result of thermal agitation. Under most physiological conditions, however, rupture is triggered by stretching or mechanical forces. During stretching of a molecule, the energy landscape of the bond gradually tilts and elongation proceeds by internal ruptures and subsequent unfolding events (Oroudjev et al. 2002). Full rupture occurs when the stretching of the bond exceeds a critical level, or a critical bond strain, beyond which the likelihood of rebinding is negligible. The force at rupture (failure force) is a random variable with a distribution that is similar to a Gaussian. The average failure force depends on the temperature and is a complicated logarithmic function of the loading rate (Dias et al. 2005). Thus, the rupture of a single molecular bond can generate a distribution of failure forces. Alternatively, studying the rate dependence of rupture forces can reveal the complex energy landscape of molecular bonds (Evans 2001).

21.2.2 Failure of ECM Components

The mechanism of failure in more complex hierarchical structures such as fibrils, fibers, cells, and entire tissues is more complicated. Among the main structural constituents of the ECM, collagen has the highest stiffness and failure stress. Indeed, to elucidate the mechanism of arterial aneurysm, it was found that pressurized vessels treated with elastase dilated, but never ruptured, whereas those treated with collagenase dilated still more and eventually ruptured in every case (Dobrin and Anidjar 1991). What structure is responsible for the failure of a collagen fiber? The subunits of type I collagen, the most abundant structural protein in tissues, form a triple helix. Several molecules assemble into fibrils and the failure properties of such fibrils depend on the length and diameter of the fibril and the cross-link density (Tang et al. 2010). Fibrils can assemble into microfibrils which show substantial hierarchical complexity with lateral packing of molecules that can exhibit fluid-like disorder (Holmes et al. 2001). The fibrils can further organize into thicker fibers through cross-linking and these fibers go on to form random (lung tissue, cartilage) or quasi-deterministic (tendon) networks within an organ. Molecular dynamic simulations suggest that the failure strain of fibrils range from 6 to 40 % and failure stresses increase from 0.5 to 5 GPa with increasing cross-link density (Tang et al. 2010). Experiments show that the ultimate tensile strength of fibrils is about 0.5 MPa nearly independent of fibril diameter in the range from 20 to 40 nm (Christiansen et al. 2000).

Elastin, another essential load bearing component of the ECM, is resilient over a large range of strains and provides elasticity to tissues. It is notable that the stiffness of elastin is at least two orders of magnitude smaller than that of collagen (Fung 1993). This is likely a result of the amorphous nature of elastin compared to the more regular organization of collagen fibers. The tropoelastin has two major types of alternating domains, the hydrophilic helical domains and the hydrophobic lysine-rich domains. These nonrandom, regularly repeating structures exhibit dominantly entropic elasticity by means of damping of internal chain dynamics on extension (Urry and Parker 2002). The failure behavior of purified elastin is similar to that of amorphous, noncrystallizing elastomers and exhibits a strong viscoelastic behavior (Lillie and Gosline 2002). In the normal range of temperatures and hydration levels, the failure stress of elastin is between 1 and 2 MPa and the failure strain is at least 200 % (Lillie and Gosline 2002).

The ECM also contains a significant amount of gel-like PGs (Hukins 1984) which change during maturation and disease (Juul et al. 1993). Critical constituents of this matrix are the glycosaminoglycans (GAGs) which are long chains of repeating disaccharide units that are variably sulfated and highly charged. Proteoglycans can also associate ionically with one another to form large aggregates that exhibit a high level of hierarchical organization. The majority of studies on proteoglycan mechanics have been carried out in cartilage. The elastic modulus of the cartilage tissue measured using the indentation atomic force microscopy at the nanometer scale is around 20 kPa and is related to the GAG content of the tissue (Stolz et al. 2004). Molecular dynamics simulations suggest that the GAG chains can take up significant strains (>800 %) and can transfer stress from one collagen fibril to a neighboring one (Redaelli et al. 2003). However, the failure force between two hyaluronan and its binding protein is very small, only 40 pN (Liu et al. 2004). Another important function of PGs is that they can oppose the folding of large and stiff fibers toward the direction of strain with their compressive resistance (Cavalcante et al. 2005). As discussed below, this mechanism plays an important role in the failure of complex tissues.

21.2.3 Failure of Cells and Tissues

The second large class of tissue constituents is the adherent cells. Most cells in the body are continuously exposed to various forms of mechanical forces which deform the cells and trigger biological responses (Ingber 2006). Under abnormal conditions, mechanical forces can exceed the failure threshold of the cells and they either release their attachments to each other and/or the ECM or they rupture. There are little experimental data on the mechanisms of cell rupture. Under compressive loading, the membrane bulges out, and depending on the cell type, bursting failure occurs around 40–80 % strain (Nguyen et al. 2009; Peeters et al. 2005; Stenson et al. 2011). The corresponding bursting force in mouse myoblast is around 9 μN (Peeters et al. 2005) whereas the circumferential stress at failure in yeast cells is 115 MPa (Stenson et al. 2011). Molecular dynamics simulations suggest that biaxial tension in the membrane causes pore formation at membrane forces between 150 and 400 pN depending on the strain rate (Koshiyama and Wada 2011).

These components also interact with each other by forming bonds of varying strength. When the tissue is exposed to supraphysiological forces, it is possible that the inter-constituent bonds start to break before the constituents themselves fail. Figure 21.1a demonstrates the rich features of the failure stress–strain curve in an engineered tissue sheet. In diseases, the tissues can undergo remodeling and enzymatic digestion weakening the structural integrity of the constituents as well as eliminating some of the bonds among them. In such cases, normal physiological forces might be sufficient to produce damage and ultimately failure. Figure 21.1b compares the failure of normal lung tissue with that of emphysematous tissue. In the next section, we will consider several network models where the failure process of tissues with varying constituent types and amounts is simulated.

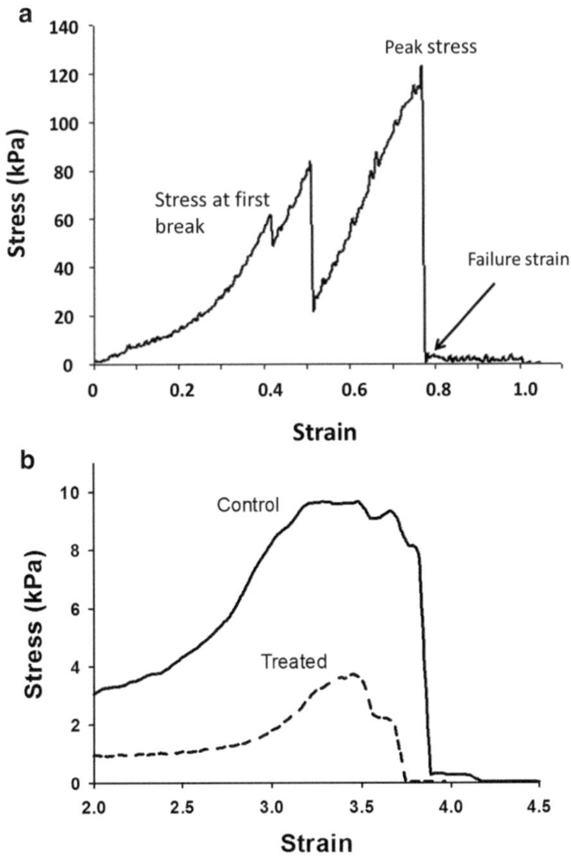

Fig. 21.1 Failure stress–strain curves of an engineered elastin sheet (**a**) and a lung tissue piece obtained from a control mouse and a mouse that had been treated with porcine pancreatic elastase (**b**). With permission from Black et al. (2005) and Ito et al. (2005)

21.3 Homogeneous Network Models of Failure

21.3.1 Percolation

In this section, we introduce the concept of percolation (Stauffer and Aharony 1992) and apply it to tissue failure. Consider a simple square lattice in which each pair of neighboring nodes are either connected by a bond with probability p or disconnected with probability $(1-p)$. A cluster is defined as the set of connected bonds. Percolation is concerned with the properties of these clusters, including their size and organization. As p increases from 0 (disconnected lattice) to 1 (fully connected lattice), the mean size of the clusters increases, and there is a point at which a large cluster spans the lattice providing full connectivity from one side to the other. The transition from a system of disconnected clusters to a system with a connected cluster spanning the system occurs when p crosses a critical percolation threshold p_c. The structure of the incipient percolation cluster at $p=p_c$ is a self-similar fractal. Percolation has been applied to the gelation of bronchial mucus

(McCullagh et al. 1995), to explain airway opening in the lung (Barabasi et al. 1996), the progression of symptoms in pulmonary emphysema and fibrosis (Bates et al. 2007), and the mechanics and failure of the ECM (Black et al. 2008; Ritter et al. 2009).

With regard to damage and failure, let us consider a model of the tissue as a lattice of elastic springs. As the lattice is diluted from its intact state ($p = 1$) by decreasing p, springs are gradually removed, and as p approaches p_c, fewer and fewer bonds hold the network together. Just under p_c, the network loses its integrity and falls apart or depercolates corresponding to tissue failure. Alternately, consider the process of stretching a network of springs. If each spring represents a fiber, then the failure of individual fibers can be thought of as a stochastic process with a failure probability of $(1 - p)$ which increases as the local strain on the fiber increases. This process can mimic the rupture of the tissue, implying that there will be a critical level of strain or stress, corresponding $p = p_c$, at which the network suddenly separates into two subnetworks representing tissue failure. This is indeed often seen in experiments as the stress suddenly drops to zero (Fig. 21.1). We note that p_c for an elastic network is larger than p_c for pure topological percolation due to the folding or alignment of the bonds before they begin to stretch. By including a torsional spring, called bond bending, that resist folding, the elastic and geometric p_c will be the same (Stauffer and Aharony 1992).

As another application of percolation to tissue failure, consider the coalescence of cracks or voids, rather than the connectivity of fibers. An intact tissue lacks voids and each fiber failure introduces a void. If the failures are random and independent of each other, the voids are mostly small, comprising single failures, until the network is sufficiently diluted such that the voids begin to coalesce. On the other hand, if the failures occur near an existing void with a high probability, clusters of voids grow quickly until a percolating cluster of voids physically separates the tissue.

The concept of percolation provides a rich scaling description of the system near the percolation threshold which as we have seen corresponds to tissue failure. The percolation threshold p_c depends on the underlying network topology, especially its local and long-range connectivity. The behavior of the network is critically dependent on the distance from p_c. For example, the elastic modulus G scales as $(p - p_c)^f$ while the typical size of a cluster ξ scales as $(p - p_c)^\nu$ where f and ν are scaling exponents. These exponents describe the mechanical and structural behavior of the tissue as it approaches failure and can be calculated for networks with or without bond bending forces (Sahimi and Arbabi 1993).

21.3.2 Network Damage: Effects of Mechanical Forces

The simplest process of depercolation is a purely random removal of bonds, possibly mimicking enzymatic digestion of an unstressed tissue comprising of identical fibers. However, most tissues in the body are under a prestress. Consequently, as loading continues, failure of one fiber can result in the redistribution of the stress the fiber carried among the neighboring fibers causing a spatially correlated failure process (Parameswaran et al. 2011). In our first example, we compare the random and the correlated degradation of lung tissue during the progression of emphysema. The lung parenchyma is composed of myriads of alveolar walls each composed of elastin, collagen PGs, and cells. At scales larger than an alveolus and away from major airways, the tissue can be thought of as a network of septal walls with nearly identical properties. During the progression of emphysema, alveolar walls are lost leading to a heterogeneous structure with a reduced surface area for gas exchange. The question is whether the progression of emphysema can be modeled as a depercolation process and if so what can we learn about the disease from the modeling exercise.

In a recent study, respiratory compliance C was correlated with biochemical and structural parameters of the mouse lung before and after elastase-induced emphysema (Hamakawa et al. 2011). Interestingly, C did not correlate with bulk measures of soluble type I collagen, type III collagen, or elastin. There was, however, a strong association between C and the mean equivalent diameter of airspaces (D) obtained from histology, and an even stronger relation between C and an index representing the area weighted mean diameter (D_2) with r^2 values of 0.675 ($p<0.01$) and 0.933 ($p<0.001$), respectively (Fig. 21.2a). Since D_2 includes higher order moments of the distribution of D, it is highly sensitive to heterogeneities. The strong correlations provide evidence that it is not the mean airspace size, but its heterogeneity that primarily determines function as characterized by C. Further insight can be gained by examining how function emerges from the underlying structure, when heterogeneity is generated in various ways. To do this, a 2D rectangular network of linear elastic springs was used to predict the above structure–function relations (Hamakawa et al. 2011). A uniform negative pressure was applied at the boundaries of the network to mimic pleural pressure, which generated tensile forces on the springs. Damage in the tissue during emphysema progression was then simulated in 3 different ways at subsequent iterations: (1) springs were uniformly weakened, (2) springs were randomly eliminated, and (3) springs carrying high force were eliminated based on previous findings (Kononov et al. 2001). The network configuration was obtained by minimizing the total energy of the system using simulated annealing (Cavalcante et al. 2005). The 2D compliance C of the network was calculated and the iterations were repeated until C increased by 2, the increase seen in the data (Fig. 21.2a). Following proper normalization, the simulations can be compared to the data. Each of these mechanisms produces an approximately linear relation between C and D_2 but the slopes of these relations are different (Fig. 21.2b). The simulations corresponding to the force-based destruction approximates best the experimental data. These results suggest that damage in the lung is produced by a process similar to correlated percolation due to the effects of mechanical forces and this has been corroborated using 3D networks (Parameswaran et al. 2011). Furthermore, producing a few large defect holes has a much stronger effect on function than uniformly weakening the alveolar walls.

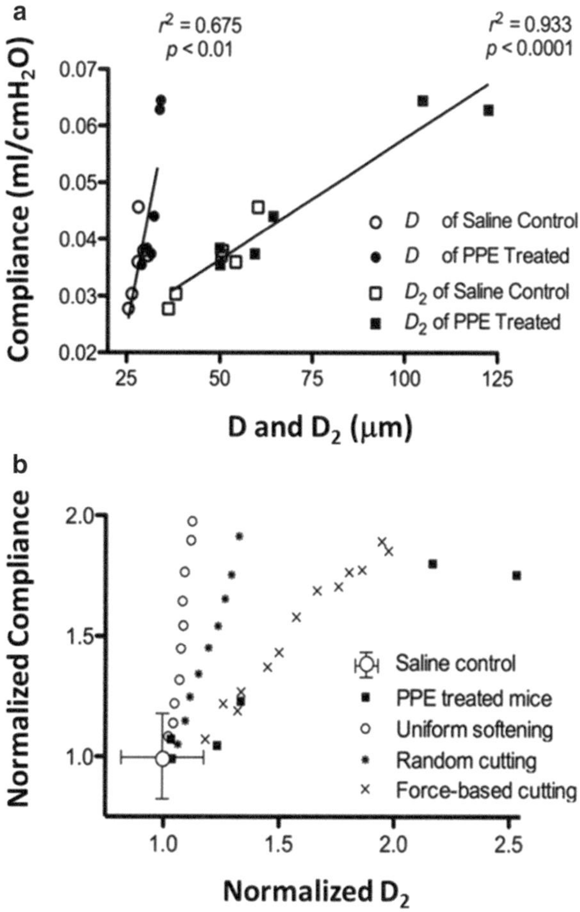

Fig. 21.2 Structure–function relations obtained in saline control and porcine pancreatic elastase-treated mice. (**a**) Respiratory compliance, C, as a function of the mean equivalent diameter (D) or the area weighted mean equivalent diameter (D_2). (**b**) Comparison of structure–function relation from a homogeneous network model with various digestion processes and from experiments. Experimental data were normalized with the mean value of C and D_2 of control mice. Each *black square* corresponds to data of a single mouse. The symbols corresponding to the various model cases represent model outputs normalized by the output of the base-line model. The data and simulations are from Hamakawa et al. (2011) with permission

21.4 Two-Phase Networks

In this section, we will consider forced rupture tests applied to various networks. First, we note that the model in the previous section was a network with homogeneous constituents. At a smaller scale than the whole lung, however, it may be necessary to account for the different properties of the constituents of the tissue. For simplicity, we consider a network of two constituents only, a soft element such as PGs and a stiff element such as elastin or collagen. We also assume that the failure strain of the two components differ significantly. The full network corresponding to $p = 1$ is now a mixture of the two constituents and the relative fraction of each is characteristic of the particular tissue we wish to model. In such a system, we need to consider whether one of the constituents percolates before any rupture is induced by the failure test. The reason is that the failure process and the failure stress and strain will critically depend on the organization of the network. For example, if the constituent with a low failure strain percolates through the network, then upon stretching the network, these elements will eventually rupture along their connected cluster. However, if neither constituent percolates, the failure process cannot be predicted from the properties of the individual components. Instead, a failure threshold emerges as a network phenomenon (Ritter et al. 2009).

21.4.1 Networks with Percolating Constituents

Consider a tissue with layers that have different constituents. For example, an arterial wall segment that has a layer of stiff collagen that fails at a low strain sandwiched between two layers of identical but softer constituents that have a much higher failure strain such as elastin. For simplicity, we assume that the stiffness or spring constant of the two components are the same but the middle layer fails at a mean strain of 0.1 (uniform distribution between 0.09 and 0.11) whereas the two outer layers have failure strain of 0.6. In the unstretched configuration, the initial lengths of the springs are set to have a narrow distribution and the network is submitted to a uniaxial failure test (Fig. 21.3a). At low strains, ε, the fibers begin to fold toward the horizontal direction slightly increasing stress (top network in Fig. 21.3b). Beyond $\varepsilon = 0.1$, the fibers become more and more stretched and around $\varepsilon = 0.2$, fibers in the middle layer start to fail with a full failure at $\varepsilon = 0.3$ (second network in Fig. 21.3b). As the network is stretched further, no rupture occurs until

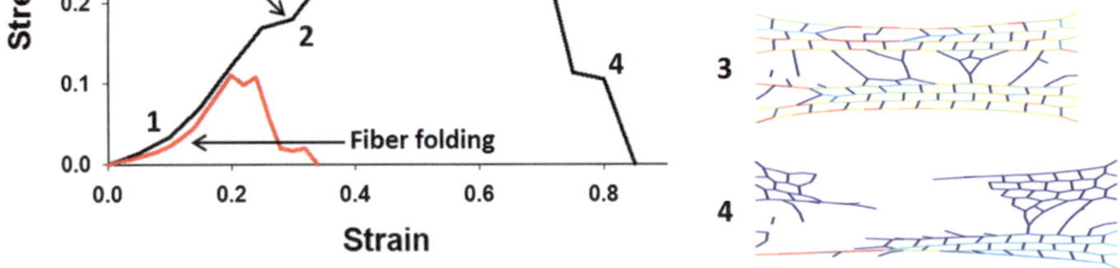

Fig. 21.3 Stress–strain failure curves and network configurations of a two-phase model with a layer of percolating springs that have a failure strain five times lower than the rest of the springs. (**a**) The *black line* corresponds to a network in which the layer is horizontal and the *red line* corresponds to a network in which the layer is vertical. Stretching proceeds in the horizontal direction. (**b**) Snapshots of the network configurations corresponding to the *black line* in (**a**) at the points marked by numbers along the stress–strain curve. The colors are proportional to strain; *blue* is low and *red* is high strain

the peak stress is reached at $\varepsilon = 0.65$ (third network). Beyond this limit, the outer layers begin to break down and the stress drops to zero (bottom network just before failure).

The conclusion from this simulation is that in a two-phase network with percolating constituents, the layer that has the higher failure strain threshold determines the ultimate strength and failure strain of the network whereas the weaker constituent can contribute to features of the stress–strain curve at lower strains. Furthermore, a higher failure strain of the stronger component generally results in a smaller fraction of the full strain range dominated by folding. This leads to a closer agreement between the strain at the peak stress and the failure strain of the individual constituent (0.65 vs. 0.6). However, the failure strain of the network can substantially differ from that of the individual components, which is 0.85 in this particular network. It is also interesting to note that our network is anisotropic. If the network was stretched in the vertical direction, then the failure strain of the weaker element would have dominated the stress–strain curve. Indeed, simulations show that in this case, the peak stress occurs at $\varepsilon = 0.2$ and the network fails at $\varepsilon = 0.32$ (red curve in Fig. 21.3a). As we shall see, in the absence of percolation, the situation is drastically different.

21.4.2 Networks with Non-percolating Constituents

In this section, we consider a two-phase network with non-percolating elements as a model of a sheet of elastin embedded in PGs. Before introducing the model, we note that two groups reported similar results related to the failure of tissues composed primarily of elastin and PGs using tissue-engineered ECM constructs (Black et al. 2005) and the inferior mesenteric vein of patients with abdominal aortic aneurysm (Goodall et al. 2002). In both tissues, elastin degradation led to a significant decrease in peak stress during a failure test. Surprisingly, the failure strains were in the range of 60–120 %, much lower than the 200 % failure strain of elastin. Furthermore, in both studies elastin degradation had no effect on the failure strain. In the previous section we saw that if the elastin fibers percolate, we would expect the failure strain of the composite to be near that of elastin. On the other hand, if elastin does not form a percolating network, the failure of PG bridges between the elastin fibers should reduce the global failure strain of the ECM below that of elastin. To test this hypothesis, we developed a novel spring network model, the Zipper Network Model (ZNM) (Ritter et al. 2009).

The model is a hexagonal network of linear springs mimicking the stiffness of elastin and PGs. The network also contains torsional springs, which resist angular rotation of the springs around their pin joints. The total elastic energy, E, of the network is given by $E = \frac{1}{2} \Sigma \ (k\Delta l^2 + b\Delta\theta^2)$ where Δl is the change in length of the spring, k is the spring constant, b is the angular spring or bond bending constant, and $\Delta\theta$ is the change in angle between two springs from the stress-free initial angle.

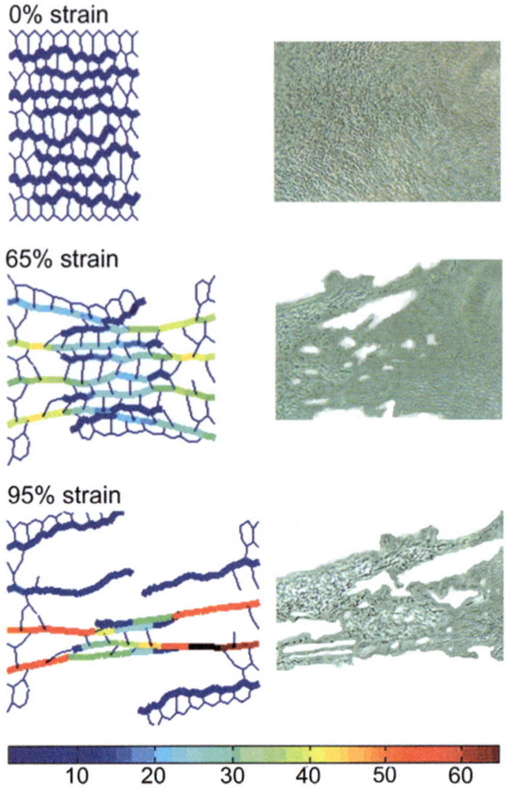

Fig. 21.4 *Left*: Images of the ZNM being stretched to various strains. The elastin fibers are drawn as thick lines and the PGs as thin lines. Note that the elastin does not percolate across the network. The color scale shows the relative forces on each spring. *Right*: Phase contrast images of a region of an engineered tissue construct sample undergoing failure taken at strains comparable to those in the network model on the left panel. With permission from Ritter et al. (2009)

The summation runs through all linear and bond bending springs. Bond bending resists the springs from aligning with the direction of strain and is related to the compressibility of the PGs (Cavalcante et al. 2005).

The springs corresponding to elastin and PG have a k value of 6 and 2 (arbitrary units) and strain failure threshold of 200 % and 0.5 %, respectively. The bond bending constant is $b = 0.1$. The springs are arranged such that there are long fibers of elastin springs which are embedded in a network of PG springs (Fig. 21.4). Note that the elastin fibers do not percolate across the tissue; instead, they reach about ¾ of the way across and overlap from opposite sides creating a "zipper-like" formation. A small amount of variability was introduced in the initial length of the springs. Failure of the ZNM was simulated by stretching it in steps of 5 % until failure and solving the network by minimizing E at each step.

The ZNM was also used to test a phenomenon referred to as avalanching. This occurs when the breaking of one spring leads to a condition that causes other springs to break without the network being further stretched. Without avalanching, increasing strain step sizes leads to increasing peak stress and failure strain. During avalanching, however, the model was insensitive to step size. Since the peak stress and failure strain should not depend on how the stress–strain curve is recorded, the more realistic avalanching was kept throughout. We note that avalanching is a dynamic phenomenon that reduces the overall stress in the material. For example, the sudden drop in stress at 0.5 strain in Fig. 21.1a is due to avalanching. In reality, the redistribution of stress after the failure of a fiber depends on the loading rate of the test. As the loading rate increases, the sequence of ruptures in an avalanche is interrupted by the next avalanche. Consequently, the failure gradually becomes similar to a depercolation process in which the weakest bonds are sequentially eliminated each opening an isolated void in the network.

Network configurations of the ZNM during a failure test are shown in Fig. 21.4 together with images obtained during the digestion of an engineered tissue sheet. During stretch, the elastin fibers gradually become straight while carrying most of the forces in the network as the colors demonstrate. Above 35 % strain, the maximum strain on PGs fluctuates near 0.5 % (data not shown) as more and more PGs reach failure. It is thus the failure of the PG bridges that eventually leads to a complete network failure at 100 % macroscopic strain. To mimic elastin digestion, half of the elastin fibers were cut into two fragments. This resulted in an elimination of only 3.1 % of the elastin springs from the network. Figure 21.5 compares the stress–strain curves for a control and an elastin-digested failure simulation. The peak stress significantly decreased by about 40 % ($p < 0.001$) while the failure strain of the network was 109 ± 9 % and did not change with digestion similar to experimental

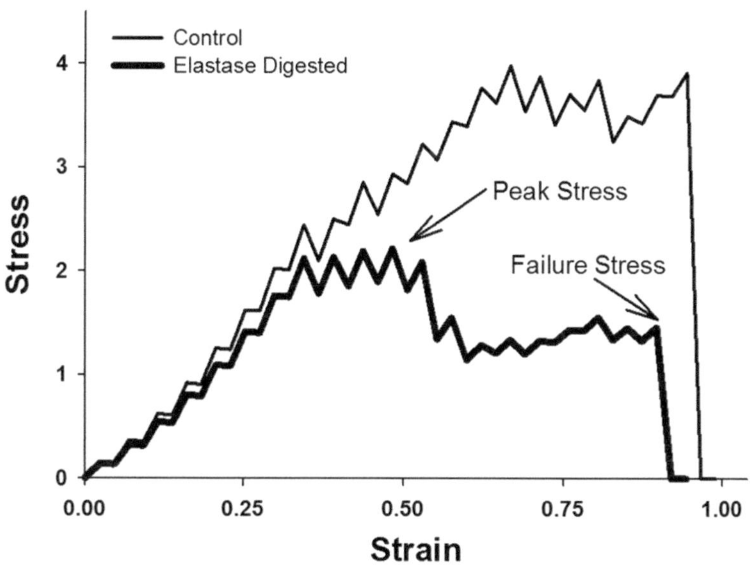

Fig. 21.5 Stress–strain curves from the ZNM with and without simulated elastase digestion. Notice that the peak stress and the failure stress can be different. Additionally, the two curves show many small drops in stress that run parallel because these correspond to the failure of PG bridges and initial ZNM was the same in both cases except that in the digestion simulation the elastin was fragmented. With permission from Ritter et al. (2009)

data (Black et al. 2005). The effects of graded digestion of PGs on the stress–strain curves of the network were also calculated. The model predicted that the removal of 60 % of PG springs decreased the peak stress and the failure strain by only 29 % and 15 %, respectively ($p < 0.05$). These simulations were then compared digestion of the sheets with trypsin that eliminates the core protein of PG without digesting elastin. After 30 min, the stiffness in trypsin-digested samples dropped significantly to about 75–80 % of baseline. The trypsin digestion led to 31 and 29 % drop in peak stress and failure strain. Furthermore, biochemical analysis showed that the trypsin digestion eliminated 65 % of chondroitin sulfate and 45 % of heparin sulfate from the samples (Ritter et al. 2009).

In summary, simulations using the two-phase ZNM with non-percolating elastin fibers resulted in the following main conclusions. First, the mechanism of failure is a gradual straightening of the elastin fibers followed by pulling them out of the PG matrix by successively breaking the PG bridges, a novel network phenomenon. Second, the ZNM correctly reproduced the failure properties of the ECM sheets containing elastin and PGs, such that the network fails at approximately 100 % strain while keeping elastin at its known failure strain of 200 %. The ZNM also accounted for previous experimental results that elastin digestion decreases peak stress but not failure strain in engineered ECM sheets (Black et al. 2005) and native vessels (Goodall et al. 2002). Finally, an important result of the network analysis is that elastin carries the load and determines the peak stress, whereas PGs reduce the failure strain of the network.

21.4.3 Analytical Modeling

Analytical modeling of failure processes of multi-phase tissues is difficult. Some results can be obtained by simplifying the ZNM as a block consisting of a single zipper of N staggered elastic fibers (e.g., elastin) with spring constant k_F, held together by short bridges (e.g., PGs) with spring constant k_B (Fig. 21.6). Let us assume that the bridges are stiffer than the fibers ($k_B > k_F$), since the fibers are much longer than the bridges. This arrangement ensures that neither the elastic fibers nor the bridges percolate. As the network is stretched, part of the macroscopic strain is transferred to the bridges. A bridge ruptures once its local strain exceeds the strain threshold, ε_B, which is much smaller than that of the fibers. With increasing macroscopic strain, a sequence of bridge ruptures eventually leads to complete failure.

We analytically calculate the strain on each bridge in a block for increasing N as a function of the macroscopic strain and the ratio of spring constants ($\lambda = k_B/k_F$). When the maximum strain among the bridges exceeds ε_B, it ruptures. Once a rupture occurs, a block either separates into two smaller blocks which are mechanically

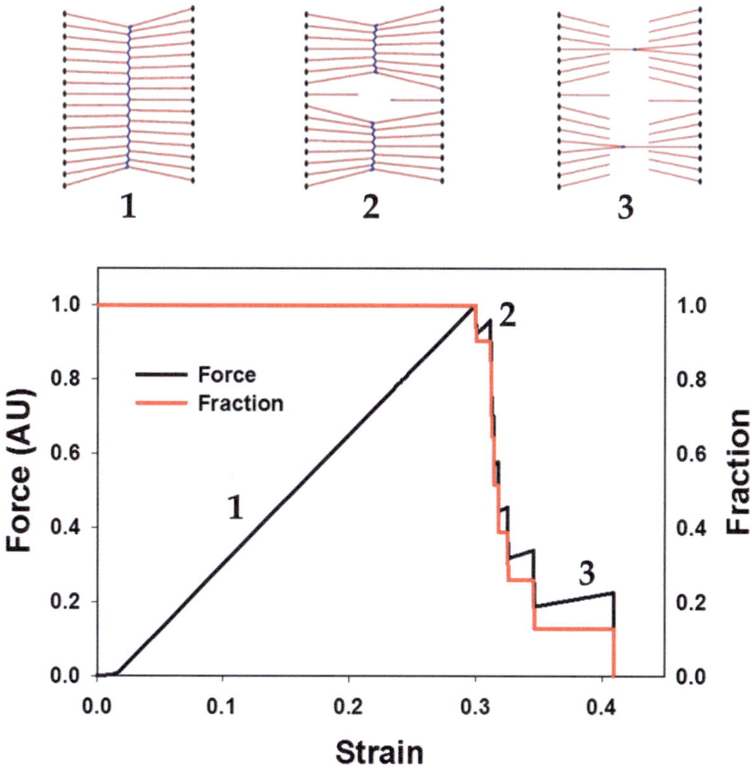

Fig. 21.6 Force–strain curve and the fraction of surviving bridges of the network undergoing failure when stretched along its fiber orientation

independent or becomes a single smaller block if the rupture occurs at the edge of the block. Ruptures can occur in avalanches as in the ZNM leading to multiple bridges breaking simultaneously.

The failure strain of a network of N bridges can be written as a ratio of polynomials in λ, suggesting that the stiffness ratio of bridges and fibers is the primary determinant of the failure pattern. We find that for small N, the maximum extension is always seen by bridges at the edges of the block and thus the sizes of blocks tend to decrease by one. Blocks with even number of fibers N tend to have higher maximum bridge extension than blocks with $(N+1)$ fibers and consequently tend to break down immediately. The force–length relationship during the breakdown shows sharp drops at distinct macroscopic strains, consistent with experimental and simulation results (Figs. 21.1 and 21.5). The sequence of events preceding complete failure can thus be linked to the mechanical relation among the constituents.

21.5 Network Repair

In this section we discuss the possibility of targeted network repair and demonstrate its possible utility to treat pulmonary emphysema. As demonstrated in Fig. 21.2, percolation can link microscopic alterations in lung parenchymal structure (D or D_2) to measurable clinical parameters (C), which in turn is closely related to symptoms. Since percolation deals with geometric connectivity within a structure, the macroscopic properties of the lung that contribute to physiologic function, and hence quality of life, are to a large extent determined by how the emphysematous lesions are linked together across the parenchyma. Let us suppose that we are able to repair or regenerate damaged tissue that we model by adding back a spring into network which would decrease C and increase its inverse, the bulk modulus B. There are a large number of ways in which the individual springs could be repaired or replaced, not all of which will lead to the same time course of symptomatic relief. This opens up the possibility of identifying optimal strategies for targeted delivery of therapies. We will assume that the repair is ideal in the sense that it returns a local region of tissue to normality without adverse biological side effects. In the case of emphysema it means that we can either regenerate alveolar walls or insert engineered parenchymal tissue into the lung at desired locations. We will assess the functional consequences of these in silico therapies in terms of their effects on B during both random and targeted repair or replacement of springs (Suki et al. 2007).

The comparison of random and targeted repair of an emphysematous network is shown in Fig. 21.7. We begin with a network containing 312 holes of area $A = 1810 \pm 12650$ (Fig. 21.7a) and $B = 0.622$. By inserting 8 springs at random locations (Fig. 21.7b), we increase recoil and reduce heterogeneity to obtain $B = 0.824$ and $A = 1751 \pm 6254$. Adding 3 or 4 springs at targeted sites designed to divide large holes into smaller ones gives only modest improvements of $B = 0.778$ and $B = 0.802$, respectively, which is still not as good as the 8 random springs. We do better with 5 targeted replacements to obtain $B = 0.851$ and $A = 1766 \pm 4909$ (Fig. 21.7c). Thus, we gain improvement over the random approach with smaller amount of tissue repair. It can be shown that in the case of fibrosis, a similar targeted approach provides a spectacular improvement over the random repair (Suki et al. 2007). Of course, there are multiple ways one can bridge a large hole with a single spring, and initial modeling suggests that the best approach may be to maximize both the distance that an added spring bridges and the increase it causes in local strain energy within the network. Selecting a site for repair may, in the future, also be guided by macro- or microimaging. For example, specific features of low attenuation areas on lung CT images may help in the optimization process. In principle, it should be possible to map a CT image into an elastic network which could subsequently be used to test the effectiveness of various repair procedures without having to carry them out in a patient. As a general concept, we propose that in order to optimally repair the functionality of a damaged tissue or organ, the design of local interventions must also take into account the redistribution of forces in the global network.

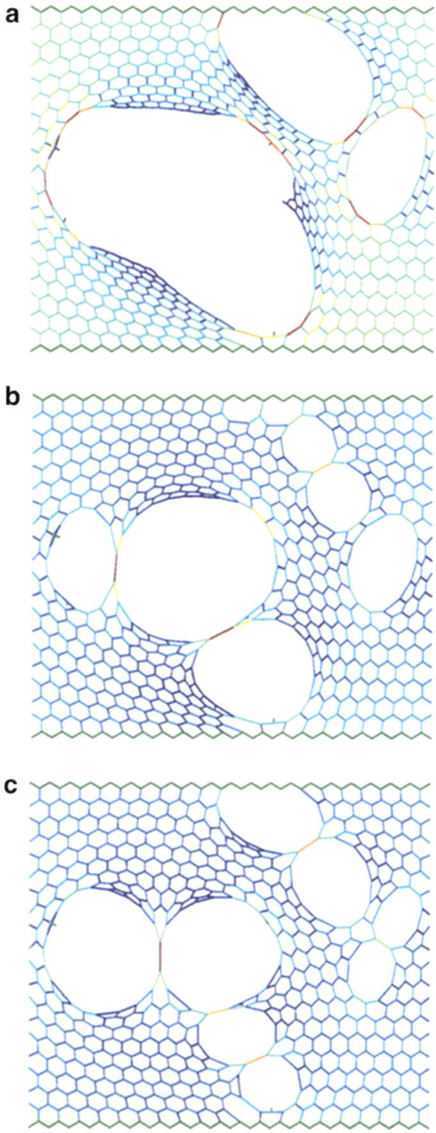

Fig. 21.7 Modeling the repair of the emphysematous lung. The colors are related to the force the elements carry as in Fig. 21.3. (**a**) Shows the initial configuration with the fraction c of missing springs $c = 0.034$ and the bulk modulus $B = 0.635$. (**b**) Shows the configuration of the network following random repair with $c = 0.027$ and $B = 0.824$. (**c**) Shows the configuration of the network following targeted repair with $c = 0.03$ and $B = 0.851$. With permission from Suki et al. (2007)

21.6 Conclusions

In this chapter, we summarized recent modeling efforts using network approaches to the biomechanical problem of tissue failure. However, we did not consider several issues that may influence the failure process. For example, boundaries that resemble the shapes of tissues and organs likely alter the failure stress and strain due to stress concentration at locations of small radius of curvature. Incorporation of more realistic heterogeneous structural elements such as blood vessels and airways could also change the failure properties. The extent to which the presence of such structures influence failure depends on whether or not they percolate, leading to yet another percolation-like phenomenon. If the tissue undergoes cyclic stretching such as the lung or blood vessels, several other mechanisms may influence failure such as fatigue, viscoelastic hysteresis, and the presence of interstitial fluid. Additionally, since peak stretches of tidal breathing as well as systolic blood pressure fluctuate on a cycle by cycle basis, the corresponding variations in peak stresses may accelerate fatigue and hence failure in an already damaged tissue. We also note that local failure inside the tissue substantially alters the regional distribution of strain which in turn has a significant impact on subsequent cellular remodeling of the tissue via mechanotransduction (Ingber 2006) potentially leading to pathological tissue or organ function. Finally, more advanced multiscale models could incorporate further details at the fiber and fibril levels as well as include the effects of enzyme diffusion, binding, and cleaving as a function of the tensile force along the fibers (Araujo et al. 2011).

Network models do have advantages over the more traditional biomechanical approaches using continuum models and strain energy functions in that they allow us to easily incorporate structural heterogeneity and anisotropy, important characteristics of biological tissues, without assuming affine deformations. Despite the limitations of the network approach that analytical solutions may not exist and simulations can require significant computational efforts, the results have significant implications for the understanding of failure processes in normal and diseased tissues. For example, network modeling suggests a key role for the matrix and PGs in failure biomechanics. Additionally, in designing tissue-engineered materials, it is important to match the mechanical properties of the native tissue for the replacement tissue to function properly in the body. Furthermore, the replacement tissue should be able to withstand the naturally occurring stresses and strains even during diseases with proteolytic activity. Network modeling provides an understanding of what components and organizational features of the ECM determine the failure properties and hence may offer a rational basis for future design of engineered materials with target functional mechanical properties.

Acknowledgement This work was supported by NIH grants HL090757 and HL-098976.

References

Angele P, Abke J, Kujat R, Faltermeier H, Schumann D, Nerlich M, Kinner B, Englert C, Ruszczak Z, Mehrl R, Mueller R. Influence of different collagen species on physico-chemical properties of crosslinked collagen matrices. Biomaterials. 2004;25(14):2831–41.

Araujo AD, Majumdar A, Parameswaran H, Yi E, Spencer JL, Nugent MA, Suki B. Dynamics of enzymatic digestion of elastic fibers and networks under tension. Proc Natl Acad Sci U S A. 2011;108(23):9414–9. doi:10.1073/pnas.1019188108.

Arroyo LH, Lee RT. Mechanisms of plaque rupture: mechanical and biologic interactions. Cardiovasc Res. 1999;41(2):369–75.
Barabasi AL, Buldyrev SV, Stanley HE, Suki B. Avalanches in the lung: a statistical mechanical model. Phys Rev Lett. 1996;76(12):2192–5.
Bates JH, Davis GS, Majumdar A, Butnor KJ, Suki B. Linking parenchymal disease progression to changes in lung mechanical function by percolation. Am J Respir Crit Care Med. 2007;176(6):617–23.
Billiar KL, Throm AM, Frey MT. Biaxial failure properties of planar living tissue equivalents. J Biomed Mater Res A. 2005;73(2):182–91.
Black LD, Brewer KK, Morris SM, Schreiber BM, Toselli P, Nugent MA, Suki B, Stone PJ. Effects of elastase on the mechanical and failure properties of engineered elastin-rich matrices. J Appl Physiol. 2005;98(4):1434–41.
Black LD, Allen PG, Morris SM, Stone PJ, Suki B. Mechanical and failure properties of extracellular matrix sheets as a function of structural protein composition. Biophys J. 2008;94(5):1916–29.
Bland RD, Carlton DP, Scheerer RG, Cummings JJ, Chapman DL. Lung fluid balance in lambs before and after premature birth. J Clin Invest. 1989;84(2):568–76. doi:10.1172/JCI114200.
Carey DJ. Biological functions of proteoglycans: use of specific inhibitors of proteoglycan synthesis. Mol Cell Biochem. 1991;104(1–2):21–8.
Cavalcante FS, Ito S, Brewer KK, Sakai H, Alencar AM, Almeida MP, Andrade Jr JS, Majumdar A, Ingenito EP, Suki B. Mechanical interactions between collagen and proteoglycans: implications for the stability of lung tissue. J Appl Physiol. 2005;98:672–9.
Chen HY, Chu YP. Theoretical determination of the strength of soft noncovalent molecular bonds. Phys Rev E Stat Nonlin Soft Matter Phys. 2005;71(1 Pt 1):010901.
Christian Gasser T. An irreversible constitutive model for fibrous soft biological tissue: a 3-D microfiber approach with demonstrative application to abdominal aortic aneurysms. Acta Biomater. 2011;7(6):2457–66. doi:10.1016/j.actbio.2011.02.015.
Christiansen DL, Huang EK, Silver FH. Assembly of type I collagen: fusion of fibril subunits and the influence of fibril diameter on mechanical properties. Matrix Biol. 2000;19(5):409–20.
Dias CL, Dube M, Oliveira FA, Grant M. Scaling in force spectroscopy of macromolecules. Phys Rev E Stat Nonlin Soft Matter Phys. 2005;72(1 Pt 1):011918.
Dobrin PB, Anidjar S. Pathophysiology of arterial aneurysms. Arch Mal Coeur Vaiss. 1991;84(Spec No 3):57–62.
Donovan DL, Schmidt SP, Townshend SP, Njus GO, Sharp WV. Material and structural characterization of human saphenous vein. J Vasc Surg. 1990;12(5):531–7.
Evans E. Probing the relation between force–lifetime–and chemistry in single molecular bonds. Annu Rev Biophys Biomol Struct. 2001;30:105–28. doi:10.1146/annurev.biophys.30.1.105.
Fung YC. Biomechanics: mechanical properties of living tissues. 2nd ed. New York: Springer; 1993.
Fung YC. Biomechanics: circulation. 2nd ed. New York: Springer; 1997.
Goodall S, Crowther M, Bell PR, Thompson MM. The association between venous structural alterations and biomechanical weakness in patients with abdominal aortic aneurysms. J Vasc Surg. 2002;35(5):937–42.
Hamakawa H, Bartolak-Suki E, Parameswaran H, Majumdar A, Lutchen KR, Suki B. Structure-function relations in an elastase-induced mouse model of emphysema. Am J Respir Cell Mol Biol. 2011;45(3):517–24. doi:10.1165/rcmb.2010-0473OC. pii: 2010-0473OC.
Holmes DF, Gilpin CJ, Baldock C, Ziese U, Koster AJ, Kadler KE. Corneal collagen fibril structure in three dimensions: Structural insights into fibril assembly, mechanical properties, and tissue organization. Proc Natl Acad Sci U S A. 2001;98(13):7307–12.
Hukins DWL. Connective tissue matrix. Topics in molecular and structural biology, vol. 5. London: Macmillan; 1984.
Humphrey JD, Delange SL. An introduction to biomechanics: solids and fluids, analysis and design. New York: Springer; 2004.

Ingber DE. Cellular mechanotransduction: putting all the pieces together again. Faseb J. 2006;20(7):811–27.

Ito S, Ingenito EP, Brewer KK, Black LD, Parameswaran H, Lutchen KR, Suki B. Mechanics, nonlinearity, and failure strength of lung tissue in a mouse model of emphysema: possible role of collagen remodeling. J Appl Physiol. 2005;98:503–11.

Ito D, Tanaka E, Yamamoto S. A novel constitutive model of skeletal muscle taking into account anisotropic damage. J Mech Behav Biomed Mater. 2010;3(1):85–93. doi:10.1016/j.jmbbm.2009.05.001.

Jesudason R, Black L, Majumdar A, Stone P, Suki B. Differential effects of static and cyclic stretching during elastase digestion on the mechanical properties of extracellular matrices. J Appl Physiol. 2007;103(3):803–11.

Joyce EM, Moore JJ, Sacks MS. Biomechanics of the fetal membrane prior to mechanical failure: review and implications. Eur J Obstet Gynecol Reprod Biol. 2009;144 Suppl 1:S121–7. doi:10.1016/j.ejogrb.2009.02.014.

Juul SE, Kinsella MG, Wight TN, Hodson WA. Alterations in nonhuman primate (M. nemestrina) lung proteoglycans during normal development and acute hyaline membrane disease. Am J Respir Cell Mol Biol. 1993;8(3):299–310.

Kielty CM, Sherratt MJ, Shuttleworth CA. Elastic fibres. J Cell Sci. 2002;115(Pt 14):2817–28.

Kononov S, Brewer K, Sakai H, Cavalcante FS, Sabayanagam CR, Ingenito EP, Suki B. Roles of mechanical forces and collagen failure in the development of elastase-induced emphysema. Am J Respir Crit Care Med. 2001;164(10 Pt 1):1920–6.

Koshiyama K, Wada S. Molecular dynamics simulations of pore formation dynamics during the rupture process of a phospholipid bilayer caused by high-speed equibiaxial stretching. J Biomech. 2011;44(11):2053–8. doi:10.1016/j.jbiomech.2011.05.014.

Lanir Y. Constitutive equations for fibrous connective tissues. J Biomech. 1983;16(1):1–12.

Lee TC, Midura RJ, Hascall VC, Vesely I. The effect of elastin damage on the mechanics of the aortic valve. J Biomech. 2001;34(2):203–10.

Lillie MA, Gosline JM. The viscoelastic basis for the tensile strength of elastin. Int J Biol Macromol. 2002;30(2):119–27.

Liu X, Sun JQ, Heggeness MH, Yeh ML, Luo ZP. Direct quantification of the rupture force of single hyaluronan/hyaluronan binding protein bonds. FEBS Lett. 2004;563(1–3):23–7. doi:10.1016/S0014-5793(04)00232-7.

Lokshin O, Lanir Y. Viscoelasticity and preconditioning of rat skin under uniaxial stretch: microstructural constitutive characterization. J Biomech Eng. 2009;131(3):031009. doi:10.1115/1.3049479.

Marque V, Kieffer P, Gayraud B, Lartaud-Idjouadiene I, Ramirez F, Atkinson J. Aortic wall mechanics and composition in a transgenic mouse model of Marfan syndrome. Arterioscler Thromb Vasc Biol. 2001;21(7):1184–9.

McCullagh CM, Jamieson AM, Blackwell J, Gupta R. Viscoelastic properties of human tracheobronchial mucin in aqueous solution. Biopolymers. 1995;35(2):149–59. doi:10.1002/bip.360350203.

Nguyen BV, Wang Q, Kuiper NJ, El HaJ AJ, Thomas CR, Zhang Z. Strain-dependent viscoelastic behaviour and rupture force of single chondrocytes and chondrons under compression. Biotechnol Lett. 2009;31(6):803–9. doi:10.1007/s10529-009-9939-y.

Oroudjev E, Soares J, Arcdiacono S, Thompson JB, Fossey SA, Hansma HG. Segmented nanofibers of spider dragline silk: atomic force microscopy and single-molecule force spectroscopy. Proc Natl Acad Sci U S A. 2002;99 Suppl 2:6460–5. doi:10.1073/pnas.082526499.

Parameswaran H, Majumdar A, Suki B. Linking microscopic spatial patterns of tissue destruction in emphysema to macroscopic decline in stiffness using a 3D computational model. PLoS Comput Biol. 2011;7(4), e1001125. doi:10.1371/journal.pcbi.1001125.

Peeters EA, Oomens CW, Bouten CV, Bader DL, Baaijens FP. Mechanical and failure properties of single attached cells under compression. J Biomech. 2005;38(8):1685–93. doi:10.1016/j.jbiomech.2004.07.018.

Redaelli A, Vesentini S, Soncini M, Vena P, Mantero S, Montevecchi FM. Possible role of decorin glycosaminoglycans in fibril to fibril force transfer in relative mature tendons–a computational study from molecular to microstructural level. J Biomech. 2003;36(10):1555–69.
Ritter MC, Jesudason R, Majumdar A, Stamenovic D, Buczek-Thomas JA, Stone PJ, Nugent MA, Suki B. A zipper network model of the failure mechanics of extracellular matrices. Proc Natl Acad Sci U S A. 2009;106(4):1081–6. doi:10.1073/pnas.0808414106. pii: 0808414106.
Sahimi M, Arbabi S. Mechanics of disordered solids. 3. Fracture properties. Phys Rev B. 1993;47(2):713–22.
Silver FH, Freeman JW, Seehra GP. Collagen self-assembly and the development of tendon mechanical properties. J Biomech. 2003;36(10):1529–53.
Stauffer D, Aharony A. Introduction to percolation theory. 2nd ed. London: Taylor & Francis; 1992.
Stenson JD, Hartley P, Wang C, Thomas CR. Determining the mechanical properties of yeast cell walls. Biotechnol Prog. 2011;27(2):505–12. doi:10.1002/btpr.554.
Stolz M, Raiteri R, Daniels AU, VanLandingham MR, Baschong W, Aebi U. Dynamic elastic modulus of porcine articular cartilage determined at two different levels of tissue organization by indentation-type atomic force microscopy. Biophys J. 2004;86(5):3269–83. doi:10.1016/S0006-3495(04)74375-1. pii: S0006-3495(04)74375-1.
Suki B, Lutchen KR, Ingenito EP. On the progressive nature of emphysema: roles of proteases, inflammation, and mechanical forces. Am J Respir Crit Care Med. 2003;168(5):516–21.
Suki B, Ito S, Stamenovic D, Lutchen KR, Ingenito EP. Biomechanics of the lung parenchyma: critical roles of collagen and mechanical forces. J Appl Physiol. 2005;98(5):1892–9.
Suki B, Majumdar A, Nugent MA, Bates JH. In silico modeling of interstitial lung mechanics: implications for disease development and repair. Drug Discov Today Dis Models. 2007;4(3):139–45.
Tanaka R, Al-Jamal R, Ludwig MS. Maturational changes in extracellular matrix and lung tissue mechanics. J Appl Physiol. 2001;91(5):2314–21.
Tang Y, Ballarini R, Buehler MJ, Eppell SJ. Deformation micromechanisms of collagen fibrils under uniaxial tension. J R Soc Interface. 2010;7(46):839–50. doi:10.1098/rsif.2009.0390.
Urry DW, Parker TM. Mechanics of elastin: molecular mechanism of biological elasticity and its relationship to contraction. J Muscle Res Cell Motil. 2002;23(5–6):543–59.
Volokh KY. Fung's model of arterial wall enhanced with a failure description. Mol Cell Biomech. 2008;5(3):207–16.
Vorp DA. Biomechanics of abdominal aortic aneurysm. J Biomech. 2007;40(9):1887–902.
Vorp DA, Schiro BJ, Ehrlich MP, Juvonen TS, Ergin MA, Griffith BP. Effect of aneurysm on the tensile strength and biomechanical behavior of the ascending thoracic aorta. Ann Thorac Surg. 2003;75(4):1210–4.
West JB. Invited review: pulmonary capillary stress failure. J Appl Physiol. 2000;89(6):2483–9; discussion 2497.
Wilson K, Bradbury A, Whyman M, Hoskins P, Lee A, Fowkes G, McCollum P, Ruckley CV. Relationship between abdominal aortic aneurysm wall compliance and clinical outcome: a preliminary analysis. Eur J Vasc Endovasc Surg. 1998;15(6):472–7.

Chapter 22
A Microstructurally Based Multi-Scale Constitutive Model of Active Myocardial Mechanics

Adarsh Krishnamurthy, Benjamin Coppola, Jared Tangney, Roy C.P. Kerckhoffs, Jeffrey H. Omens, and Andrew D. McCulloch

Abstract Contraction of cardiac muscle cells provides the work for ventricular pumping. The primary component of this contractile stress development in myocardium acts along the axis of the myofilaments; however, there may be a component directed transversely as well. Biaxial testing of tonically activated cardiac tissue has shown that myocardium can generate active stresses in the transverse direction that are as high as 50 % of those developed along the fiber axis. The microstructural basis for this is not clear. We hypothesized that transverse active stresses are generated at the crossbridge and myofilament lattice scales and transmitted via the myocardial laminar sheets as plane stress objects. To test this hypothesis, we developed a multi-scale constitutive model accounting for crossbridge and myofilament lattice structures as well as multicellular myofiber and sheet angle dispersions. Integrating these properties in a finite element model of an actively contracting myocardial tissue slice suggested that these mechanisms may be sufficient to explain the results of biaxial tests in contracted myocardium.

22.1 Introduction

It is well known that cardiac muscle fibers develop active force along the longitudinal myofibril axis of the myocyte. Both the actin and myosin filaments are oriented along the myofibrils, and it is their relative motions that lead to muscle fiber shortening and thickening. However, the acto-myosin crossbridges are not oriented

---

Invited Book Chapter in "Structure-Based Mechanics of Tissues and Organs: A Tribute to Yoram Lanir"

A. Krishnamurthy (✉) • B. Coppola • J. Tangney • R.C.P. Kerckhoffs
• J.H. Omens • A.D. McCulloch
Departments of Bioengineering and Medicine, Cardiac Biomedical Science and Engineering Center, University of California, San Diego, CA, USA
e-mail: amcculloch@ucsd.edu

G.S. Kassab, M.S. Sacks (eds.), *Structure-Based Mechanics of Tissues and Organs*,
DOI 10.1007/978-1-4899-7630-7_22

parallel to the myofilaments. Structural studies suggest that the actin-binding region of myosin when the crossbridge is in the strongly bound state forms an acute angle between the binding site on actin and the backbone of the thick filament (Huxley 1985; Huxley and Kress 1985). Theoretical analysis suggests that this may give rise to a significant component of force radial to the myofilament long axis (Schoenberg 1980a, b; Zahalak 1996). The magnitude of this radial component likely depends on this binding angle, the filament spacing, and the sarcomere length. In biaxial tests of an isolated myocardial tissue preparation, Lin and Yin (1998) showed that the multicellular myocardium can generate significant systolic transverse stresses (greater than 40% of those in the fiber direction). They concluded that fiber angle variations within the specimens alone would be insufficient to explain transverse stresses of this magnitude, thus implicating an active cellular mechanism for transverse tension generation. Finite element models of the heart have traditionally used uniaxial active stress models (Guccione and McCulloch 1993; Hunter et al. 1998). However, it has been shown that the inclusion of transverse active stress in models of ventricular mechanics significantly improves the agreement between predicted systolic wall strains and experimentally measured deformations in the intact heart (Usyk et al. 2000).

Thus, there is a need for microstructurally derived constitutive models to link crossbridge models of tension development in sarcomeres to tissue-scale models of systolic myocardial wall stress development. Here we consider structural mechanisms at four different scales of myocardial organization of multi-axial systolic stress development and derive a hierarchical multi-scale microstructural model of anisotropic systolic myocardial stress–strain relations. We assume that the input to such a model is a lumped parameter model of calcium-dependent myofilament activation and crossbridge interactions such as Markov model described recently (Campbell et al. 2010). This model in turn could be activated by a model of dynamic myocyte depolarization and excitation–contraction coupling such as the model by Campbell et al. (2009). The twitch tension developed in these models depends on processes at the crossbridge, sarcomere, and whole cell scales. However, we can use the computed tension to derive the force in a single average crossbridge, for the purposes of deriving a microstructural model of three-dimensional myocardial active stresses.

We consider mechanisms at four scales: (1) In the single crossbridge scale, we consider the two-dimensional static equilibrium of a strongly bound crossbridge to resolve the crossbridge stiffness into longitudinal and transverse components, using a similar approach to that proposed by Schoenberg (1980a, b) and accounting for changes in lattice spacing with sarcomere length; (2) At the intracellular scale, we consider the hexagonal arrangements of thick and thin filaments in the organized myofilament lattice to derive how active stresses are developed anisotropically within myocytes; (3) At the multicellular single laminar sheet scale, we integrate these anisotropic stress tensors within a laminar sheet to derive the tissue-scale effects of dispersion of myofibers about the mean fiber orientation (Karlon et al. 1998); and (4) finally we consider the effects of distributions of myocardial laminae and their orientations within the myocardium on orthotropic systolic stress development (LeGrice et al. 1995).

In order to test the resulting multi-scale microstructural constitutive law, we integrated it into a three-dimensional finite element model. The model includes measurements of fiber-sheet distributions in one dog. The stresses developed in the model were then compared with those in the biaxial tests performed in tonically activated rabbit myocardium by Lin and Yin (1998).

22.2 Methods

22.2.1 Histological Measurements

The histological measurements used in the current model were taken from a canine heart used in a previous study in our laboratory (Coppola et al. 2007). All animal studies were performed according to the National Institutes of Health *Guide for the Care and Use of Laboratory Animals*. All protocols were approved by the Animal Subjects Committee of the University of California, San Diego, which is accredited by the American Association for Accreditation of Laboratory Animal Care. An adult mongrel dog was instrumented as described in detail in Coppola et al. (2007). The heart was fixed in situ at end-diastolic pressure with 2.5% gluteraldehyde and stored in 10% formalin. The heart was then sectioned for histology as previously described (Ashikaga et al. 2004). These sections are cut perpendicular to the mean fiber direction so that the sheet angle β can be visualized directly. Sheet angles were measured at each transmural depth using the method of Karlon et al. (1998). Ten 10-μm sections were analyzed (50–70 sheet angles per section) for a total of about 600 measurements of the sheet angle, β, across the wall thickness. These sheet angle populations were incorporated into the finite element model.

22.2.2 Crossbridge Mechanics

The static equilibrium of the strongly bound myosin molecule based on a 2D simplification of the model originally proposed by Schoenberg (1980a, b) was used to resolve crossbridge tension into axial and transverse (radial) components. The axial and radial stiffnesses of the structure were then derived and used to compute the resulting axial and radial stresses in the hexagonal sarcomere lattice model.

Tension in the elastic S2 segment is resolved into axial and transverse components using a two-dimensional force and moment balance derived from the geometry shown in Fig. 22.1 and assuming no net moment at the attachment of the S1 head to the thin filament. To derive crossbridge stiffness components, the lattice spacing of the crossbridge model is displaced by a small value and the resulting change in the axial and radial forces is obtained (Fig. 22.1). This change in force is then used

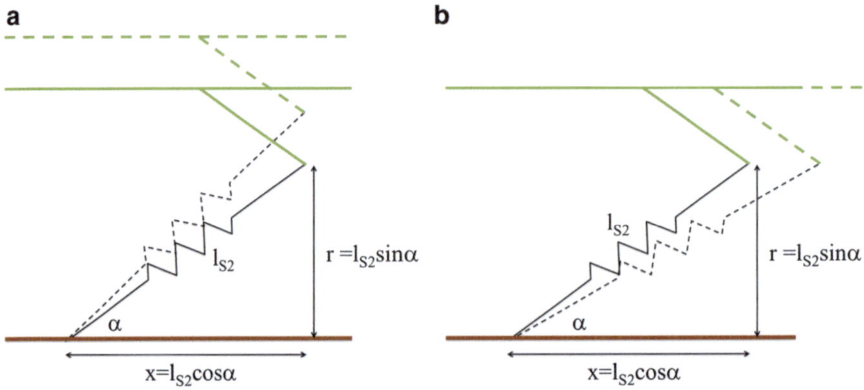

Fig. 22.1 Computing the crossbridge stiffness in the axial and radial direction based on the geometry of the myosin S2 segment. The myosin S2 segment makes an angle α with the thick filament of myosin

to derive the instantaneous stiffnesses in the two directions (Tangney et al. 2013). It can be shown that the transverse to fiber crossbridge stiffness-ratio is

$$\frac{K_{\mathrm{t}}}{K_{\mathrm{f}}} = \frac{l_{\mathrm{S2}} - l_0 \cos^2\alpha}{l_{\mathrm{S2}} - l_0 \sin^2\alpha} \tag{22.1}$$

where l_{S2} is the length of the S2 segment of the crossbridge, l_0 is the resting length of the crossbridge considering the S2 segment to be a linear spring, and α is the angle between the S2 segment and the thick filament of the myosin molecule.

22.2.3 *Lattice Model*

A hexagonal lattice model of the sarcomeres was used to derive transverse and axial stresses as a function of the lattice spacing and crossbridge stiffness components. The lateral force interactions in a myofilament lattice due to crossbridge formation between thick and thin filaments can be analyzed in two perpendicular planes: one parallel to the axis of the sarcomeres, and one perpendicular to the axis of the sarcomeres. Figure 22.2a shows the cross-section in this perpendicular plane. We make use of energy conservation in a hexagonal unit cell of width Δ_0 and axial length δ_0 consisting of three pairs of crossbridges as shown in Fig. 22.2b to derive the resultant axial and radial stress components. For this analysis, the three pairs of crossbridges are assumed to be in the same plane.

The complete derivation for the strain analysis is given in Tangney et al. 2013. In terms of the ratio of the transverse to the radial stiffness ($K_{\mathrm{t}}/K_{\mathrm{f}}$) derived from

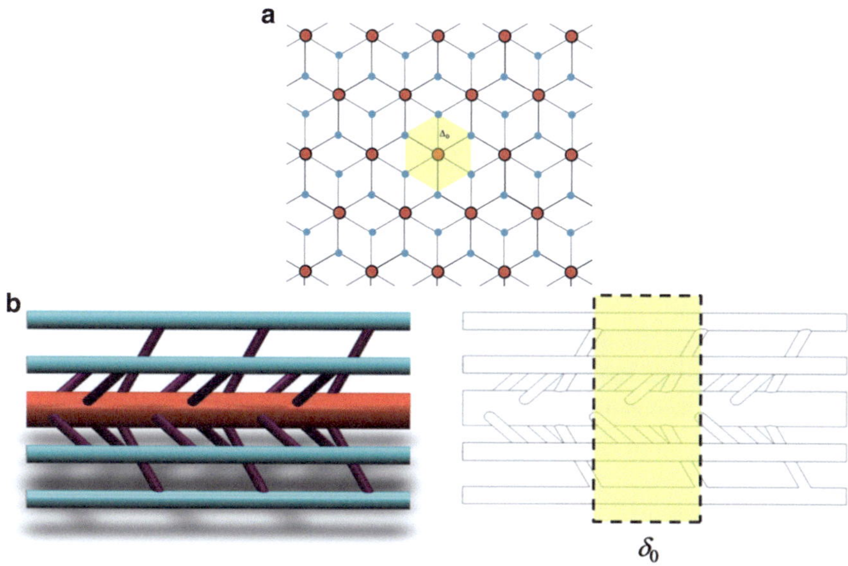

Fig. 22.2 (**a**) Lattice structure showing the cross-section of the thick myosin filaments (*red*) and the thin actin filaments (*blue*) in a 2D view. The section consists of three pairs of crossbridges. A single unit cell used for the analysis with six crossbridges is highlighted in *yellow*. (**b**) Axial view of sarcomeres showing three sets of three pairs of crossbridges spaced 120° apart. The unit cell consisting of three pairs of crossbridges and axial length δ_0 is marked in *yellow*

the crossbridge analysis, the ratio of the stresses in the radial direction to the axial direction derived for the lattice is

$$\nu = \frac{S_t}{S_f} = \frac{1}{2}\left(\frac{\Delta}{\delta_0}\right)\frac{K_t}{K_f} \tag{22.2}$$

22.2.4 Active Systolic Stress

We use the stress ratio derived from the previous section to derive the active stress tensor at the tissue level. At each integration point of the finite element model, the active fiber stress of a myocyte is calculated from the Guccione activation model (Guccione and McCulloch 1993) as a function of sarcomere length and time. From this fiber stress, S_f, we use (22.2) to obtain the transverse stress $S_t = \nu \bullet S_f$ at the single myocyte scale. We assume the resulting myocyte systolic active stress is transversely isotropic. Now considering myocytes distributed within a planar sheet assumed to behave as a plane stress element due to weak sheet-sheet coupling, then there can be no stress acting in the sheet-normal direction. Therefore, contribution of a single myocyte directed parallel to the mean fiber orientation within a sheet is

$$\boldsymbol{T} = \begin{bmatrix} S_{\mathrm{f}} & 0 & 0 \\ 0 & \nu \cdot S_{\mathrm{f}} & 0 \\ 0 & 0 & 0 \end{bmatrix} \quad \text{with respect to} : \{\mathbf{e}_{\mathrm{f}}, \mathbf{e}_{\mathrm{s}}, \mathbf{e}_{\mathrm{n}}\} \tag{22.3}$$

where $\mathbf{e}_{\mathrm{f}}$ represents the fiber direction, $\mathbf{e}_{\mathrm{s}}$ represents the within-sheet direction, and $\mathbf{e}_{\mathrm{n}}$ represents the cross-sheet direction associated with this individual sheet.

We assume that all myofibers within a sheet are parallel to the plane of the sheet but are distributed with an angular distribution $f(\varphi)$ with respect to the mean fiber direction (see Fig. 22.3a). Taking into account this distribution, the stresses in this single sheet can be obtained by integrating:

$$\boldsymbol{T}_{\text{sheet}} = \int_{-\pi/2}^{\pi/2} \boldsymbol{R}_{\varphi} \cdot \boldsymbol{T} \cdot \boldsymbol{R}_{\varphi}^{T} \cdot f(\varphi) \cdot d\varphi$$
$$\boldsymbol{R}_{\varphi} = \begin{bmatrix} \cos(\varphi) & \sin(\varphi) & 0 \\ -\sin(\varphi) & \cos(\varphi) & 0 \\ 0 & 0 & 1 \end{bmatrix} \quad \text{with respect to} : \{\mathbf{e}_{\mathrm{f}}, \mathbf{e}_{\mathrm{s}}, \mathbf{e}_{\mathrm{n}}\} \tag{22.4}$$

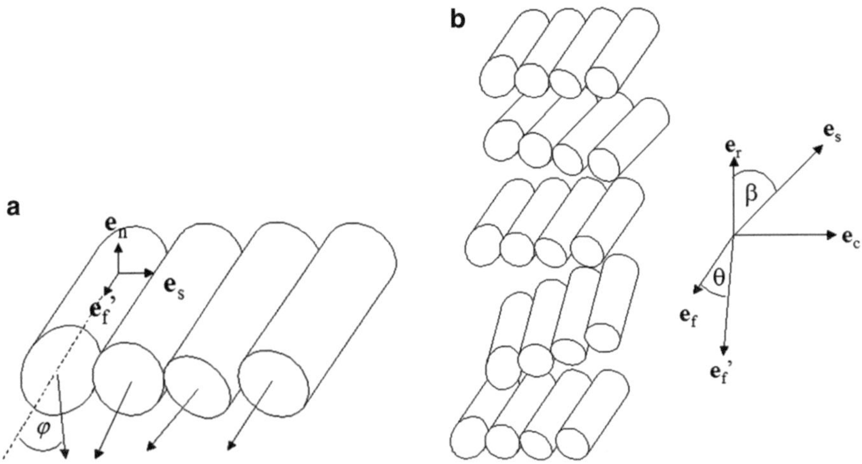

Fig. 22.3 Description of angles in model. (**a**) Schematic representation of a single sheet. $\mathbf{e}_{\mathrm{f}}'$ represents the sheet mean fiber axis, $\mathbf{e}_{\mathrm{s}}$ represents the direction transverse to the fiber direction but within the sheet, and $\mathbf{e}_{\mathrm{n}}$ represents the sheet-normal direction (across the thickness of the sheet). φ represents the deviation of a single myocytes fiber axis relative to the mean fiber axis. φ is measured in the ($\mathbf{e}_{\mathrm{f}}'$, $\mathbf{e}_{\mathrm{s}}$) plane. (**b**) Schematic representation of several sheets. $\mathbf{e}_{\mathrm{f}}$ represents the tissue mean fiber axis, $\mathbf{e}_{\mathrm{r}}$ represents the radial direction, and $\mathbf{e}_{\mathrm{c}}$ is perpendicular to both ($\mathbf{e}_{\mathrm{c}} = \mathbf{e}_{\mathrm{r}} \times \mathbf{e}_{\mathrm{f}}$). Each sheet has a sheet mean fiber axis at an angle θ to the tissue mean fiber axis, in the ($\mathbf{e}_{\mathrm{f}}$, $\mathbf{e}_{\mathrm{r}}$) plane. Each sheet also has an angle β, which is in the ($\mathbf{e}_{\mathrm{c}}$, $\mathbf{e}_{\mathrm{r}}$) plane

where $f(\varphi)$ represents the distribution of φ, which we define to have a mean of zero. $\boldsymbol{T}_{\text{sheet}}$ remains a plane stress, but it is no longer a diagonal tensor in general. In other words, there may be a shear stress component due to the dispersion of φ within the sheet. However, if the distribution is symmetric, the shear terms cancel out making the tensor diagonal.

In the finite element formulation, stresses are integrated at the tissue scale. At this scale, there are many sheets. Each sheet has an orientation associated with it, which can be described as a function of two angles: β, which is the histologically measured sheet angle (Costa et al. 1999) and θ, which is the angle relative to the mean fiber direction (about the sheet direction). For an illustration of these angles, see Fig. 22.3b. Now the stress at the tissue level can be integrated:

$$\begin{gathered}
\boldsymbol{T}_{\text{tissue}} = \iint \boldsymbol{R}_{\theta}\boldsymbol{R}_{\beta}\boldsymbol{T}_{\text{sheet}}\boldsymbol{R}_{\beta}^{\mathrm{T}}\boldsymbol{R}_{\theta}^{\mathrm{T}} f(\theta) f(\beta)\, d\theta d\beta \\
\boldsymbol{R}_{\beta} = \begin{bmatrix} 1 & 0 & 0 \\ 0 & \sin(\beta) & \cos(\beta) \\ 0 & -\cos(\beta) & \sin(\beta) \end{bmatrix}, \quad \boldsymbol{R}_{\theta} = \begin{bmatrix} \cos(\theta) & 0 & \sin(\theta) \\ 0 & 1 & 0 \\ -\sin(\theta) & 0 & \cos(\theta) \end{bmatrix} \\
\text{with respect to}: \{\mathbf{e}_{\mathrm{f}}, \mathbf{e}_{\mathrm{c}}, \mathbf{e}_{\mathrm{r}}\}
\end{gathered} \tag{22.5}$$

where $\mathbf{e}_{\mathrm{f}}$ represents the fiber direction, $\mathbf{e}_{\mathrm{r}}$ represents the radial (transmural) direction, and $\mathbf{e}_{\mathrm{c}}$ represents the cross-fiber direction ($\mathbf{e}_{\mathrm{c}} = \mathbf{e}_{\mathrm{r}} \times \mathbf{e}_{\mathrm{f}}$) associated with the bulk tissue; $f(\theta)$ and $f(\beta)$ represent the distributions of θ and β, respectively. Note that the different format of $\mathbf{R}_{\beta}$ is consistent with Costa's definition (Costa et al. 1999).

The angles φ, θ, and β change through time as the heart deforms. In other words, they are functions of Lagrangian strain (**E**), as shown in the Appendices 1 and 2. Because these quantities vary through time, the integration has to be performed at each time step of the finite element solver. However, if we assume there is no interaction between the angles, the integration terms can be separated and can be evaluated by evaluating the following definite integrals:

$$\begin{aligned}
I_{\cos^2\gamma} &= \int_{-\pi/2}^{\pi/2} \cos^2\gamma f(\gamma)\, d\gamma \\
I_{\cos\gamma} &= \int_{-\pi/2}^{\pi/2} \cos\gamma f(\gamma)\, d\gamma \\
I_{\sin^2\gamma} &= \int_{-\pi/2}^{\pi/2} \sin^2\gamma f(\gamma)\, d\gamma
\end{aligned} \tag{22.6}$$

where f represents a Von-Mises distribution for the angles and γ represents θ, β, or φ. In addition, as shown in the Appendix 2, the effect of the strain on these distributions is not significant in the range of strains experienced by a typical myocardial tissue. Hence, these functions can be pre-computed before attempting to solve the finite element problem.

Finally, (22.5) can be expanded in terms of the distributions of the angles in the reference configuration. The equations for all six independent terms of the stress tensor look similar and are of the following form:

$$T_{\text{tissue}}(j,k) = S_f.F_{jk}(I_i(\gamma), \nu) \\ \text{where } i = \cos^2\gamma, \cos\gamma, \sin^2\gamma \quad \gamma = \beta, \varphi, \theta \tag{22.7}$$

F_{jk} are pre-computable functions of the angle distributions. The actual form of these functions is given in the Appendix 1.

22.2.5 Simplifying Assumptions

In addition to the assumptions inherent to the construction of the model, a few additional simplifying assumptions were made. Because we have no detailed measurements of the angles φ and θ, these angles were replaced with a Von-Mises distribution centered about 0° with a fiber dispersion standard deviation of 12° (Karlon et al. 1998). This simplifies the resulting model due to the symmetry of these distributions, making the active stress tensor diagonal.

22.2.6 Finite Element Computational Model

The crossbridge and lattice models were derived analytically with MATLAB code utilizing a symbolic library. The resulting code was then implemented into the laboratory's custom finite element modeling environment (Continuity 6.3, www.continuity.ucsd.edu).

The active contraction in a continuum tissue was simulated with a nonlinear finite element model of a tissue slab. The chosen model includes passive material properties and a biophysically based tension generation. It is a 27-node, 8-element mesh of a tissue sample, which is synchronously activated.

Myocardial stresses are determined at each integration point within the finite element mesh by a summation of passive stress (due to distension) and active stress (due to crossbridge cycling). The passive stress model described by Guccione et al. (1991) is used as is. The active stresses are determined by the model described in the previous section.

To simulate equi-biaxial tests, the tissue was first activated to maximum active tension corresponding to the strong attachment of all crossbridges. This gives a new geometry where the passive stresses generated in the tissue are in equilibrium with the generated active stresses (Fig. 22.4). This geometry was then stretched equi-biaxially to generate curves similar to those reported by Lin and Yin (1998). Volume conservation was enforced using a semi-incompressible penalty formulation (Doll and Schweizerhof 2000).

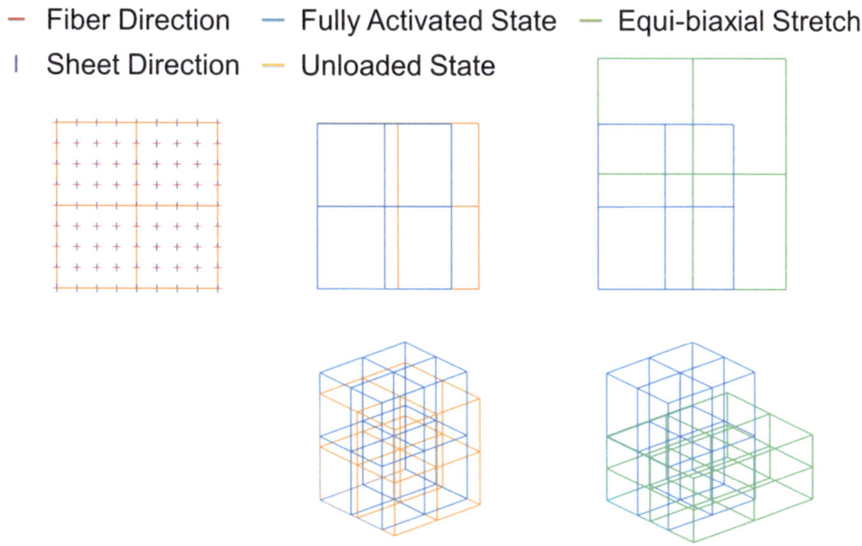

Fig. 22.4 Simulation of equi-biaxial stretch. The sample is first activated fully to obtain the geometry corresponding to the fully activated state (*middle*). The geometry is then stretched equally to simulate an equi-biaxial stretch

22.3 Results

22.3.1 Sheet Angle Measurements

Automated measurements of sheet angle were performed on 10 μm sections at every 1 mm depth through the ventricular wall (ten depths). Figure 22.5 shows an example of one section, as well as the results of the automated processing scheme, which was performed as described by Karlon et al. (1998). Results from the subendocardial and subepicardial regions are shown in Fig. 22.6. Note that there is substantial dispersion about the mean sheet angle ($\sigma > 10°$), particularly deeper in the wall.

For our simulations, we used the average dispersion data from these distributions. The average standard deviation of the dispersion is found to be about 30°. This corresponds to a concentration parameter (κ) value of 4 in the Von-Mises distribution. However, we did not include the bimodal distribution of sheets in our simulations since we were interested only in biaxial tests in isolated myocardial tissue.

22.3.2 Lattice Model

The effect of the lattice spacing on the transverse to fiber stress ratio was computed using the crossbridge and lattice model at typical lattice-spacing values (Julian et al.

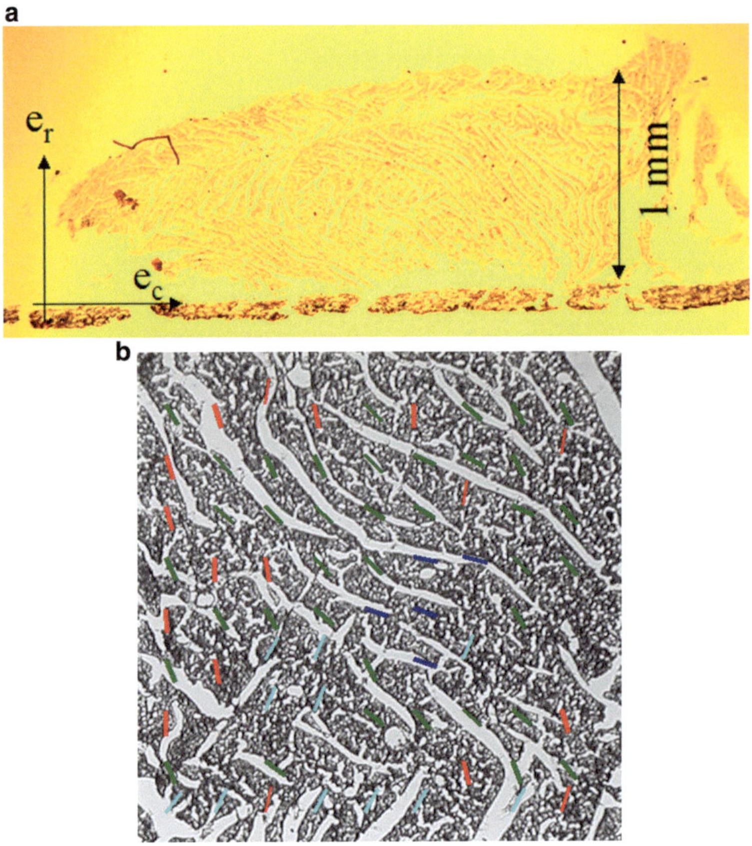

Fig. 22.5 Automated measurements of the sheet angle β. (**a**) 10 μm section of myocardium cut perpendicular to fiber angles. Gaps in tissue represent cleavage planes between myocardial sheets, which have opened up as tissue was allowed to desiccate for 10 min. In this image, two distinct populations of sheets are present. (**b**) Enlarged view of tissue section showing automated measurements of sheet angle. The region of interest for each measurement was 76 μm^2

1978; Schoenberg 1980a, b; Rayment et al. 1993). The lattice spacing is measured as the distance between adjacent actin and myosin filaments. The parameter values used for the crossbridge model are tabulated in Table 22.1. It can be seen that the radial to axial stress ratio is nonlinearly dependent on the lattice spacing.

It can be seen from Fig. 22.7 that this ratio of stresses depends on the length of the myosin S2 segment at lattice spacing corresponding to the unloaded sarcomere length. This length determines the angle the S2 segment makes with the myosin thick filament and hence in turn mediates the transverse force generated by the crossbridge.

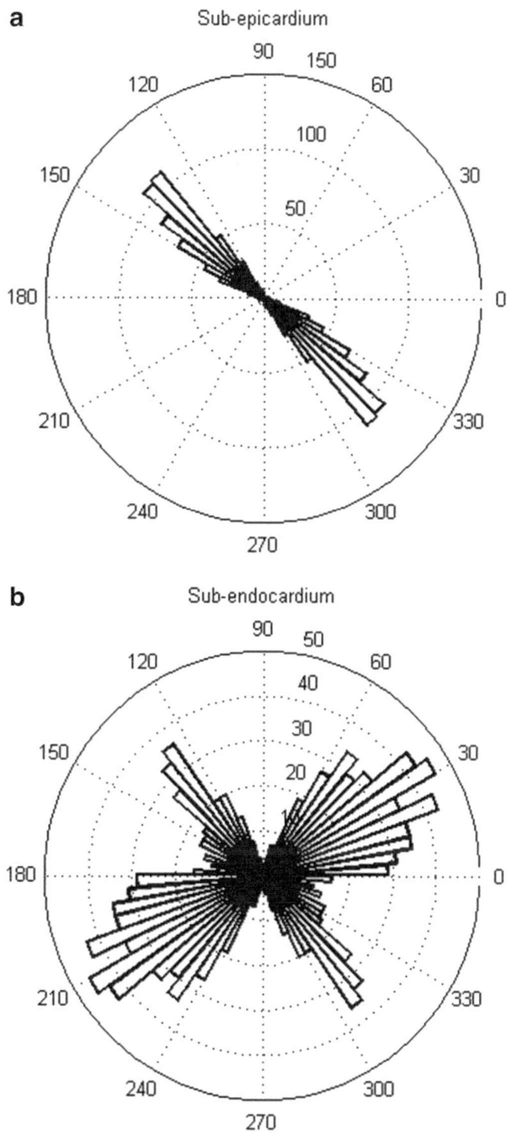

Fig. 22.6 Experimental distributions of the sheet angle β. Rose plots (circular histograms) showing the distribution of the sheet angle β in the sub-epicardium (**a**) and sub-endocardium (**b**). 0°/180° indicates that the sheet lie along the radial direction. It is clear that this animal has a second population of sheet angles in the sub-endocardium

22.3.3 *Finite Element Model*

The effect of fiber dispersion was tested using a finite element computational model of a rectangular slab of myocardium. The dimensions of the slab relative to the actual wall thickness of the heart are small such that sheet angle does not vary within the slab. The final form of the active stress-coupling model is given in Appendix 1. Equi-biaxial stretch in a fully activated myocardial tissue was simulated. Three

Table 22.1 Parameter values for the crossbridge and lattice model

Parameter	Description	Value	Reference
l_0	Resting S2 segment length	12 nm	Williams et al. (2010)[a]
l_{S20}	S2 segment length at reference lattice spacing	16–20 nm	
l_{S1}	S1 segment length	11 nm	Schoenberg (1980a, b)[a]
α_{S1}	Angle of S1 attachment	45°	Julian et al. (1978)
δ_0	Axial distance of three myosin head pairs	43.5 nm	Craig and Woodhead (2006)
Δ_0	Lattice spacing at unloaded sarcomere length	19 nm	

[a]Values projected to 2D from a 3D model

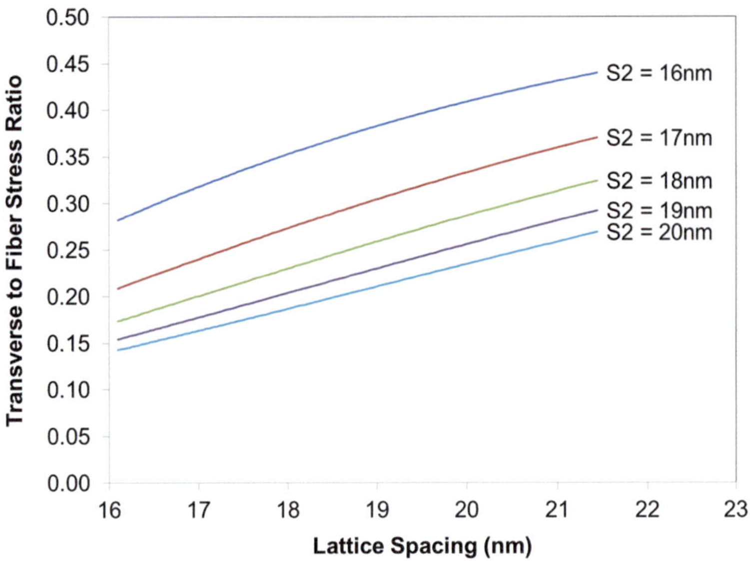

Fig. 22.7 Plot of the transverse to axial stress ratio as a function of lattice spacing between the actin and myosin filaments. This ratio is also a function of the length of the S2 myosin segment at reference lattice spacing as shown by the family of curves

simulations at sheet angles 0°, 45°, and 90° with respect to the second stretch direction were performed. Figure 22.8 shows the total stresses in the fiber and cross-fiber direction for the three sheet angles. It can be seen that the ratio of the cross-fiber to fiber stress varies depending on the sheet angle orientation. These may explain some of the variations in the experimental measurements by Lin and Yin (1998). In addition, this shows that the angle of the sheet relative to the applied stretch has a large effect on the total generated stress.

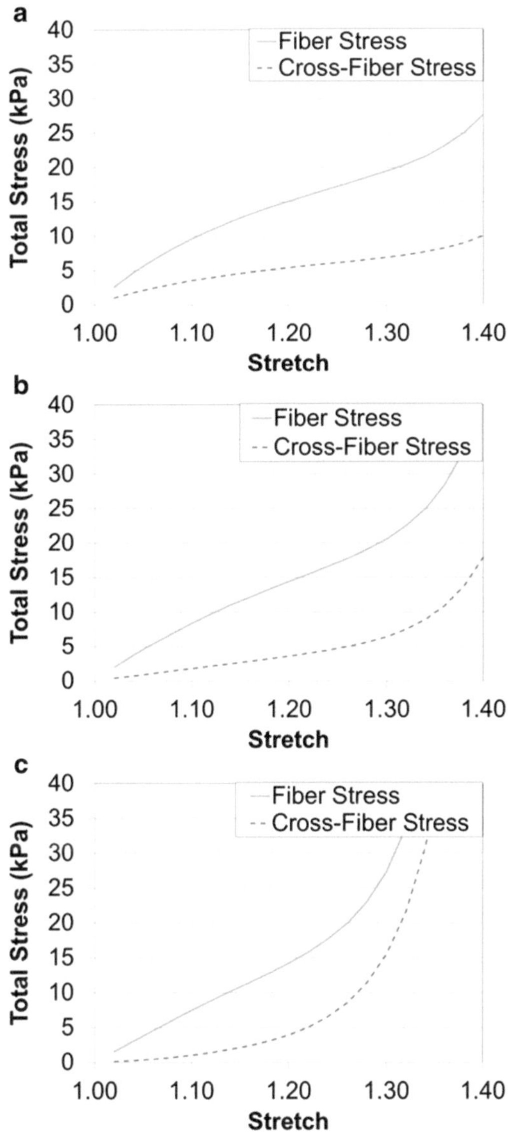

Fig. 22.8 Total fiber and cross-fiber stresses for three sheet angle orientations: 0° (**a**), 45° (**b**), and 90° (**c**). It can be seen that the ratio of the transverse to fiber stresses varies with the sheet angles

Figure 22.9 shows the active normal stresses generated in sheet coordinates as a function of the equi-biaxial stretch with the sheets parallel to the stretch plane. The experimental measurements given by Lin and Yin (1998) at one equi-biaxial stretch level are shown as points. It can be seen that on an average, the cross-fiber (sheet) stresses were around 40% of the fiber stresses and the stresses in the sheet-normal direction were around 10% of the fiber stresses. The active stress generated in the sheet-normal direction is only due to dispersion in the sheet angles.

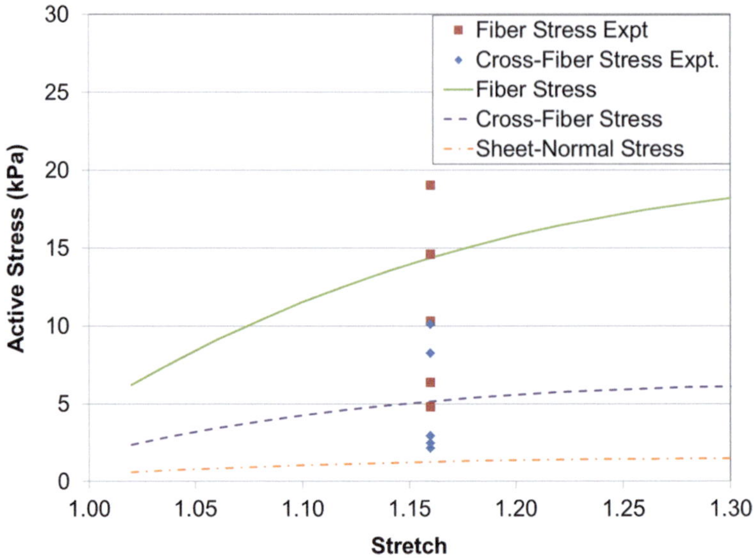

Fig. 22.9 Active fiber, cross-fiber, and sheet-normal stress generated in the tissue during equi-biaxial stretch. The experimental values from Lin and Yin (1998) at an equi-biaxial stretch of 1.16 are marked with *dots*

22.4 Discussion

In this study, we derive a multi-scale mathematical model to investigate the relationship between active force development within the sarcomere of a cardiac myocyte and stress transverse to the fiber orientation at the tissue level. The model incorporates structural dispersion including histological measurements of sheet orientation, and incorporates crossbridge and sarcomere lattice geometry. The results of the finite element model are compared with measured experimental stress in biaxial deformation tests. The results suggest that these mechanisms can explain the source of forces generated transverse to the fiber direction in myocardial tissue.

The transverse force generation in the crossbridge model is sensitive to the parameters of the model, such as the length of the S1 and S2 segments. While estimates for these quantities vary between publications and muscle types and species, they are measurable microstructural properties rather than arbitrary parameters.

Our analysis suggests that the strain dependence of fiber and sheet dispersion is very small and unlikely to affect the analysis. However, the strain dependence on lattice spacing gives rise to larger transverse stresses at larger lattice spacing. In the current model, we assume, based on electron microscopy and X-ray crystallography, that lattice spacing is only determined by fiber strain because the lattice isotropically expands in the transverse direction as sarcomeres shorten to maintain approximately constant sarcomere volume. This implies that anisotropic macroscopic strains in the myocardium must be accommodated either by rearrangement of myofibrils within myocytes, myocytes within sheets, or sheets within the tissue.

Since we consider the sheet to be a plane stress object, transverse stresses are not transmitted in the sheet-normal direction. The only mechanism of active stress generation in the sheet-normal direction in our model is through sheet dispersion. This is probably not completely accurate there is some form of inter-sheet coupling that can transmit active stresses in the sheet-normal direction. Myocardial sheets have also been shown to have unique passive material properties. For instance, they are stiffer within the plane of the sheet than across it (Dokos et al. 2002). Despite this, it has been shown that simulations of systole are insensitive to changes in parameters controlling passive sheet properties (Usyk et al. 2000).

In conclusion, we have developed a mathematical model linking scales from the myofilament crossbridge up to the tissue-scale myocardial continuum. The stress developed transverse to the myofilaments, in combination with dispersions of the muscle fibers and sheets, leads to significant transverse stress at the tissue level found in previous experimental tests. The transverse active stress development in the tissue depends on structural geometry at multiple scales in the tissue. The orientation of the sheets relative to tissue deformation plays an important role in the total stress that is measured experimentally. The strain dependence of the transverse stress developed at the crossbridge level is significant while the strain dependence on the dispersion is found to be small as shown in the Appendix 2. Thus, we have developed a microstructurally based multi-scale model of active myocardial mechanics that takes into account the crossbridge and sarcomere lattice geometry and the myocardial sheet structure. Such a theoretical model can be easily incorporated into realistic ventricular geometry to simulate cardiac function that match closely with experimental observations.

Acknowledgement Supported by NIH grants 5P01HL46345, GM103426, 1R01HL96544, GM094503, 1RO1HL091036, and 1R01HL105242.

A.1 Appendix 1: Fiber-Sheet Dispersion Effects on Active Stress

Here we give details of the derivation of the fiber-sheet dispersion effects on active stress from (22.5). We used a Von-Mises distribution for the three angles. The probability density of a Von-Mises distribution is given by the following equation:

$$f(\gamma) = \frac{e^{\kappa \cos \gamma}}{2\pi I_0(\kappa)} \tag{22.8}$$

where I_0 is the modified Bessel function of order 0 and κ is called the concentration parameter that controls the standard deviation of the distribution. The components of the stress tensor can be computed to be

$$\begin{aligned}
\boldsymbol{T}_{11} &= S_{\mathrm{f}}\Big[\nu\left(I_{\sin^2\varphi}I_{\cos^2\theta} + I_{\cos^2\beta}I_{\cos^2\varphi}I_{\sin^2\theta}\right) \\
&\quad + \left(I_{\cos^2\varphi}I_{\cos^2\theta} + I_{\cos^2\beta}I_{\sin^2\varphi}I_{\sin^2\theta}\right)\Big] \\
\boldsymbol{T}_{22} &= S_{\mathrm{f}}\Big[\nu\left(I_{\sin^2\beta}I_{\cos^2\varphi}\right) + \left(I_{\sin^2\beta}I_{\sin^2\varphi}\right)\Big] \\
\boldsymbol{T}_{33} &= S_{\mathrm{f}}\Big[\nu\left(I_{\sin^2\varphi}I_{\sin^2\theta} + I_{\cos^2\beta}I_{\cos^2\varphi}I_{\cos^2\theta}\right) \\
&\quad + \left(I_{\cos^2\varphi}I_{\sin^2\theta} + I_{\cos^2\beta}I_{\sin^2\varphi}I_{\cos^2\theta}\right)\Big]
\end{aligned} \tag{22.9}$$

where the integrals I can be computed numerically from the distribution. For a standard dispersion of 12° for φ and θ, and a 30° for b, we get the active stress components to be given by

$$\begin{aligned}
\boldsymbol{T}_{11} &= S_{\mathrm{f}}\,[0.067\,\nu + 0.924] \\
\boldsymbol{T}_{22} &= S_{\mathrm{f}}\,[0.201\,\nu + 0.008] \\
\boldsymbol{T}_{33} &= S_{\mathrm{f}}\,[0.724\,\nu + 0.067]
\end{aligned} \tag{22.10}$$

These equations were then used in the finite element model and the k computed from the lattice model is used as the input to these models.

A.2 Appendix 2: Strain Dependence of Angle Distributions

In continuum mechanics, deformations of bodies create changes in angles. For example, consider the two-dimensional example in Fig. A.1. Suppose the fibers in this tissue are originally oriented at an angle γ_0. After undergoing deformation, this angle is represented by γ. The relationship between γ and γ_0 can be derived from continuum mechanics principles (Fung 1993), and is given by:

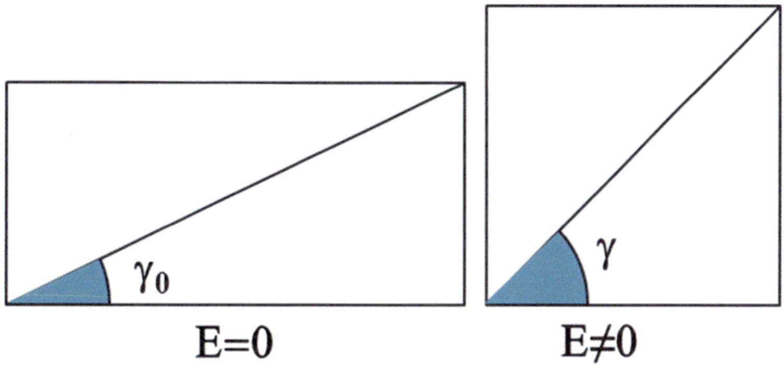

Fig. A.1 Schematic diagram representing the change in angle γ as a body deforms. In this example, due to horizontal shortening and vertical lengthening, $\gamma > \gamma_0$. The angle would also be affected by shearing deformation (*not shown*)

$$
\begin{aligned}
\cos\gamma &= \frac{\mathbf{u}_1\mathbf{C}\mathbf{u}_2}{\sqrt{\mathbf{u}_1\mathbf{C}\mathbf{u}_1}\sqrt{\mathbf{u}_2\mathbf{C}\mathbf{u}_2}} \\
&= \frac{\begin{bmatrix}1\\0\end{bmatrix}\begin{bmatrix}C_{11} & C_{12}\\ C_{12} & C_{22}\end{bmatrix}\begin{bmatrix}\cos\gamma_0\\ \sin\gamma_0\end{bmatrix}}{\sqrt{C_{11}}\sqrt{\begin{bmatrix}\cos\gamma_0\\ \sin\gamma_0\end{bmatrix}\begin{bmatrix}C_{11} & C_{12}\\ C_{12} & C_{22}\end{bmatrix}\begin{bmatrix}\cos\gamma_0\\ \sin\gamma_0\end{bmatrix}}} \\
&= \frac{C_{11}\cos\gamma_0 + C_{12}\sin\gamma_0}{\sqrt{C_{11}}\sqrt{C_{11}\cos^2\gamma_0 + 2C_{12}\sin\gamma_0\cos\gamma_0 + C_{22}\sin^2\gamma_0}}
\end{aligned}
\tag{22.11}
$$

In terms of the strain components $\mathbf{E}$, the $\cos(\gamma)$ can be computed from the equation,

$$
\cos\gamma = \frac{(2E_{11}+1)\cos\gamma_0 + 2E_{12}\sin\gamma_0}{\sqrt{(2E_{11}+1)}\sqrt{\Big(2E_{11}\cos^2\gamma_0 + 4E_{12}\sin\gamma_0\cos\gamma_0 + 2E_{22}\sin^2\gamma_0 + 1}}
\tag{22.12}
$$

In order to understand the strain dependence of the fiber dispersion functions, several numerical experiments were performed. Samples of 5000 angles were drawn from a Von-Mises distribution of known κ, the concentration parameter, which gives a standard deviation of 12°. The change in the angle γ is computed for different values of biaxial strains, and the new standard deviation and the κ parameter were computed for the resulting distribution (Fig. A.2). This was then compared with directly computing the change in the standard deviation angle using (22.12). It can

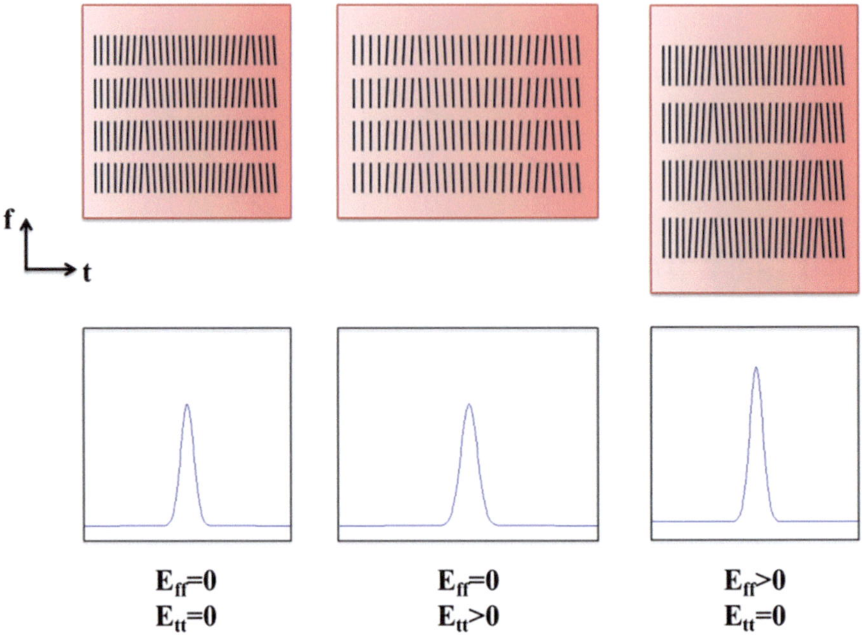

Fig. A.2 Effect of strain on fiber distribution. A positive transverse strain increases the standard deviation of the angle distribution while a positive fiber strain decreases the standard deviation

be seen from Figs. A.3 and A.4 that the predicted standard deviations are within few degrees of the predicted values. Under shear strain, the mean is not zero, but this deviation in the mean is <2° for reasonable shear strains.

Next, the strain dependence of the active stress components was computed. The concentration parameter was varied from 10 to 40 for φ and θ, and from 2 to 10 for the sheet angle β. These correspond to a standard deviation of 18°–9° for φ and θ, and 48°–18° for β, respectively. It can be seen from Fig. A.5 that the strain dependence is very small for practical values of standard deviation of fiber dispersion and strains. Consequently, the strain dependence can be ignored for typical strains in a myocardium. In addition, if the strain values are extreme, the strain dependence can be incorporated by computing the new standard deviation of the distribution and using the concentration parameter that corresponds to this standard deviation value in the simulations.

Fig. A.3 Comparison of actual standard deviation with predicted values for different strains

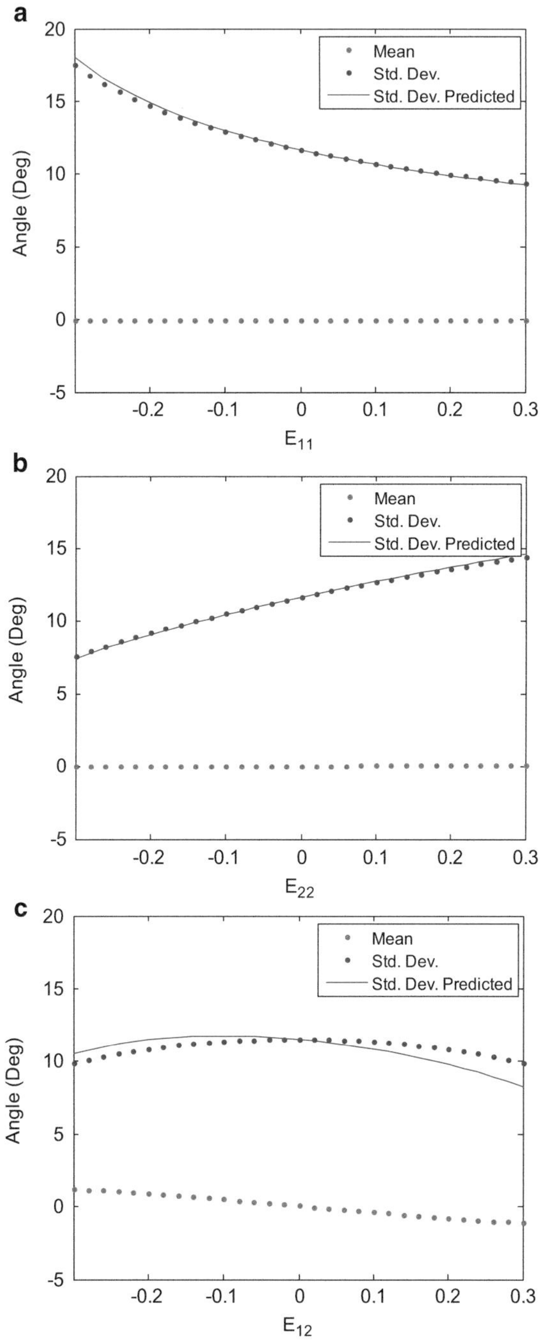

Fig. A.4 Effect of combined biaxial strains on standard deviation and its prediction

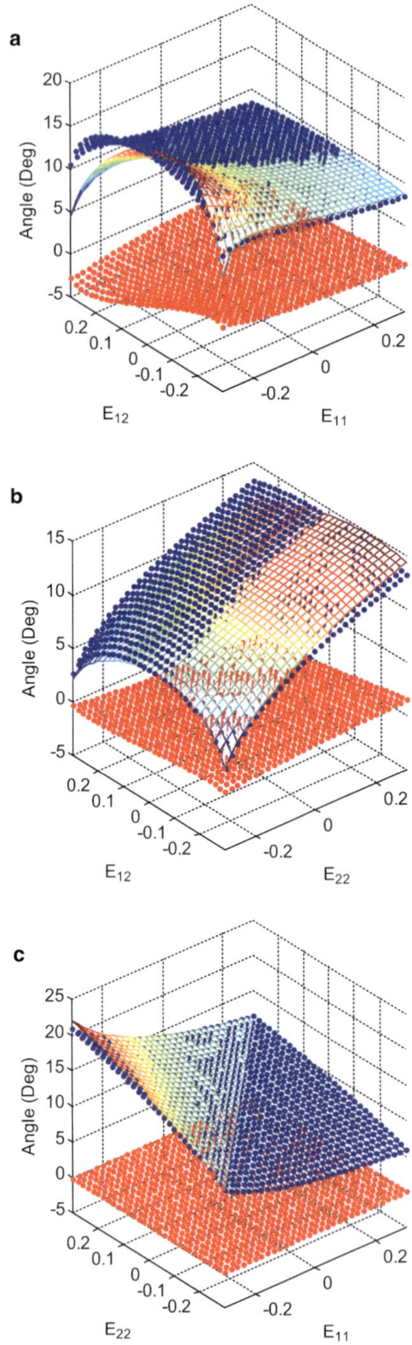

Fig. A.5 Effect of the concentration parameter κ on the diagonal components of the active stress tensor. It can be seen that the strain dependence is very small and we can ignore it for practical simulations

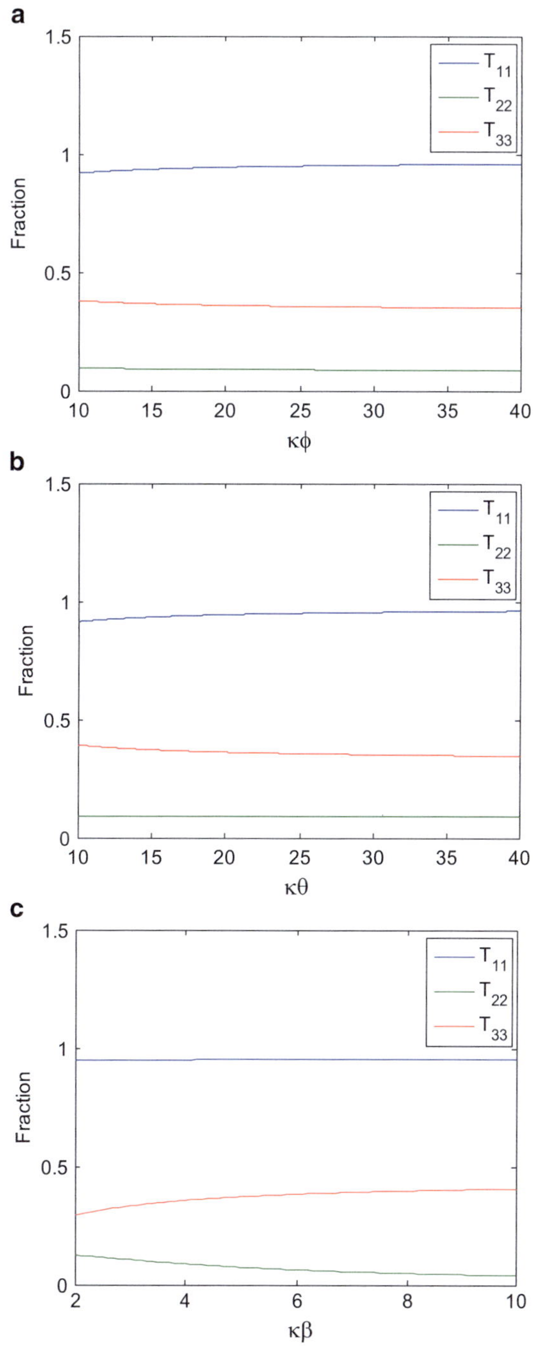

References

Ashikaga H, Omens JH, et al. Transmural mechanics at left ventricular epicardial pacing site. Am J Physiol Heart Circ Physiol. 2004;286:H2401–7.

Campbell SG, Howard E, et al. Effect of transmurally heterogeneous myocyte excitation-contraction coupling on canine left ventricular electromechanics. Exp Physiol. 2009;94: 541–52.

Campbell SG, Lionetti FV, et al. Coupling of adjacent tropomyosins enhances cross-bridge-mediated cooperative activation in a Markov model of the cardiac thin filament. Biophys J. 2010;98:2254–64.

Tangney JR, Chuang JS, Janssen MS, Krishnamurthy A, Liao P, Hoshijima M, Wu X, et al. Novel role for vinculin in ventricular myocyte mechanics and dysfunction. Biophys J. 2013;104(7):1623–33.

Coppola BA, Covell JW, et al. Asynchrony of ventricular activation affects magnitude and timing of fiber stretch in late-activated regions of the canine heart. Am J Physiol Heart Circ Physiol. 2007;293:H754–61.

Costa KD, Takayama Y, et al. Laminar architecture and three-dimensional systolic mechanics in canine ventricular myocardium. Am J Physiol. 1999;276:H595–607.

Craig R, Woodhead JL. Structure and function of myosin filaments. Curr Opin Struct Biol. 2006;16(2):204–12.

Dokos S, Smaill BH, et al. Shear properties of passive ventricular myocardium. Am J Physiol Heart Circ Physiol. 2002;283:2650–9.

Doll S, Schweizerhof K. On the development of volumetric strain energy functions. J Appl Mech. 2000;67:17.

Fung YC. A first course in continuum mechanics. Englewood Cliffs: Prentice-Hall; 1993.

Guccione JM, McCulloch AD. Mechanics of active contraction in cardiac muscle: Part I—constitutive relations for fiber stress that describe deactivation. J Biomech Eng. 1993;115: 72–81.

Guccione JM, McCulloch AD, et al. Passive material properties of intact ventricular myocardium determined from a cylindrical model. J Biomech Eng. 1991;113:42–55.

Hunter PJ, McCulloch AD, et al. Modelling the mechanical properties of cardiac muscle. Prog Biophys Mol Biol. 1998;69(2/3):289–331.

Huxley HE. The crossbridge mechanism of muscular contraction and its implications. J Exp Biol. 1985;115:17–30.

Huxley HE, Kress M. Crossbridge behaviour during muscle contraction. J Muscle Res Cell Motil. 1985;6(2):153–61.

Julian FJ, Moss RL, et al. The mechanism for vertebrate striated muscle contraction. Circ Res. 1978;42(1):2–14.

Karlon WJ, Covell JW, et al. Automated measurement of myofiber disarray in transgenic mice with ventricular expression of ras. Anat Rec. 1998;252(4):612–25.

LeGrice IJ, Smaill BH, et al. Laminar structure of the heart: ventricular myocyte arrangement and connective tissue architecture in the dog. Am J Physiol. 1995;269:H571–82.

Lin DH, Yin FC. A multiaxial constitutive law for mammalian left ventricular myocardium in steady-state barium contracture or tetanus. J Biomech Eng. 1998;120(4):504–17.

Rayment I, Holden HM, et al. Structure of the actin-myosin complex and its implications for muscle contraction. Science. 1993;261(5117):58–65.

Schoenberg M. Geometrical factors influencing muscle force development. I. The effect of filament spacing upon axial forces. Biophys J. 1980a;30(1):51–67.

Schoenberg M. Geometrical factors influencing muscle force development. II. Radial forces. Biophys J. 1980b;30:69–77.

Usyk TP, Mazhari R, et al. Effect of laminar orthotropic myofiber architecture on regional stress and strain in the canine left ventricle. J Elas. 2000;61:143–64.

Williams CD, Regnier M, et al. Axial and radial forces of cross-bridges depend on lattice spacing. PLoS Comput Biol. 2010;6(12):e1001018.

Zahalak GI. Non-axial muscle stress and stiffness. J Theor Biol. 1996;182:59–84.

Index

G.S. Kassab, M.S. Sacks (eds.), *Structure-Based Mechanics of Tissues and Organs*,
DOI 10.1007/978-1-4899-7630-7

D

E

T

U

V

MIX
Papier aus verantwortungsvollen Quellen
Paper from responsible sources
FSC® C105338

If you have any concerns about our products,
you can contact us on
ProductSafety@springernature.com

In case Publisher is established outside the EU,
the EU authorized representative is:
Springer Nature Customer Service Center GmbH
Europaplatz 3, 69115 Heidelberg, Germany

Printed by Libri Plureos GmbH
in Hamburg, Germany